JEFFREY BENNETT
University of Colorado at Boulder

MEGAN DONAHUE
Space Telescope Science Institute

NICHOLAS SCHNEIDER
University of Colorado at Boulder

MARK VOIT
Space Telescope Science Institute

The Cosmic Perspective

BRIEF EDITION

An Imprint of Addison Wesley Longman, Inc.
San Francisco • Reading, Massachusetts • New York • Harlow, England •
Don Mills, Ontario • Sydney • Mexico City • Madrid • Amsterdam

Acquisitions Editor: *Sami Iwata*
Publisher: *Robin J. Heyden*
Marketing Manager: *Jennifer Schmidt*
Publishing Associates: *Nancy Gee, Tony Asaro*
Production Coordination: *Joan Marsh*
Production: *Mary Douglas*
Photo Research: *Myrna Engler*
Technical Artist: *John Goshorn/Techarts*
Copyeditor: *Mary Roybal*
Text and Cover Designer: *Andrew Ogus*
Cover Photos: *Galaxy in Andromeda/Jerry Lodriguss © Photo-
    Disc, Inc. Canoeists/Alan and Sandy Carey*
Composition: *Thompson Type*
Film: *H&S Graphics/Tom Anderson, Michelle Kessel*
Cover Printer: *Phoenix Color Corporation*
Printer and Binder: *Von Hoffmann Press*

Library of Congress Cataloging-in-Publication Data
The cosmic perspective / Jeffrey Bennett . . . [et al.].—Brief ed.
    p.   cm.
  Includes index.
  ISBN 0-201-69734-3
  1. Astronomy.   I. Bennett, Jeffrey O.
QB45.C67   2000
520—dc21   99-056692

1 2 3 4 5 6 7 8 9 10—VHP—02 01 00 99

BRIEF CONTENTS

DETAILED CONTENTS

PREFACE

We humans have gazed into the sky for countless generations, wondering how our lives are connected to the Sun, Moon, planets, and stars that adorn the heavens. Today, through the science of astronomy, we know that these connections go far deeper than our ancestors ever imagined. This book tells the story of modern astronomy and the new perspective—*The Cosmic Perspective*—with which it allows us to view ourselves and our planet. It is written for anyone who is curious about the universe, but it is designed primarily as a textbook for college students who are not planning to major in mathematics or science.

The Cosmic Perspective, Brief Edition, is designed primarily for one-term courses in introductory astronomy. The brief edition focuses on the major conceptual ideas of the cosmic perspective and does not involve complex mathematics. In preparing the brief edition, we have incorporated many of the suggestions and feedback we have received from the companion full version, *The Cosmic Perspective.*

Approach

This book grew out of our experience teaching astronomy both to college students and to the general public over the past 20 years. During this time, a flood of new discoveries fueled a revolution in our understanding of the cosmos, but the basic organization and approach of most astronomy textbooks remained unchanged. We felt that the time had come to rethink how to organize and teach the major concepts in astronomy. This book is the result. Several major innovations in organization set this book apart.

A big picture approach. To help students achieve a lasting appreciation of astronomy, we have tried to ensure that astronomical facts are always expressed in the context of the big picture of the universe. The first chapter sets the stage by providing a broad overview of modern astronomy, ensuring that students understand the structural hierarchy and general history of the universe, the scale of space and time, and the motions of Earth in space. The rest of the book builds upon this framework, and each chapter ends with a section called "The Big Picture," which reinforces how the concepts covered fit into the broader context of astronomy.

True comparative planetology. Departing from the traditional planet-by-planet approach to studying our solar system, we focus on the physical processes that determine the similarities and differences among planets. This approach helps students see the relevance of planetary science to their own lives by enabling them to gain a far deeper appreciation of our unique world. We begin our section on comparative planetology (Part III of the text) by discussing solar system formation (Chapter 7), which provides students with the context needed to make planetary comparisons. We then turn to the study of the terrestrial worlds (Chapter 8), focusing on planetary geology and planetary atmospheres. Our chapter on the jovian worlds (Chapter 9) similarly focuses on processes, with emphasis on the atmospheric processes of the jovian planets themselves and on the "ice geology" of the major jovian moons. In Chapter 10, we focus on asteroids, comets, and Pluto (because it is now known to have much more in common with the comets of the Kuiper belt than with the other planets of our solar system). We also discuss the important role of impacts in Earth's past and the threat of possible impacts in the future. Finally, in Chapter 11, we bring all the previous lessons from other planets to bear on the study of our home world—and then use what we learn about Earth to investigate the prospects for finding life elsewhere in the universe.

An evolving vision of galaxies and dark matter. Traditionally, galaxies have been covered from a classification standpoint, rather like the way that stars were studied in the early part of the 20th century. In just the past decade or two, our understanding of galaxy evolution has grown enormously. As a result, it is now possible to discuss the entire subject of galaxies and cosmology (Part V of the text) within a unifying evolutionary context. Chapter 16 begins this discussion by focusing on the Milky Way as a dynamic system that cycles gas through stars, chemically enriching the galaxy as a whole. We also discuss some of the key clues that our galaxy offers about galactic evolution, such as the differences between stars in the disk and halo and the evidence that dark matter dominates the total mass of the galaxy. Chapter 17 emphasizes how we study galaxy evolution. We introduce the cosmic distance scale, Hubble's law, and the concepts of lookback time and the cosmological horizon. We place starburst galaxies and quasars in their proper context as phases in the evolution of otherwise normal galaxies, rather than as separate classes of abnormal galaxies. We devote Chapter 18 to the important topic of dark matter, which is integral to the process of galaxy formation and evolution as well as to questions of the fate of the universe. With the evolutionary context well established, students are better able to understand the importance of studying the Big Bang itself in Chapter 19.

A focus on the universality of physics. We believe that one of the primary goals of any astronomy course should be to help students understand how we are able to learn so much about the universe. We therefore emphasize the "universality of physics"—the idea that knowing a few key principles that govern matter, energy, light, and motion can help them better understand both the phenomena of their daily lives and the mysteries of the cosmos. Each of our chapters covering fundamental ideas of science (Part II of the text) begins with a section on science in everyday life in which we use everyday experiences to help students build a scientific view of the world. We then develop formal concepts of physics by building upon these familiar experiences. Also, we devote a chapter (Chapter 3) to the development of science and how it functions in the modern world and another entire chapter (Chapter 4) to the concepts of matter and energy, with emphasis on conservation of energy. In addition, because student difficulties often stem from prior misconceptions, we address many misconceptions head on with our boxed features called "Common Misconceptions," which are found throughout the text.

Art That Teaches

Textbook art and photos are a balance of beauty and teaching. The art and photo program must not only engage the eye; most importantly it must illustrate and reinforce the material within the text discussion. With this goal in mind, we have worked side by side with our artists, weaving together the connection between the book's words and illustrations. All photos have been chosen not only for their eye-catching

quality but for how well they complement the text. Our aim was to achieve a new level of integration between content and art that would help students better understand astronomy.

Additional Features

Time Out to Think. Appearing throughout every chapter, the "Time Out to Think" features pose a short conceptual question designed to help students reflect on important new ideas. They also serve as excellent starting points for classroom discussions.

Common Misconceptions. These boxes address and correct popularly held but incorrect ideas related to the chapter material. Examples of topics covered in these boxes include misconceptions about the seasons (in Chapter 1), about why astronauts are weightless in orbit (in Chapter 5), and about what would happen to a spaceship passing near a black hole (in Chapter 15).

Thinking About. These boxes contain supplementary discussion of topics related to the chapter material but not prerequisite to the continuing discussion. Examples of topics covered in these boxes include the work of Aristarchus (in Chapter 2) and the history of the cosmological constant (in Chapter 19).

End-of-Chapter Questions and Problems. The following three sections appear at the end of each chapter:

- ■ *Review Questions*: Every important concept in a chapter is covered by at least one review question, and the questions can be answered by rereading relevant sections of the chapter. For students, these questions are more useful than a standard chapter summary because they require interaction, as opposed to passive reading. Instructors can help students focus their studying by telling them which questions they are responsible to know for exams.

- ■ *Discussion Questions*: These questions are meant to be particularly thought-provoking and generally do not have objective answers. As such, they are ideal for in-class discussion.

- ■ *Problems*: Each chapter includes several problems of varying levels of difficulty that are designed to be assigned as written homework. Problem titles help explain the topic of a problem and explicitly state when problems require outside research or activity.

Cross-References. Cross-references in brackets help students find the relevant discussion and create connections between chapters.

SkyGazer, College Edition. This CD-ROM brings backyard astronomy into the classroom. Based on *Voyager,* one of the world's most popular planetarium programs, *SkyGazer, College Edition,* is guaranteed to capture students' attention and introduce them to the wonders of the night sky. Intended for students in introductory astronomy, *SkyGazer, College Edition,* makes it easy for novice users to explore the sky and learn more about their place in the universe. Exercises for the software appear on the book's web site, Astrospot.

Here are just a few of the exciting activities students can experience using *SkyGazer, College Edition.*

- Navigate with easy-to-use chart controls to access deep-sky objects such as clusters, nebulae, and galaxies.

- View the changing patterns of the constellations.

- Explore phenomena such as meteor showers, the seasons, and the cycle of day/night.

- Display and print detailed sky charts for any time and season and from any location on the planet. Multiple sky charts permit viewing different areas of the sky simultaneously.

- Explore the motion of comets and asteroids without getting lost in the details of orbital elements.

- Customize the horizon area to match the local landscape.

SkyGazer, College Edition, is part of Carina Software's *Voyager III* suite of products, which includes *Voyager III, Voyager III with SkyPilot,* and *SkyGazer.*

Stay Current with Our Web Site: *www.astrospot.com*

New discoveries in astronomy are occurring at a remarkable rate. To stay current, visit our accompanying web site at www.astrospot.com. Astrospot includes many features for both students and instructors including:

- chapter-by-chapter study aids including chapter summaries, immediate feedback quizzes, and a hyperlinked glossary.

- current articles, weekly updates on new discoveries, and interviews with scientists

- animations and interactive simulations to clarify key concepts

- weekly sky calendar

- links to other on-line astronomical resources, sorted and organized to help users make sense of the vast amount of available information

- additional resources for instructors, including an on-line version of the Instructor's Guide

- A complimentary subscription to Astrospot is packaged with each new copy of the text.

Supplements for Instructors

In addition to the Instructor's Resource Area at www.astrospot.com, the following resources are available to qualified adopters; contact your Addison Wesley Longman sales representative for more information.

- *Instructor's Guide (0-8053-8015-9).* Includes an overview of the philosophy of the text, interactive teaching methods, a chapter-by-chapter commentary, and solutions to end-of-chapter problems.

- *Addison Wesley Longman Course Management System.* Allows the instructor to offer on-line quizzing, create individualized syllabi, conduct threaded discussion groups, and provide a format for administering on-line content.

- *Instructor's Presentation CD-ROM (0-8053-8014-0).* Contains all of the full-color line art figures and hundreds of photos from the text for use in lecture presentations. Images may be exported into other programs, such as PowerPoint.

- *Transparency Acetates (0-8053-8013-2).* A selection of over 200 figures from the text printed on full-color transparency acetates.

- *Test Bank (0-8053-8016-7).* Includes multiple-choice, true/false, and free-response questions for each chapter.

- *TestGen-EQ with QuizMaster-EQ (Windows/Mac: 0-8053-8017-5).* Allows the instructor to view and edit questions, add questions, and create and print different versions of tests. A built-in question editor creates graphs, imports graphics, and inserts text. QuizMaster-EQ automatically grades exams, stores results on disk, and allows the instructor to view or print a variety of reports.

- *Supplementary Texts for Courses or Labs.* For students who wish to learn more about the sky, *Norton's Star Atlas* provides an unmatched resource with extensive star charts, information about what to see in the sky, and hints on observing techniques, use of telescopes, and astrophotography. It is also an excellent resource for astronomy labs. Contact your bookstore or your Addison Wesley Longman sales representative for ordering information. *Norton's Star Atlas and Reference Handbook,* 19th edition, edited by Ian Ridpath. ISBN: 0-582-31283-3 (hardcover), ISBN: 0-582-35655-5 (softcover).

Acknowledgments

It has taken the intense efforts of many people to bring this book to publication. We could not possibly list everyone who has helped, but we would like to call attention to a few people who have played particularly important roles. First, we thank the people who helped turn our writing into an actual book: our editors and friends at Addison Wesley Longman who have stuck with us through thick and thin, including Bill Poole, Robin Heyden, Linda Davis, Joan Marsh, Jennifer Schmidt, Catherine Flack, Nancy Gee, and especially Sami Iwata; our production team, including Mary Douglas, Myrna Engler, Mary Roybal, and Karen Stough; our designer, Andrew Ogus; our web team, including Claire Masson and Rebecca Lindell Adrian; and the artists who have created the outstanding illustrations, Joe Bergeron, John Goshorn, and Stan Maddock.

We've also been fortunate to have an outstanding group of reviewers, whose extensive comments and suggestions helped us shape the book. We thank the following people who reviewed drafts of this or previous editions in various stages:

John Beaver, University of Wisconsin at Fox Valley
Dipak Chowdhury, Indiana University–Purdue University at Fort Wayne
Robert Egler, North Carolina State University at Raleigh
Robert A. Fesen, Dartmouth College
Sidney Freudenstein, Metropolitan State College of Denver
Richard Gray, Appalachian State University
David Griffiths, Oregon State University
David Grinspoon, University of Colorado
Bruce Jakosky, University of Colorado
Kristine Larsen, Central Connecticut State University
Larry Lebofsky, University of Arizona
Michael LoPresto, Henry Ford Community College
William R. Luebke, Modesto Junior College
Marie Machacek, Massachusetts Institute of Technology
Steven Majewski, University of Virginia
John Safko, University of South Carolina
James A. Scarborough, Delta State University
Joslyn Schoemer, Challenger Center for Space Science Education
James Schombert, University of Oregon
Dale Smith, Bowling Green State University

John Spencer, Lowell Observatory
John Stolar, West Chester University
Jack Sulentic, University of Alabama
C. Sean Sutton, Mount Holyoke College
J. Wayne Wooten, Pensacola Junior College
Dennis Zaritsky, University of California, Santa Cruz

In addition, we thank the following colleagues who helped us clarify technical points or checked the accuracy of technical discussions:

Thomas Ayres, University of Colorado
Cecilia Barnbaum, Valdosta State University
Rick Binzel, Massachusetts Institute of Technology
Humberto Campins, University of Florida
Robin Canup, Southwest Research Institute
Mark Dickinson, Space Telescope Science Institute
Harry Ferguson, Space Telescope Science Institute
Andrew Hamilton, University of Colorado
Todd Henry, Harvard–Smithsonian Center for Astrophysics
Dave Jewitt, University of Hawaii
Hal Levison, Southwest Research Institute
Mario Livio, Space Telescope Science Institute
Mark Marley, New Mexico State University
Bob Pappalardo, Brown University
Michael Shara, Space Telescope Science Institute
Glen Stewart, University of Colorado
John Stolar, West Chester University
Dave Tholen, University of Hawaii
Nick Thomas, MPI/Lindau (Germany)
Don Yeomans, Jet Propulsion Laboratory

Finally, we thank the many people who have greatly influenced our outlook on education and our perspective on the universe over the years, including Tom Ayres, Fran Bagenal, Forrest Boley, Robert A. Brown, George Dulk, Erica Ellingson, Timothy Ferris, Katy Garmany, Jeff Goldstein, David Grinspoon, Don Hunten, Catherine McCord, Dick McCray, Dee Mook, Cheri Morrow, Charlie Pellerin, Carl Sagan, Mike Shull, John Spencer, and John Stocke.

Jeff Bennett
Megan Donahue
Nick Schneider
Mark Voit

HOW TO SUCCEED IN
YOUR ASTRONOMY COURSE

Most readers of this book are enrolled in a college course in introductory astronomy. If you are one of these readers, we offer you the following hints to help you succeed in your astronomy course.

Using This Book

Before we address general strategies for studying, here are a few guidelines that will help you use *this* book most effectively.

- Read assigned material twice.
  - Make your first pass *before* the material is covered in class. Use this pass to get a "feel" for all the material and to identify concepts that you may want to ask about in class.
  - Read the material for the second time shortly after it is covered in class. This will help solidify your understanding and allow you to make notes that will help you study for exams later.
- Take advantage of the features that will help you study.
  - Always read the *Time Out to Think* features, and use them to help you absorb the material and to identify areas where you may have questions.
  - Use the *cross-references* indicated in brackets in the text and the *glossary* at the back of the book to find more information about terms or concepts that you don't recall.
  - Go to the text web site at **www.astrospot.com** to find additional study aids, including chapter summaries and quizzes with answers.
- It's your book, so don't be afraid to make notes in it that will help you study later.

- Don't highlight—<u>underline</u>! Using a pen or pencil to <u>underline</u> material requires greater care than highlighting and therefore helps keep you alert as you study. And be selective in your underlining—for purposes of studying later, it won't help if you underlined everything.
- There's plenty of "white space" in the margins and elsewhere, so use it to make notes as you read. Your own notes will later be very valuable when you are doing homework or studying for exams.

- After you complete the reading, and again when you study for exams, make sure you can answer the *Review Questions* at the end of each chapter. If you are having difficulty with a review question, reread the relevant portions of the chapter until the answer becomes clear.

Budgeting Your Time

One of the easiest ways to ensure success in any college course is to make sure you budget enough time for studying. A general rule of thumb for college classes is that you should expect to study about 2 to 3 hours per week *outside* class for each unit of credit. For example, based on this rule of thumb, a student taking 15 credit hours should expect to spend 30 to 45 hours each week studying outside of class. Combined with time in class, this works out to a total of 45 to 60 hours spent on academic work—not much more than the time a typical job requires, and you get to choose your own hours. Of course, if you are working while you attend school, you will need to budget your time carefully.

As a rough guideline, your studying time in astronomy might be divided as follows:

If Your Course Is:	Time for Reading the Assigned Text (per week)	Time for Homework Assignments (per week)	Time for Review and Test Preparation (average per week)	Total Study Time (per week)
3 credits	2 to 4 hours	2 to 3 hours	2 hours	6 to 9 hours
4 credits	3 to 5 hours	2 to 4 hours	3 hours	8 to 12 hours
5 credits	3 to 5 hours	3 to 6 hours	4 hours	10 to 15 hours

If you find that you are spending fewer hours than these guidelines suggest, you can probably improve your grade by studying more. If you are spending more hours than these guidelines suggest, you may be studying inefficiently; in that case, you should talk to your instructor about how to study more effectively.

General Strategies for Studying

■ Don't miss class. Listening to lectures and participating in discussions is much more effective than reading someone else's notes. Active participation will help you retain what you are learning.

■ Budget your time effectively. An hour or two each day is more effective, and far less painful, than studying all night before homework is due or before exams.

■ If a concept gives you trouble, do additional reading or studying beyond what has been assigned. And if you still have trouble, ask for help: You surely can find friends, colleagues, or teachers who will be glad to help you learn.

■ Working together with friends can be valuable in helping you understand difficult concepts. However, be sure that you learn *with* your friends and do not become dependent on them.

■ Be sure that any work you turn in is of *collegiate quality:* neat and easy to read, well organized, and demonstrating mastery of the subject matter. Although it takes extra effort to make your work look this good, the effort will help you solidify

your learning and is also good practice for the expectations that future professors and employers will have.

Preparing for Exams

■ Study the review questions, and rework problems and other assignments; try additional questions to be sure you understand the concepts. Study your performance on assignments, quizzes, or exams from earlier in the term.

■ Try the quizzes available on the text web site at **www.astrospot.com.**

■ Study your notes from lectures and discussions. Pay attention to what your instructor expects you to know for an exam.

■ Reread the relevant sections in the textbook, paying special attention to notes you have made on the pages.

■ Study individually *before* joining a study group with friends. Study groups are effective only if every individual comes prepared to contribute.

■ Don't stay up too late before an exam. Don't eat a big meal within an hour of the exam (thinking is more difficult when blood is being diverted to the digestive system).

■ Try to relax before and during the exam. If you have studied effectively, you are capable of doing well. Staying relaxed will help you think clearly.

We shall not cease from exploration
And the end of all our exploring
Will be to arrive where we started
And know the place for the first time.
  T. S. ELIOT

Dedication

TO ALL WHO HAVE EVER WONDERED about the mysteries
of the universe. We hope this book will answer some of your
questions—and that it will also raise new questions in your
mind that will keep you curious and interested in the ongoing
human adventure of astronomy.

And, especially, to the members of the "baby boom" that
has occurred among the authors and editors during the writing of this book: Michaela, Emily, Rachel, Sebastian, Elizabeth,
and Grant (and one more on the way). The study of the universe begins at birth, and we hope that you will grow up in a
world with far less poverty, hatred, and war so that all people
will have the opportunity to contemplate the mysteries of the
universe into which they are born.

ABOUT THE AUTHORS

Jeffrey Bennett received a B.A. in biophysics from the University of California at San Diego in 1981 and a Ph.D. in astrophysics from the University of Colorado in 1987. His thesis research focused on Sun-like stars, but he now specializes in mathematics and science education. He has extensive teaching experience at the elementary and secondary levels and has taught more than fifty college courses in astronomy, physics, mathematics, and education. During a 2-year term as a Visiting Senior Scientist at NASA headquarters, he guided the creation of NASA's Initiative to Develop Education through Astronomy (IDEA). He is also the author of a mathematics textbook (Bennett and Briggs, *Using and Understanding Mathematics,* Addison Wesley Longman, 1999) and is currently working with the Smithsonian Institution and the Challenger Center to build a scale-model solar system near the National Mall. When not working, he enjoys participating in Masters Swimming and hiking the trails of Boulder, Colorado, with his wife, Lisa, son, Grant, and dog, Max.

Megan Donahue is an astronomer at the Space Telescope Science Institute in Baltimore, Maryland. After growing up in rural Nebraska, she obtained an S.B. in physics from the Massachusetts Institute of Technology in 1985 and a Ph.D. in astrophysics from the University of Colorado in 1990. Her thesis on intergalactic gas and clusters of galaxies won the Robert J. Trumpler Award (1993). She continued her research as a Carnegie Fellow at the Observatories of the Carnegie Institution in Pasadena, California, and later was an Institute Fellow at the Space Telescope Science Institute. She is an active observer, using ground-based telescopes, the Hubble Space Telescope, and orbiting X-ray telescopes. Her research focuses on questions of galaxy evolution, the nature of intergalactic space, large-scale structure formation, and dark matter and the fate of the universe. She married Mark Voit while in graduate school and they are the parents of two children, Michaela and Sebastian.

Nicholas Schneider is an associate professor in the Department of Astrophysical and Planetary Sciences at the University of Colorado and a researcher in the Laboratory for Atmospheric and Space Physics. He received his B.A. in physics and astronomy from Dartmouth College in 1979 and his Ph.D. in planetary science from the University of Arizona in 1988. In 1991, he received the National Science Foundation's Presidential Young Investigator Award. His research interests include planetary atmospheres and planetary astronomy, with a focus on the odd case of Jupiter's moon Io. He enjoys teaching at all levels and is active in efforts to improve undergraduate astronomy education. Off the job, he enjoys exploring the outdoors with his family and figuring out how things work.

Mark Voit is an astronomer in the Office of Public Outreach at the Space Telescope Science Institute. He earned his A.B. in physics at Princeton University in 1983 and his Ph.D. in astrophysics at the University of Colorado in 1990. He continued his studies at the California Institute of Technology, where he was Research Fellow in theoretical astrophysics. NASA then awarded him a Hubble Fellowship, under which he conducted research at the Johns Hopkins University. His research interests range from interstellar processes in our own galaxy to the clustering of galaxies in the early universe. Occasionally he escapes to the outdoors, where he and his wife, Megan Donahue, enjoy running, hiking, orienteering, and playing with their children.

ABOUT THE ARTIST

Joe Bergeron is a space artist who collaborated with the authors on many of the illustrations in this text. His decades as an amateur astronomer, combined with his imagination and ability to analyze the way things look, form his unique mixture of artistic skills. His science fiction work has been included in books by Isaac Asimov, Piers Anthony, and Gregory Benford. He was a contributor to the Time-Life Books series *Voyage Through the Universe.*

Bergeron's artwork is influenced by his background as an amateur astronomer. Many of his most memorable sightings were revealed by small telescopes, or even the naked eye. He has been privileged to see the sky painted in bold colors by the aurora borealis, the glory of the solar corona revealed during an eclipse, a golden star flickering behind the rings of Saturn, and pepper-black spots deposited on the clouds of Jupiter by a rain of comets—and, of course, light streaming like water from comets Hyakutake and Hale–Bopp.

Bergeron still paints, and he also creates digital images using a Macintosh computer. All of his illustrations that appear in this text were created on the computer. His work is not restricted to space themes but includes wildlife, landscapes, fantasy, and figures. For more about him, visit his web site at: http://members.aol.com/jabergeron

PART I

DEVELOPING A PERSPECTIVE

Fig. III.

Our Place in the Universe

1

FAR FROM CITY LIGHTS ON A CLEAR NIGHT, YOU CAN GAZE UPWARD AT A sky filled with stars. If you lie back and watch for a few hours, you will observe the stars marching steadily across the sky. Confronted by the seemingly infinite heavens, you might wonder how the Earth and the universe came to be. With these thoughts, you will be sharing an experience common to humans around the world and in thousands of generations past.

Remarkably, modern science offers answers to many fundamental questions about the universe and our place within it. We now know the basic content and scale of the universe. We know the age of the Earth and the approximate age of the universe. And, although much remains to be discovered, we are rapidly learning how the simple constituents of the early universe developed into the incredible diversity of life on Earth.

In this first chapter, we will survey the contents, history, scales, and motions of our universe. We'll thereby develop a "big picture" perspective on our place in the universe that will provide a base upon which we can build a deeper understanding throughout the rest of this book.

the Local Supercluster

the Local Group

GURE 1.1 Our place in the universe.

Earth

the Milky Way Galaxy

the solar system

1.1 A Modern View of the Universe

If you observe the sky carefully, you can see why most of our ancestors believed that the heavens revolved about the Earth. The Sun, Moon, planets, and stars *appear* to circle around our sky each day, and we cannot feel the constant motion of the Earth as it rotates on its axis and orbits the Sun. The ancient belief in an Earth-centered, or **geocentric,** universe was shattered only about 400 years ago, when the work of Copernicus, Kepler, and Galileo finally established that the Earth is just one of many planets orbiting the Sun. We have learned far more about the cosmos since that time. (In astronomy, the term *cosmos* is synonymous with *universe*.) Let's begin our overview of the universe by examining what modern astronomy has to say about our cosmic location and our cosmic origins.

Our Cosmic Address

If we are not the center of the universe, then what *is* our place? Figure 1.1 shows the answer in general terms. Earth is the third (in order of distance) of nine planets that orbit the Sun. Our **solar system** consists of the Sun, the planets and their moons, and the myriad of smaller objects, such as asteroids and comets, that orbit the Sun.

Our Sun is a star, just like the countless stars in our night sky. The Sun and all the stars you can see with your naked eye make up only a small part of a huge, disk-shaped collection of stars called the **Milky Way Galaxy.** A galaxy is a great island of stars in space, containing from a few hundred million to a trillion or more stars. The Milky Way Galaxy is relatively large, containing more than 100 billion stars. Our solar system is located a little over halfway from the galactic center to the edge of the galactic disk.

Many galaxies congregate in small **groups** or larger **clusters** that may contain hundreds or thousands of galaxies. The Milky Way Galaxy belongs to a group of 30 or so galaxies called the **Local Group.**

On the largest scales, groups, clusters, and isolated individual galaxies appear to be loosely associated into giant chains and sheets that span great distances across the universe. Between these vast structures lie huge **voids** containing few if any galaxies. Some of the galaxy structures appear more tightly grouped than others and are called **superclusters.** The supercluster to which our Local Group belongs is called, not surprisingly, the **Local Supercluster.**

Finally, the **universe** is the sum total of all matter and energy. That is, it encompasses the superclusters and voids, and everything within them. As you progress through this text, you will learn much more about the different levels of structure in the universe. To keep the hierarchy straight, you might imagine how a faraway friend would address a postcard to someone here on Earth (Figure 1.2).

TIME OUT TO THINK *Some people think that our tiny physical size in the vast universe makes us insignificant. Others think that our ability to learn about the wonders of the universe gives us significance despite our small size. What do you think?*

Our Cosmic Origins

Neither humanity nor the Earth has existed forever, and people have always wondered how we came to be. Much of the rest of this text discusses the scientific evidence concerning our cosmic origins, and we'll see that humans are relative newcomers in an old universe. For now, let's look at a quick overview of the scientific story of creation, as summarized in Figure 1.3 (p. 8–9).

As we'll discuss shortly, telescopic observations of distant galaxies show that the entire universe is **expanding.** That is, average distances between galaxies are increasing with time. If the universe is expanding, we can logically conclude that everything was closer together in the past. From the observed rate of expansion, we estimate that the expansion must have started somewhere between 10 and 16 billion years ago. This beginning is commonly referred to as the **Big Bang.**

Basic Astronomical Definitions

Star A large, glowing ball of gas that generates energy through nuclear fusion in its core. The term *star* is sometimes applied to objects that are in the process of becoming true stars (e.g., protostars) and to the remains of stars that have died (e.g., neutron stars).

Planet An object that orbits a star and that, while much smaller than a star, is relatively large in size. There is no "official" minimum size for a planet, but the nine planets in our solar system are all at least 2,000 kilometers in diameter. Planets may be rocky, icy, or gaseous in composition, and they shine primarily by reflecting light from their star.

Moon (or *satellite*) An object that orbits a planet. The term *satellite* is also used more generally to refer to *any* object orbiting another object.

Asteroid A relatively small, rocky object that orbits a star. Asteroids are sometimes called *minor planets* because they are similar to planets but smaller.

Comet A relatively small, icy object that orbits a star.

The universe has continued to expand ever since the Big Bang, but not always at the same rate. In particular, the force of gravity that attracts all objects to all other objects has presumably slowed the overall expansion, and in some relatively small regions of the universe gravity has halted the expansion altogether. Within a few billion years after the Big Bang, gravity caused local concentrations of matter to collapse into the galaxies we see today, including our own Milky Way. Gravity also drives the collapse of smaller clumps of gas and dust within galaxies, thereby forming stars and planets.

All the matter in the universe was created in the Big Bang, but some of it has changed form since that time. In particular, the early universe contained only the two simplest elements: hydrogen and helium (and trace amounts of lithium). All other elements were manufactured by stars through **nuclear fusion,** in which light elements fuse (i.e., their nuclei join together) to form heavier ones.

several billion years by the time our solar system formed about 4.6 billion years ago. The cloud that gave birth to our solar system was about 98% hydrogen and helium, but the other 2% contained all the other chemical elements. The small rocky planets of our solar system, including the Earth, were made from a small part of this 2%. We do not know exactly how the elements on the Earth's surface developed into the first forms of life, but life was already flourishing on Earth more than 3 billion years ago. Biological evolution took over once life arose, leading to the great diversity of life on Earth today—and to us.

In summary, all the material from which we and the Earth are made (except hydrogen and most helium) was created inside stars that died before the birth of our Sun. We are intimately connected to the stars because we are the products of stars. In the words of astronomer Carl Sagan (1934–96), we are "star stuff."

FIGURE 1.2 A postcard from a distant friend.

The energy released by nuclear fusion is what makes stars shine. Stars shine for most of their lives by fusing hydrogen into helium, but near the ends of their lives advanced fusion reactions in the more massive stars produce elements such as carbon, oxygen, nitrogen, and iron. Many of these massive stars die in titanic explosions that release the heavy elements into space, where they mix with other gas and dust in the galaxy. Thus, galaxies function as giant recycling plants, recycling material expelled from dying stars into subsequent generations of stars and planets.

The processes of heavy element production and cosmic recycling had already been taking place for

Images of Time

We study the universe by studying light from distant stars and galaxies. Light travels extremely fast by earthly standards: The speed of light is 300,000 kilometers per second, a speed at which it would be possible to circle the Earth nearly eight times in just 1 second. Nevertheless, even light takes substantial amounts of time to travel the vast distances in space. For example, light takes about 1 second to reach the Earth from the Moon and about 8 minutes to reach the Earth from the Sun.

Light from the stars takes many years to reach us, so we measure distances to the stars in units called

The universe has been expanding
ever since its hot and dense beginning in
the Big Bang. Each of the three cubes represents
the same region of the universe, showing how the region
expands with time.

FIGURE 1.3 Our cosmic origins. All the matter and
energy in the universe was created in the Big Bang.
This sequence of paintings shows the progression
of that matter and energy from the Big Bang to
human life. Note that the elements from which we
are made were produced in stars that shined long
ago, and these elements formed the Earth, thanks to
the recycling role played by our galaxy.

The Earth was built with elements
produced in stars that lived and died
in the Milky Way before our solar system
formed.

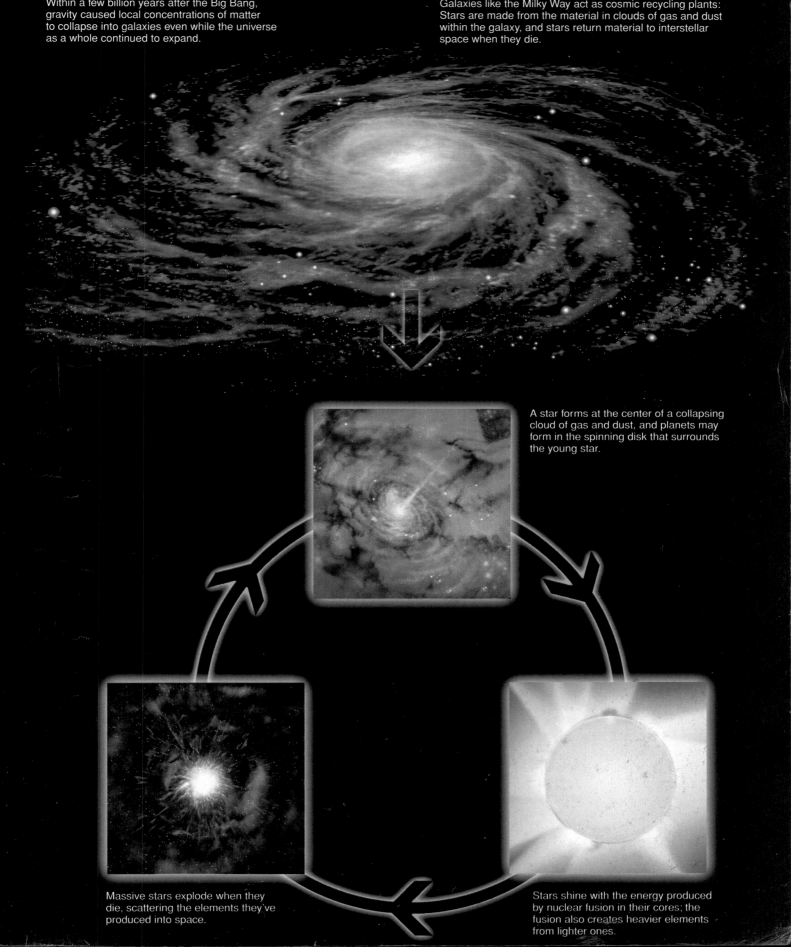

Within a few billion years after the Big Bang, gravity caused local concentrations of matter to collapse into galaxies even while the universe as a whole continued to expand.

Galaxies like the Milky Way act as cosmic recycling plants: Stars are made from the material in clouds of gas and dust within the galaxy, and stars return material to interstellar space when they die.

A star forms at the center of a collapsing cloud of gas and dust, and planets may form in the spinning disk that surrounds the young star.

Massive stars explode when they die, scattering the elements they've produced into space.

Stars shine with the energy produced by nuclear fusion in their cores; the fusion also creates heavier elements from lighter ones.

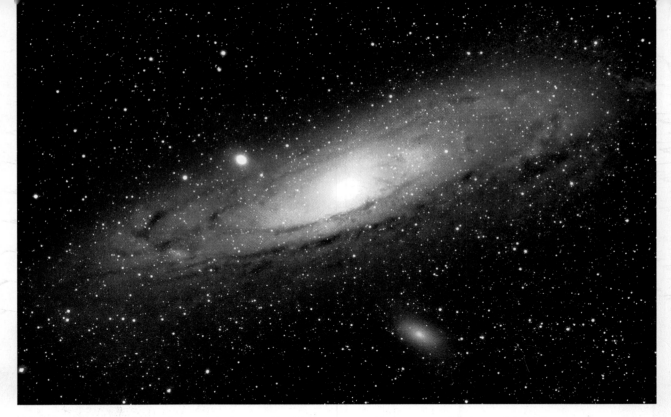

FIGURE 1.4 M31, the Great Galaxy in Andromeda, is about 2.5 million light-years away, so this photo captures light that traveled through space for 2.5 million years to reach us. Because the galaxy is 100,000 light-years in diameter, the photo also captures 100,000 years of time in M31: We see the galaxy's near side as it looked 100,000 years later than the time at which we see the far side.

light-years. One light-year is the distance that light can travel in 1 year, which is about 10 trillion kilometers, or 6 trillion miles. (You can calculate this distance by multiplying the speed of light, 300,000 km/s, by the number of seconds in 1 year.) Note that a light-year is a unit of *distance*, not of time.

The brightest star in the night sky, Sirius, is about 8 light-years from our solar system, which means that it takes light from Sirius about 8 years to reach us. Thus, when we look at Sirius, we see light that left the star about 8 years ago. The Orion Nebula, a star-forming region visible to the naked eye as a small, cloudy patch in the sword of the constellation Orion, lies about 1,500 light-years from Earth. Thus, we see the Orion Nebula as it looked about 1,500 years ago—at about the time of the fall of the Roman Empire. If any major events have occurred in the Orion Nebula since that time, we cannot yet know about them because the light from these events has not yet reached us.

Because it takes time for light to travel through space, *the farther away we look in distance, the further back we look in time.* If we look to a galaxy that lies 10 million light-years away, we are seeing it as it was 10 million years ago. If we observe a distant cluster of galaxies that lies 1 billion light-years away, we are seeing the cluster as it was 1 billion years ago.

Ultimately, the speed of light limits the portion of the universe that we can see. For example, if the universe is 12 billion years old, then light from galaxies more than 12 billion light-years away would not have had time to reach us. We would therefore say that the **observable universe** extends 12 billion light-years in all directions from Earth. (As we'll discuss in Chapter 17, this explanation is somewhat oversimplified, but it captures the basic idea that

Common Misconceptions: Light-Years

A recent advertisement illustrated a common misconception by claiming, "It will be light-years before anyone builds a better product." This advertisement makes no sense, because light-years are a unit of *distance*, not a unit of time. If you are unsure whether the term *light-years* is being used correctly, try testing the statement by remembering that 1 light-year is approximately 10 trillion kilometers, or 6 trillion miles. The advertisement then reads, "It will be 6 trillion miles before anyone builds a better product," which clearly does not make sense.

our *observable* universe may not include the *entire* universe.)

It is also amazing to realize that any "snapshot" of a distant galaxy or cluster of galaxies is a picture of both space and time. For example, the Great Galaxy in Andromeda, also known as M31, actually lies about 2.5 million light-years from Earth. Figure 1.4 therefore is a picture of how M31 looked about 2.5 million years ago, when early humans were first walking the Earth. Moreover, the diameter of M31 is about 100,000 light-years, so light from the far side of the galaxy required 100,000 years longer to reach us than light from the near side. Thus, the picture of M31 shows 100,000 years of time: This single photograph captured light that left the near side of the galaxy some 100,000 years later than the light it captured from the far side. When we study the universe, it is impossible to separate space and time.

1.2 The Scale of the Universe

We've already given numerical answers to questions about the size and age of the universe, but these numbers may have little meaning for you—after all, they are literally astronomical. In this section, we try to give meaning to incredible cosmic distances and times, beginning with our own solar system.

A Walking Tour of the Solar System

Let's imagine a walk through a model solar system built on a scale of 1 to 10 billion (that is, diameters and distances in the model are *one ten-billionth*, or 10^{-10}, of actual diameters and distances). Table 1.1 lists the diameters and distances of the planets on this scale (along with their real values). Figure 1.5 shows the sizes of the Sun and planets on this scale: The model Sun is about the size of a grapefruit, and

FIGURE 1.5 The sizes of the Sun and the planets on the 1-to-10-billion scale.

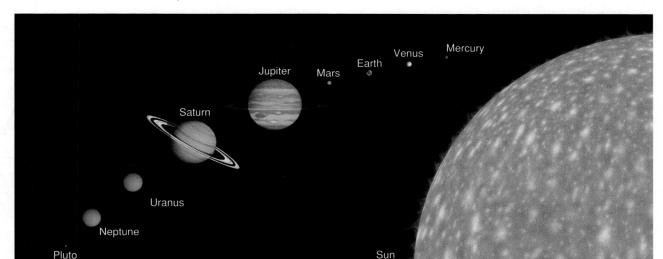

Table 1.1 Solar System Sizes and Distances, 1-to-10-Billion Scale

Object	Real Diameter	Real Distance from Sun (average)	Model Diameter	Model Distance from Sun
Sun	1,392,500 km	—	139 mm = 13.9 cm	—
Mercury	4,880 km	57.9 million km	0.5 mm	6 m
Venus	12,100 km	108.2 million km	1.2 mm	11 m
Earth	12,760 km	149.6 million km	1.3 mm	15 m
Mars	6,790 km	227.9 million km	0.7 mm	23 m
Jupiter	143,000 km	778.3 million km	14.3 mm	78 m
Saturn	120,000 km	1,427 million km	12.0 mm	143 m
Uranus	52,000 km	2,870 million km	5.2 mm	287 m
Neptune	48,400 km	4,497 million km	4.8 mm	450 m
Pluto	2,260 km	5,900 million km	0.2 mm	590 m

FIGURE 1.6 Pedestals for planet locations in a 1-to-10-billion scale model solar system.

the model planets range in size from dust speck–size Pluto to marble-size Jupiter. Figure 1.6 shows the layout of such a model. Note that all the model planets lie within 0.6 kilometer (less than ½ mile) of the model Sun, so our walk will take only a few minutes. Along the way, we'll pause to get acquainted with the objects we encounter. Keep in mind that, although our model lines up all the planets in a straight line, such an arrangement is extremely rare because each planet orbits the Sun independently.

TIME OUT TO THINK *To help yourself picture the scale, make your own quick model. Use a grapefruit or similar-size object to represent the Sun and a pinhead to represent the Earth. In a long hallway or outside, place your Sun on a chair or table and take your tiny Earth about 15 meters from your Sun. Where would the rest of the planets be located?*

We start our walk at the Sun, the central object of our solar system (Figure 1.7). The Sun contains more than 99.9% of the solar system's total mass; that is, it outweighs everything else in the solar system combined by more than a thousand times. Like most stars, the Sun is essentially a great ball of hydrogen and helium gas that generates energy through nuclear fusion in its core. A small fraction of this energy reaches the Earth, where it provides the light and heat that sustain life.

The innermost planet, Mercury, lies just a few steps from the model Sun. Mercury is the smallest planet except for Pluto, and in our model it is only about as big as the period at the end of this sentence. Mercury is a desolate, cratered world that looks much like our Moon (Figure 1.8). Temperatures on its daylight side are hot enough to melt lead. But with no atmosphere to retain heat during its long nights (which last about 3 months), temperatures

FIGURE 1.7 A photograph of the Sun. The dark splotches are sunspots.

plummet far below freezing. Thus, Mercury has the most extreme temperature range of any planet.

We reach the model Venus in just a few steps, and a few more steps take us to Earth and then to Mars. If we go back and forth among them, we see that Venus and Earth are nearly identical in size—pinheads in our model solar system—while Mars is about half as big. These three planets probably once were very similar, but they are very different today. Venus is covered in dense clouds (Figure 1.9). Because it is not that much closer to the Sun than is Earth, science fiction writers once speculated that Venus might be a lush, tropical paradise. Instead, an extreme *greenhouse effect* bakes its surface to temperatures hotter than in a pizza oven. The thick atmosphere bears down on the surface with a pressure equivalent to that found nearly a kilometer (0.6 mile) beneath the ocean surface on Earth. Besides the crushing pressure and searing temperature, a visitor to Venus would feel the corrosive effects of sulfuric acid and other toxic chemicals in its atmosphere. Far from being a tropical paradise, Venus resembles a traditional view of hell.

Earth is striking in its beauty. Blue oceans cover nearly three-fourths of the surface, broken by the continental land masses and scattered islands. The polar caps are white with snow and ice, and white clouds are scattered above the surface. Earth is the first planet on our tour to have a moon. On our model scale, the Moon is a barely visible speck that orbits the Earth at a distance of about 4 centimeters. Thus, your thumb could easily cover both the Earth and the Moon. The Moon is a world with its own interesting history and geology. It is also the only world besides Earth on which humans have stepped. Looking back from the barren Moon, we begin to get a sense of the Earth as a rare and precious oasis of life in the cosmos (Figure 1.10).

FIGURE 1.8 Mercury. This picture is a mosaic of photographs from the Mariner 10 spacecraft (1974). The areas labeled "no data" were not photographed.

FIGURE 1.9 Venus. This picture was taken by the *Pioneer Venus Orbiter* with cameras sensitive to ultraviolet light; with visible light, cloud features cannot be distinguished from the general haze.

FIGURE 1.10 This photo shows astronaut Harrison Schmitt on the Moon during the Apollo 17 mission, with a welcoming Earth in the black lunar sky. Apollo 17, the last of six Apollo missions to land on the Moon (Apollo 11, 12, and 14–17), visited the Moon in December 1972. No humans have been back since.

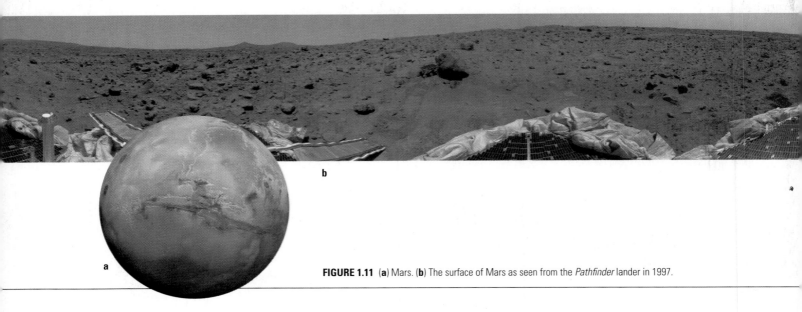

b

a

FIGURE 1.11 (**a**) Mars. (**b**) The surface of Mars as seen from the *Pathfinder* lander in 1997.

Mars is an intriguing world (Figure 1.11), with extinct volcanoes that dwarf the largest mountains on Earth, a great canyon that runs nearly one-fifth of the way around the planet, and polar caps made of water ice and frozen carbon dioxide ("dry ice"). Although all the water on Mars is frozen today, the presence of dried-up riverbeds indicates that water once flowed on the Martian surface. Thus, Mars may once have been hospitable for life, though so far we have no evidence of any life surviving today. Mars has two tiny moons, Phobos and Deimos, but they are so small as to be microscopic on our 1-to-10-billion scale.

As we walk beyond Mars, we notice one of the most important characteristics of our solar system. In contrast to the few steps separating the four inner planets in the model solar system, the walk from Mars to Jupiter covers more than 50 meters, or more than half the length of a football field. As we look ahead, we see that increasingly large distances separate the remaining planets. Thus, the solar system has two major regions: the **inner solar system,** where Mercury, Venus, Earth, and Mars orbit the Sun relatively closely; and the **outer solar system,** where the remaining planets are widely separated. In

FIGURE 1.12 This NASA photo, taken by the Voyager 1 spacecraft, shows a portion of Jupiter's southern hemisphere and the moons Io and Europa. Io is about 350,000 km (217,000 mi) above Jupiter's Great Red Spot, and Europa is about 600,000 km (372,000 mi) above Jupiter's clouds. Io is a bizarre world on which volcanoes erupt incessantly. Europa is an ice-crusted world that may have a subsurface ocean deeper than any ocean on Earth.

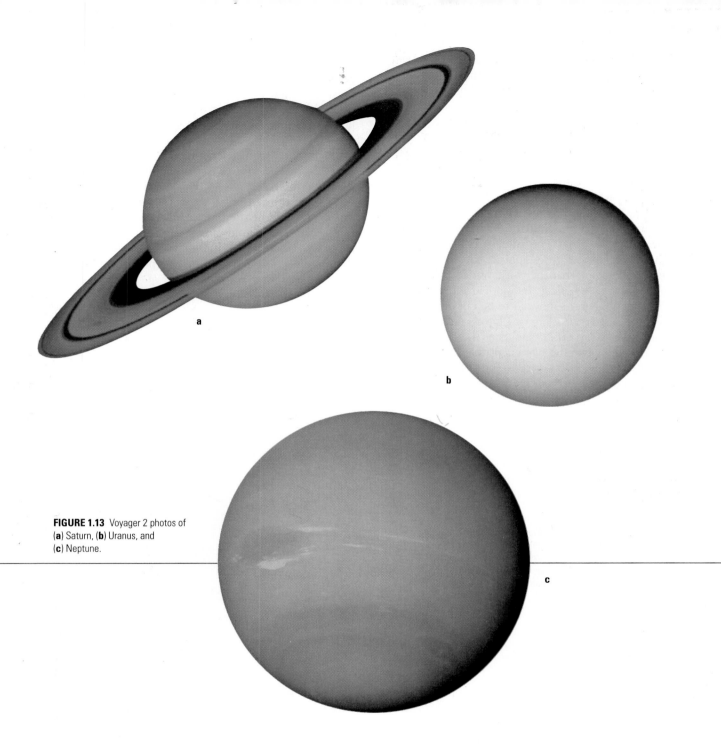

FIGURE 1.13 Voyager 2 photos of (**a**) Saturn, (**b**) Uranus, and (**c**) Neptune.

between we find the **asteroid belt,** where thousands of asteroids orbit the Sun. Despite the large number of asteroids, their total mass is less than that of any of the inner planets, and most of them are no bigger than the small Martian moons.

Upon reaching the model Jupiter, we immediately see that it is a giant in comparison to the planets of the inner solar system (Figure 1.12). Indeed, Jupiter is more massive than all the rest of the planets combined. At least 16 moons orbit Jupiter. The four

largest (Io, Europa, Ganymede, and Callisto) were discovered by Galileo and are fascinating worlds in their own right—easily visible on the scale of the model solar system. Together, Jupiter and its moons look much like a solar system in miniature.

Walking the larger distances to Saturn, Uranus, and Neptune (Figure 1.13), we find that these planets bear many similarities to Jupiter. Like Jupiter, all these worlds are giants compared to the planets of the inner solar system and are orbited by their own sets of intriguing moons. Each is made largely of hydrogen and helium and has no solid surface to speak of. If you descended into the atmospheres of any of these planets, you would be crushed by the growing gas pressure long before you reached any solid core. All are also orbited by *rings* that are actually made up of countless individual particles ranging in size from dust specks to large boulders. However, only Saturn's rings are easily visible from Earth, and no one knew of the rings around the other planets until the 1970s and 1980s.

Leaving Neptune behind, we must walk some 140 meters to reach the model Pluto (at its average distance from the Sun). The grapefruit-size Sun, more than $\frac{1}{2}$ kilometer away, appears tiny from here. It is easy to imagine that the world of Pluto must be cold and dark. One look at the model Pluto, along with its moon Charon, shows it to be out of character with the rest of the planets. Pluto is neither large and gaseous like the other planets of the outer solar system nor rocky like the planets of the inner solar system. Instead, it is very small—the smallest planet by far—and made mostly of ices. Pluto's orbit is also unusual (Figure 1.14). While all the other planets

Onward to the Stars

Having completed our walk through the model solar system, let's briefly look back at the trip. Perhaps the most striking feature of the solar system is its *emptiness*. Although we imagined walking to model planets laid out in a straight line from the Sun, a better model would show their complete orbits. On the 1-to-10-billion scale, such a model would require an area of some 1.5 km² (0.6 mi²), equivalent to more than 300 football fields. Aside from the grapefruit-size Sun at the center, all we would find in the rest of this area would be the nine planets and their moons, plus scattered asteroids and comets—no object larger than marble-size Jupiter, and most far smaller than pinhead-size Earth. With this perspective on our solar system in mind, the next question is: Where are the stars on this scale?

TIME OUT TO THINK *Before reading on, make a guess. Look at the grapefruit-size model Sun that you placed nearby. Where do you think you'll find the next grapefruit-size star? Within a few blocks? Within your city? Farther away?*

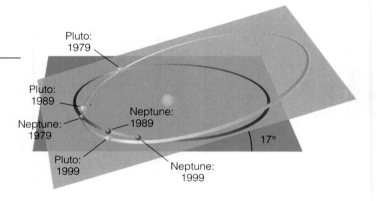

FIGURE 1.14 Pluto's orbit is highly elongated compared to the nearly circular orbits of the other planets. As a result, Pluto is closer to the Sun than Neptune for 20 years out of each of its 248-year orbits of the Sun; this was the case between 1979 and 1999. Pluto's orbit is also inclined by 17° to the ecliptic plane. Thus, it does not actually cross Neptune's orbit, and there is no danger of a collision between Neptune and Pluto.

orbit the Sun along nearly circular paths, Pluto's orbit is highly elongated and substantially inclined relative to the orbits of the other planets.

Although these characteristics of Pluto make it a "misfit" among the planets, they give it much in common with the only other inhabitants of the outskirts of the solar system: the balls of ice and dust that we call *comets*. You are probably familiar with the occasional appearance of a comet in the inner solar system, where the heat of the Sun evaporates some of its ice and it grows a beautiful tail. But these spectacular comets are rare visitors from among perhaps a trillion comets in our solar system, many of which orbit the Sun far beyond the orbit of Pluto.

The nearest star system to our own, called *Alpha Centauri*, contains three stars. The largest is about the same size as our Sun—another grapefruit on the 1-to-10-billion scale. Alpha Centauri is about 4.4 light-years away. Recall that a light-year is about 10 trillion kilometers, which becomes 1,000 kilometers on the 1-to-10-billion scale (because 10 trillion ÷ 10 billion = 1,000). The 4.4-light-year distance to Alpha Centauri therefore becomes about 4,400 kilometers (2,700 miles) on this scale. Thus, simply trying to *see* the largest star of Alpha Centauri from the Earth is equivalent to trying to look from New York at a grapefruit in Los Angeles (neglecting the problems introduced by the curvature of the Earth). And the other stars in our night sky are even farther away. No wonder the stars appear as mere points of light in the sky even when viewed through powerful telescopes.

Now consider the difficulty of seeing planets orbiting other stars. Imagine looking from New York and trying to see a pinhead-size model Earth orbiting a grapefruit-size model Sun in Los Angeles. You probably won't be surprised to learn that no Earth-size planets have yet been detected around any other star. Indeed, the bigger surprise may be that we *have* discovered Jupiter-size planets around other stars (by their gravitational effects, not by direct imaging). With planet-detecting technology improving rapidly, we will probably know whether Earth-size planets orbit nearby stars within a couple of decades.

The Milky Way Galaxy and Beyond

The vast separation between our solar system and Alpha Centauri is typical of the separations among star systems here in the outskirts of the Milky Way Galaxy. Thus, the 1-to-10-billion scale is useless for modeling even just a few dozen of the nearest stars, because they could not all be spaced properly on the surface of the Earth. Visualizing the entire galaxy is even more difficult and clearly requires a new scale.

Let's reduce our scale further by a factor of 1 billion (making it a scale of 1 to 10^{19}). On this new scale, each light-year becomes 1 millimeter; for example, the 4.4-light-year separation between the Sun and Alpha Centauri becomes just 4.4 millimeters on this scale—smaller than the width of your little finger. Aside from the fact that the stars themselves become about the size of individual atoms, this new scale makes it easy to visualize the Milky Way Galaxy. Its 100,000-light-year diameter becomes 100 meters on this scale, or about the length of a football field.

TIME OUT TO THINK *Visualize (or go to) a football field. Imagine our scale model of the galaxy centered over midfield. Standing on the 20-yard line puts you at about the correct location for our solar system. Hold your thumb and forefinger about 4 millimeters apart to represent the separation between the Sun and Alpha Centauri. Now think about how this scale compares to our earlier scale of the solar system. How do these scale models affect your perspective on our place in the universe?*

Another way to put the galaxy in perspective is to consider its number of stars, which is more than 100 billion. Imagine that, tonight, you are having difficulty falling asleep (perhaps because you are contemplating the enormity of the universe). Instead of counting sheep, you decide to count stars. If you are

able to count about one star each second, how long would it take you to count the 100 billion stars in the Milky Way?

Clearly, the answer is 100 billion (10^{11}) seconds. But how long is that? Amazingly, 100 billion seconds turns out to be more than 3,000 years. (You can confirm this by dividing 100 billion by the number of seconds in 1 year.) Thus, you would need thousands of years just to *count* the stars in the Milky Way Galaxy. And that assumes that you never take a break—no sleeping, no eating, and absolutely no dying!

As incredible as the scale of our galaxy may seem, the Milky Way is only one of some 100 billion galaxies in the observable universe. If we assume 100 billion stars per galaxy, then the total number of stars in the observable universe is 10,000,000,000,000,000,000,000 (10^{22}). How big is this number? Visit a beach. Run your hands through the fine-grained sand. Try to imagine counting every one of the tiny grains of sand as they slip through your fingers. Then imagine counting every grain of sand on the beach where you are sitting. Next think about counting *all* the grains of sand on *all* the beaches everywhere on Earth. The number you would count would be less than the number of stars in the observable universe.

The Scale of Time

Now that we have developed some perspective on the scale of space, we can do the same for the scale of time. Imagine the entire history of the universe, from the Big Bang to the present, compressed into a single year. We can represent this history with a *cosmic calendar,* on which the Big Bang takes place at

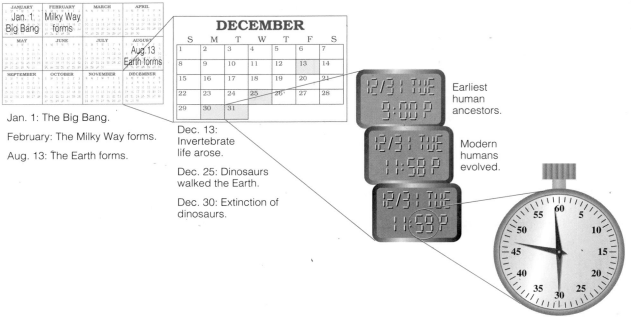

FIGURE 1.15 The cosmic calendar compresses the history of the universe into 1 year; this version assumes that the universe is 12 billion years old. The stopwatch shows the last minute of the last hour of the last day (December 31); it is only here that human civilization has taken shape.

Jan. 1: The Big Bang.

February: The Milky Way forms.

Aug. 13: The Earth forms.

Dec. 13: Invertebrate life arose.

Dec. 25: Dinosaurs walked the Earth.

Dec. 30: Extinction of dinosaurs.

Earliest human ancestors.

Modern humans evolved.

30 seconds: Agriculture arose.

47 seconds: Pyramids built.

59 seconds: Kepler and Galileo proved Earth a planet.

the first instant of January 1 and the present day is just before the stroke of midnight on December 31 (Figure 1.15). If we assume that the universe is about 12 billion years old (recall that we think the age is between 10 and 16 billion years), then each month on the cosmic calendar represents about 1 billion years.

On this scale, the Milky Way Galaxy probably formed sometime in February. Many generations of stars lived and died in the subsequent cosmic months, enriching the galaxy with the heavier elements from which we and the Earth are made. Not until about August 13, which represents a time 4.6 billion years ago, did our solar system form from a cloud of gas and dust in the Milky Way.

No one knows exactly when the earliest life arose on Earth, but it certainly was within the first billion years, or by mid-September on the cosmic calendar. However, for most of the Earth's history, living organisms remained relatively primitive and microscopic in size. Larger invertebrate life arose only about 600 million years ago, or about December 13 on the cosmic calendar. The age of the dinosaurs began shortly after midnight on Christmas (December 25) on the cosmic calendar and came to a sud-

den end around 2 A.M. on December 30 (65 million years ago in real time).

With the dinosaurs gone, small furry mammals inherited the Earth. Some 60 million years later, or around 9 P.M. on December 31 of the cosmic calendar, the earliest hominids (human ancestors) walked upright. Most of the major events of human history took place within the final seconds of the final minute of the final day on the cosmic calendar. Agriculture arose about 30 seconds ago. The Egyptians built the Pyramids about 13 seconds ago. It was only about 1 second ago on the cosmic calendar that Kepler and Galileo first proved that the Earth is a planet. The average college student was born less than 0.1 second ago, around 11:59:59.9 P.M. on December 31. On this scale of cosmic time, the human species is the youngest of infants, and a human lifetime is a mere blink of an eye.

1.3 Spaceship Earth

Now that we have discussed the scale of space and time, the next step in our "big picture" overview is understanding motion in the universe. Wherever you are as you read this book, you probably have the feeling that you're "just sitting here." Nothing could be further from the truth.

In fact, you are being spun in circles as the Earth rotates, you are racing around the Sun in the Earth's orbit, and you are careening with the Milky Way Galaxy through the cosmos. In the words of the noted inventor and philosopher R. Buckminster Fuller (1895–1983), you are a traveler on *spaceship Earth*.

Rotation and Revolution

The Earth **rotates** on its axis once each day. (The axis is a line joining the North and South poles and passing through the center of the Earth.) As a result, you are whirling around the Earth's axis at a speed of 1,000 kilometers per hour (600 miles per hour) or more—faster than most airplanes travel (Figure 1.16). It is this rotation that makes the Sun and stars appear to make their daily circles around the sky. You experience sunrise when the rotating Earth first allows you to glimpse the Sun in the east, noon when your location faces directly toward the Sun, and sunset just as your location rotates to the Earth's night side, where the Sun is hidden from view. (Because the Earth orbits the Sun at the same time it rotates, our 24-hour day is slightly longer—by about 4 minutes—than the Earth's actual rotation period.)

The Earth **revolves** around the Sun once each year, following an orbit that is not quite circular (Figure 1.17). The *average* distance of the Earth from the Sun is called an **astronomical unit,** or **AU** (more technically, an astronomical unit is the *semimajor axis* of the Earth's orbit):

$$1 \text{ AU} \approx 150 \text{ million km (93 million mi)}$$

The Earth travels slightly faster in its orbit when it is nearer to the Sun and slightly slower when it is farther from the Sun. But, at all times, the Earth is carrying you around the Sun at a speed in excess of 100,000 kilometers per hour (60,000 miles per hour).

The plane of the Earth's orbit around the Sun is called the **ecliptic plane** (Figure 1.18). The Earth's rotation axis happens to be tilted by $23\frac{1}{2}°$ from a line *perpendicular* to the ecliptic plane, pointing nearly in the direction of Polaris, the North Star. Viewed from above the North Pole, the Earth both rotates and revolves counterclockwise.

The combination of the Earth's rotation and its revolution around the Sun explains why we have seasons. Because the Earth's rotation axis remains pointed in the same direction (toward Polaris) throughout the year, its orientation *relative to the Sun* changes as it orbits the Sun, as shown in Figure 1.19. The seasonal changes that we generally notice—longer and warmer days in summer, and

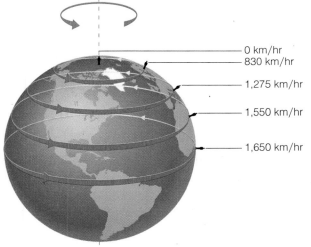

FIGURE 1.16 As the Earth rotates, your speed around the Earth's axis depends on your latitude. Unless you live very near the North or South Pole, your speed is over 1,000 km/hr.

0 km/hr
830 km/hr
1,275 km/hr
1,550 km/hr
1,650 km/hr

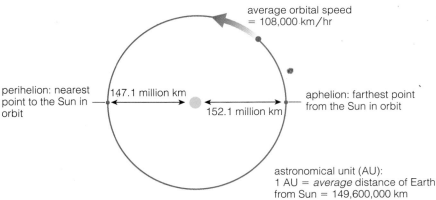

FIGURE 1.17 Basic properties of the Earth's orbit.

average orbital speed = 108,000 km/hr

perihelion: nearest point to the Sun in orbit

147.1 million km

152.1 million km

aphelion: farthest point from the Sun in orbit

astronomical unit (AU):
1 AU = *average* distance of Earth from Sun = 149,600,000 km

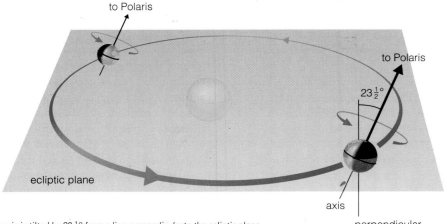

FIGURE 1.18 The Earth's axis is tilted by $23\frac{1}{2}°$ from a line *perpendicular* to the ecliptic plane. Note that, as viewed from above the North Pole, the Earth rotates *and* revolves counterclockwise.

1. Spring Equinox
Spring begins in the Northern Hemisphere, fall in the Southern Hemisphere.

2. Summer Solstice
Summer begins in the Northern Hemisphere, winter in the Southern Hemisphere.

4. Winter Solstice
Winter begins in the Northern Hemisphere, summer in the Southern Hemisphere.

3. Fall Equinox
Fall begins in the Northern Hemisphere, spring in the Southern Hemisphere.

FIGURE 1.19 Seasons occur because, even though the Earth's axis remains pointed toward Polaris throughout the year, the orientation of the axis *relative to the Sun* changes as the Earth orbits the Sun. Around the time of the summer solstice, the Northern Hemisphere has summer because it is tipped toward the Sun, and the Southern Hemisphere has winter because it is tipped away from the Sun. The situation is reversed around the time of the winter solstice when the Northern Hemisphere has winter and the Southern Hemisphere has summer. At the equinoxes, both hemispheres receive equal amounts of light.

shorter and cooler days in winter—arise because of changes in how directly the Sun's rays strike a particular location on Earth. When the rays are fairly direct, as is the case for most locations in the Northern Hemisphere on one side of the orbit and most locations in the Southern Hemisphere on the opposite side of the orbit, the Sun's path through the sky is long and high and the days are warm. (By analogy, think of how you sunburn more easily on parts of your body that face the Sun more directly.) Similarly, when the Sun's rays strike less directly, which again occurs for most places in the Northern and Southern hemispheres on opposite sides of the orbit, the Sun's path is short and low through the sky and the days are cool. Thus, the seasons are opposite in the Northern and Southern hemispheres.

Four points in the Earth's orbit have special names related to the seasons. The Earth is at point 1 in Figure 1.19 on about March 21 each year, which represents the **spring equinox** (or *vernal equinox*). Both hemispheres receive equal amounts of sunlight at this time; it is the beginning of spring for the Northern Hemisphere but the beginning of fall for the Southern Hemisphere. The Earth reaches point 2, which represents the **summer solstice,** around June 21—the day on which the Northern Hemisphere receives its most direct sunlight and the Southern Hemisphere receives its least direct sunlight. Point 3 represents the **fall equinox** (or *autumnal equinox*), which occurs around September 21; both hemispheres again receive the same amount of sunlight, but it is the beginning of fall in the Northern Hemisphere and the beginning of spring in the Southern Hemisphere. Finally, point 4 represents the **winter solstice,** which occurs around December 21, the day on which the Northern Hemisphere receives its least direct sunlight and the Southern Hemisphere receives its most direct sunlight.

TIME OUT TO THINK *By studying Figure 1.19, explain why regions near the Earth's equator do not experience seasons in the same way as midlatitudes (such as North America, Europe, and Australia). When is the Sun's path highest in the sky for equatorial regions?*

When asked the cause of the seasons, many people mistakenly answer that they are caused by variations in Earth's distance from the Sun. Indeed, in a famous survey, even many Harvard graduates gave this incorrect answer. By knowing that the Northern and Southern hemispheres experience opposite seasons, you can immediately realize that Earth's varying distance from the Sun *cannot* be the cause of the seasons; if it were, both hemispheres would have summer at the same time. Although Earth's distance from the Sun *does* vary slightly over the course of a year (Earth is closest to the Sun in January and farthest away in July), this effect is greatly overwhelmed by the way the tilt of the rotation axis causes the Northern and Southern hemispheres to alternately receive more or less direct sunlight. That is, Earth's varying distance from the Sun has no noticeable effect on our seasons. (Interestingly, this is not necessarily the case for other planets; seasons on Mars, for example, *are* affected by its varying distance from the Sun.)

Although the Earth's axis will remain pointed toward Polaris throughout our lifetimes, it has not always been so and will not always remain so. The reason is a gradual motion called **precession.** Like the axis of a spinning top, the Earth's axis sweeps out a circle (precesses)—but much more slowly (Figure 1.20). Each cycle of the Earth's precession takes

You probably know your astrological "sign." When astrology began a few thousand years ago, your sign was supposed to represent the constellation in which the Sun appeared on your birth date. However, for most people, this no longer is the case. For example, if your birthday is the spring equinox, March 21, a newspaper horoscope will show that your sign is Aries, but the Sun appears in Pisces on that date. In fact, your astrological sign corresponds to the constellation in which the Sun *would have appeared* on your birth date if you had lived about 2,000 years ago. Why? Because the astrological signs are based on the positions of the Sun among the stars as described by the Greek scientist Ptolemy in his book *Tetrabiblios*, which was written in about A.D. 150.

about 26,000 years. Today, the axis points toward Polaris, which is what makes it our North Star. In about 13,000 years, the axis will point nearly in the direction of the star Vega, making it our North Star. During most of the precession cycle, the axis does not point very near any bright star. Precession also changes the location in the Earth's orbit where the equinoxes and solstices occur. For example, the spring equinox today occurs when the Sun appears in the direction of the constellation Pisces, but 2,000 years ago it occurred when the Sun appeared in Aries. (That is why the spring equinox is sometimes called "the first point of Aries.")

FIGURE 1.20 (**a**) A spinning top wobbles, or *precesses,* more slowly than it spins. (**b**) The Earth's axis also precesses. Each precession cycle takes about 26,000 years. Note that the angle of the axis tilt remains about the same throughout the cycle, but the changing orientation of the axis means that Polaris is only a temporary North Star.

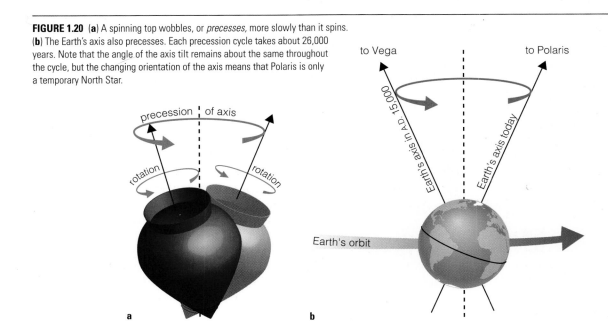

Traveling in the Milky Way Galaxy

In addition to moving within our solar system, we are also traveling with our solar system on a journey through the Milky Way Galaxy. For example, we are moving relative to nearby stars at an average speed of about 70,000 kilometers per hour (40,000 miles per hour)—about three times faster than the Space Shuttle orbits the Earth. In fact, all stars are in constant motion, and it is equally valid to say that the stars are moving relative to us as it is to say that we are moving relative to them.

If nearby stars move at such high speeds, why don't we see them racing around the sky? The answer lies with their vast distances from us. You've probably noticed that a distant airplane appears to move through your sky more slowly than one flying close overhead. If we extend this idea to the stars, we find that even at speeds of 70,000 kilometers per hour stellar motions would be noticeable to the naked eye only if we watched for thousands of years. Thus, the patterns in the constellations seem to remain fixed. Nevertheless, in 10,000 years the constellations will

be noticeably different from those we see today. In 500,000 years, they will be unrecognizable. If you could watch a time-lapse movie made over millions of years, you would see the stars of our *local solar neighborhood* (the Sun and nearby stars) racing around in seemingly random directions (Figure 1.21).

If you look closely at leaves floating in a stream, their motions relative to one another might appear random, just like the motions of stars in the local solar neighborhood. As you widen your view, you see that all the leaves are being carried in the same general direction by the downstream current. In the same way, if you widened your view beyond the local solar neighborhood, you would see that all the stars are moving in a simple and general way: The entire Milky Way Galaxy is rotating. Stars at different distances from the galactic center take different amounts of time to complete an orbit. Our solar system, located about 28,000 light-years from the galactic center, completes one orbit of the galaxy in about 230 million years (Figure 1.22). Even if you could watch from outside our galaxy, this motion would be unnoticeable to your naked eye. However, if you calculate the speed of our solar system as we orbit the center of the galaxy, you will find that it is close to 1 million kilometers per hour (600,000 miles per hour).

Before we discuss larger-scale motion, it is worth noting that the galactic rotation reveals one of the greatest mysteries in science—one that we will study in depth later in the book. The speeds at which stars orbit the galactic center depend on the strength of gravity, and the strength of gravity depends on how mass is distributed throughout the galaxy. Thus, careful study of the galactic rotation allows us to determine the distribution of mass in the galaxy. Remarkably, we find that the stars we see in the disk of the galaxy represent only the "tip of the iceberg" compared to the mass of the entire galaxy. That is, most of the mass of the galaxy is located outside the visible disk, in the galaxy's *halo* (Figure 1.23). We don't know the nature of this mass because we have not detected any light coming from it, and we therefore call it **dark matter.** Studies of other galaxies show that they are also made mostly of dark matter. In fact, most of the mass in the universe appears to be made of this mysterious dark matter, even though we do not know what it is.

The Expanding Universe

The billions of galaxies in the universe also move relative to one another. Within the Local Group (see Figure 1.1), the motions

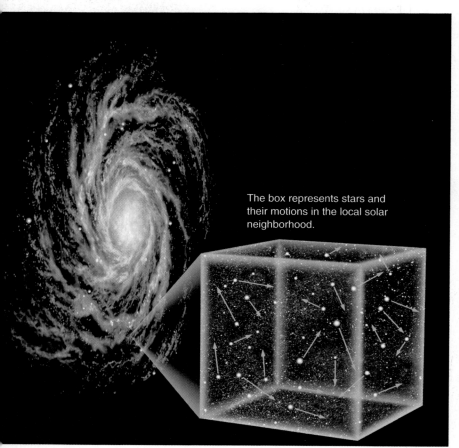

FIGURE 1.21 As shown in this painting, the local solar neighborhood is only a tiny portion of the Milky Way Galaxy. The stars in the local solar neighborhood actually move quite fast relative to our solar system, but the enormous distances between stars make this motion barely detectable on human time scales. If you could watch a time-lapse movie made over a time period of millions of years, you would see the stars in the local solar neighborhood racing around in seemingly random directions.

The box represents stars and their motions in the local solar neighborhood.

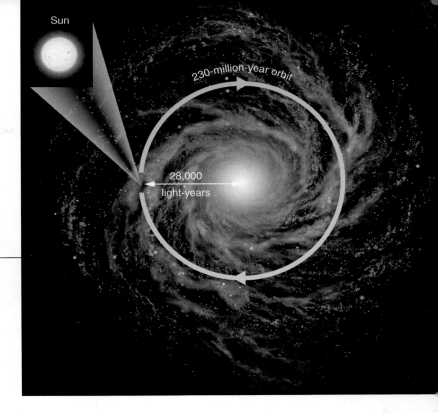

FIGURE 1.22 The entire Milky Way Galaxy rotates (in a direction that would tend to wind up the spiral arms). Our solar system is located about 28,000 light-years from the galactic center. At this distance, each orbit takes about 230 million years.

Sun

230-million-year orbit

28,000 light-years

FIGURE 1.23 Careful study of galactic rotation indicates that most of the mass in the galaxy is in the form of mysterious *dark matter*. (The dark matter may actually extend quite far out; see Figure 18.1.)

Most of the galaxy's light comes from stars and gas in the galactic disk...

...but most of the galaxy's mass lies above and below the disk in the *halo*.

are about as we might expect: Some of the galaxies move toward the Milky Way Galaxy, some move away, and some move in more complex ways. (For example, two small galaxies—called the *Large Magellanic Cloud* and the *Small Magellanic Cloud*—apparently orbit the Milky Way.) Again, the speeds are enormous by earthly standards but unnoticeable to our eyes. For example, the Milky Way is moving toward the Great Galaxy in Andromeda (M31) at a speed of about 300,000 kilometers per hour (180,000 miles per hour). At that rate, it will take us more than 90 million years to move just 1% of the 2.5-million-light-year distance to M31.

When we look outside the Local Group, however, we find two astonishing facts that were first discovered in the 1920s by Edwin Hubble, for whom the Hubble Space Telescope was named:

1. Virtually every galaxy outside the Local Group is moving *away* from us.

2. The more distant the galaxy, the faster it appears to be racing away from us.

Upon first learning of these two facts, you may be tempted to conclude that our Local Group suffers a cosmic case of chicken pox. However, there is a much more natural explanation: *The entire universe is expanding.* Although discussion of the details of this expansion will have to wait until later in the book, we can understand the basic idea by using the analogy of baking a raisin cake.

Imagine that you make a raisin cake and that the distance between adjacent raisins is 1 centimeter. You place the cake in the oven, where it expands as it bakes. After 1 hour, you remove the cake, which has expanded so that the distance between adjacent raisins is 3 centimeters (Figure 1.24).

Pick any raisin (it doesn't matter which one), call it the Local Raisin, and identify it in the pictures of the cake both before and after baking; Figure 1.24 shows one possible choice for the Local Raisin, with several other nearby raisins labeled. Note that, before baking, Raisin 1 is 1 centimeter away from the Local Raisin, Raisin 2 is 2 centimeters away, and Raisin 3 is 3 centimeters away. After baking, Raisin 1 is 3 centimeters away from the Local Raisin, Raisin 2 is 6 centimeters away, and Raisin 3 is 9 centimeters away. Nothing should seem surprising so far, as the new distances simply reflect the fact that the cake expanded uniformly everywhere.

Now suppose you live *inside* the Local Raisin. From your vantage point, the other raisins appear to move away from you as the cake expands. For example, Raisin 1 is 1 centimeter away before baking and 3 centimeters away after baking. It therefore appears to move 2 centimeters during the hour, so you'll see it moving away from you at a speed of 2 centimeters

FIGURE 1.24 An expanding raisin cake illustrates basic principles of the expansion of the universe.

Table 1.2 Distances and Speeds As Seen from the Local Raisin in Figure 1.24

Raisin Number	Distance Before Baking	Distance After Baking (1 hour later)	Speed
1	1 cm	3 cm	2 cm/hr
2	2 cm	6 cm	4 cm/hr
3	3 cm	9 cm	6 cm/hr
4	4 cm	12 cm	8 cm/hr
⋮	⋮	⋮	⋮
10	10 cm	30 cm	20 cm/hr
⋮	⋮	⋮	⋮

per hour. Raisin 2 is 2 centimeters away before baking and 6 centimeters away afterward; thus, it appears to move 4 centimeters during the hour of baking, giving it a speed of 4 centimeters per hour away from you. Table 1.2 lists the distances of other raisins before and after baking, along with their speeds as seen from the Local Raisin. Note that, although the entire cake expanded uniformly, from the vantage point of the Local Raisin each subsequent raisin moved away at increasingly faster speeds. Furthermore, because selection of the Local Raisin was arbitrary, you will find the same results no matter which raisin you choose to represent your Local Raisin.

If you now imagine the Local Raisin to represent our Local Group of galaxies, and the other raisins to represent more distant galaxies, you have a basic picture of the expansion of the universe. That is, *space* itself is growing, and more distant galaxies move away from us faster because they are carried along with this expansion like raisins in an expanding cake. The same effects of expansion would be seen from any place in the universe.

Let's summarize the motions we have covered. We spin around the Earth's axis as we revolve around the Sun. Our solar system moves among the stars of the local solar neighborhood, while this entire neighborhood orbits the center of the Milky Way Galaxy. Our galaxy, in turn, moves among the other galaxies of the Local Group, while the Local Group is carried along with the overall expansion of the universe. Spaceship Earth is indeed carrying us on a remarkable journey.

How vast those Orbs must be, and how inconsiderable this Earth, the Theatre upon which all our mighty Designs, all our Navigations, and all our Wars are transacted, is when compared to them. A very fit consideration, and matter of Reflection, for those Kings and Princes who sacrifice the Lives of so many People, only to flatter their Ambition in being Masters of some pitiful corner of this small Spot.

CHRISTIAAN HUYGENS, DUTCH ASTRONOMER AND SCHOLAR (c. 1690)

THE BIG PICTURE

In this first chapter, we developed a broad overview of our place in the universe. It is not yet necessary for you to remember all the details—everything presented in this chapter will be covered in greater detail later in the book. However, you should understand enough so that the following "big picture" ideas are clear:

- The Earth is not the center of the universe but instead a planet orbiting a rather ordinary star in the Milky Way Galaxy. The Milky Way Galaxy, in turn, is one of billions of galaxies in our observable universe.

- We are "star stuff." The atoms from which we are made began as hydrogen and helium in the Big Bang and were later fused into heavier elements in massive stars. When these stars died, they released these atoms into space, where our galaxy recycled them into our solar system when it formed some 4.6 billion years ago.

- Cosmic distances are literally astronomical, but we can put them in perspective with the aid of scale models and other scaling techniques. When you think about these enormous scales, don't forget that every star is a sun and every planet a unique world.

- We are latecomers on the scale of cosmic time. The universe was more than halfway through its history by the time our solar system formed, and then it took billions of years before humans arrived on the scene.

- All of us are being carried through the cosmos on spaceship Earth. Although we cannot feel this motion in our everyday lives, the associated speeds are surprisingly high. Learning about the motions of spaceship Earth not only gives us a new perspective on the cosmos but also helps us understand the nature and history of our universe.

Review Questions

1. What do we mean by a *geocentric* universe? About how long ago was the geocentric viewpoint shown to be incorrect?

2. Describe the major levels of structure in the universe.

3. What do we mean when we say that the universe is *expanding*? Why does an expanding universe imply a beginning called the *Big Bang*?

4. Summarize our cosmic origins and explain what Carl Sagan meant when he said that we are "star stuff."

5. Explain the statement *The farther away we look in distance, the further back we look in time.* What do we mean by the *observable universe*?

6. Describe key features of our solar system as they would appear on a scale of 1 to 10 billion.

7. How do the distances to the stars compare to distances within our solar system?

8. Describe a scale on which we could view the Milky Way Galaxy. How long would it take to count the stars in the Milky Way Galaxy?

9. How many galaxies are in the universe? How many stars are in the universe? Put these numbers in perspective.

10. Imagine that you are describing the cosmic calendar to friends. In your own words, give them a feel for how the human race fits into the scale of time.

11. Briefly summarize the major motions of the Earth through the universe.

12. What is an *astronomical unit?* What is the *ecliptic plane?*

13. Describe the cause of the Earth's seasons and explain why the seasons are opposite in the Northern and Southern hemispheres.

14. Given the relatively high speeds of stars in the local solar neighborhood, why don't we notice them racing around the sky?

15. Describe the raisin cake model of the universe, and explain how it shows that a uniform expansion would lead us to see more distant galaxies moving away from us at higher speeds.

Discussion Questions

1. *Vast Orbs.* The end-of-chapter quotation from Christiaan Huygens suggests that humans might be less inclined to war if everyone appreciated the Earth's place in the universe. Do you agree? Defend your opinion.

2. *Infant Species.* In the last few tenths of a second before midnight on December 31 of the cosmic calendar, we have developed an incredible civilization and learned a great deal about the universe, but we also have developed technology through which we could destroy ourselves. The midnight bell is striking, and the choice for the future is ours. How far do you think our civilization will survive into the next cosmic year? Defend your opinion.

Problems

1. *Sensible Statements?* Evaluate each of the following statements, and decide whether it is sensible. Explain.

   Example: I walked east from our base camp at the North Pole. *Solution:* The statement does not make sense because *east* has no meaning at the North Pole—all directions are south from the North Pole.

   a. The universe is between 10 and 16 billion years old.

b. The universe is between 10 and 16 billion light-years old.

c. It will take me light-years to complete this homework assignment!

d. Someday, we may build spaceships capable of traveling at a speed of 1 light-minute per hour. (Think carefully about this one!)

e. Astronomers recently discovered a moon that does not orbit a planet.

f. No one has ever seen a photograph of the Milky Way Galaxy taken from outside it.

g. The observable universe is the same size today as it was a few billion years ago.

2. *Tour Report.* Imagine that, following your visit to a scale model solar system, a friend asks you the following questions. Answer each question in one paragraph.

   a. Is the Sun really much bigger than the Earth?

   b. Is it true that the Sun uses nuclear energy?

   c. Would it be much harder to send humans to Mars than to the Moon?

   d. In elementary school, I heard that Neptune is farther from the Sun than Pluto. Is this true?

   e. I read that Pluto is not really a planet. What's the story?

   f. Why weren't there any stars besides the Sun in the scale model?

   g. What was the most interesting thing you learned during your tour?

3. *No Axis Tilt.* Suppose that the Earth's axis had no tilt. Would we still have seasons? Why or why not?

4. *Raisin Cake Universe.* Suppose that before baking all the raisins in a cake are 1 cm apart and after baking they are all 4 cm apart.

   a. Draw diagrams to represent your cake before and after baking.

   b. Identify one raisin as the Local Raisin on your diagrams. Construct a table similar to Table 1.2 showing the distances and speeds of other raisins as seen from the Local Raisin.

   c. In your own words, explain how your expanding cake is similar to the expansion of the universe.

5. *Light-Year.* Calculate the distance that light travels in 1 year. Show your work. (*Hint:* Multiply the speed of light, 300,000 km/s, by the number of seconds in a minute, the number of minutes in an hour, and so on until you have the distance light travels in a year.)

6. *Project: Make Your Own Time-Line.* Make a 1-meter-long time-line of the history of the universe. Show the locations of important events in cosmic history and in the Earth's history.

Discovering the ^{2} Universe for Yourself

It is an exciting time in the history of astronomy. A new generation of telescopes is probing the depths of the universe. Increasingly sophisticated space probes are collecting new data about the planets and other objects in our solar system. And rapid advances in computing technology are helping scientists analyze the vast amounts of new data and model the processes that occur in planets, stars, galaxies, and the universe.

One of the goals of this book is to help *you* share in the ongoing human adventure of astronomical discovery. One of the best ways to become a part of this adventure is to discover the universe for yourself by doing what other humans have done for thousands of generations: Go outside, observe the sky around you, and contemplate the awe-inspiring universe of which you are a part. In this chapter, we'll discuss a few key ideas that will help you understand what you see in the sky.

FIGURE 2.1 This photograph shows stars in the constellation Orion. Betelgeuse and Bellatrix represent the shoulders, while Saiph and Rigel are the feet. The three central stars are the belt, and the downward extension containing the Orion Nebula is the sword. All stars are points of light in the sky but appear different sizes in photos because brighter stars are overexposed. The crosses on bright stars are an artifact of the telescope used to take the photograph.

FIGURE 2.2 Red lines mark official borders of several constellations near Orion. Yellow lines connect recognizable patterns of stars within constellations. Sirius, Procyon, and Betelgeuse form a pattern spanning several constellations and called the Winter Triangle; it is easy to find on clear winter evenings.

2.1 Patterns in the Sky

Shortly after sunset, as daylight fades to darkness, the sky appears to fill slowly with stars. On clear, moonless nights far from city lights, as many as 2,000–3,000 stars may be visible to your naked eye. As you look at the stars, your mind might group them into many different patterns. If you observe the sky night after night or year after year, you will recognize the same patterns of stars.

People of nearly every culture have given names to patterns in the sky. The pattern that the Greeks named Orion, the hunter (Figure 2.1), was seen as a supreme warrior called _Shen_ by the ancient Chinese. Hindus in ancient India also saw a warrior, called _Skanda,_ who rode a peacock as the general of a great celestial army. The three stars of Orion's belt were seen as three fishermen in a canoe by Aborigines of northern Australia. As seen from southern California, these three stars climb almost straight up into the sky as they rise in the east, which may explain why the Chemehuevi Indians of the California desert saw them as a line of three sure-footed mountain sheep. These are but a few of the many names, each accompanied by a rich folklore, given to the pattern of stars that we call Orion.

Constellations

The patterns of stars seen in the sky are usually called _constellations_. In astronomy, however, the term **constellation** refers to a _region_ of the sky. Any place you point in the sky belongs to some constellation; familiar patterns of stars merely help locate particular constellations. For example, the constellation Orion includes all the stars in the familiar pattern of the hunter, _along with_ the region of the sky in which these stars are found (Figure 2.2).

The official borders of the constellations were agreed on in 1928 by members of the International Astronomical Union (IAU), an association of astronomers from around the world. The IAU divided the sky into 88 constellations whose borders correspond roughly to the star patterns recognized by Europeans. Thus, despite the wide variety of names given to patterns of stars by different cultures, the "official" names of constellations visible from the Northern Hemisphere can be traced back to the ancient Greeks and to other cultures of southern Europe, the Middle East, and northern Africa. No one knows exactly when these constellations first were named, although it was probably at least 5,000 years ago. The official names of the constellations visible from the

Southern Hemisphere are primarily those given by seventeenth-century European explorers.

Learning your way around the constellations is no more difficult than learning the way around your neighborhood, and recognizing the patterns of just 20–40 constellations is enough to make the entire sky seem familiar. The best way to learn the constellations is to go out and practice; it may help to start with a few visits to a planetarium and to purchase an inexpensive star chart to help you identify the patterns you see.

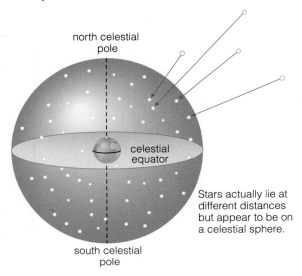

FIGURE 2.3 Our lack of depth perception when we look into space creates the illusion that the Earth is surrounded by a *celestial sphere*. Thus, stars that *appear* very close to one another in our sky may actually lie at very different distances from Earth.

The Celestial Sphere

The stars in a particular constellation may appear to lie close to one another, but this is an illusion: Each star in a particular constellation lies in a similar direction from Earth, but all do not necessarily lie at a similar distance (Figure 2.3). However, because we lack depth perception when we look into space, we get the illusion that all the stars lie at the same distance. Ancient people mistook this illusion for reality, imagining the Earth to be surrounded by a great **celestial sphere.** (Some ancient thinkers even imagined it to be made of glass.) Today, the illusion of a celestial sphere is still useful for learning about the sky, even though we know that the Earth does not really lie in the center of a giant ball. We give names to special locations on the imaginary celestial sphere. The point directly above the Earth's North Pole is called the **north celestial pole** (NCP), and the point directly above the Earth's South Pole is called the **south celestial pole.** The **celestial equator** represents an extension of the Earth's equator into space.

The stars form the fixed patterns of the constellations on the celestial sphere, while the Sun, Moon, and planets slowly wander among the stars. As we'll discuss shortly, the apparent wandering of the Moon and planets is fairly complex. The Sun, however, appears to follow a simple path around the celestial sphere called the **ecliptic,** which is the projection of the ecliptic plane [Section 1.3] into space. A model of the celestial sphere typically shows the patterns of the stars, the borders of the 88 official constellations, the ecliptic, and the celestial equator and poles (Figure 2.4).

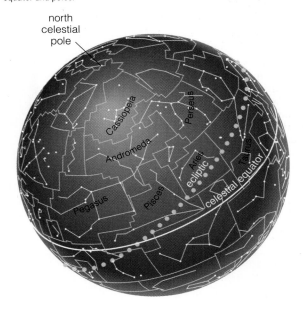

FIGURE 2.4 A model of the celestial sphere shows the patterns of the stars, the borders of the 88 official constellations, the ecliptic, and the celestial equator and poles.

Common Misconceptions: Stars in the Daytime

Because we don't see stars in the daytime, some people believe that the stars disappear in the daytime and "come out" at night. In fact, the stars are always present. The only reason you cannot see stars in the daytime is that their dim light is overwhelmed by the bright daytime sky. You *can* see bright stars in the daytime with the aid of a telescope, and you may see stars in the daytime if you are fortunate enough to observe a total eclipse of the Sun. Astronauts also see stars in the daytime: Above the Earth's atmosphere, where no air is present to scatter sunlight through the sky, the Sun is a bright disk against a dark sky filled with stars.

The Milky Way

As your eyes adapt to darkness at a dark site, you'll begin to see the whitish band of light called the *Milky Way*. It is from this band of light that our Milky Way Galaxy gets its name. You can see only part of the Milky Way at any particular time, but it stretches all the way around the celestial sphere. If you look carefully, you will notice that the Milky Way varies in width and has dark fissures running through it. The widest and brightest parts of the Milky Way are most easily seen from the Southern Hemisphere (Figure 2.5), which probably explains why the Aborigines of Australia gave names to the dark patterns they saw within the Milky Way in the same way that other cultures gave names to patterns of stars.

The band of light called the Milky Way bears an important relationship to the Milky Way Galaxy: *It traces the galactic plane as it appears from our location in the outskirts of the galaxy* (Figure 2.6). Recall that the Milky Way Galaxy is shaped like a thin pancake with a bulge in the middle (see Figure 1.23). We view the universe from our location more than halfway outward from the center

FIGURE 2.5 A "fish eye" view of the Milky Way in the Australian sky. Comet Hyakutake is also visible in this 1996 photo.

looking out of plane: clear view to distant universe

location of our solar system

looking into plane: the Milky Way band of light

looking out of plane: clear view to distant universe

FIGURE 2.6 Artist's conception of the Milky Way galaxy from afar, showing how the galaxy's structure affects our view from Earth. When we look *into* the galactic plane in any direction, our view is blocked by stars, gas, and dust. Thus, we see the galactic plane as the band of light we call the Milky Way, stretching a full 360° around our sky (i.e., around the celestial sphere). We have a clear view to the distant universe only when we look away from the galactic plane.

of this "pancake." When we look in any direction *within* the plane of the galaxy, we see countless stars, along with interstellar gas and dust. These stars and glowing clouds of gas form the band of light we call the Milky Way. The dark fissures appear in regions where particularly dense interstellar clouds obscure our view of stars behind them. The central bulge of the galaxy makes the Milky Way slightly wider in the direction of the galactic center, which is the direction of the constellation Sagittarius in our sky.

When we look in directions pointing *outward* from our location within the galactic plane, we see fewer stars, and there is relatively little gas and dust to obscure our view. Thus, we have a clear view to the far reaches of the universe, limited only by what our eyes or instruments allow. If you are fortunate enough to have a very dark sky, you may see a fuzzy patch in the constellation Andromeda (Figure 2.7). Although this patch may look like nothing more than a small cloud, you are actually seeing the Great Galaxy in Andromeda (M 31)—some 2.5 million light-years away.

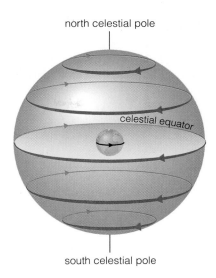

FIGURE 2.8 Because the Earth rotates daily from west to east (black arrow), the celestial sphere *appears* to rotate around us from east to west. Viewed from outside, the stars (and the Sun, Moon, and planets) therefore appear to make simple daily circles around us. The red circles represent the apparent daily paths of a few selected stars.

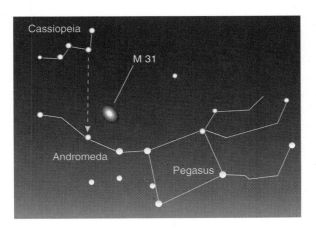

FIGURE 2.7 Location of M 31 in the night sky.

TIME OUT TO THINK *Contemplate the fact that light from M 31 journeyed through space for some 2.5 million years to reach you and is the combined light of hundreds of billions of stars. Do you think it possible that among some of those stars students like yourself are gazing outward in amazement at the Milky Way Galaxy?*

2.2 The Circling Sky

If you spend a few hours out under a starry sky, you'll see most stars rise and set much like the Sun and Moon. In reality, it is not the stars that are moving, but us. The Sun and the stars appear to rise in the east and set in the west because the Earth rotates in the opposite direction, from west to east. In fact, the Earth's rotation makes the entire celestial sphere appear to rotate around us each day. If you could view the celestial sphere from outside, the daily circles of the stars would appear very simple (Figure 2.8).

However, because we live on the Earth, the ground beneath us always blocks our view of half of the celestial sphere. As a result, we see only *half* of the celestial sphere at any one moment. The particular half that we see depends on the time, the date, and our location on Earth. In addition, the simple daily paths of the stars look somewhat more complex in our sky than they do when viewed from outside the celestial sphere. Let's investigate.

The Dome of the Sky

Standing in a flat, open field, your **horizon** marks a clear boundary that cuts the celestial sphere in half (Figure 2.9). The half that lies above your horizon makes your sky appear to take the shape of a dome. The point directly overhead is your **zenith.** Your **meridian** is an imaginary half-circle stretching from your horizon due south, through your zenith, to your horizon due north. You can pinpoint the position of any object in your sky by stating its **direction** along your horizon (more technically measured as *azimuth*) and its **altitude** above your horizon. For

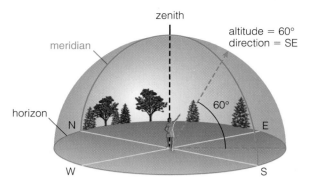

FIGURE 2.9 Definitions in your local sky.

example, Figure 2.9 shows a person pointing to a star located in a southeasterly direction at an altitude of 60°. (We cannot define a direction for the zenith, but it is uniquely identified as the only point with altitude 90°.)

Because of our lack of depth perception in the sky, we cannot tell the true sizes of objects or the true distances between objects just by looking. For example, the Sun and Moon look about the same size in our sky, but the Sun is actually about 400 times larger

in diameter than the Moon. We can, however, measure *angles* in the sky. The **angular size** of an object such as the Sun or the Moon is the angle that it appears to span in our field of view. The **angular distance** between a pair of objects is the angle that appears to separate them. For example, Figure 2.10 shows that the angular size of the Moon is about $\frac{1}{2}°$ (about the same as that of the Sun), while the angular distance between the pointer stars of the Big Dipper is about 5°. Note that you can make rough estimates of angles in the sky by using your outstretched hand as a simple measuring device.

For more precise astronomical measurements, we subdivide each degree into 60 **arcminutes** and subdivide each arcminute into 60 **arcseconds.** Thus, there are 60 arcseconds in 1 arcminute, 60 arcminutes in 1°, and 360° in a full circle.

Variation with Latitude

The location of any point on Earth can be described by its north or south **latitude** and its east or west **longitude** (Figure 2.11). Latitude is defined to be 0° at the equator, so the North Pole and the South Pole have latitude 90°N and 90°S, respectively. Note that "lines of latitude" actually are circles running parallel to the equator, while "lines of longitude" are semicircles extending from the North Pole to the South Pole. The line of longitude passing through Greenwich, England, is defined to have a longitude of 0° (this line is sometimes called the *prime meridian*); the choice of Greenwich was made by international treaty in 1884. Stating a latitude and a longitude

a

b

c

FIGURE 2.10 We see angular sizes or distances, rather than actual sizes or distances, when we look at objects in the sky. (**a**) The angular size of the Moon is about $\frac{1}{2}°$ (**b**) The angular distance between the pointer stars of the Big Dipper is about 5°. (**c**) You can estimate angular sizes or distances with your outstretched hand.

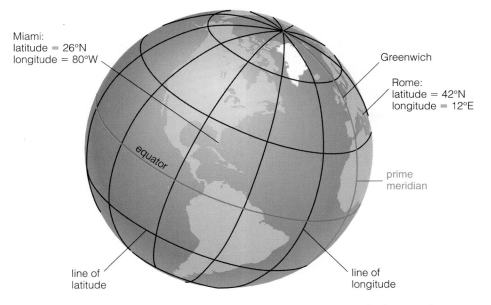

Miami:
latitude = 26°N
longitude = 80°W

Greenwich

Rome:
latitude = 42°N
longitude = 12°E

equator

prime
meridian

line of
latitude

line of
longitude

FIGURE 2.11 We can locate any place on the Earth's surface by its latitude and longitude. Latitude measures angular distance north or south of the equator. Longitude measures angular distance east or west of the prime meridian (which passes through Greenwich, England).

pinpoints a location. For example, Figure 2.11 shows that Rome lies at about 42°N latitude and 12°E longitude and that Miami lies at about 26°N latitude and 80°W longitude.

If we combine the idea of the sky's dome-shaped appearance on the ground with the idea that stars make simple daily circles around the celestial sphere, we can see why the apparent paths of stars depend on *latitude.* Your latitude determines your orientation on Earth relative to the celestial sphere—"down" is toward the center of the Earth, and "up" (toward your zenith) is directly away from the center of the Earth. Figure 2.12 shows sample orientations relative to the celestial sphere for mid-latitudes in each hemisphere. If you study the diagrams carefully, you'll discover the following key features:

- In the Northern Hemisphere, the daily circles of stars relatively near the north celestial pole remain constantly above the horizon. We say that such stars are **circumpolar.** They never rise or set but instead make daily, counterclockwise circles around the north celestial pole (Figure 2.13). In the Southern Hemisphere, circumpolar stars make daily clockwise circles around the south celestial pole.

- In the Northern Hemisphere, the daily circles of stars relatively near the south celestial pole remain constantly below the horizon, so these stars can never be seen. In the Southern Hemisphere, stars near the north celestial pole are never visible. Thus, different sets of constellations are visible in northern and southern skies.

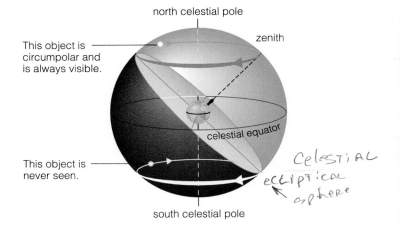

north celestial pole

zenith

This object is circumpolar and is always visible.

celestial equator

This object is never seen.

Celestial

ecliptical
sphere

south celestial pole

a Northern Hemisphere

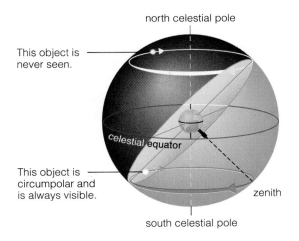

north celestial pole

This object is never seen.

celestial equator

This object is circumpolar and is always visible.

zenith

south celestial pole

b Southern Hemisphere

FIGURE 2.12 Your latitude determines the portion of the celestial sphere visible in your sky. (**a**) A Northern Hemisphere sky. (**b**) A Southern Hemisphere sky.

FIGURE 2.13 This time-exposure photograph, taken at Arches National Park in Utah, shows how the Earth's rotation causes stars to make daily circles around the sky. The north celestial pole lies at the center of the circles, and we say that nearby stars are circumpolar. For stars farther from the north celestial pole, we see only portions of their daily circles, which is why they appear to rise in the east and set in the west.

■ All other stars (those that are neither circumpolar nor never visible) rise daily in the east and set daily in the west. Their precise paths depend on their precise locations on the celestial sphere.

■ If you study the geometry of the diagrams carefully, you'll see that in the Northern Hemisphere *the altitude of the north celestial pole in your sky is equal to your latitude.* For example, if the north celestial pole appears in your sky 40° above your horizon, your latitude is 40°N. Similarly, in the Southern Hemisphere the altitude of the south celestial pole in your sky is equal to your latitude.

The last feature is very useful for navigation, since it allows you to determine your latitude just by finding the celestial pole in your sky. Finding the north celestial pole is fairly easy because it lies very close to the star Polaris (Figure 2.14a). In the Southern Hemisphere, you can find the south celestial pole with the aid of the Southern Cross (Figure 2.14b).

TIME OUT TO THINK *Answer the following questions for* your *latitude: Where is the north (or south) celestial pole in your sky? Where should you look to see circumpolar stars? What portion of the celestial sphere is never visible in your sky?*

Common Misconceptions:
What Makes the North Star Special?

Most people are aware that the North Star, Polaris, is a special star. Contrary to a relatively common belief, however, it is *not* the brightest star in the sky; more than 50 other stars are either considerably brighter or comparable in brightness. Polaris is special because it so closely marks the direction of due north and because its altitude in your sky is nearly equal to your latitude on Earth, making it very useful in navigation.

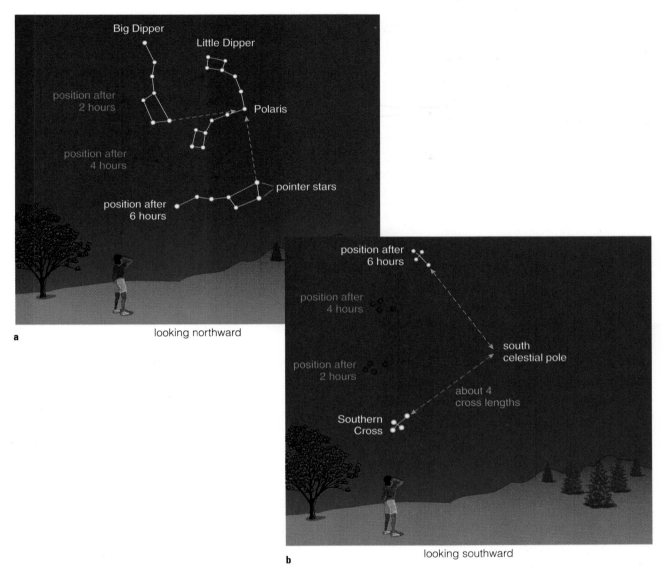

FIGURE 2.14 The altitude of the celestial pole in your sky is equal to your latitude. (**a**) In the Northern Hemisphere, the pointer stars of the Big Dipper point to Polaris, which lies within 1° of the north celestial pole. This diagram shows how the orientation of these stars changes over a period of 6 hours. Note that the sky appears to turn counterclockwise around the north celestial pole. (**b**) In the Southern Hemisphere, the Southern Cross points to the south celestial pole, which is not marked by any bright star. Note that the sky appears to turn clockwise around the south celestial pole.

Seasonal Changes in the Sky

The basic patterns of motion in the sky remain the same from one day to the next: The Sun, Moon, planets, and stars trace daily circles around the sky, and those bodies that are not circumpolar from your latitude rise in the east and set in the west. However, if you observe the sky night after night, you will notice changes that cannot be seen in a single night. One pattern that becomes obvious after a few months is that the constellations visible at a particular time of night change with the seasons. For example, Orion is prominent in the evening sky in February, but in September it can be seen only shortly before dawn.

The nighttime constellations change with the seasons because of the Earth's orbit around the Sun. Viewed from Earth, the Sun appears to move steadily *eastward* around the celestial sphere over the course

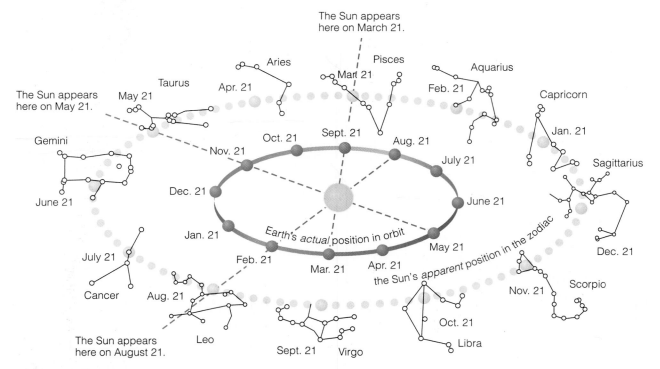

FIGURE 2.15 This diagram shows why the Sun appears to move steadily eastward around the ecliptic each year, passing through the constellations that make up the zodiac. If you look at the Earth's position in its orbit on May 21, for example, you'll see that the Sun appears to be in the constellation Taurus. Similarly, from the Earth's position on August 21, the Sun appears to be in Leo. You can use this diagram to determine the constellation in which the Sun appears on various dates of the year.

of the year (Figure 2.15). We've already identified the Sun's apparent path around the celestial sphere as the _ecliptic_ (see Figure 2.4). The constellations along this path are called the constellations of the **zodiac.** (Tradition places 12 constellations along the zodiac, but the official constellation borders include a thirteenth constellation, Ophiuchus.) The Sun's apparent location along the ecliptic determines which constellations we see at night. In late August, for example, the Sun appears to be in the constellation Leo, so you won't see the stars of Leo at all—they're above your horizon only during the daytime. Aquarius, however, will be visible on your meridian at midnight because it is opposite Leo on the celestial sphere. Six months later, in February, it is Aquarius that cannot be seen and Leo that is on the meridian at midnight.

TIME OUT TO THINK _Based on Figure 2.15 and today's date, where does the Sun currently appear among the zodiac constellations? What zodiac constellation will be on your meridian at midnight? What zodiac constellation will you see in the west shortly after sunset? Go outside at night to confirm your answers._

The Sun's path through the daytime sky also changes with the seasons. Recall that the Earth's orientation _relative to the Sun_ changes during the year because of the $23\frac{1}{2}°$ tilt of the axis (see Figure 1.19). This is also why the ecliptic crosses the celestial equator at angles of $23\frac{1}{2}°$ on the celestial sphere (see Figure 2.4). When the Sun appears north of the celestial equator, the Sun's daily path is long and high for the Northern Hemisphere (making the long, hot days of summer) but short and low for the Southern Hemisphere (making the short, cool days of winter). The situation for the two hemispheres is reversed when the Sun appears south of the celestial equator. At any particular location, you can notice this annual change by observing the Sun's position in your sky at the same time each day (Figure 2.16).

Moreover, just as the daily path of any star through your sky depends on your latitude, so does the daily path of the Sun at any time of year. During the summer, the Sun's daily path is longer at higher latitudes than at latitudes nearer the equator. The Sun's daily path is correspondingly shorter at higher latitudes in the winter. That is why higher latitudes have longer summer days and longer winter nights. In fact, the Sun becomes circumpolar at very high latitudes (within the _Arctic_ and _Antarctic circles_) in the summer. The Sun never sets during these summer days, making what we call the _land of the midnight Sun_ (Figure 2.17). Of course, the name "land of noon darkness" is more appropriate in the winter, when the Sun never rises above the horizon at these high latitudes.

FIGURE 2.16 This composite photograph shows images of the Sun snapped at 10-day intervals over an entire year, always from the same location and at the same mean solar time. The three bright streaks show the path of the Sun's rise on three particular dates.

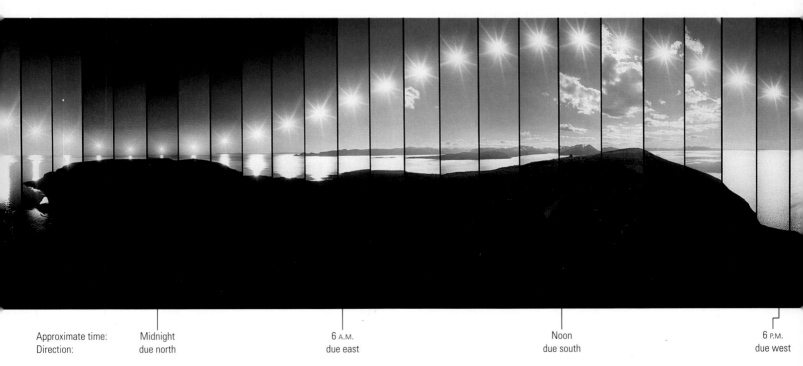

Approximate time:	Midnight	6 A.M.	Noon	6 P.M.
Direction:	due north	due east	due south	due west

FIGURE 2.17 This sequence of photos shows the progression of the Sun around the horizon on the summer solstice at the Arctic Circle.

Timekeeping

The Sun's motion through the sky forms the basis of our systems for keeping track of time. For example, you are probably aware that the Sun rises in the east, crosses the meridian around noon, and sets in the west. Thus, we call the morning hours A.M. for *ante meridian* (*ante* means "before") and the afternoon hours P.M. for *post meridian*. Note that the Sun generally does not cross the meridian *precisely* at noon, for several reasons. One reason is that we divide the world into time zones, so that everyone within a particular time zone keeps the same time regardless of the precise position of the Sun in their sky. Another is that many states and countries switch to *daylight saving time* in the spring and summer, which generally makes the Sun's crossing of the meridian come closer to 1 P.M. than to noon.

The origins of our calendar go back to ancient Egypt. By 4200 B.C., the Egyptians were using a calendar that counted 365 days in a year. Because the length of a year actually is closer to $365\frac{1}{4}$ days, the Egyptian calendar drifted out of synchronization with the seasons by about 1 day every 4 years. To keep the seasons and the calendar synchronized, in 46 B.C. Julius Caesar decreed the adoption of a new calendar. This **Julian calendar** introduced the concept of a **leap year:** Every fourth year has 366 days, rather than 365, so that the average length of the calendar year is $365\frac{1}{4}$ days.

Unfortunately, the Julian calendar also slowly drifted out of synchronization with the seasons because a year is *not exactly* $365\frac{1}{4}$ days. Today, we use a modification of the Julian calendar called the **Gregorian calendar,** which was introduced by Pope Gregory XIII in 1582. On this calendar, there is an exception to the pattern of having a leap year every 4 years: A century year is not a leap year unless that year is evenly divisible by 400. For example, 1700, 1800, and 1900 are *not* leap years because none of these numbers is divisible by 400. But 2000 *is* a leap year because 2000 is divisible by 400.

TIME OUT TO THINK *Will the year 2100 be a leap year? Explain.*

2.3 The Moon, Our Constant Companion

Like all objects on the celestial sphere, each day the Moon rises in the east and sets in the west. Like the Sun, the Moon also moves gradually eastward through the constellations of the zodiac. However, it takes the Moon only about a month to make a complete circuit around the celestial sphere. If you carefully observe the Moon's position relative to bright stars, you can notice the Moon's drift among the constellations in just a few hours.

As the Moon moves through the sky, its appearance and the times at which it rises and sets all change with the cycle of **lunar phases** (Figure 2.18). Each complete cycle from one new moon to the next takes about $29\frac{1}{2}$ days—hence the origin of the word *month* (think of "moonth"). The easiest way to understand the lunar phases is with a simple demonstration, using a small ball to represent the Moon while your head represents the Earth. If it's daytime and the Sun is shining, take your ball outside and watch how you see phases as you move the ball around your head; if it's dark or cloudy, you can place a flashlight a few meters away to represent the Sun. As you hold your ball at various places in its "orbit" around your head, you'll observe that phases result from just two basic facts:

1. At any particular time, half of the ball faces the Sun (or flashlight) and therefore is bright, while the other half faces away from the Sun and therefore is dark.

2. As you look at the ball, you see some combination of its bright and dark faces. This combination is the phase of the ball.

We see lunar phases for the same reason: Half of the Moon is always illuminated by the Sun, but the amount of this illuminated half that we see from Earth depends on the Moon's position in its orbit. As indicated in Figure 2.18, the time of day during which the Moon is visible also depends on its phase. For example, full moon occurs when the Moon is opposite the Sun in the sky, so the full moon rises around sunset, reaches the meridian at midnight, and sets around sunrise.

TIME OUT TO THINK *Suppose you go outside in the morning and notice that the visible face of the Moon is half light and half dark. Is this a first-quarter or third-quarter moon? How do you know? (Hint: Study Figure 2.18.)*

Common Misconceptions: Moon on the Horizon

You've probably noticed that the full moon appears far larger when it is near the horizon than when it is high in your sky. However, this appearance is an illusion. If you measure the angular size of the full moon on a particular night, you'll find it is the same whether it is near the horizon or high in the sky. In fact, the Moon's angular size in the sky depends only on its distance from the Earth. Although this distance varies over the course of the Moon's monthly orbit, it does not change enough to cause a noticeable effect on a single night. (You can eliminate the illusion by viewing the Moon upside down between your legs when it is on the horizon.)

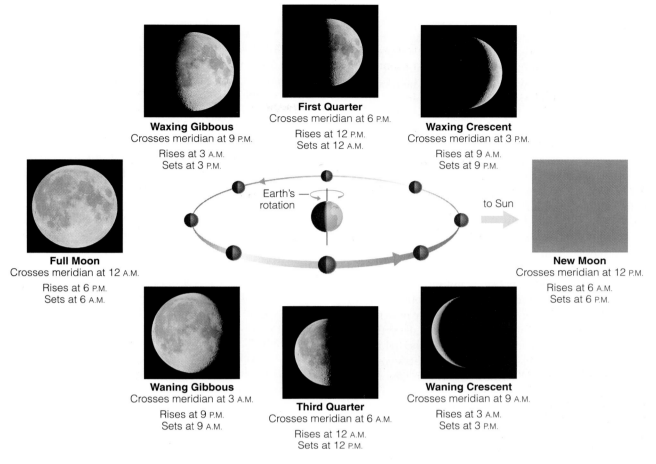

Waxing Gibbous
Crosses meridian at 9 P.M.

Rises at 3 A.M.
Sets at 3 P.M.

First Quarter
Crosses meridian at 6 P.M.
Rises at 12 P.M.
Sets at 12 A.M.

Waxing Crescent
Crosses meridian at 3 P.M.

Rises at 9 A.M.
Sets at 9 P.M.

Earth's rotation

to Sun

Full Moon
Crosses meridian at 12 A.M.

Rises at 6 P.M.
Sets at 6 A.M.

New Moon
Crosses meridian at 12 P.M.

Rises at 6 A.M.
Sets at 6 P.M.

Waning Gibbous
Crosses meridian at 3 A.M.

Rises at 9 P.M.
Sets at 9 A.M.

Third Quarter
Crosses meridian at 6 A.M.

Rises at 12 A.M.
Sets at 12 P.M.

Waning Crescent
Crosses meridian at 9 A.M.

Rises at 3 A.M.
Sets at 3 P.M.

FIGURE 2.18 The phases of the Moon depend on its location relative to the Sun in its orbit. To understand rise and set times, imagine yourself standing on the rotating Earth and looking at the Moon in a particular phase. (The photo for new moon shows only a blue sky because a new moon is above the horizon in daylight, but cannot be seen as it is very close to the Sun in the sky.)

Although we see many *phases* of the Moon, we do not see many *faces*. In fact, from the Earth we always see (nearly) the same face of the Moon. This tells us that the Moon must rotate once on its axis in the same time that it makes a single orbit of the Earth, as you can see with another simple demonstration. Place a ball on a table to represent the Earth, while you represent the Moon. Start by facing the ball. If you do not rotate as you orbit the ball, you'll be looking away from it by the time you are halfway around your orbit (Figure 2.19a). The only way you can face the ball at all times is by completing exactly one rotation while you complete one orbit (Figure 2.19b). (We'll see why the Moon's periods of rotation and orbit are the same in Section 5.5.)

FIGURE 2.19 You can demonstrate why the Moon's rotational period must be the same as its orbital period by walking around a model of Earth. (**a**) If you do not rotate while walking around the model, you will not always face it. (**b**) You can face the model at all times only if you rotate exactly once during each orbit.

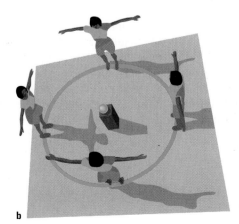

a

b

The View from the Moon

A good way to solidify your understanding of the lunar phases is to imagine that you live on the side of the Moon that faces the Earth. Look again at Figure 2.18. Note that at new moon you would be facing the day side of the Earth. Thus, you would see *full earth* when people on Earth see new moon. Similarly, at full moon you would be facing the night side of the Earth. Thus, you would see *new earth* when people on Earth see full moon. In general, you'd always see the Earth in a phase opposite the phase of the Moon that people on Earth see. Moreover, because your "day" lasts about a month, you'd have about 2 weeks of daylight and 2 weeks of darkness as you watched the Earth go through its cycle of phases.

Thinking about the view from the Moon clarifies another interesting feature of the lunar phases: The dark portion of the lunar face is not *totally* dark. Imagine that you are standing on the Moon when it is in a crescent phase. Because it's nearly new moon as seen from Earth, you will see nearly full earth in your sky. Just as we can see at night by the light of the Moon, the light of the Earth illuminates your night moonscape. (In fact, Earth is much larger and brighter in the lunar sky than the full moon is in Earth's sky.) This faint light illuminating the "dark" portion of the Moon's face is often called the *ashen light* or *earthshine*; it is the light that enables us to see the outline of the full face of the Moon even when the Moon is not full.

Eclipses

Look once more at Figure 2.18. If this figure told the whole story of the lunar phases, a new moon would always block our view of the Sun, and the Earth would always prevent sunlight from reaching a full moon. More precisely, this figure makes it look as if the Moon's shadow should fall on Earth during new moon and that Earth's shadow should fall on the Moon during full moon. Any time one astronomical object casts a shadow on another, we say that an **eclipse** is occurring. Thus, Figure 2.18 makes it look as if we should have an eclipse with every new moon and every full moon—but we don't.

The missing piece of the story in Figure 2.18 is that the Moon's orbit is inclined to the ecliptic plane by about 5°. An easy way to visualize this inclination is to imagine the ecliptic plane as the surface of a pond, as shown in Figure 2.20. Because of the inclination of its orbit, the Moon spends most of its time either above or below this surface. It crosses *through* this surface only twice during each orbit: once coming out and once going back in. The two points in each orbit at which the Moon crosses the surface (the ecliptic plane) are called the **nodes** of the Moon's orbit.

Figure 2.20 shows the position of the Moon's orbit at several different times of year. Note that the nodes are aligned the same way in each case (diagonally on the page). As a result, the nodes lie in a straight line with the Earth and the Sun only about twice each year. (For reasons we'll discuss shortly, it is not *exactly* twice each year.) Because an eclipse can occur only when the Earth and the Moon lie in a straight line with the Sun, two conditions must be met simultaneously for an eclipse to occur:

1. The nodes of the Moon's orbit must be nearly aligned with the Earth and the Sun.

2. The phase of the Moon must be either new or full.

Types of Eclipse There are two basic types of eclipse. A **lunar eclipse** occurs when the Moon passes through the Earth's shadow and therefore can occur only at *full moon*. A **solar eclipse** occurs when the Moon's shadow falls on the Earth and therefore can occur only at *new moon*. But the full story of eclipse types is more complex, because the shadow of the Moon or the Earth consists of two distinct regions: a central **umbra,** where sunlight is completely blocked, and a surrounding **penumbra,** where sunlight is only partially blocked (Figure 2.21).

A lunar eclipse begins at the moment that the Moon's orbit first carries it into the Earth's penumbra. After that, we will see one of three types of lunar eclipse (Figure 2.22). If the Sun, the Earth, and the Moon are nearly perfectly aligned, the Moon will pass through the Earth's umbra, and we will see a

---

**Common Misconceptions:
The "Dark Side" of the Moon**

The term *dark side of the Moon* really should be used to mean the side facing away from the Sun. Unfortunately, *dark side* traditionally meant what would better be called the *far side*—the hemisphere that never can be seen from the Earth. Many people still refer to the far side as the "dark side," even though this side is not necessarily dark. For example, during new moon the far side faces the Sun and hence is completely sunlit; the only time the far side actually is completely dark is at full moon, when it faces away from both the Sun and the Earth.

FIGURE 2.20 This painting represents the ecliptic plane as the surface of a pond. The Moon's orbit is slightly tilted to the ecliptic plane, thus, in this illustration, the Moon spends half of each orbit above the pond surface and half below the surface. Eclipses occur only when the Moon is both crossing the ecliptic plane (splashing through the pond surface) *and* has a phase of either new moon or full moon—as is the case with the lower left and top right orbits shown. At all other times new moons and full moons occur above or below the ecliptic plane, so no eclipse is possible.

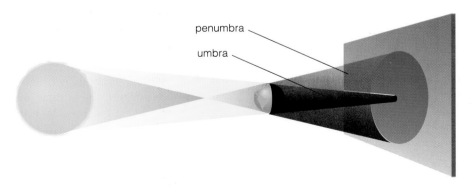

penumbra

umbra

FIGURE 2.21 The shadow cast by an object in sunlight: Sunlight is fully blocked in the umbra and partially blocked in the penumbra.

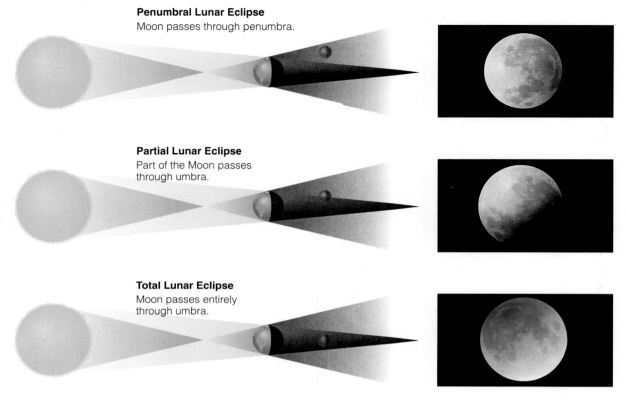

Penumbral Lunar Eclipse
Moon passes through penumbra.

Partial Lunar Eclipse
Part of the Moon passes through umbra.

Total Lunar Eclipse
Moon passes entirely through umbra.

FIGURE 2.22 The three types of lunar eclipse.

total lunar eclipse. If the alignment is somewhat less perfect, only part of the full moon will pass through the umbra (with the rest in the penumbra), and we will see a **partial lunar eclipse.** If the Moon passes *only* through the Earth's penumbra, we will see a **penumbral lunar eclipse.**

Penumbral eclipses are the most common type of lunar eclipse, but they are difficult to notice because the full moon darkens only slightly. Partial lunar eclipses are easier to see because the Earth's umbral shadow clearly darkens part of the Moon's face. Note that the Earth's umbra casts a curved shadow on the Moon, a fact that helped convince the ancient Greeks that the Earth is round. A total lunar eclipse is particularly spectacular, because the Moon becomes dark and eerily red during **totality** (the time during which the Moon is entirely engulfed in the umbra)—dark because it is in shadow, and red because the Earth's atmosphere bends some of the red light from the Sun around the Earth and toward the Moon.

We can also see three types of solar eclipse (Figure 2.23). The Moon's distance from the Earth varies in its orbit, and if a solar eclipse occurs when the Moon is relatively close to the Earth, the Moon's umbra touches a small area of the Earth (no more than about 270 kilometers in diameter). Anyone

within this area will see a **total solar eclipse.** Surrounding this region of totality is a much larger area (typically about 7,000 kilometers in diameter) that falls within the Moon's penumbral shadow; anyone within this region will see a **partial solar eclipse.** If the eclipse occurs when the Moon is relatively far from the Earth, the umbra may not reach the Earth at all. In that case, anyone in the small region of the Earth directly behind the umbra will see an **annular eclipse,** in which a ring of sunlight surrounds the disk of the Moon. (Again, anyone in the surrounding penumbral shadow will see a partial solar eclipse.) During any solar eclipse, the combination of the Earth's rotation and the orbital motion of the Moon causes the circular umbral and penumbral shadows to race across the face of the Earth at a typical speed of about 1,700 kilometers per hour (relative to the ground). As a result, the umbral (or annular) shadow traces a narrow path across the Earth, and totality (or annularity) never lasts more than a few minutes in any particular place.

A total solar eclipse is a spectacular sight. It begins when the disk of the Moon first appears to touch the Sun. Over the next couple of hours, the Moon appears to take a larger and larger "bite" out of the Sun. As totality approaches, the sky darkens and

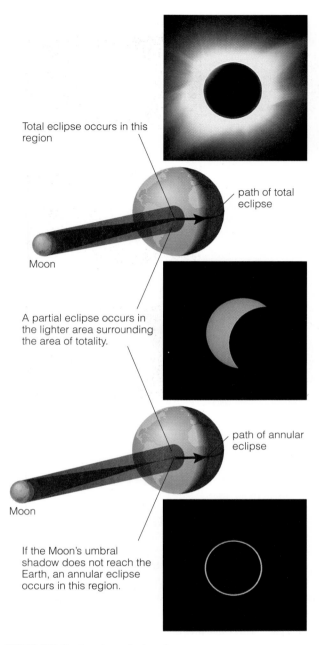

Total eclipse occurs in this region

path of total eclipse

Moon

A partial eclipse occurs in the lighter area surrounding the area of totality.

path of annular eclipse

Moon

If the Moon's umbral shadow does not reach the Earth, an annular eclipse occurs in this region.

FIGURE 2.23 The three types of solar eclipse.

FIGURE 2.24 This multiple-exposure photograph shows the progression of a total solar eclipse. Totality (central image) lasts only a few minutes.

temperatures fall. Birds head back to their nests, and crickets begin their nighttime chirping. During the few minutes of totality, the Moon completely blocks the normally visible disk of the Sun, allowing the faint *corona* to be seen (Figure 2.24). The surrounding sky takes on a twilight glow, and planets and bright stars become visible in the daytime. As totality ends, the Sun slowly emerges from behind the Moon over the next couple of hours. However, because your eyes have adapted to the darkness, totality *appears* to end far more abruptly than it began.

Predicting Eclipses Few phenomena have so inspired and humbled humans throughout the ages as eclipses. For many cultures, eclipses were mystical events associated with fate or the gods, and countless stories and legends surround eclipses. Much of the mystery of eclipses probably stems from the fact that they are relatively difficult to predict. Look again at Figure 2.20. The two periods each year when the nodes of the Moon's orbit are nearly aligned with the Sun are called **eclipse seasons.** Each eclipse season lasts a few weeks, so some type of lunar eclipse occurs during each eclipse season's full moon, and some type of solar eclipse occurs during its new moon. If Figure 2.20 told the whole story, eclipse seasons would occur every 6 months and predicting eclipses would be easy. For example, if eclipse seasons always occurred in January and July, eclipses would occur only on the dates of new and full moons in those months. But Figure 2.20 does not show one important fact about the Moon's orbit: The nodes slowly precess around the orbit. As a result, eclipse seasons actually occur slightly less than 6 months apart (about 173 days apart) and therefore do not recur in the same months year after year.

The combination of the changing dates of eclipse seasons and the $29\frac{1}{2}$-day cycle of lunar phases turns out to make eclipses recur in a cycle of about 18 years $11\frac{1}{3}$ days. For example, if a solar eclipse occurs today, another will occur 18 years $11\frac{1}{3}$ days from now. This roughly 18-year cycle is called the **saros cycle.**

Astronomers in many ancient cultures identified the saros cycle and therefore could predict *when* eclipses would occur. However, the saros cycle does

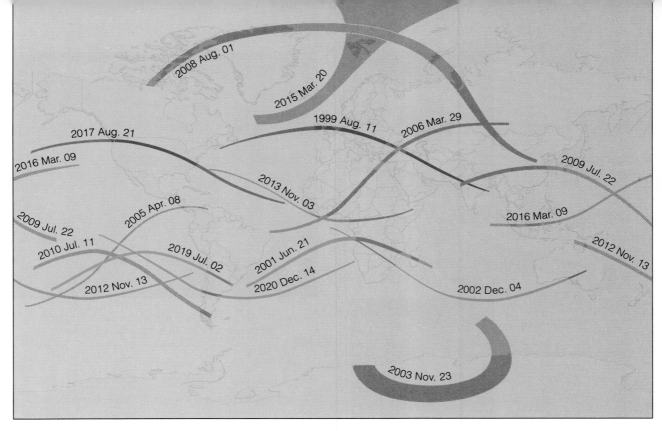

FIGURE 2.25 This map shows the paths of totality for solar eclipses from 1999 to 2020. Paths of the same color represent eclipses occurring in successive saros cycles, separated by 18 years 11 days. For example, the 2017 eclipse occurs 18 years 11 days after the 1999 eclipse.

Table 2.1 Total Lunar Eclipses 2000–2004

January 21, 2000
July 16, 2000
January 9, 2001
May 16, 2003
November 9, 2003
May 4, 2004
October 28, 2004

not account for all the complications involved in predicting eclipses. If a solar eclipse occurs today, the one that occurs 18 years $11\frac{1}{3}$ days from now will not be visible from the same places on the Earth and may not be of the same type (e.g., one may be total and the other only partial). As a result, no ancient culture achieved the ability to predict eclipses in every detail. Today, eclipses *can* be predicted because we know the precise details of the orbits of the Earth and the Moon. In fact, *you* can predict the dates of eclipses far into the future with the aid of one of many inexpensive astronomical software packages. Table 2.1 lists upcoming total lunar eclipses, and Figure 2.25 shows paths of totality for upcoming total solar eclipses. It is well worth your while to plan a trip to see a total solar eclipse.

2.4 The Ancient Mystery of the Planets

Five planets are easy to find with the naked eye: Mercury, Venus, Mars, Jupiter, and Saturn. Mercury can be seen only infrequently, and then only just after sunset or just before sunrise because it is so close to the Sun. Venus often shines brightly in the early evening in the west or before dawn in the east; if you see a very bright "star" in the early evening or early morning, it probably is Venus. Jupiter, when it is visible at night, is the brightest object in the sky besides the Moon and Venus. Mars is recognizable by its red color, but be careful not to confuse it with a bright red star. Saturn also is easy to see with the naked eye, but because many stars are just as bright as Saturn, it helps to know where to look. (Also, planets tend not to twinkle as much as stars.) Sometimes several planets and the Moon may appear close together in the sky, offering a particularly beautiful sight (Figure 2.26).

Like the Sun and the Moon, the planets appear to move slowly through the constellations of the zodiac. (The word *planet* comes from the Greek for "wandering star.") However, while the Sun and the Moon always appear to move eastward relative to the stars, the planets occasionally reverse course and

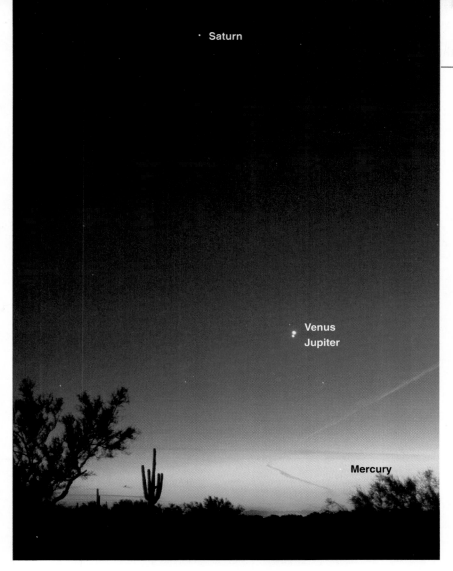

FIGURE 2.26 This photograph shows Mercury, Venus, Jupiter, and Saturn appearing close together on February 23, 1999.

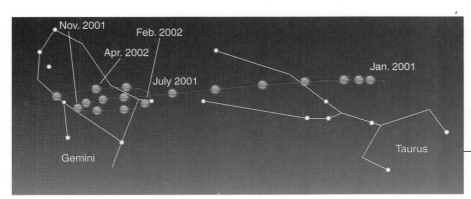

Dots represent Jupiter's position at 1-month intervals.

FIGURE 2.27 This diagram shows Jupiter's position among the stars in our sky during 2001–2002. Jupiter generally appears to drift eastward among the stars, but for about 4 months each year it turns back toward the west. Here, this apparent retrograde motion occurs between about November 2001 and February 2002. Note that Jupiter moves about $\frac{1}{12}$ of the way through the zodiac each year because it takes 12 years to complete one orbit of the Sun.

appear to move *westward* through the zodiac. A period during which a planet appears to move westward relative to the stars is called a period of **apparent retrograde motion** (*retrograde* means "backward"). Figure 2.27 shows a period of apparent retrograde motion for Jupiter.

Ancient astronomers could easily "explain" the daily paths of the stars through the sky by imagining that the celestial sphere was real and that it really rotated around the Earth each day. But the apparent retrograde motion of the planets posed a far greater mystery: What could cause the planets sometimes to go backward? As we'll discuss in Chapter 3, the ancient Greeks came up with some very clever ways to "explain" the occasional backward motion of the planets. But these ideas never allowed them to predict planetary positions in the sky with great accuracy. The problem was that they were wedded to the incorrect idea of an Earth-centered universe.

In contrast, it's very easy to explain apparent retrograde motion in a Sun-centered solar system. You can demonstrate it for yourself with the help of a friend (Figure 2.28). Pick a spot in an open field to represent the Sun. You can represent the Earth, walking counterclockwise around the Sun, while your friend represents a more distant planet (e.g., Mars, Jupiter, or Saturn) by walking counterclockwise around the Sun at a greater distance. Your friend should walk more slowly than you because more distant planets orbit the Sun more slowly. As you walk, watch how your friend appears to move relative to buildings or trees in the distance. Although both of you always walk the same way around the Sun, your friend will *appear* to move backward against the background during the part of your "orbit" at which you catch up with and pass him or her. (To understand the apparent retrograde motions of Mercury and Venus, which are closer to the Sun than is the Earth, simply switch places with your friend and repeat the demonstration.)

The apparent retrograde motion demonstration applies directly to the planets. For example, because Mars takes about 2 years to orbit the Sun (actually 1.88 years), it covers about half its orbit during the 1 year in which Earth makes a complete orbit. If you trace lines of sight from Earth to Mars from different points in their orbits, you will see that the line of sight usually moves eastward relative to the stars but moves westward during the time when Earth is passing Mars in its orbit (Figure 2.29). Like your friend in the demonstration, Mars never really changes direction; it only *appears* to change direction from our perspective on Earth.

If the apparent retrograde motion of the planets is so readily explained by recognizing that the Earth is a planet, why wasn't this idea accepted in ancient times? In fact, the idea that the Earth goes

FIGURE 2.28 The retrograde motion demonstration. Watch how your friend usually appears to you to move forward against the background of buildings in the distance but appears to move backward as you catch up with and pass him or her in your "orbit."

FIGURE 2.29 The explanation for apparent retrograde motion. Follow the lines of sight from Earth to Mars in numerical order. The period where the lines of sight shift *westward* relative to the distant stars is the period during which we observe apparent retrograde motion for Mars.

Aristarchus

Until the early 1600s, nearly everyone believed that the Earth was the center of the universe. Yet Aristarchus (c. 310–230 B.C.) argued otherwise almost 2,000 years earlier. How did he reach his conclusion?

In about 260 B.C., Aristarchus pointed out that the apparent retrograde motion of the planets could be explained much more simply if Earth and other planets orbited the Sun. To account for the lack of detectable stellar parallax, Aristarchus suggested that the stars were extremely far away.

He further strengthened his argument by estimating the sizes of the Moon and the Sun. By observing the shadow of the Earth on the Moon during a lunar eclipse, Aristarchus estimated the Moon's diameter to be about one-third of the Earth's diameter—only slightly higher than the actual value. He then used a geometric argument, based on measuring the angle between the Moon and the Sun at first- and third-quarter phases, to conclude that the Sun must be larger than the Earth. (His measurements were imprecise, so he estimated the Sun's diameter to be about 7 times the Earth's, rather than the correct value of about 100 times.) Because the Sun is larger than the Earth, he argued that it was more natural for the Earth to go around the Sun than vice versa.

Like most scientific work, Aristarchus's work built upon the work of others. In particular, Heracleides (c. 388–315 B.C.) had suggested that the Earth rotates, an idea that Aristarchus drew from to explain the apparent daily rotation of the stars in our sky. Heracleides also was the first to suggest that *not all* heavenly bodies circle the Earth; based on the fact that Mercury and Venus always are close to the Sun in the sky, he argued that these two planets must orbit the Sun. Thus, in suggesting that all the planets orbit the Sun, Aristarchus was extending the ideas of Heracleides and others before him. Alas, Aristarchus's arguments were not widely accepted in ancient times and were revived only with the work of Copernicus some 1,800 years later.

---

around the Sun was suggested as early as 260 B.C. by the Greek astronomer Aristarchus but did not gain wide acceptance until the early 1600s. Although there were many reasons for the historic reluctance to abandon the idea of an Earth-centered universe, perhaps the most prominent involved the inability of ancient peoples to detect something called _stellar parallax_.

Extend your arm and hold up one finger. If you keep your finger still but alternately close your left eye and right eye, your finger will *appear* to jump back and forth against the background. This apparent shifting in the position of your finger demonstrates _parallax_ and occurs simply because your two eyes view your finger from opposite sides of your nose. Note that if you move your finger closer to your face, the parallax increases. In contrast, if you look at a distant tree or flagpole instead of your finger, you probably cannot detect any parallax by alternately closing your left eye and right eye. Thus, parallax depends on distance, with nearer objects exhibiting greater parallax than more distant objects.

Now imagine that your two eyes represent the Earth at opposite sides of its orbit around the Sun and that your finger represents a distant star—this is the idea of **stellar parallax.** That is, because we view the stars from different places in our orbit at different times of year, nearby stars should *appear* to shift back and forth against distant stars in the background during the course of the year (Figure 2.30).

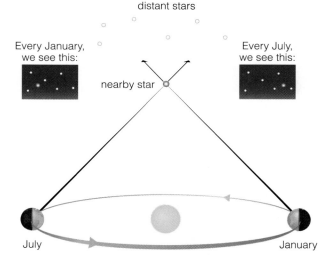

distant stars

Every January, we see this:

nearby star

Every July, we see this:

July

January

FIGURE 2.30 Stellar parallax is an apparent shift in the position of a nearby star as we look at it from different places in the Earth's orbit. This figure is greatly exaggerated; in reality, the amount of shift is far too small to detect with the naked eye.

However, no matter how hard they searched, ancient astronomers could find no sign of stellar parallax. They therefore concluded that one of the following must be true:

1. The stars are so far away that stellar parallax is undetectable to the naked eye.

2. There is no stellar parallax because the Earth is the center of the universe.

Unfortunately, with notable exceptions such as Aristarchus, ancient astronomers rejected the correct answer (1) because they could not imagine that the stars could be *that* far away. Today, we can detect stellar parallax with the aid of telescopes, thereby providing direct proof that the Earth really does orbit the Sun. (Careful measurements of stellar parallax also provide the most reliable means of measuring distances to nearby stars [Section 13.2].)

TIME OUT TO THINK *How far apart are opposite sides of the Earth's orbit? How far away are the nearest stars? Describe the challenge of detecting stellar parallax. It may help to visualize the Earth's orbit and the distance to stars on the 1-to-10-billion scale used in Chapter 1.*

Thus, we see that the ancient mystery of the planets drove much of the historical debate over the Earth's place in the universe. Indeed, in many ways, the modern technological society that we take for granted today can be traced directly back to the scientific revolution that began because of the quest to explain the slow wandering of the planets among the stars in our sky.

Have you noticed that the astronomers and mathematicians are much the most cheerful people of the lot? I suppose that perpetually contemplating things on so vast a scale makes them feel either that it doesn't matter a hoot anyway, or that anything so large and elaborate must have some sense in it somewhere.

DOROTHY L. SAYERS (1893-1957)

THE BIG PICTURE

In this chapter, we surveyed the phenomena of our sky. Keep the following "big picture" ideas in mind as you continue your study of astronomy:

■ You can enhance your enjoyment of learning astronomy by spending time outside observing the sky. The more you learn about the appearance and apparent motions of the sky, the more you will appreciate what you can see in the universe.

■ From our vantage point on Earth, it is convenient to imagine that we are at the center of a great celestial sphere—even though we really are on a planet orbiting a star in a vast universe.

■ Most of the phenomena of the sky are relatively easy to observe and understand. But the more complex phenomena, particularly eclipses and planetary motion, challenged our ancestors for thousands of years and helped drive the development of mathematics and science.

Review Questions

1. What is a *constellation*? How is a constellation related to a pattern of stars in the sky?

2. What is the *celestial sphere*? Describe the major features shown on a model of the celestial sphere.

3. What is the *Milky Way* in our sky, and how is it related to the Milky Way Galaxy?

4. Why does your sky look like a dome? Define your *horizon, zenith,* and *meridian*.

5. Explain why we can measure only *angular* sizes and distances for objects in the sky.

6. Briefly describe how and why the sky varies with latitude.

7. Briefly describe how and why the sky changes with the seasons.

8. How are our methods of timekeeping based on astronomical phenomena?

9. Describe the Moon's cycle of phases and explain why we see phases of the Moon.

10. Why don't we see an eclipse at every new and full moon? Describe the conditions that must be met for us to see a solar or lunar eclipse.

11. Why are eclipses so difficult to predict? Explain the importance of *eclipse seasons* and the *saros cycle* to eclipse prediction.

12. What is *apparent retrograde motion* of the planets? Why was it difficult for ancient astronomers to explain, but easy for us to explain?

13. What is *stellar parallax*? Describe the role it played in making ancient astronomers believe in an Earth-centered universe.

Discussion Questions

1. *Geocentric Language.* Many common phrases reflect the ancient Earth-centered view of our universe. For example, the phrase "the Sun rises each day" implies that the Sun is really moving over the Earth. In fact, the Sun only *appears* to rise as the rotation of the Earth carries us to a place where we can see the Sun in our sky. Identify other common phrases that imply an Earth-centered viewpoint.

2. *Round Earth.* As early as 500 B.C., observations of lunar eclipses and the fact that the sky varies with latitude convinced the ancient Greeks that the Earth is round. Discuss how these pieces of evidence point to a round Earth.

Problems

1. *View from Afar.* Describe how the Milky Way Galaxy would look in the sky of someone observing from a planet around a star in M 31.

2. *Your View.*
   a. Find your latitude and longitude, and state the source of your information.
   b. Describe the altitude and direction in your sky at which the north or south celestial pole appears.
   c. Is Polaris a circumpolar star in your sky? Explain.

   d. Describe the path of the meridian in your sky.
   e. Describe the path of the celestial equator in your sky. (*Hint:* Study Figure 2.12.)

3. *View from the Moon.* Suppose you lived on the Moon, near the center of the face that we see from Earth.
   a. During the phase of full moon, what phase would you see for Earth? Would it be daylight or dark where you live?
   b. On Earth, we see the Moon rise and set in our sky each day. If you lived on the Moon, would you see the Earth rise and set? Why or why not? (*Hint:* Remember that the Moon always keeps the same face toward Earth.)
   c. What would you see when people on Earth were experiencing an eclipse? Answer for both solar and lunar eclipses.

4. *A Farther Moon.* Suppose the distance to the Moon were twice its actual value. Would it still be possible to have a total solar eclipse? An annular eclipse? A total lunar eclipse? Explain.

5. *A Smaller Earth.* Suppose the Earth were smaller in size. Would solar eclipses be any different? If so, how? What about lunar eclipses? Explain.

6. *Project: Sky Information.* Look for sources of daily information about sky phenomena (such as lunar phases, times of sunrise and sunset, and unusual sky events) in newspapers and magazines and on the Web. Identify and briefly describe your favorite source.

7. *Observing Project: Apparent Planetary Motion.* Find out what planets are currently visible in your evening skies. At least once a week, observe the planets and draw a diagram showing the position of each visible planet relative to stars in a zodiac constellation. From week to week, note how the planets are moving relative to the stars. Can you see any of the apparently "erratic" features of planetary motion? Explain.

PART II
KEY CONCEPTS FOR ASTRONOMY

T³he Science of Astronomy

TODAY WE KNOW THAT THE EARTH IS A PLANET ORBITING A RATHER ordinary star, in a galaxy of a hundred billion or more stars, in an incredibly vast universe. We know that the Earth, along with the entire cosmos, is in constant motion. We know that, on the scale of cosmic time, human civilization has existed for only the briefest moment. Yet we have acquired all this knowledge only recently in human history. How have we managed to learn these things?

It wasn't easy. Astronomy is the oldest of the sciences, with roots extending as far back as recorded history allows us to see. But while our current understanding of the universe rests on foundations laid long ago, the most impressive advances in knowledge have come in just the last few centuries.

In this chapter, we will study the development and nature of the science of astronomy. We'll explore what we mean by *science* and the *scientific method* and trace how modern astronomy arose from its roots in ancient observations.

3.1 Everyday Science

A common stereotype holds that scientists are men and women in white lab coats who somehow think differently than other people. In reality, scientific thinking is a fundamental part of human nature.

Think about how a baby behaves. By about a year of age, she notices that objects fall to the ground when she drops them. She lets go of a ball; it falls. She pushes a plate of food from her high chair; it falls too. She continues to drop all kinds of objects, and they all plummet to Earth. Through her powers of observation, the baby has learned something about the physical world: Things fall when they are unsupported. Eventually, she becomes so certain of this fact that, to her parents' delight, she no longer needs to test it continually.

Then one day somebody gives the baby a helium balloon. She releases it; to her surprise, it rises to the ceiling! Her rudimentary understanding of physics must be revised. She now knows that the principle "all things fall" does not represent the whole truth, although it still serves her quite well in most situations. It probably will be years before she learns enough about the atmosphere, the force of gravity, and the concept of density to understand *why* the balloon rises when most other objects fall. For the moment, the baby probably will be delighted that she has learned something new and unexpected.

The baby's experience with falling objects and balloons exemplifies scientific thinking, which, in essence, is a way of learning about nature through careful observation and trial-and-error experiments. Rather than thinking differently than other people, modern scientists simply are trained to organize this everyday thinking in a way that makes it easier for them to share their discoveries with others and thereby employ the collective wisdom of many people in the search for new knowledge. This organizational scheme is what we call the *scientific method;* it is the method by which humanity has acquired its present knowledge of the universe.

TIME OUT TO THINK *When was the last time you used trial and error to learn something? Describe a few cases where you have learned by trial and error in cooking, participating in sports, fixing something, or any other situation.*

3.2 Ancient Observations

We will discuss the modern scientific method shortly, but first we will explore how it arose from the observations of ancient peoples. Our exploration begins in central Africa, where people of many indigenous societies predict the weather with reasonable accuracy by making careful observations of the Moon. The Moon begins its monthly cycle as a crescent in the western sky just after sunset. Through long traditions of sky watching, central African societies learned that the orientation of the crescent "horns" relative to the horizon is closely tied to rainfall patterns (Figure 3.1).

No one knows when central Africans first developed the ability to predict weather using lunar cycles. But the earliest known written astronomical record comes from central Africa, near the border of modern-day Congo and Uganda. It consists of an animal bone (known as the *Ishango bone*) etched with patterns that appear to represent part of a lunar calendar, probably carved around 6500 B.C.

Why did ancient Africans, and other people around the world, bother to make such careful and detailed observations of the sky? Ancient peoples were just as curious and intelligent as we are today. In

FIGURE 3.1 This diagram shows how central Africans used the Moon to predict the weather. The graph depicts the annual rainfall pattern in central Nigeria, characterized by a wet season and a dry season. The Moon diagrams represent the orientation of a waxing crescent moon relative to the western horizon at different times of year. The angle of the crescent "horns" allows observers to determine the time of year and hence the expected rainfall. (The orientation is measured in degrees, where 0° means the crescent horns are parallel to the horizon.)

the daytime, they surely recognized the importance of the Sun to their lives. At night, without electric light, they were much more aware of the starry sky than we are today. Thus, it's not surprising that they paid attention to patterns of motion in the sky and developed ideas and stories to explain what they saw.

Astronomy played a practical role in ancient societies by enabling them to keep track of time and seasons, a crucial skill for people who depended on agriculture for survival. This ability may seem quaint today, when digital watches tell us the precise time and date, but it required considerable knowledge and skill in ancient times, when the only clocks and calendars were in the sky.

Modern measures of time still reflect their ancient astronomical roots. Our 24-hour day is the time it takes for the Sun to circle our sky. The length of a month comes from the lunar cycle, and our calendar year is based on the cycle of the seasons. The days of the week are named after the seven naked-eye objects that appear to move among the constellations: the Sun, the Moon, and the five planets recognized in ancient times (Table 3.1).

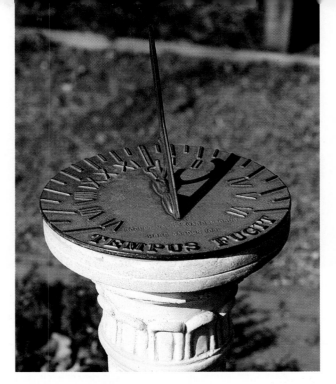

FIGURE 3.2 Sundials similar to this one were probably used by many ancient cultures.

Determining the Time of Day

In the daytime, ancient peoples could tell time by observing the Sun's path through their sky. They knew it was morning when the Sun was in the east, and afternoon when the Sun was in the west. For greater precision, many ancient cultures probably used sundials (Figure 3.2). Sundials remain popular today both for their decorative value and as reminders that the sky once was our only guide to time.

At night, the Moon's position and phase give an indication of the time. (See Figure 2.18.) For example, a first-quarter moon sets around midnight, so it is not yet midnight if the first-quarter moon is still above the western horizon. The positions of the stars also indicate the time if you know the approximate date. For example, in December the constellation Orion rises around sunset, reaches the meridian around midnight, and sets around sunrise. Hence, if it is winter and Orion is setting, dawn must be approaching. Most ancient peoples probably were adept at estimating the time of night, although written evidence is sparse.

Determining the Time of Year

Many cultures built structures to help them mark the seasons. One of the oldest standing human-made structures served such a purpose: Stonehenge in southern England, which was constructed in stages

Table 3.1 The Seven Days of the Week and the Astronomical Objects They Honor

The correspondence between objects and days is easy to see in French and Spanish. In English, the correspondence becomes clear when we look at the names of the objects used by the Teutonic tribes who lived in the region of modern-day Germany.

Object	Teutonic Name	English	French	Spanish
Sun	Sun	Sunday	dimanche	domingo
Moon	Moon	Monday	lundi	lunes
Mars	Tiw	Tuesday	mardi	martes
Mercury	Woden	Wednesday	mercredi	miércoles
Jupiter	Thor	Thursday	jeudi	jueves
Venus	Fria	Friday	vendredi	viernes
Saturn	Saturn	Saturday	samedi	sábado

a

b

FIGURE 3.3 (**a**) Stonehenge today. (**b**) A sketch showing how archaeologists believe Stonehenge looked when construction was completed around 1550 B.C. Note, for example, that observers standing in the center would see the Sun rise directly over the Heel Stone on the summer solstice.

FIGURE 3.4 This scale model shows the Templo Mayor and the surrounding plaza as they looked before Aztec civilization was destroyed by the Conquistadors.

from about 2750 B.C. to about 1550 B.C. (Figure 3.3). Observers standing in its center see the Sun rise directly over the Heel Stone only on the summer solstice. Stonehenge also served as a social gathering place and probably as a religious site; no one knows whether its original purpose was social or astronomical. Perhaps it was built for both: In ancient times, social rituals and practical astronomy probably were deeply intertwined.

One of the most spectacular structures used to mark the seasons was the Templo Mayor in the Aztec city of Tenochtitlán, located on the site of modern-day Mexico City (Figure 3.4). Twin temples surmounted a flat-topped, 150-foot-high pyramid. From the location of a royal observer watching from the opposite side of the plaza, the Sun rose directly through the notch between the twin temples on the equinoxes. Like Stonehenge, the Templo Mayor

FIGURE 3.5 The Sun Dagger is an arrangement of rocks that shape the Sun's light into a dagger, with a spiral carved on a rock face behind them. The dagger pierces the center of the carved spiral each year at noon on the summer solstice.

served important social and religious functions in addition to its astronomical role.

Many ancient cultures aligned their buildings with the cardinal directions (north, south, east, and west), enabling them to mark the rising and setting of the Sun relative to the building orientation. Some even created monuments with a single special astronomical purpose. Perhaps in a practical form of ancient art, someone among the ancient Anasazi people carved a spiral known as the Sun Dagger on a vertical cliff face in Chaco Canyon, New Mexico (Figure 3.5). There the Sun's rays form a dagger of sunlight that pierces the center of a carved spiral only once each year—at noon on the summer solstice.

Lunar Cycles

Many ancient civilizations paid particular attention to the lunar cycle, often using it as the basis for lunar calendars. The months on lunar calendars generally have either 29 or 30 days, chosen to make the average agree with the approximately $29\frac{1}{2}$-day lunar cycle. A 12-month lunar calendar has only 354 or 355 days, or about 11 days less than a calendar based on the tropical year. Such a calendar is still used in the Muslim religion, which is why the month-long fast of Ramadan (the ninth month) begins about 11 days earlier with each subsequent year.

Other lunar calendars take advantage of the fact that 19 years is almost precisely 235 lunar months. Because this fact was discovered in 432 B.C. by the Babylonian astronomer Meton, the 19-year period is called the **Metonic cycle.** A lunar calendar can be synchronized to the Metonic cycle by adding a thirteenth month to 7 of every 19 years (making exactly 235 months in each 19-year period), thereby ensuring that "new year" comes on approximately the same date every nineteenth year. The Jewish calendar follows the Metonic cycle, adding a thirteenth month in the third, sixth, eighth, eleventh, fourteenth, seventeenth, and nineteenth years of each cycle. This also explains why the date of Easter changes each year: The New Testament ties the date of Easter to the Jewish festival of Passover, which has its date set by the Jewish lunar calendar. In a slight modification of the original scheme, most Christians now celebrate Easter on *the first Sunday after the first full moon after March 21;* if the full moon falls on Sunday, Easter is the following Sunday.

Some ancient cultures learned to predict eclipses by recognizing the 18-year saros cycle [Section 2.3]. In the Middle East, the ancient Babylonians achieved remarkable success in predicting eclipses more than 2,500 years ago. Perhaps the most successful predictions of eclipses prior to modern times were made by the Mayans in Central America. The Mayan calendar featured a sacred cycle that almost certainly was related to eclipses. This Mayan cycle, called the *sacred round,* lasted 260 days—almost exactly $1\frac{1}{2}$ times the 173.3 days between successive eclipse seasons [Section 2.3].

FIGURE 3.6 The Moon rising between two stones of the 4,000-year-old sacred stone circle at Callanish, Scotland (on the Isle of Lewis in the Scottish Hebrides). The full moon rises in this position only once every 18.6 years.

FIGURE 3.7 The ruins of the Mayan observatory at Chichén Itzá.

The complexity of the Moon's orbit leads to other long-term patterns in the Moon's appearance. For example, the full moon rises at its most southerly point along the eastern horizon only once every 18.6 years, a phenomenon that can be vividly observed from the 4,000-year-old sacred stone circle at Callanish, Scotland (Figure 3.6).

Observations of Planets and Stars

Many cultures also made careful observations of the planets and stars. Mayan observatories in Central America, such as the one still standing at Chichén Itzá (Figure 3.7), had windows strategically placed for observations of Venus.

More than 2,000 years ago, the early Aztec people marked the beginning of their year when the group of stars called the Pleiades first rose in the east after having been hidden from view for 40 days by the glare of the Sun. On the other side of the world, southern Greeks also used the Pleiades to time their planting and harvests, as described in this excerpt from an epic poem (*Works and Days*) composed by the Greek farmer Hesiod in about 800 B.C.:

> *When you notice the daughters of Atlas, the Pleiades, rising, start on your reaping, and on your sowing when they are setting. They are hidden from your view for a period of forty full days, both night and day, but then once again, as the year moves round, they reappear at the time for you to be sharpening your sickle.*

FIGURE 3.8 Hundreds of lines and patterns are etched in the sand of the Nazca desert in Peru. This aerial photo shows a large, etched figure of a hummingbird.

Observational aids could also mark the rising and setting of the stars. More than 800 lines, some stretching for miles, are etched in the dry desert sand of Peru between the Ingenio and Nazca rivers. Many of these lines may simply have been well-traveled pathways, but others are aligned in directions that point to places where bright stars or the Sun rose at particular times of year. In addition to the many straight lines, the desert also features many large figures of animals, which may be representations of constellations made by the Incas who lived in the region (Figure 3.8)

TIME OUT TO THINK *The animal figures show up clearly only when seen from above. As a result, some UFO enthusiasts argue that the patterns must have been created by aliens. What do you think of this argument? Defend your opinion.*

The great Incan cities of South America clearly were built with astronomy in mind. Entire cities appear to have been designed so that particular arrangements of roads, buildings, or other human-made structures would point to places where bright stars rise or set, or where the Sun rises and sets at particular times of year.

Structures for astronomical observation also were popular in North America. Lodges built by the Pawnee people in Kansas featured strategically placed holes for observing the passage of constellations that figured prominently in their folklore. In the northern plains of the United States, Native American "Medicine Wheels" probably were designed for astronomical observations. The "spokes" of the Medicine Wheel at Big Horn, Wyoming, were aligned with the rising and setting of bright stars, as well as with the rising and setting of the Sun on the equinoxes and solstices (Figure 3.9). The 28 spokes of the Medicine Wheel probably relate to the month of the Native

FIGURE 3.9 Aerial and ground views of the Big Horn Medicine Wheel in Wyoming. Note the 28 "spokes" radiating out from the center. These spokes probably relate to the month of the Native Americans.

Americans, which they measured as 28 days (rather than 29 or 30) because they did not count the day of the new moon.

Perhaps the people most dependent on knowledge of the stars were the Polynesians, who lived and traveled among the many islands of the mid- and South Pacific. Because the next island in a journey usually was too distant to be seen, poor navigation meant becoming lost at sea. As a result, the most esteemed position in Polynesian culture was that of the Navigator, a person who had acquired the detailed knowledge necessary to navigate great distances among the islands. The Navigators employed a combination of detailed knowledge of astronomy and equally impressive knowledge of the patterns of waves and swells around different islands (Figure 3.10). The stars provided their broad navigational sense, pointing them in the correct direction of their intended destination. As they neared a destination, the wave and swell patterns would guide them to their precise landing point.

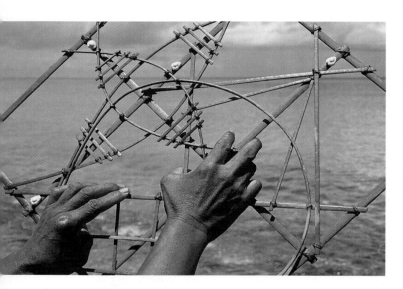

FIGURE 3.10 A traditional Polynesian navigational instrument.

From Observation to Science

Before a structure such as Stonehenge could be built, careful observations had to be made and repeated over and over to ensure their validity. Careful, repeatable observations also underlie the modern scientific method. To this extent, elements of modern science were present in many early human cultures.

The degree to which scientific ideas developed in different societies depended on practical needs, social and political customs, and interactions with other cultures. Because all of these factors can change, it should not be surprising that different cultures were more scientifically or technologically advanced than others at different times in history. The ancient Chinese kept remarkably detailed records of astronomical observations beginning at least 5,000 years ago, and these records enabled them to discover many patterns of sky motion hundreds of years before their counterparts in other parts of the world. The Chinese also developed advanced instrumentation to aid their observations, which may have

FIGURE 3.11 This photo shows a model of the celestial sphere and other instruments at the ancient astronomical observatory in Beijing.

spurred their general technological development (Figure 3.11).

Yet, by the end of the 1600s, Chinese science and technology had fallen behind that of Europe. Some historians argue that a primary reason for this decline was that the Chinese tended to regard their science and technology as state secrets. This secrecy may have slowed Chinese scientific development by preventing the broad-based collaborative science that fueled the European advance beginning in the Renaissance.

In Central America, the ancient Mayans also were ahead of their time in many ways. For example, their system of numbers and mathematics looks distinctly modern; their invention of the concept of zero preceded by some 500 years its introduction in the Eurasian world (by Hindu mathematicians, around A.D. 600). The Aztecs, Incas, Anasazi, and other ancient peoples of the Americas may also have been quite advanced in many other areas, but few written records survive to tell the tale.

It appears that virtually all cultures employed scientific thinking to varying degrees. Had the circumstances of history been different, any one of these many cultures might have been the first to develop what we consider to be modern science. But, in the end, history takes only one of countless possible paths, and the path that led to modern science emerged from the ancient civilizations of the Mediterranean and the Middle East.

3.3 The Modern Lineage

By 3000 B.C., civilization was well established in two major regions of the Middle East: Egypt and Mesopotamia (Figure 3.12). Their geographical location

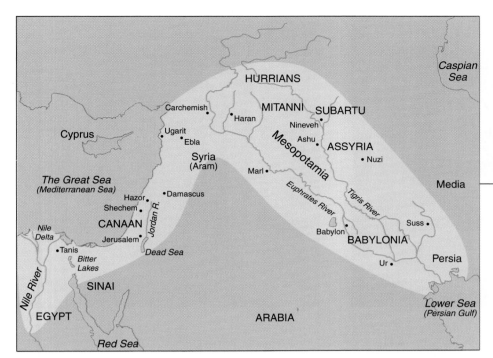

FIGURE 3.12 This map shows the Middle East in ancient times. Mesopotamia was the ancient Greek name for the region between the Tigris and Euphrates rivers, but modern historians use the name for the entire region today occupied by Iraq. Three major cultures thrived at various times in ancient Mesopotamia: Sumer, Babylonia, and Assyria.

THINKING ABOUT . . .

Eratosthenes Measures the Earth

In a remarkable ancient feat, the Greek astronomer and geographer Eratosthenes (c. 276–196 B.C.) estimated the size of the Earth in about 240 B.C. He did it by comparing the altitude of the Sun on the summer solstice in the Egyptian cities of Syene (modern-day Aswan) and Alexandria.

Eratosthenes knew that the Sun passed directly overhead in Syene on the summer solstice. He also knew that, in the city of Alexandria to the north, the Sun came within only 7° of the zenith on the summer solstice. He therefore concluded that Alexandria must be 7° of latitude to the north of Syene (Figure 3.13). Because 7° is $\frac{7}{360}$ of a circle, he concluded that the north-south distance between Alexandria and Syene must be $\frac{7}{360}$ of the circumference of the Earth.

Eratosthenes estimated the north-south distance between Syene and Alexandria to be 5,000 stadia (the *stadium* was a Greek unit of distance). Thus, he concluded that

$$\frac{7}{360} \times \text{circumference of Earth} = 5,000 \text{ stadia}$$

from which he found the Earth's circumference to be about 250,000 stadia.

Today, we don't know exactly what distance a stadium meant to Eratosthenes; however, based on the actual sizes of Greek stadiums, it must have been about $\frac{1}{6}$ kilometer. Thus, Eratosthenes estimated the circumference of the Earth to be about $\frac{250,000}{6}$ kilometers, or about 42,000 kilometers—remarkably close to the modern value of just over 40,000 kilometers.

FIGURE 3.13 On the summer solstice, the Sun reaches 7° higher in the sky in Syene than in Alexandria, so 7° of latitude must separate Syene from Alexandria.

FIGURE 3.14 These renderings show an artist's reconstruction, based on scholarly research, of how (**a**) the Great Hall and (**b**) a scroll room of the Library of Alexandria might have looked.

placed these civilizations at a crossroads for travelers, merchants, and armies of Europe, Asia, and Africa. This mixing of cultures fostered creativity, and the broad interactions among peoples ensured that new ideas spread throughout the region. Over the next 2,500 years, numerous great cultures arose. For example, the ancient Egyptians built the Great Pyramids between 2700 and 2100 B.C., using their astronomical knowledge to orient the Pyramids with the cardinal directions. They also invented papyrus scrolls and ink-based writing. The Babylonians invented methods of writing on clay tablets and developed arithmetic to serve in commerce and later in astronomical calculations. Many more of our modern principles of commerce, law, religion, and science originated with the cultures of Egypt and Mesopotamia.

The development of principles of modern science accelerated with the rise of Greece as a power in the Middle East, beginning around 500 B.C. In 330 B.C., Alexander the Great led the expansion of the Greek empire throughout the Middle East, absorbing all the former empires of Egypt and Mesopotamia. Alexander had a keen interest in science and education, perhaps fueled by his association with Aristotle [Section 5.3], who was his personal tutor. He encouraged the pursuit of knowledge and respect for foreign cultures. On the delta of the Nile in Egypt, he founded the city of Alexandria, which he hoped would become a center of world culture.

Although Alexander died at the age of 35, his dream came true. His successors continued to build Alexandria, including the construction of a great library and research center around 300 B.C. (Figure 3.14). The Library of Alexandria was the world's preeminent center of research for the next 700 years. Its end is associated with the death of Hypatia, perhaps the most prominent female scholar of the ancient world. Hypatia was a resident scholar of the Library of Alexandria, director of the observatory in Alexandria, and one of the leading mathematicians and astronomers of her time. Unfortunately, she lived during a time of rising sentiment against free inquiry and was murdered by anti-intellectual mobs in A.D. 415. The final destruction of the Library of Alexandria followed not long after her death.

At its peak, the Library of Alexandria held more than a half million books, hand-written on papyrus scrolls. As the invention of the printing press was far in the future, most of the scrolls probably were original manuscripts or the single copies of original manuscripts. When the library was destroyed, most of its storehouse of ancient wisdom was lost forever.

TIME OUT TO THINK *How old is the school that you attend? Compare its age to the 700 years that the Library of Alexandria survived. Estimate the number of books you've read in your life. Compare this number to the half million books that once were housed in the Library of Alexandria.*

FIGURE 3.15 This diagram depicts the heavenly spheres, an ancient Greek model of the universe (c. 200 B.C.).

The Ptolemaic Model of the Universe

Perhaps the most important scientific idea developed by the ancient Greeks was that of creating **models** of nature. Just as a model airplane is a representation of a real airplane, a scientific model seeks to represent some aspect of nature. However, scientific models usually are conceptual models rather than miniature representations. The purpose of a scientific model is to explain and predict real phenomena, without any need to invoke myth, magic, or the supernatural. The ancient Greek idea of a celestial sphere is an example of a scientific model: The celestial sphere is the model, and it can be used to explain and predict the apparent motions of stars in our sky.

By combining the concept of modeling with advances in logic and mathematics, the ancient Greeks practiced a pursuit of knowledge very much like modern science. They used their models to make predictions, such as predicting where the planets should appear among the constellations of the night sky. If the predictions proved inaccurate, they changed and refined their models. In this way, the Greeks soon learned that the simple idea of a celestial

sphere could not explain all the phenomena they saw in the sky (particularly the apparent retrograde motion of the planets). They therefore added spheres for the Sun, the Moon, and each of the planets and then continued to change their ideas of how the spheres moved in an ongoing attempt to make the model agree with reality (Figure 3.15).

The culmination of this ancient modeling came with the work of Claudius Ptolemy (c. A.D. 100–170), pronounced *tol-e-mee*. Ptolemy's model is a tribute to human ingenuity in that it made reasonably accurate predictions despite being built on the flawed premise of an Earth-centered universe. Following a Greek tradition dating back to Plato (428–348 B.C.), Ptolemy imagined that all heavenly motions must be in perfect circles. To explain the apparent retrograde motion of planets, Ptolemy used an idea first suggested by Apollonius (c. 240–190 B.C.) and further developed by Hipparchus (c. 190–120 B.C.). This idea held that each planet moved along a small circle that, in turn, moved around a larger circle (Figure 3.16). As seen from Earth, this "circle upon circle" motion meant that a planet usually moved eastward relative to the stars but sometimes moved westward in apparent retrograde motion.

Ptolemy selected the sizes of the circles and the rates of motion of the planets along the circles so that his model reproduced the apparent retrograde motion

FIGURE 3.16 Ptolemy's model of the motions of the Sun, Moon, and planets around the Earth.

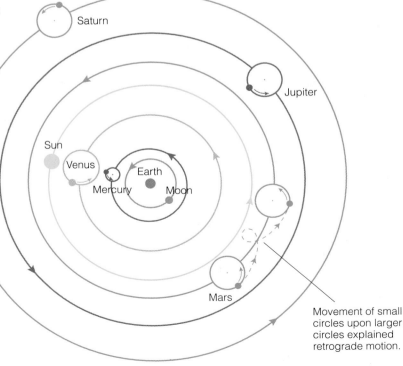

Movement of small circles upon larger circles explained retrograde motion.

of all the known planets. He also incorporated more complex ideas into his model, such as allowing the Earth to be slightly off-center from the circles of a particular planet, so that it could forecast future planetary positions to within a few degrees—which was considered quite accurate at that time. Ptolemy's model remained in use for some 1,500 years.

The Islamic Role

Most of the great scholarly work of the ancient Greeks was lost with the fall of the Roman Empire and the destruction of the Library of Alexandria in the fifth century A.D. Much more would have been lost had it not been for the rise of a new center of intellectual achievement in Baghdad (Iraq). While European civilization fell into the period of intellectual decline known as the Dark Ages, scholars of the new religion of Islam sought knowledge of mathematics and astronomy in hopes of better understanding the wisdom of Allah. During the eighth and ninth centuries, the Muslim empire centered in Baghdad translated and thereby saved many of the ancient Greek works.

Around A.D. 800, an Islamic leader named Al-Mamun (A.D. 786–833) established a "House of Wisdom" in Baghdad comparable in scope to the destroyed Library of Alexandria. Founded in a spirit of great openness and tolerance, the House of Wisdom employed Jews, Christians, and Muslims, all working together in scholarly pursuits. Using the translated Greek scientific manuscripts as building blocks, these scholars developed algebra and many new instruments and techniques for astronomical observation. Most of the official names of constellations and stars come from Arabic because of the work of the scholars at Baghdad. If you look at a star chart, you will see that the names of many bright stars begin with *al* (e.g., Aldebaran, Algol), which simply means "the" in Arabic. Ptolemy's 13-volume manuscript describing his astronomical system is also known by its Arabic name, *Almagest,* from Arabic words meaning "the greatest compilation."

The Islamic world of the Middle Ages had frequent contact with Hindu scholars from India, who in turn brought knowledge of ideas and discoveries from China. Hence, the intellectual center in Baghdad achieved a synthesis of the surviving work of the ancient Greeks with that of the Indians and Chinese. The accumulated knowledge of the Arabs spread throughout the Byzantine Empire (the eastern part of the former Roman Empire). When the Byzantine capital of Constantinople (modern-day Istanbul) fell to the Turks in 1453, many Eastern scholars headed west to Europe, bringing with them the knowledge that helped ignite the European Renaissance.

3.4 Modern Science and the Scientific Method

As we have seen, many different cultures developed scientific modes of thinking, and the ancient Greeks developed principles of modeling very similar to those underlying the modern scientific method. But there are important differences between these ancient ideas and modern science. For example, Ptolemy's writings suggest that he designed his model to aid in calculations, with little concern over whether it reflected reality. While modern scientists also employ purely computational models when necessary, understanding the underlying reality of nature is an important goal of modern science.

The principles that govern modern science coalesced during the European Renaissance. Within 100 years after the fall of Constantinople, Copernicus began the process of overturning the Earth-centered, Ptolemaic model, setting into motion a burst of scientific discovery that continues to this day. Before we can fully appreciate the Copernican revolution, which we'll discuss in Chapter 5, we must investigate a more basic question: What is modern science?

The word **science** comes from the Latin *scientia,* meaning "knowledge"; it is also the root of the words *conscience* (related to the idea of self-knowledge or self-awareness) and *omniscience* (the quality of being all-knowing). Today, science connotes a special kind of knowledge: the kind that makes predictions that can be confirmed by rigorous observations and experiments. In brief, the modern **scientific method** is an organized approach to explaining observed facts with a model of nature, subject to the constraint that any proposed model must be testable and the provision that the model must be modified or discarded if it fails these tests.

In its most idealized form, the scientific method begins with a set of observed facts. A fact is supposed to be a statement that is objectively true. For example, we consider it a fact that the Sun rises each morning, that the planet Mars appeared in a particular place in our sky last night, and that the Earth rotates. Facts are not always obvious, as illustrated by the case of the Earth's rotation—for most of human history, the Earth was assumed to be stationary at the center of the universe. In addition, our interpretations of facts often are based on beliefs about the world that others might not share. For example, when we say that the Sun rises each morning, we assume that it is the *same* Sun day after day—an idea that might not have been accepted by ancient Egyptians, whose mythology held that the Sun died with every sunset and was reborn with every sunrise. Nevertheless, facts are the raw material that scientific models seek to explain, so it is important that scientists agree on the facts. In the context of science, a fact must therefore be something that anyone can verify for himself or herself, at least in principle.

Once the facts have been collected, a model can be proposed to explain them. A useful model must also make predictions that can be tested through experiments or further observations. Ptolemy's model of the universe was useful because it predicted future locations of the Sun, Moon, and planets in the sky. However, although the Ptolemaic model remained in use for nearly 1,500 years, it eventually became clear that its predictions did not quite match actual observations—a key reason why the Earth-centered model of the universe finally was discarded.

In summary, the idealized scientific method proceeds as follows:

- *Observation:* The scientific method begins with the collection of a set of observed facts.

- *Hypothesis:* A model is proposed to explain the observed facts and to make new predictions. A proposed model is often called a **hypothesis,** which essentially means an *educated guess.*

- *Experiment:* The model's predictions are tested through experiments or further observations. When a prediction is verified, we gain confidence that the model truly represents nature. When a prediction fails, we recognize that the model is flawed, and we therefore must refine or discard the model.

- *Theory:* A model must be continually challenged with new observations and experiments by many different scientists. A model achieves the status of a **scientific theory** only after a broad range of its predictions have been repeatedly verified. Note that, while we can have great confidence that a scientific theory truly represents nature, we can never prove a theory to be true *beyond all doubt.* Therefore, even well-established theories must be subject to continuing challenges through further observations and experiments.

In reality, scientific discoveries rarely are made by a process as mechanical as the idealized scientific method described here. For example, Johannes Kepler, who discovered the laws of planetary motion in the early 1600s, tested his model against observations that had been made previously, rather than verifying new predictions of his model [Section 5.3]. Moreover, like most scientific work, Kepler's work involved intuition, collaboration with others, moments of insight, and luck. Nevertheless, with hindsight we can look back at Kepler's theory and see that other scientists eventually made plenty of observations to verify the planetary positions predicted by his model. In that sense, the scientific method represents an ideal prescription for judging objectively whether a proposed model of nature is close to the truth.

TIME OUT TO THINK *Look up* theory *in a dictionary. How does its definition differ from that given here for a* scientific theory?

Science, Nonscience, and Pseudoscience

People often seek knowledge in ways that do not follow the basic tenets of the scientific method and therefore are not science. When we leave the realm of science, claims of knowledge become far more subjective and difficult to verify. Nevertheless, it can be useful to categorize such claims of knowledge as either *pseudoscience* or *nonscience.*

By **pseudoscience** we mean attempts to search for knowledge in ways that may at first seem scientific but do not really adhere to the testing and verification requirements of the scientific method; the prefix *pseudo* means "false." For example, at the beginning of each year you can find tabloid newspapers offering predictions made by people who claim to be able to "see" the future. Because they make specific predictions, we can test their claims by checking whether their predictions come true, and numerous studies have shown that their predictions come true no more often than would be expected by pure chance. But these seers seem unconcerned with such studies. The fact that they make testable claims but then ignore the results of the tests marks their claimed ability to see the future as pseudoscience.

By **nonscience** we mean knowledge sought through pure intuition, societal traditions, ancient scriptures, and other processes that make no claim to adhere to the scientific method. In nonscience, experimental testing is irrelevant because nonscientific beliefs are based on things such as faith, political conviction, and tradition. Because nonscience makes no pretense of following the scientific method, science can say nothing about the validity of nonscience.

TIME OUT TO THINK *Can the scientific method be applied in any way to questions of faith, prayer, emotion, or values? Defend your view.*

Is Science Objective?

The boundaries between science, nonscience, and pseudoscience are sometimes blurry. In particular, because science is practiced by human beings, individual scientists carry their personal biases and beliefs with them in their scientific work. These biases can influence how a scientist proposes or tests a model, and in some cases scientists have been known to cheat—either deliberately or subconsciously—to obtain the results they desire. For example, in the late nineteenth and early twentieth centuries, some astronomers claimed to see networks of "canals" in their blurry telescopic images of Mars, and they

Astrology

Although the terms *astrology* and *astronomy* sound very similar, today they describe very different practices. In ancient times, however, astrology and astronomy often went hand in hand, and astrology played an important role in the historical development of astronomy. Indeed, astronomers and astrologers were usually one and the same.

In brief, the basic tenet of astrology is that human events are influenced by the apparent positions of the Sun, Moon, and planets among the stars in our sky. The origins of this idea are easy to understand. After all, there is no doubt that the position of the Sun in the sky influences our lives—it determines the seasons and hence the times of planting and harvesting, of warmth and cold, and of daylight and darkness. Similarly, the Moon determines the tides, and the cycle of lunar phases coincides with many biological cycles. Because the planets also appear to move among the stars, it seemed reasonable to imagine that planets also influence our lives, even if these influences were much more difficult to discover.

Ancient astrologers hoped that they might learn *how* the positions of the Sun, Moon, and planets influence our lives. They charted the skies, seeking correlations with events on Earth. For example, if an earthquake occurred when Saturn was entering the constellation Leo, might Saturn's position have been the cause of the earthquake? If the king became ill when Mars appeared in the constellation Gemini and the first-quarter moon appeared in Scorpio, might it mean another tragedy for the king when this particular alignment of the Moon and Mars next recurred? Surely, the ancient astrologers thought, the patterns of influence eventually would become clear. Thus, the astrologers hoped that they might someday learn to forecast human events with the same reliability with which astronomical observations of the Sun could forecast the coming of spring.

This hope was never realized. Although many astrologers still attempt to predict future events, scientific tests have shown that their predictions come true no more often than would be expected by pure chance. Moreover, in light of our current understanding of the universe, the original ideas behind astrology no longer make sense. For example, today we use ideas of gravity and energy to explain the influences of the Sun and the Moon, and these same ideas tell us that the planets are too far from Earth to have a similar influence on our lives.

Of course, many people continue to practice astrology, perhaps because of its ancient and rich traditions. Scientifically, we cannot say anything about such traditions, because traditions are not testable predictions. But if you want to understand the latest discoveries about the cosmos, you'll need a science that can be tested and refined—and astrology can't meet these requirements.

hypothesized that Mars was home to a dying civilization that used the canals to transport water to thirsty cities [Section 8.6]. Because no such canals actually exist, these astronomers apparently were allowing their beliefs to influence how they interpreted blurry images. It was, in essence, a form of cheating—even though it certainly was not intentional.

Sometimes, bias can show up even in the thinking of the scientific community as a whole. Thus, some valid ideas may not be considered by any scientist because the ideas fall too far outside the general patterns of thought, or the **paradigm,** of the time. One such example is the development of Einstein's theory of relativity. Many other scientists had gleaned hints of this theory in the decades before Einstein but did not investigate them, at least in part because they seemed too outlandish.

The beauty of the scientific method is that it encourages continued testing by many people. Thus, even if personal biases affect some results, tests by others eventually will discover the mistakes. Simi-

larly, even when a new idea falls outside the accepted paradigm, sufficient testing and verification of the idea will eventually force a change in the paradigm. Thus, although individual scientists rarely follow the scientific method in its idealized form, the collective action of many scientists over many years generally *does* follow the basic tenets of the scientific method. That is why, despite the biases of individual scientists, science *as a whole* is objective.

3.5 Astronomy Today

The modern scientific method has fueled a dramatic rise in human knowledge. It took humanity some 6,000 years to progress from the early lunar astronomy recorded in Africa to the ancient Greek idea of creating models of the universe, and another 2,000 years to finally recognize that the Earth orbits the Sun. In contrast, the past 400 years have seen discovery upon discovery, and we have come to know that the universe is filled with wonders far beyond the wildest imagination of our ancestors.

Today, the science of astronomy continues to develop at an ever-growing rate. New telescopes are fueling a renaissance in astronomy. In the 1990s alone, we made the first discoveries of planets around other stars, witnessed the birth of galaxies near the edge of the observable universe, and accumulated strong evidence for the existence of black holes. If we measure information as a computer does, in *bytes*—where 1 byte represents the information content of one typed character and 1 megabyte represents the information in a 500-page book—then the information recorded at a single major observatory in a single night exceeds the total amount of astronomical information recorded by all ancient peoples combined.

The abundance of new data is disseminated worldwide via computer networks, and new data are collected at international observatories in many different countries and in space. The result is that, after a few centuries during which advances in astronomy were made primarily in Europe and the United States, astronomy is once again a worldwide endeavor. Just as the ancient civilizations of Egypt and Mesopotamia benefited from the cross-fertilization of ideas brought by people of many different cultures, astronomy today is benefiting from the great variety of perspectives brought by scientists from all over the world.

We especially need imagination in science.

MARIA MITCHELL (1818–1889), Astronomer and first woman elected to the American Academy of Arts and Sciences

THE BIG PICTURE

In this chapter, we turned from the big picture of the universe presented in Chapters 1 and 2 to the topic of how we have learned so much about the universe. Key "big picture" concepts from this chapter include the following:

- The basic ingredients of scientific thinking—careful observation and trial-and-error testing—are a part of everyone's experience. The modern scientific method simply provides a way of organizing this everyday thinking to facilitate the learning and sharing of new knowledge.

- Although knowledge about the universe is growing rapidly today, each new piece rests upon foundations of older discoveries. The foundations of astronomy reach far back into history and are intertwined with the general development of human culture and civilization.

- The concept of making models of nature lies at the heart of modern science. Models are created by generalizing from many specific facts. They then yield further predictions that allow them to be tested. Models that withstand repeated testing rise to the status of scientific theories, while models that fail must be modified or discarded.

Review Questions

1. In what ways is scientific thinking a fundamental part of human nature? What characteristics does the modern scientific method possess that go beyond this everyday type of thinking?

2. Explain how the people of central Africa predicted the weather by observing the Moon.

3. How are the names of the seven days of the week related to astronomical objects?

4. Briefly describe each of the following, and give a brief description of its astronomical uses: Stonehenge, the Templo Mayor, the Sun Dagger, the Mayan observatory at Chichén Itzá, lines in the Nazca desert, Pawnee lodges, the Big Horn Medicine Wheel.

5. Explain why the Muslim fast of Ramadan occurs earlier with each subsequent year.

6. What is the *Metonic cycle?* How does the Jewish calendar follow the Metonic cycle? How does this influence the date of Easter?

7. What role did the Navigator play in Polynesian society?

8. What was the Library of Alexandria? What role did it play in the ancient empires of Greece and Rome?

9. What do we mean by a *model* of nature? Explain how the celestial sphere represents such a model.

10. Who was Ptolemy? Briefly describe how the Ptolemaic model of the universe explains apparent retrograde motion while preserving the idea of an Earth-centered universe.

11. Briefly describe the role played by Islamic scholars in preserving ancient Greek knowledge and laying the groundwork for the European Renaissance.

12. Describe the basic process of the scientific method: observation, hypothesis, experiment, and theory. What is the role of a *model* in this method? What do we mean by *facts* when discussing science?

13. Briefly describe how science differs from *pseudoscience* and *nonscience*.

14. Briefly explain how it is possible for science as a whole to be objective despite the fact that all individual scientists have personal biases and beliefs.

15. Briefly describe the international character of modern astronomy.

Discussion Questions

1. *The Impact of Science.* The modern world is filled with ideas, knowledge, and technology that developed through science and application of the scientific method. Discuss some of these things and how they affect our lives. Which of these impacts do you think are positive? Which are negative? Overall, do you think that science has benefited the human race? Defend your opinion.

2. *Secrecy and Science.* The text mentions that some historians believe that Chinese science and technology fell behind that of Europe because of the Chinese culture of secrecy. Do you agree that secrecy can hold back the advance of science? Why or why not? For the past 200 years, the United States has allowed a greater degree of free speech than most other countries in the world. How much of a role do you think this has played in making the United States the world leader in science and technology? Defend your opinion.

Problems

1. *Cultural Accomplishments in Astronomy.* Choose a particular culture of interest to you, and research the astronomical knowledge and accomplishments of that culture. Write a two- to three-page summary.

2. *Astronomical Structures.* Choose an ancient astronomical structure of interest to you (e.g., Stonehenge, Nazca lines, Pawnee lodges, etc.), and research its history. Write a two- to three-page summary.

3. *Venus and the Mayans.* The planet Venus apparently played a particularly important role in Mayan society. Research the evidence, and write a one- to two-page summary of current knowledge about the role of Venus in Mayan society.

4. *Easter.* Find out when Easter is celebrated by different Christian churches. Then research how and why different groups set different dates for Easter. Summarize your findings in a one- to two-page written report.

5. *Greek Astronomers.* Many ancient Greek scientists had ideas that, in retrospect, seem well ahead of their time. Choose one or more of the following ancient Greek scientists, and learn enough about his or her work in science and astronomy to write a one- to two-page "scientific biography."

Thales	Seleucus	Eudoxus
Democritus	Pythagoras	Eratosthenes
Callipus	Plato	Hypatia
Hipparchus	Archimedes	Empedocles
Anaximander	Ptolemy	Aristotle
Meton	Anaxagoras	Apollonius
Aristarchus		

6. *Project: Devise Your Own Test.* Devise your own scientific test of astrology. Clearly define the methods you will use in your test and how you will evaluate the results. Then carry out the test.

A Universe of Matter and Energy

4

IN THIS AND THE NEXT TWO CHAPTERS, WE TURN OUR ATTENTION TO THE
scientific concepts that lie at the heart of modern astronomy. We begin
by investigating the nature of matter and energy, the fundamental stuff from which the universe is made.

The history of the universe essentially is a
story about the interplay between matter and energy since the beginning of time. Interactions
between matter and energy govern everything
from the creation of hydrogen and helium in
the Big Bang to the processes that make Earth
a habitable planet. Understanding the universe
therefore depends on familiarity with how matter responds to the ebb and flow of energy.

The concepts of matter and energy
presented in this chapter will enable you to understand most of the topics in this book. Some of the concepts and terminology may already be familiar to you. If not,
don't worry. We will go over them again as they arise in different
contexts, and you can refer back to this chapter as needed during the
remainder of your studies.

4.1 Matter and Energy in Everyday Life

The meaning of **matter** is obvious to most people, at least on a practical level. Matter is simply material, such as rocks, water, or air. You can hold matter in your hand or put it in a box. The meaning of **energy** is not quite as obvious, although we certainly talk a lot about it. We pay energy bills to the power companies, we use energy from gasoline to run our cars, and we argue about whether nuclear energy is a sensible alternative to fossil fuels. On a personal level, we often talk about how energetic we feel on a particular day. But what *is* energy?

Broadly speaking, energy is what makes matter move. For Americans, the most familiar way of measuring energy is in units of Calories, which we use to describe how much energy our bodies can draw from food. A typical adult uses about 2,500 Calories of energy each day. Among other things, this energy keeps our hearts beating and our lungs breathing, generates the heat that maintains our 37°C (98.6°F) body temperature, and allows us to walk and run.

Just as there are many different units for measuring height, such as inches, feet, and meters, there are many alternatives to Calories for measuring energy. If you look closely at an electric bill, you'll probably find that the power company charges you for electrical energy in units called *kilowatt-hours.* If you purchase a gas appliance, its energy requirements may be labeled in *British thermal units,* or BTUs. In science and internationally, the favored unit of energy is the **joule,** which is equivalent to about 1/4,000 of a Calorie. Thus, the 2,500 Calories used daily by a typical adult is equivalent to about 10 million joules. For purposes of comparison, some energies given in joules are listed in Table 4.1.

Although energy can always be measured in joules, it comes in many different forms. Already we have talked about food energy, electrical energy, the energy of a beating heart, heat energy represented by our body temperature, and more. Fortunately, the many forms of energy can be grouped into three basic categories.

First, whenever matter is moving, it has energy of motion, or **kinetic energy** (*kinetic* comes from a Greek word meaning "motion"). Falling rocks, the moving blades on an electric mixer, a car driving down the highway, and the molecules moving in the air around us are all examples of objects with kinetic energy.

The second basic category of energy is **potential energy,** or energy being stored for later conversion into kinetic energy. A rock perched on a ledge has *gravitational* potential energy because it will fall if it slips off the edge. Gasoline contains *chemical* potential energy, which a car engine converts to the kinetic energy of the moving car. Power companies supply *electrical* potential energy, which we use to run dishwashers and other appliances.

The third basic category is energy carried by light, or **radiative energy** (the word *radiation* often is used as a synonym for *light*). Plants directly convert the radiative energy of sunlight into chemical potential energy through the process called *photosynthesis.* Radiative energy is fundamental to astronomy, because telescopes collect the radiative energy of light from distant stars.

TIME OUT TO THINK *We buy energy in many different forms. For example, we buy chemical potential energy in the form of food to fuel our bodies. Describe several other forms of energy that you commonly buy.*

Understanding how energy changes from one form to another helps us understand many common phenomena. For example, a diver standing on a 10-meter platform has gravitational potential energy owing to her height above the water and chemical potential energy stored in her body tissues. She uses the chemical potential energy to flex her muscles in such a way as to initiate her dive and then to help her execute graceful twists and spins (Figure 4.1). Meanwhile, her gravitational potential energy becomes kinetic energy of motion as she falls toward the water.

Following how energy changes from one form to another also helps us understand the universe. But before we study astronomical phenomena, we first need ways to describe energy quantitatively.

Table 4.1 Energy Comparisons

Item	Energy (joules)
Average daytime solar energy striking Earth, per m² per second	1×10^3
Energy released by metabolism of one average candy bar	1×10^6
Energy needed for 1 hour of walking (adult)	1×10^6
Kinetic energy of average car traveling at 60 mi/hr	1×10^6
Daily energy needs of average adult	1×10^7
Energy released by burning 1 liter of oil	1.2×10^7
Energy released by fission of 1 kg of uranium-235	5.6×10^{13}
Energy released by 1-megaton H-bomb	5×10^{15}
Energy released by major earthquake (magnitude 8.0)	2.5×10^{16}
U.S. annual energy consumption	10^{20}
Annual energy generation of Sun	10^{34}
Energy released by supernova (explosion of a star)	$10^{44}-10^{46}$

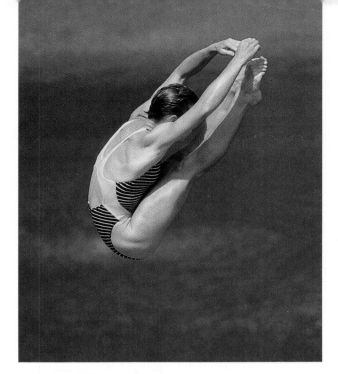

FIGURE 4.1 Understanding energy can help us understand the graceful movements of a diver and the story of the universe.

4.2 A Scientific View of Energy

In this section, we discuss a few ways to quantify kinetic and potential energy; we'll discuss radiative energy in Chapter 6, in which we study properties of light.

Kinetic Energy

You can calculate the kinetic energy of any moving object with a very simple formula:

$$\text{kinetic energy} = \tfrac{1}{2}mv^2$$

where m is the mass of the object and v is its speed (v for *velocity*). If you measure the mass in kilograms and the speed in meters per second, the resulting answer will be in joules. The kinetic energy formula is easy to interpret. The m tells us that kinetic energy is proportional to mass: A 5-ton truck has 5 times the kinetic energy of a 1-ton car moving at the same speed. The v^2 tells us that kinetic energy increases with the *square* of the velocity: If you double your speed (e.g., from 30 km/hr to 60 km/hr), your kinetic energy becomes $2^2 = 4$ times greater.

Thermal Energy
Suppose we want to know about the kinetic energy of the

countless tiny particles (atoms and molecules) inside a rock, or of the countless particles in the air or in a distant star. Each of these tiny particles has its own motion relative to surrounding particles, and these motions constantly change as the particles jostle one another. The result is that the particles inside a substance appear to move randomly: Any individual particle may be moving in any direction with any of a wide range of speeds. Despite the seemingly random motion of particles within a substance, it's easy to measure the *average* kinetic energy of the particles—which we call **temperature** (Figure 4.2). A higher temperature simply means that, on average, the particles have more kinetic energy and hence are moving faster (Figure 4.3). The speeds of particles within a substance can be surprisingly fast. For example, the air molecules around you move at typical speeds of about 500 meters per second (about 1,000 miles per hour).

The energy contained *within* a substance as measured by its temperature often is called **thermal energy.** Thus, thermal energy represents the collective kinetic energy of the many individual particles moving within a substance.

Temperature and Heat The concepts of *temperature* and *heat* are not quite the same. To understand the difference, imagine the following experiment (but don't try it!). Suppose you heat your oven to 500°F. Then you open the oven door, quickly thrust your

FIGURE 4.2 Temperature is a measure of the average kinetic energy of the particles in a substance. This diagram compares three common temperature scales. The Fahrenheit scale is used in the United States, but nearly all other countries use the Celsius scale. Scientists prefer the Kelvin scale because 0 K represents *absolute zero,* the coldest possible temperature. (The degree symbol ° is not usually used with the Kelvin scale.)

Longer arrows mean higher speed.

FIGURE 4.3 The particles in the box on the right have a higher temperature because their average speeds are higher (assuming that both boxes contain particles of the same mass).

FIGURE 4.4 Both boxes have the same temperature, but the box on the right contains more *heat* because it contains more particles.

arm inside (without touching anything), and immediately remove it. What will happen to your arm? Not much. Now suppose you boil a pot of water. Although the temperature of boiling water is only 212°F, you would be badly burned if you put your arm in the pot, even for an instant. Thus, the pot of water transfers more heat to your arm than does the air in the oven, even though it has a lower temperature. That is, your arm heats more quickly in the water. This is because heat content depends on both the temperature and the total number of particles, which is much larger in a pot of water (Figure 4.4).

Let's look at what is happening on the molecular level. If air or water is hotter than your body, molecules striking your skin transfer some of their thermal energy to your arm. The high temperature in a 500°F oven means that the air molecules strike your skin harder, on average, than the molecules in a 212°F pot of boiling water. However, because the *density* is so much higher in the pot of water, many more molecules strike your skin each second. Thus, while each individual molecular collision transfers a little less energy in the boiling water than in the oven, the sheer number of collisions in the water transfers so much heat that your skin burns rapidly.

Potential Energy

Potential energy can be stored in many different forms and is not always easy to quantify. Fortunately, it is easy to describe two types of potential energy that we use frequently in astronomy.

Gravitational Potential Energy Gravitational potential energy is extremely important in astronomy. The conversion of gravitational potential energy into kinetic (or thermal) energy helps explain everything from the speed at which an object falls to the ground to the formation processes of stars and planets. The mathematical formula for gravitational potential energy can take a variety of forms, but in words the idea is simple: *The amount of gravitational potential energy released as an object falls depends on its mass, the strength of gravity, and the distance it falls.*

This statement explains the obvious fact that falling from a 10-story building hurts more than falling out of a chair. Your gravitational potential en-

ergy is much greater on top of the 10-story building than in your chair because you can fall much farther. Because your gravitational potential energy will be converted to kinetic energy when you fall, you'll have a lot more kinetic energy by the time you hit the ground after falling from the building than from the chair, which means you'll hit the ground much harder.

Gravitational potential energy also helps us understand how the Sun became hot enough to sustain nuclear fusion. Before the Sun formed, it was a large, cold, diffuse cloud of gas. Most of the individual gas particles were far from the center of this large cloud and therefore had considerable amounts of gravitational potential energy. As the cloud collapsed under its own gravity, the gravitational potential energy of these particles was converted to thermal energy, eventually making the center of the cloud so hot that it could ignite nuclear fusion.

Mass-Energy Although matter and energy seem very different in daily life, they are intimately connected. Einstein showed that mass itself is a form of potential energy, often called **mass-energy.** The mass-energy of any piece of matter is given by the formula

$$E = mc^2$$

where E is the amount of potential energy, m is the mass of the object, and c is the speed of light. If the mass is measured in kilograms and the speed of light in meters per second, the resulting mass-energy has units of joules. Note that the speed of light is a large number ($c = 3 \times 10^8$ m/s) and the speed of light squared is much larger still ($c^2 = 9 \times 10^{16}$ m^2/s^2). Thus, Einstein's formula implies that a relatively small amount of mass represents a huge amount of mass-energy.

Mass-energy can be converted to other forms of energy, but noticeable amounts of mass become other forms of energy only under special but important circumstances. The process of nuclear fusion in

FIGURE 4.5 The energy from this H-bomb comes from converting only about 0.1 kg of mass into energy, in accordance with $E = mc^2$.

the core of the Sun converts some of the Sun's mass into energy, ultimately generating the sunlight that makes life on Earth possible. On Earth, nuclear reactors and nuclear bombs also work in accord with Einstein's formula. In nuclear reactors, the splitting (fission) of elements such as uranium or plutonium converts some of the mass-energy of these materials into heat, which is then used to generate electrical power. In an H-bomb, nuclear fusion similar to that in the Sun converts a small amount of the mass-energy in hydrogen into the bomb's destructive energy. Incredibly, a 1-megaton H-bomb that could destroy a major city requires the conversion of only about 0.1 kilogram (about 3 ounces) of mass into energy (Figure 4.5).

Just as $E = mc^2$ tells us that mass can be converted into other forms of energy, it also tells us that energy can be transformed into mass. In *particle accelerators,* scientists accelerate subatomic particles to extremely high speeds. When these particles collide with one another or with a barrier, some of the energy released in the collision spontaneously turns into mass, appearing as a shower of subatomic particles. These showers of particles allow scientists to test theories about how matter behaves at extremely

high temperatures such as those that prevailed in the universe during the first fraction of a second after the Big Bang. Among the most powerful particle accelerators in the world are Fermilab in Illinois, the Stanford Linear Accelerator in California, and CERN in Switzerland.

TIME OUT TO THINK *Einstein's formula $E = mc^2$ is probably the most famous physics equation of all time. Considering its role as described in the preceding paragraphs, do you think its fame is well deserved?*

Conservation of Energy

A fundamental principle in science is that, regardless of how we change the *form* of energy, the total quantity of energy never changes. This principle is called the **law of conservation of energy.** It has been carefully tested in many experiments, and it is a pillar upon which modern theories of the universe are built. Because of this law, the story of the universe is a story of the interplay of energy and matter: All actions in the universe involve exchanges of energy or the conversion of energy from one form to another.

For example, imagine that you've thrown a base-ball so that it is moving and hence has kinetic energy. Where did this kinetic energy come from? The base-ball got its kinetic energy from the motion of your arm as you threw it; that is, some of the kinetic energy of your moving arm was transferred to the base-ball. Your arm, in turn, got its kinetic energy from the release of chemical potential energy stored in your muscle tissues. Your muscles got this energy from the chemical potential energy stored in the foods you ate. The energy stored in the food came from sun-light, which plants convert into chemical potential energy through photosynthesis. The radiative energy of the Sun was generated through the process of nuclear fusion, which releases some of the mass-energy stored in the Sun's supply of hydrogen. Thus, the ultimate source of the energy of the moving base-ball is the mass-energy stored in hydrogen—which was created in the Big Bang.

TIME OUT TO THINK *Where does the kinetic energy of the ball go after it hits the ground? Remember that energy must always be conserved.*

4.3 The Material World

Now that we have seen how energy animates the matter in the universe, it's time to consider matter itself in greater detail. You are familiar with two basic properties of matter on Earth from everyday experience. First, matter can exist in different **phases:** as a **solid,** such as ice or a rock; as a **liquid,** such as flowing water or oil; or as a **gas,** such as air. Second, even in a particular phase, matter comes in a great variety of different substances.

But what *is* matter, and why does it have so many different forms? Today, we know that all ordinary matter is composed of **atoms** and that each different type of atom corresponds to a different chemical **element.** Among the most familiar chemical elements are hydrogen, helium, carbon, oxygen, silicon, iron, gold, silver, lead, and uranium.

The number of different material substances is far greater than the number of chemical elements because atoms can combine to form **molecules.** The chemical properties of a molecule are different from those of its individual atoms. For example, water behaves very differently than pure hydrogen or pure oxygen, even though water molecules are composed of two hydrogen atoms and one oxygen atom, as indicated by the familiar symbol H_2O.

Atomic Structure

Atoms are incredibly small: Millions could fit end-to-end across the period at the end of this sentence,

and the number in a single drop of water (10^{22} to 10^{23} atoms) exceeds the number of stars in the observable universe. Yet atoms are composed of even smaller particles. Nearly all of an atom's mass is contained in its **nucleus,** which is surrounded by one or more **electrons.** The nucleus, in turn, is composed of two types of particles: **protons** and **neutrons.**

The properties of an atom depend mainly on the amount of **electrical charge** in its nucleus. Electrical charge is a fundamental physical property that is always conserved, just as energy is always conserved. Electrons carry a negative electrical charge of -1, protons carry a positive electrical charge of $+1$, and neutrons are electrically neutral. Oppositely charged particles attract one another, and similarly charged particles repel one another. The attraction between the positively charged protons in the nucleus and the negatively charged electrons that surround it is what holds an atom together. Ordinary atoms have identical numbers of electrons and protons, making them electrically neutral overall.

Although electrons can be thought of as tiny particles, they are not quite like tiny grains of sand, and they don't really orbit the nucleus as planets orbit the Sun. Instead, the electrons in an atom are "smeared out," forming a kind of cloud that surrounds the nucleus and gives the atom its apparent size. The electrons aren't really cloudy; it's just that it is impossible to pinpoint their positions. An atomic nucleus is very tiny, even compared to the atom itself: If we imagine an atom on a scale on which its nucleus is the size of your fist, its electron cloud would be many miles wide. Figure 4.6 shows the structure of a typical atom.

**Common Misconceptions:
The Illusion of Solidity**

Bang your hand on a table. Although the table feels solid, it is made almost entirely of empty space! Nearly all the mass of the table is contained in the nuclei of its atoms. But the volume of an atom is more than a trillion times the volume of its nucleus, so relatively speaking the nuclei of adjacent atoms are nowhere near to touching one another. The solidity of the table comes about from a combination of electrical interactions between the charged particles in its atoms and the strange quantum laws governing the behavior of electrons. If we could somehow pack all the table's nuclei together, the table's mass would fit into a microscopic speck. Although *we* cannot pack matter together in this way, nature can and does—in *neutron stars,* which we will study in Chapter 15.

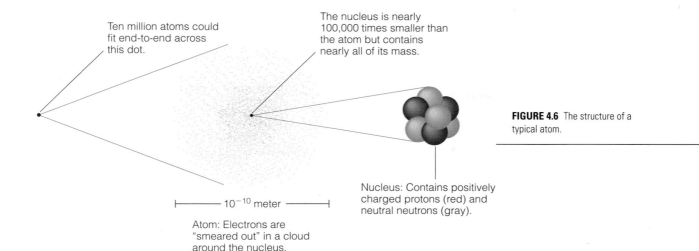

Ten million atoms could fit end-to-end across this dot.

The nucleus is nearly 100,000 times smaller than the atom but contains nearly all of its mass.

FIGURE 4.6 The structure of a typical atom.

10^{-10} meter

Atom: Electrons are "smeared out" in a cloud around the nucleus.

Nucleus: Contains positively charged protons (red) and neutral neutrons (gray).

Each different chemical element contains a different number of protons in its nucleus called its **atomic number.** For example, a hydrogen nucleus contains just one proton, so its atomic number is 1; a helium nucleus contains two protons, so its atomic number is 2.

The *combined* number of protons and neutrons in an atom is called its **atomic weight** (or **atomic mass**). The atomic weight of ordinary hydrogen is 1 because its nucleus is just a single proton. Helium usually has two neutrons in addition to its two protons, giving it an atomic weight of 4. Carbon usually has six protons and six neutrons, giving it an atomic weight of 12.

Sometimes, the same element can have varying numbers of neutrons. For example, in rare cases a hydrogen nucleus contains a neutron in addition to its proton, making its atomic weight 2; hydrogen with atomic weight 2 is often called *deuterium*. The dif-

ferent forms of hydrogen are called **isotopes** of one another. Each isotope of an element has the *same* number of protons but a *different* number of neutrons.

We usually denote different isotopes by showing the atomic weight to the upper left of the chemical symbol. For example, most carbon atoms contain six neutrons in addition to their six protons, making their atomic weight $6 + 6 = 12$; we therefore represent this isotope of carbon by writing ^{12}C. The symbol ^{14}C represents a rarer isotope of carbon that has atomic weight 14 and hence eight neutrons rather than six. Figure 4.7 illustrates some atomic terminology.

TIME OUT TO THINK *The symbol ^{4}He represents helium with an atomic weight of 4; this is the most common form, containing two protons and two neutrons. What does the symbol ^{3}He represent?*

FIGURE 4.7 Terminology of atoms.

atomic number = number of protons
atomic weight = number of protons + neutrons

Hydrogen (^{1}H)
atomic number = 1
atomic weight = 1
(1 electron)

Helium (^{4}He)
atomic number = 2
atomic weight = 4
(2 electrons)

Carbon (^{12}C)
atomic number = 6
atomic weight = 12
(6 electrons)

The number of electrons in a neutral atom equals its atomic number.

Isotopes of Hydrogen
hydrogen
^{1}H
(1 proton)

deuterium
^{2}H
(1 proton + 1 neutron)

Isotopes of Carbon
carbon-12
^{12}C
(6 protons + 6 neutrons)

carbon-14
^{14}C
(6 protons + 8 neutrons)

Phases of Matter

Everyday experience tells us that, depending on the temperature, the same substance exists in different phases. The main difference between different phases of matter is how tightly neighboring particles are bound together. As a substance is heated, the average kinetic energy of its particles increases, enabling the particles to break the bonds holding them to their neighbors. Each change in phase corresponds to the breaking of a different kind of bond. Although phase changes occur in all substances, let's consider what happens when we heat water, starting from ice (Figure 4.8):

■ Below 0°C (32°F), water molecules have a relatively low average kinetic energy, and each molecule is bound tightly to its neighbors, making the *solid* structure of ice.

■ As the temperature increases (but remains below freezing), the rigid arrangement of the molecules in ice vibrates more and more. At 0°C, the molecules have enough energy to break the solid bonds of ice. The molecules can then move relatively freely among one another, allowing the water to flow as a *liquid*. Even in liquid water, a type of bond between adjacent molecules still keeps them close together.

■ When a water molecule breaks free of all bonds with its neighbors, we call it a molecule of water vapor, which is a *gas*. Molecules in the gas phase move independently of other molecules. Even at low temperatures at which water is a solid or liquid, a few molecules will have enough energy to enter the gas phase. We call the process **evaporation** when molecules escape from a liquid and **sublimation** when they escape from a solid. As temperatures rise, the rates of sublimation and evaporation increase, eventually changing all the solid and liquid water into water vapor.

What happens if we continue to raise the temperature of water vapor? As the temperature rises, the molecules move faster, making collisions among them more and more violent. These collisions eventually split the water molecules into their component atoms of hydrogen and oxygen. The process in which the bonds that hold the atoms of a molecule together are broken is called **molecular dissociation.** At still higher temperatures, collisions can break the bonds holding electrons around the nuclei of individual atoms, allowing the electrons to go free. The loss of one or more negatively charged electrons leaves a remaining atom with a net positive charge. Such charged atoms are called **ions,** and the process of stripping electrons from atoms is called **ionization.** Thus, at high temperatures, what once was water becomes a hot gas consisting of freely moving electrons and positively charged ions of hydrogen and

FIGURE 4.8 The general progression of phase changes.

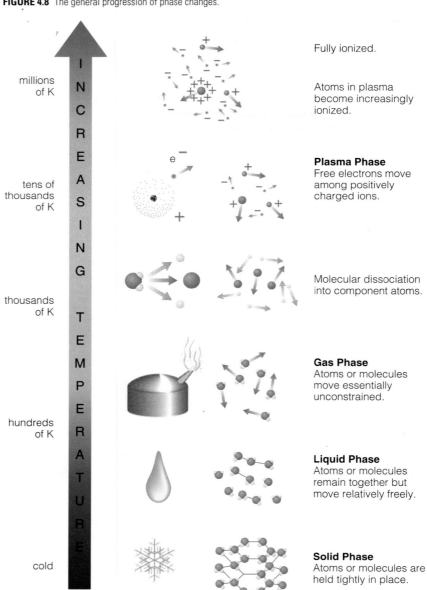

INCREASING TEMPERATURE

millions of K — Fully ionized.

Atoms in plasma become increasingly ionized.

Plasma Phase
Free electrons move among positively charged ions.

tens of thousands of K

thousands of K — Molecular dissociation into component atoms.

Gas Phase
Atoms or molecules move essentially unconstrained.

hundreds of K

Liquid Phase
Atoms or molecules remain together but move relatively freely.

Solid Phase
Atoms or molecules are held tightly in place.

cold

oxygen. This type of hot gas, in which atoms have become ionized, is called a **plasma,** sometimes referred to as "the fourth phase of matter." Other chemical substances go through similar phase changes, but the temperatures at which phase changes occur depend on the type of atom or molecule involved.

Note that neutral hydrogen contains only one electron, which balances the single positive charge of the one proton in its nucleus. Thus, hydrogen can be ionized only once, and the remaining hydrogen ion, designated H$^+$, is simply a proton. Oxygen, with atomic number 8, has eight electrons when it is neutral, so it can be ionized multiple times. *Singly ionized* oxygen is missing one electron, so it has a charge of $+1$ and is designated O$^+$. *Doubly ionized* oxygen, or O^{+2}, is missing two electrons; *triply ionized* oxygen, or O^{+3}, is missing three electrons; and so on. At extremely high temperatures, oxygen can be *fully ionized,* in which case all eight electrons are stripped away and the remaining ion has a charge of $+8$.

4.4 Energy in Atoms

So far, we've seen two different ways in which atoms have energy. First, by virtue of their mass, they possess mass-energy in the amount mc^2. Second, they possess kinetic energy by virtue of their motion.

Atoms can also contain energy in a third way: as *electric potential energy* in the distribution of their electrons around their nuclei. The simplest case is that of hydrogen, which has only one electron. Remember that an electron tends to be "smeared out" into a cloud around the nucleus. When the electron is "smeared out" to the minimum

extent that nature allows, the atom contains its smallest possible amount of electric potential energy; we say that the atom is in its **ground state** (Figure 4.9). If the electron somehow gains energy, then it becomes "smeared out" over a greater volume, and we say that the atom is in an **excited state.** If the electron gains enough energy, it can escape the atom completely, in which case the atom has been *ionized.*

ground state   excited state   ionization

FIGURE 4.9 In its ground state, an electron is "smeared out" to the minimum extent allowed by nature. Adding energy can raise the electron to an excited state that occupies a larger volume. Adding enough energy can ionize the atom.

Perhaps the most surprising aspect of atoms was discovered in the 1910s, when scientists realized that electrons in atoms can have only *particular* energies (which correspond to particular sizes and shapes of the electron cloud). A simple analogy can illuminate this idea. Suppose you're washing windows on a building. If you use an adjustable platform to reach high windows, you can stop the platform at any height above the ground (Figure 4.10a). But if you use a ladder, you can stand only at *particular* heights—the heights of the rungs of the ladder—and not at any height in between (Figure 4.10b). The possible energies of electrons in atoms are like the

FIGURE 4.10 (**a**) A window washer on an elevator can be at any height. (**b**) A window washer on a ladder can be at only particular heights. Similarly, electrons in an atom can have only particular energy levels.

a

b

possible heights on a ladder. Only a few particular energies are possible, and energies between these special few are not allowed.

The possible energy levels of the electron in hydrogen are represented like the steps of a ladder in Figure 4.11. The ground state, or level 1, is the bottom rung of the ladder; its energy is labeled zero because the atom has no excess electrical potential energy to lose. Each subsequent rung of the ladder represents a possible excited state for the electron. Each level is labeled with the electron's energy above the ground state in units of **electron-volts,** or **eV** ($1 \text{ eV} = 1.60 \times 10^{-19}$ joule). For example, the energy of an electron in energy level 2 is 10.2 eV greater than that of an electron in the ground state. That is, an electron must gain 10.2 eV of energy to "jump" from level 1 to level 2. Similarly, jumping from level 1 to level 3 requires gaining 12.1 eV of energy.

FIGURE 4.11 Energy levels for the electron in a hydrogen atom. There are many more closely spaced energy levels between level 4 and the ionization level.

Note that, unlike a ladder built for climbing, the rungs on the electron's energy ladder are closer together near the top. The top itself represents the energy of ionization—if the electron gains this much energy, or 13.6 eV above the ground state in the case of hydrogen, the electron breaks free from the atom. (Any excess energy beyond that needed for ionization becomes kinetic energy of the free-moving electron.)

Because energy is always conserved, an electron cannot jump to a higher energy level unless its atom gains the energy from somewhere else. Generally, the atom gains this energy either from the kinetic energy of another particle colliding with it or from the absorption of energy carried by light. Similarly, when an electron falls to a *lower* energy level, it either transfers its energy to another particle through a collision or emits light that carries the energy away. The key point is this: *Electron jumps can occur only with the particular amounts of energy representing differences between possible energy levels.*

The result is that electrons in atoms can absorb or emit only particular amounts of energy and not other amounts in between. For example, if you attempt to provide a hydrogen atom in the ground state with 11.1 eV of energy, the atom won't accept it because it is too high to boost the electron to level 2 but not high enough to boost it to level 3.

TIME OUT TO THINK *We will see in Chapter 6 that light comes in "pieces" called* photons *that carry specific amounts of energy. Can a hydrogen atom absorb a photon with 11.1 eV of energy? Why or why not? Can it absorb a photon with 10.2 eV of energy? Explain.*

If you think about it, the idea that electrons in atoms can jump only between particular energy levels is quite bizarre. It is as if you had a car that could go only particular speeds and not other speeds in between. How strange it would seem if your car suddenly jumped from 5 miles per hour to 20 miles per hour without gradually passing through a speed of 10 miles per hour! In scientific terminology, the electron's energy levels are said to be *quantized*, and the study of the energy levels of electrons (and other particles) is called *quantum mechanics*.

Electrons have quantized energy levels in all atoms, not just in hydrogen. Moreover, the allowed energy levels differ from element to element and even from one ion of an element to another ion of the same element. This fact holds the key to the study of distant objects in the universe. As we will see in Chapter 6 when we study light, the different energy levels of different elements allow light to carry "fingerprints" that can tell us the chemical composition of distant objects.

Common Misconceptions: Orbiting Electrons?

Most people have been taught to think of electrons in atoms as "orbiting" the nucleus like tiny planets orbiting a tiny sun. But this representation is simply not true. Electrons and other subatomic particles do not behave at all like baseballs, rocks, or planets. In fact, the behavior of subatomic particles is so strange that human minds may simply be incapable of visualizing it. Nevertheless, the *effects* of electrons are easy to observe, and one of the most important effects is that electrons give atoms their size. Thus, we say that electrons in atoms are "smeared out" into an "electron cloud" around the nucleus. This description is rather vague, but it is far more accurate than the misleading picture of electrons circling like tiny planets.

The eternal mystery of the world is its comprehensibility. The fact that it is comprehensible is a miracle.

ALBERT EINSTEIN

THE BIG PICTURE

In this chapter, we discussed the concepts of energy and matter in some detail. Key "big picture" ideas to draw from this chapter include the following:

- Energy and matter are the two basic ingredients of the universe. Therefore, understanding energy and matter, and the interplay between them, is crucial to understanding the universe.

- Energy is always conserved, and we can understand many processes in the universe by following how energy changes from one form to another in its interactions with matter. Don't forget that mass itself is a form of potential energy, called mass-energy.

- The strange laws of quantum mechanics govern the interactions of matter and energy on an atomic level. Electrons in atoms can have only particular energies and not energies in between, and each element has a different set of allowed energy levels.

Review Questions

1. Briefly describe and differentiate between *kinetic energy, potential energy,* and *radiative energy.*

2. What is the formula for the kinetic energy of an object? Based on this formula, explain why (a) a 4-ton truck moving at 100 km/hr has 4 times as much kinetic energy as a 1-ton car moving at 100 km/hr; (b) a 1-ton car moving at 100 km/hr has the *same* kinetic energy as a 4-ton truck moving at 50 km/hr.

3. What does *temperature* measure? How is it related to kinetic energy? What is *thermal energy?*

4. In your own words, briefly describe how a 500°F oven and a pot of boiling water illustrate the difference between *temperature* and *heat.*

5. What is *gravitational potential energy?* How does an object's gravitational potential energy depend on its mass, the distance it has to fall, and the strength of gravity?

6. What do we mean by mass-energy? Explain the meaning of the formula $E = mc^2$ and what it has to do with the Sun, nuclear bombs, and particle accelerators.

7. What is the law of conservation of energy? Your body is using energy right now to keep you alive. Where does this energy come from? Where does it go?

8. Briefly define *atom, element,* and *molecule.*

9. What is *electrical charge?* What type of electrical charge is carried by *protons, neutrons,* and *electrons?* Under what circumstances do electrical charges attract? Under what circumstances do they repel?

10. Briefly describe the structure of an atom. How big is an atom? How big is the *nucleus* in comparison to the entire atom?

11. Test your understanding of atomic terminology. (a) The most common form of iron has 26 protons and 30 neutrons in its nucleus. State its *atomic number,* its *atomic weight,* and its number of electrons if it is electrically neutral. (b) Consider the following three atoms: Atom 1 has seven protons and eight neutrons; atom 2 has eight protons and seven neutrons; atom 3 has eight protons and eight neutrons. Which two are *isotopes* of each other? (c) Oxygen has atomic number 8. How many times must an oxygen atom be ionized to create an O^{+5} ion? How many electrons are in an O^{+5} ion?

12. What do we mean by a *phase* of matter? Briefly describe how the idea of *bonds* between atoms (or molecules) explains the difference between the *solid, liquid,* and *gas* phases.

13. Briefly explain why a few atoms (or molecules) are always in gas phase around any solid or liquid. Then explain how *sublimation* and *evaporation* are similar and how they are different.

14. Explain why, at sufficiently high temperatures, molecules undergo *molecular dissociation* and atoms undergo *ionization.*

15. What is a *plasma?* Explain why, at very high temperatures, all matter is in this "fourth phase of matter."

16. Describe the three basic ways in which atoms can contain energy. In what way does an atom in an *excited state* contain more energy than an atom in the *ground state?*

17. How are the possible energy levels of electrons in atoms similar to the possible gravitational potential energies of a person on a ladder? How are they different?

Discussion Questions

1. *Knowledge of Mass-Energy.* Einstein's discovery that energy and mass are equivalent has led to technological developments both beneficial and dangerous. Discuss some of these developments. Overall, do you think the human race would be better or worse off if we had never discovered that mass is a form of energy? Defend your opinion.

2. *Indoor Pollution.* Given that sublimation and evaporation are very similar processes, why is sublimation generally much more difficult to notice? Discuss how sublimation, particularly from plastics and other human-made materials, can cause "indoor pollution."

Problems

1. *Gravitational Potential Energy.*

   a. Why does a bowling ball perched on a cliff ledge have more gravitational potential energy than a baseball perched on the same ledge?

   b. Why does a diver on a 10-meter platform have more gravitational potential energy than a diver on a 3-meter diving board?

   c. Why does a 100-kg satellite orbiting Jupiter have more gravitational potential energy than a 100-kg satellite orbiting the Earth, assuming both satellites orbit at the same distance from the planet centers?

2. *Einstein's Famous Formula.*

   a. What is the meaning of the formula $E = mc^2$? Be sure to define each variable.

   b. How does this formula explain the generation of energy by the Sun?

   c. How does this formula explain the destructive power of nuclear bombs?

3. *Spontaneous Human Combustion.* Suppose that all the mass in your body was suddenly converted into energy according to the formula $E = mc^2$. How much energy would be produced? Compare this to the energy of a nuclear bomb (see Table 4.1). What effect would your disappearance have on the surrounding region? (*Hint:* Multiply your mass in kilograms by c^2.)

4. *The Fourth Phase of Matter.*

   a. Explain why nearly all the matter in the Sun is in the plasma phase.

   b. Based on your answer to part (a), explain why plasma is the most common phase of matter in the universe.

   c. Given that plasma is the most common phase of matter in the universe, why is it so rare on Earth?

5. *Atomic Terminology Practice.*

   a. Consider fluorine atoms with 9 protons and 10 neutrons. What are the atomic number and atomic weight of this fluorine? Suppose we could add a proton to this fluorine nucleus. Would the result still be fluorine? Explain. What if we added a neutron to the fluorine nucleus?

   b. The most common isotope of oxygen has atomic number 8 and atomic weight 16. Another isotope of oxygen has two extra neutrons. What are the atomic number and atomic weight of this isotope?

   c. The most common isotope of gold has atomic number 79 and atomic weight 197. How many protons and neutrons does the gold nucleus contain? Assuming the gold is electrically neutral, how many electrons does it have? If the gold is triply ionized, how many electrons does it have?

   d. The most common isotope of uranium is ^{238}U, but the form used in nuclear bombs and nuclear power plants is ^{235}U. Given that uranium has atomic number 92, how many neutrons are in each of these two isotopes of uranium?

6. *Energy Level Transitions.* The labeled transitions below represent an electron moving between energy levels in hydrogen. Answer each of the following questions and explain your answers.

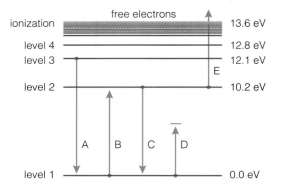

   a. Which transition could represent an electron that *gains* 10.2 eV of energy?

   b. Which transition represents an electron that *loses* 10.2 eV of energy?

   c. Which transition represents an electron that is breaking free of the atom?

   d. Which transition, as shown, is *not* possible?

   e. Describe the process taking place in transition A.

Universal Motion: From Copernicus to Newton

⁵

Everything in the universe is in constant motion, from the random meandering of molecules in the air to the large-scale drifting of galaxies in super-clusters. Remarkably, just a few physical laws govern all this motion. The elucidation of these laws over the past several centuries is surely one of the greatest scientific triumphs of all time.

We have two primary goals in this chapter: understanding the laws that govern the motion of celestial objects and understanding how those laws were discovered. Achieving these goals requires that we first explore motion and the laws that govern it more generally. Once we develop this background, you'll be able to understand the extraordinary story of how the ancient belief in an Earth-centered universe was finally overthrown.

As before, much of the subject matter of this chapter may be familiar to you already. Again, don't worry if this is not the case. The material is not difficult, and studying it carefully will greatly enhance your understanding of the astronomy in the rest of the book as well as of many everyday phenomena.

5.1 Describing Motion: Examples from Daily Life

Everything moves in the universe, so understanding motion is important to understanding the universe. We all have a great deal of experience with motion and natural intuition as to how motion works, so we begin our discussion of motion with some familiar examples. Indeed, you probably are familiar with all the terms defined in this section, although their scientific definitions may differ subtly from those you use in casual conversation.

Speed, Velocity, and Acceleration

The concepts we use in determining the trajectory of a ball, a rocket, or a planet are familiar to you from driving a car. The speedometer indicates your **speed,** usually in units of both miles per hour (mi/hr) and kilometers per hour (km/hr); for example, 100 km/hr is a speed. Your **velocity** is your speed in a certain direction; "100 km/hr going due north" describes a velocity. It is possible to change your velocity without changing your speed by, for example, holding to a steady 60 km/hr as you drive around a curve. Because your direction is changing as you round the curve, your *velocity* is also changing—even though your *speed* is constant.

Whenever your velocity is changing, you are experiencing **acceleration.** You are undoubtedly familiar with the term *acceleration* as it applies to increasing speed, such as when you accelerate away from a stop sign while driving your car. In science, we also say that you are accelerating when you slow down or turn. Slowing occurs when your acceleration is in a direction opposite to your motion; we say that your acceleration is *negative* and causes your velocity to decrease. Turning changes your *direction,* which means a change in velocity and thus involves acceleration even if your speed remains constant.

Note that you don't feel anything when you are traveling at *constant velocity,* which is why you don't feel any sensation of motion when you're traveling in an airplane on a smooth flight. In contrast, you can often feel acceleration: As you speed up in a car you feel yourself being pushed back into your seat, as you slow down you feel yourself being pulled forward from the seat, and as you drive around a curve you lean outward because of your acceleration.

The Acceleration of Gravity

One of the most important types of acceleration is that caused by gravity, which makes objects accelerate as they fall. In a famous (though probably apocryphal) experiment that involved dropping weights from the Leaning Tower of Pisa, Galileo demonstrated that gravity accelerates all objects by the same amount, regardless of their mass. This fact may be surprising because it seems to contradict everyday experience: A feather floats gently to the ground, while a rock plummets. However, this difference is caused by air resistance. If you dropped a feather and a rock on the Moon, where there is no air, both would fall at exactly the same rate.

TIME OUT TO THINK *Find a piece of paper and a small rock. Hold both at the same height, one in each hand, and let them go at the same instant. The rock, of course, hits the ground first. Next crumple the paper into a small ball and repeat the experiment. What happens? Explain how this experiment suggests that gravity accelerates all objects by the same amount.*

The acceleration of a falling object is called the **acceleration of gravity,** abbreviated *g.* On Earth, the acceleration of gravity causes falling objects to fall faster by 9.8 meters per second (m/s), or about 10 m/s, with each passing second. For example, suppose you drop a rock from a tall building. At the moment you let it go, its speed is 0 m/s. After 1 second, the rock will be falling downward at about 10 m/s. After 2 seconds, it will be falling at about 20 m/s. In the absence of air resistance, its speed will continue to increase by about 10 m/s each second until it hits the ground (Figure 5.1). We therefore say that the acceleration of gravity is about 10 *meters per second per second,* or 10 *meters per second squared,* which we write as 10 m/s². (More precisely, $g = 9.8$ m/s².)

FIGURE 5.1 On Earth, gravity causes falling objects to accelerate downward at about 10 m/s². That is, a falling object's velocity increases by about 10 m/s with each passing second. More precisely, $g = 9.8$ m/s².

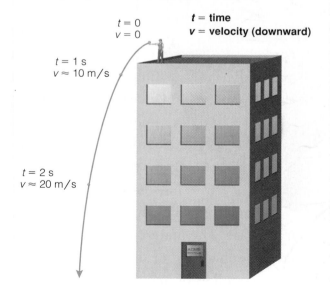

t = time
v = velocity (downward)

$t = 0$
$v = 0$

$t = 1$ s
$v \approx 10$ m/s

$t = 2$ s
$v \approx 20$ m/s

Momentum and Force

Imagine that you're stopped innocently in your car at a red light when a bug flying at a velocity of 30 km/hr due south slams into your windshield. What will happen to your car? Not much, except perhaps a bit of a mess on your windshield. Next imagine that a 2-ton truck runs the red light and hits you head-on with the same velocity as the bug. Clearly, the truck will cause far more damage.

Scientifically, we say that the truck imparts a much larger jolt than the bug because it transfers more **momentum** to you. Momentum describes a combination of mass and velocity. We can describe the momentum of the truck before the collision as "2 tons moving due south at 30 km/hr," while the momentum of the bug is perhaps "1 gram moving due south at 30 km/hr." Mathematically, momentum is defined as mass × velocity.

In transferring some of its momentum to your car, the truck (or bug) has exerted a **force** on your car. More generally, a force is anything that can cause a change in momentum. You are familiar with many types of force besides collisional force. For example, if you shift into neutral while driving along a flat stretch of road, the forces of air resistance and road friction will continually sap your car's momentum (transferring it to molecules in the air and pavement), slowing your velocity until you come to a stop. You probably are also familiar with forces arising from gravity, electricity, and magnetism.

The mere presence of a force does not always cause a change in momentum. For example, if the engine works hard enough, a car can maintain constant velocity—and hence constant momentum—despite air resistance and road friction. In this case, the force generated by the engine precisely offsets the forces of air resistance and road friction, and we say that no **net force** is acting on the car.

As long as an object is not shedding (or gaining) mass, a change in momentum necessarily means a change in velocity (because momentum is described as mass times velocity), which means an acceleration. Hence, any net force will cause acceleration, and all accelerations must be caused by a force. That is why you feel forces when you accelerate in your car.

Mass and Weight

Up until now, we've been glossing over one key term: *mass*. Your **mass** refers to the amount of matter in your body, which is different from your *weight*. Imagine standing on a scale in an elevator (Figure 5.2). When the elevator is stationary or moving at constant velocity, the scale reads your "normal" weight. When the elevator is accelerating upward, the floor exerts an additional force that makes you feel heavier, and the scale verifies your greater weight. Note that your weight is greater than its "normal" value only while the elevator is *accelerating,* not while it is moving at constant velocity. When the elevator accelerates downward, the floor and scale are dropping away, so

---

FIGURE 5.2 Mass is not the same as weight. The man's mass never changes, but his weight does.

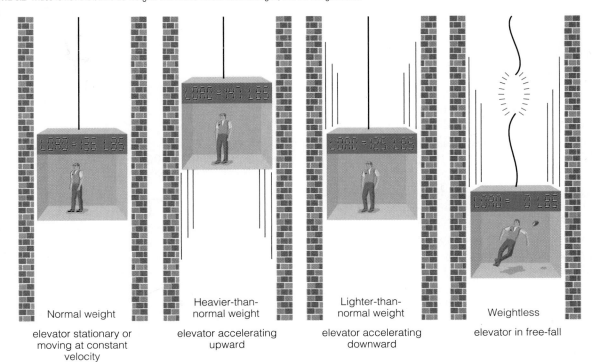

Normal weight

elevator stationary or moving at constant velocity

Heavier-than-normal weight

elevator accelerating upward

Lighter-than-normal weight

elevator accelerating downward

Weightless

elevator in free-fall

Astronauts are weightless for much longer periods because orbiting spacecraft are in a constant state of free-fall. To understand why, imagine a cannon that shoots a ball horizontally from a tall mountain. Once launched, the ball falls solely because of gravity and hence is in free-fall. Thus, the faster the cannon shoots the ball, the farther it goes (Figure 5.3). If the cannonball could go fast enough, its motion would keep it constantly "falling around" the Earth and it would never hit the ground (as long as we neglect air resistance). This is precisely what spacecraft such as the Space Shuttle and the Space Station do as they orbit the Earth. Their constant state of free-fall makes the spaceships and everything in them weightless.

FIGURE 5.3 The faster the cannonball is shot, the farther it goes before hitting the ground. If it goes fast enough, it will continually "fall around," or orbit, the Earth. With an even faster speed, it may escape the Earth's gravity altogether.

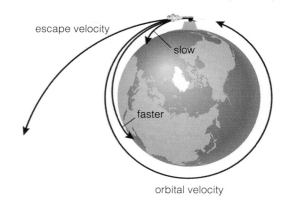

If an orbiting object travels at the **escape velocity** or faster, it can completely escape from the Earth, never to return. It will still be weightless as long as it does not fire any engines (because its motion is affected only by the gravity of the Sun, the planets, and other objects). The escape velocity from the Earth's surface is about 40,000 km/hr (25,000 mi/hr).

your weight is reduced. Thus, your weight varies with the elevator's motion, but your mass remains the same. (You can verify these facts by taking a small bathroom scale with you on an elevator.)

More precisely, your **weight** describes the *force* that acts on your mass, and it depends on the strength of gravity and other forces acting upon you (such as the force caused by the elevator's acceleration). Although your mass would be the same whether you were on the Earth, on the Moon, or in the Space Shuttle, you would weigh less on the Moon because of its weaker gravity.

Free-fall, Weightlessness, and Orbit

If the cable breaks so that the elevator is in **free-fall,** the floor drops away at the same rate that you fall. You lose contact with the scale, becoming **weightless.** In fact, you are in free-fall whenever nothing *prevents* you from falling. For example, you are in free-fall when you jump off a chair or spring from a diving board or trampoline. Surprising as it may seem, you have therefore experienced weightlessness many times in your life and can experience it right now simply by jumping off your chair. Of course, your weightlessness lasts for only the very short time until you hit the ground.

5.2 Understanding Motion: Newton's Laws

The complexity of motion in daily life might lead you to guess that the laws governing how forces affect motion are quite complex. For example, if you watch a piece of paper fall to the ground, you'll see it waft lazily back and forth in a seemingly unpredictable pattern. However, the complexity of this motion arises because the paper is affected by a variety of forces, including gravity and the changing forces caused by air currents. If you could analyze the forces individually, you'd find that each force affects the paper's motion in a simple, predictable way. Three simple laws, known as **Newton's laws of motion,** describe how forces affect motion. Let's investigate them, again using examples from everyday life.

Newton's First Law of Motion

Newton's first law of motion deals with situations in which no net force is acting on an object. It states:

In the absence of a net force, an object moves with constant velocity.

Thus, objects at rest (velocity = 0) tend to remain at rest, and objects in motion tend to remain in motion with no change in either their speed or their direction.

The idea that an object at rest should remain at rest is rather obvious; after all, if a car is parked on a flat street, it won't suddenly start moving for no reason. But what if the car is traveling along a flat, straight road? Newton's first law says that the car should keep going forever *unless* a force acts upon it. You know that the car eventually will come to a stop if you take your foot off the gas pedal, so we must conclude that some forces are stopping the car—in this case forces arising from friction and air resistance. If the car were in space, and therefore unaffected by friction or air, it would keep moving forever (though gravity would eventually alter its speed and direction). That is why interplanetary spacecraft, once launched into space, need no fuel to keep going.

Newton's first law also explains why you don't feel any sensation of motion when you're traveling in an airplane on a smooth flight. As long as the plane is traveling at constant velocity, no net force is acting on it or on you. Therefore, you feel no different from how you would feel at rest: You can walk around the cabin, play catch with a person a few rows forward, or relax and go to sleep just as though you were "at rest" on the ground.

Newton's Second Law of Motion

Newton's second law of motion tells us what happens to an object when a net force *is* present. We have already said that a net force changes an object's momentum, accelerating it in the direction of the force. Newton's second law quantifies this relationship, which can be stated in two equivalent ways:

force = rate of change in momentum

force = mass × acceleration $(F = ma)$

This law explains why you can throw a baseball farther than you can throw a shot-put. For both the baseball and the shot-put, the force delivered by your arm equals the product of mass and acceleration.

Because the mass of the shot-put is greater than that of the baseball, the same force from your arm gives the shot-put a smaller acceleration. Due to its smaller acceleration, the shot-put leaves your hand with less speed than the baseball and thus travels a shorter distance before hitting the ground.

Newton's Third Law of Motion

Think for a moment about standing still on the ground. The force of gravity acts downward on you, so if this force were acting alone Newton's second law would demand that you be accelerating downward. The fact that you are not falling means that the ground must be pushing back up on you with exactly the right amount of force to offset gravity. How does the ground know how much force to apply? The answer is given by Newton's third law of motion:

For any force, there always is an equal and opposite reaction force.

According to this law, your body exerts a gravitational force on the Earth identical to the one the Earth exerts on you, except that it acts in the opposite direction. In this mutual pull, the ground just happens to be caught in the middle. Because other forces (between atoms and molecules in the Earth) keep the ground stationary with respect to the center of the Earth, the opposite, upward force is transmitted to you by the ground, holding you in place. Newton's third law also explains how rockets work: Engines generate an explosive force driving hot gas out the back, which creates an equal and opposite force propelling the rocket forward.

Common Misconceptions: What Makes a Rocket Launch?

If you've ever watched a rocket launch, it's easy to see why many people believe that the rocket "pushes off" the ground. In fact, the ground has nothing to do with the rocket launch. The rocket takes off because of momentum conservation. Rocket engines are designed to expel hot gas with an enormous amount of momentum. To balance the explosive force driving gas out the back of the rocket, an equal and opposite force must propel the rocket forward, keeping the total momentum—gas plus rocket—unchanged. Thus, rockets can be launched horizontally as well as vertically, and a rocket can be "launched" in space (e.g., from a space station) with no need for any nearby solid ground.

Conservation of Momentum and Angular Momentum

If you look more closely at Newton's laws, you will see that they all reflect aspects of a deeper principle: the _conservation of momentum_. Like the amount of energy, the total amount of momentum in the universe is conserved—that is, it does not change. Newton's first law says that an individual object's momentum will not change at all if the object is left alone. Newton's second law says that a force can change the object's momentum, but Newton's third law says that another equal and opposite force simultaneously changes some other object's momentum by a precisely opposite amount. Total momentum always remains unchanged.

In astronomy, a special kind of momentum is particularly important. Consider an ice skater spinning in place. She certainly has some kind of momentum, but it is a little different from the momentum we've discussed previously because she's not actually going anywhere. Technically, her "spinning momentum" is called **angular momentum.** (The term _angular_ arises because each spin means turning through an _angle_ of 360°.) In fact, any object that is spinning or orbiting has angular momentum. We can write a simple formula for the angular momentum of an object moving in a circle:

$$\text{angular momentum} = m \times v \times r$$

where m is the object's mass, v is its speed around the circle, and r is the radius of the circle (Figure 5.4).

Just as momentum can be changed only by a force, the angular momentum of any object or group of objects can be changed only by a "twisting force" (more formally called a _torque_). Thus, in the absence of any such twisting forces, we find a very important principle called the **law of conservation of angular momentum:**

In the absence of a net twisting force (torque), the total angular momentum of a system remains constant.

We will see countless applications of this law throughout the book, but in everyday life it is most familiar to us in the example of the ice skater. Since there is so little friction on ice, a spinning ice skater essentially keeps a constant angular momentum. When the skater pulls in her extended arms, she effectively decreases her radius. Therefore, for the product $m \times v \times r$ to remain unchanged, her velocity of rotation must increase (Figure 5.5).

FIGURE 5.5 A spinning skater conserves angular momentum.

In the product $m \times v \times r$, extended arms mean larger radius and smaller velocity of rotation.

Bringing in her arms decreases her radius and therefore increases her rotational velocity.

FIGURE 5.4 The angular momentum of an object moving in a circle is $m \times v \times r$.

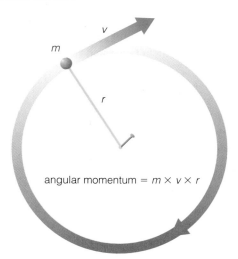

$$\text{angular momentum} = m \times v \times r$$

5.3 Planetary Motion: The Copernican Revolution

Now that we have described motion in general terms, we can turn to the important issue of planetary motion in our solar system. Although planetary motion is among the simplest motions found in nature, it remained mysterious for most of human history because of its *apparent* complexity in our sky. The prevailing view of European scholars at the dawn of the Renaissance remained that of Ptolemy, whose complex model had planets spinning on circles upon circles around the Earth [Section 3.3]. The sea change in human understanding of planetary motion is one of the most remarkable stories in the history of science. It began early in the sixteenth century, with the work of Nicholas Copernicus.

Nicholas Copernicus: The Revolution Begins

Copernicus was born in Toruń, Poland, on February 19, 1473. His family was wealthy, and he received a first-class education, studying painting, mathematics, medicine, and law. He began studying astronomy in his late 20s. By that time, tables of planetary motion based on the Ptolemaic model of the universe were noticeably inaccurate. Copernicus concluded that planetary motion could be more simply explained in a Sun-centered solar system, as had been suggested by Aristarchus some 1,800 years earlier [Section 2.4], and he began developing a Sun-centered system for predicting planetary positions.

Copernicus was hesitant to publish his work for fear that his suggestion that the Earth moved would be considered heretical. Nevertheless, he discussed his system with other scholars and generated great interest in his work. At the urging of some of these scholars, including some high-ranking officials of the Church, he finally agreed to publish a book describing his system. The book, *De Revolutionibus Orbium Caelestium*, or "Concerning the Revolutions of the Heavenly Spheres," was published in 1543. Legend has it that Copernicus saw the first printed copy on the day he died—May 24, 1543.

In addition to its aesthetic advantages, the Sun-centered system of Copernicus allowed him to discover a mathematical relationship between a planet's true orbital period around the Sun and the time between successive appearances of the planet in its "full" phase (directly opposite the Sun in the sky). He was also able to use geometrical techniques to estimate the distances of the planets from the Sun in terms of the Earth-Sun distance (i.e., distances in astronomical units). However, the actual model published by Copernicus did not predict planetary positions substantially more accurately than did the old Ptolemaic model. Moreover, it remained complex

because Copernicus held to the ancient Greek belief that all heavenly motions must follow perfect circles. Because the true orbits of the planets are *not* circles, Copernicus found it necessary to add circles upon circles to his system, just as in the Ptolemaic system. As a result, the Copernican system won relatively few converts in the 50 years after it was published. After all, why throw out thousands of years of tradition for a new system that predicted planetary motion equally poorly?

Tycho Brahe: The Greatest Naked-Eye Observer of All Time

Part of the difficulty faced by astronomers who sought to improve either the Ptolemaic or the Copernican system was a lack of quality data. The telescope had not yet been invented, and existing naked-eye observations were not very accurate. In the late 1500s, a Danish nobleman named Tycho Brahe (1546–1601) set about correcting this problem.

When Tycho was a young boy, his family discouraged his interest in astronomy. He therefore hid his passion, learning the constellations from a miniature model of a celestial sphere that he kept hidden. In 1563, Tycho decided to observe a widely anticipated conjunction of Jupiter and Saturn in the sky. To his surprise, the conjunction occurred about a month later than had been predicted. He resolved to improve the state of astronomical prediction and set about compiling careful observations of stellar and planetary positions in the sky.

Tycho often was arrogant about both his noble birth and his learned abilities. At age 20, he fought a duel with another student over which of them was the better mathematician. Part of his nose was cut off, so he designed a replacement piece made of silver and gold. Tycho's fame grew after he observed what he called a *nova*, meaning "new star," in 1572 and proved that it was at a distance much farther away than the Moon. Today, we know that Tycho saw a *supernova*—the explosion of a distant star. In 1577, Tycho observed a comet and proved that it, too, lay in the realm of the heavens; others, notably Aristotle, had argued that comets were phenomena of the Earth's atmosphere. King Frederick II of Denmark then decided to sponsor Tycho's ongoing work, providing him with money to build an unparalleled observatory for naked-eye observations (Figure 5.6). Over a period of three decades, Tycho and his assistants compiled naked-eye observations accurate to within less than 1 arcminute. Because the telescope was invented shortly after his death, Tycho's data remain the best set of naked-eye observations ever made.

Despite the quality of his observations, Tycho never succeeded in coming up with a satisfying

FIGURE 5.6 Tycho's naked-eye observatory.

of calculation, Kepler found a circular orbit that matched nearly all of Tycho's observations of Mars to within 2 arcminutes. In two cases, however, this orbit predicted a position for Mars that differed from Tycho's observations by the slightly greater margin of 8 arcminutes.

Kepler surely was tempted to ignore these two observations and attribute them to errors by Tycho. After all, 8 arcminutes is barely one-fourth the angular diameter of the full moon. But Kepler trusted Tycho's careful work, and the missed 8 arcminutes finally led him to abandon the idea of circular orbits—and to find the correct solution to the ancient riddle of planetary motion. About this event, Kepler wrote:

> If I had believed that we could ignore these eight minutes [of arc], I would have patched up my hypothesis accordingly. But, since it was not permissible to ignore, those eight minutes pointed the road to a complete reformation in astronomy.

Kepler summarized his discoveries with three simple laws that we now call **Kepler's laws of planetary motion.** He published the first two laws in 1610 and the third in 1618. (Be careful not to confuse Kepler's three laws of *planetary* motion with Newton's three laws that apply generally to *all* motion.)

Kepler's key discovery was that planetary orbits are not circles but instead are a special type of oval called an **ellipse.** You probably know how to draw a circle by putting a pencil on the end of a string, tacking the string to a board, and pulling the pencil around (Figure 5.7a). Drawing an ellipse is similar,

explanation for planetary motion. He did, however, succeed in finding someone who could: In 1600, he hired a young German astronomer named Johannes Kepler (1571–1630). Kepler and Tycho had a strained relationship while Tycho was living. But in 1601, as Tycho lay on his deathbed, he bequeathed all his notebooks of observations to Kepler and begged him to find a system that would make sense of the observations so "that it may not appear I have lived in vain."

Kepler's Reformation: The Laws of Planetary Motion

Kepler was deeply religious and believed that understanding the geometry of the heavens would bring him closer to God. Kepler, like Copernicus, believed that Earth and the other planets traveled around the Sun in circular orbits, and he worked diligently to match circular motions to Tycho's data.

Kepler worked with particular intensity to find an orbit for Mars, which posed the greatest difficulties in matching the data to a circular orbit. After years

FIGURE 5.7

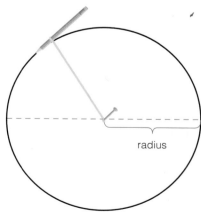

a Drawing a circle.

except that you must stretch the string around *two* tacks (Figure 5.7b). The locations of the two tacks are called the **foci** (singular, **focus**) of the ellipse. By altering the distance between the two foci while keeping the same length of string, you can draw ellipses of varying **eccentricity,** a quantity that describes how much an ellipse deviates from a perfect circle (Figure 5.7c).

Kepler's first law states that *the orbit of each planet about the Sun is an ellipse with the Sun at one focus* (Figure 5.8). (There is nothing at the other focus.) This law tells us that a planet's distance from the Sun varies during its orbit: It is closest at the point called **perihelion,** and farthest at the point called **aphelion.** The *average* of a planet's perihelion and aphelion distances is called its **semimajor axis;** we will refer to this simply as the planet's average distance from the Sun.

Kepler's second law states that, *as a planet moves around its orbit, it sweeps out equal areas in equal times.* As shown in Figure 5.9, this means that the planet moves a greater distance when it is near perihelion than it does in the same amount of time near aphelion; that is, the planet travels faster when it is nearer to the Sun and slower when it is farther from the Sun.

Kepler's third law tells us that *more distant planets move more slowly in their orbits.* More precisely, it describes how a planet's *period,* or the

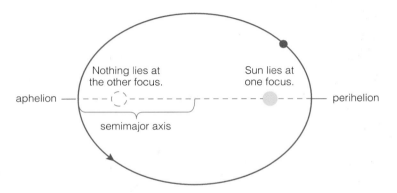

FIGURE 5.8 Kepler's first law: The orbit of each planet about the Sun is an ellipse with the Sun at one focus.

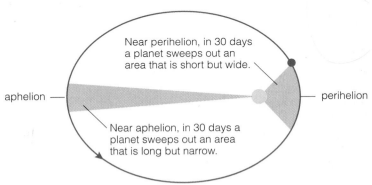

The areas swept out in 30-day periods are all equal.

FIGURE 5.9 Kepler's second law: As a planet moves around its orbit, it sweeps out equal areas in equal times.

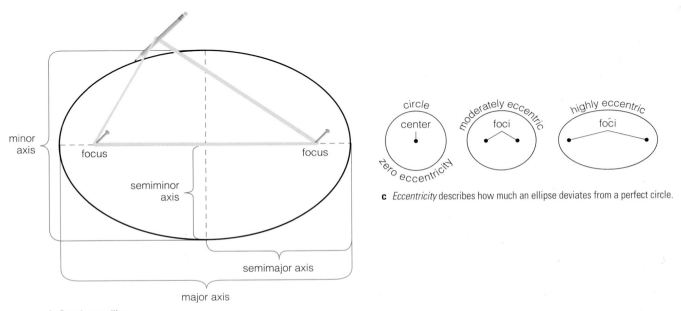

b Drawing an ellipse.

c *Eccentricity* describes how much an ellipse deviates from a perfect circle.

number of years it takes to complete one orbit of the Sun, is related to its average distance from the Sun in astronomical units (1 AU ≈ 150 million km):

$$(\text{orbital period in years})^2 = (\text{average distance in AU})^3$$

This formula is often written more simply as $p^2 = a^3$, where it is understood that p is the orbital period measured in years and a is the average distance measured in AU. Note that a planet's period does not depend on the eccentricity of its orbit: All orbits with the same semimajor axis have the same period. Nor does the period depend on the mass of the planet: Any object located an average of 1 AU from the Sun would orbit the Sun in a year (as long as the object's mass was small compared to the Sun's mass).

Galileo: The Death of the Earth-Centered Universe

The success of Kepler's laws in matching Tycho's data provided strong evidence in favor of Copernicus's placement of the Sun, rather than the Earth, at the center of the solar system. Nevertheless, many scientists still voiced reasonable objections to the Copernican view. There were three basic objections, all rooted in the 2,000-year-old beliefs of Aristotle (384–322 B.C.) and other ancient Greeks. First, Aristotle had held that the Earth could not be moving because, if it were, objects such as birds, falling stones, and clouds would be left behind as the Earth moved along its way. Second, the idea of noncircular orbits contradicted the ancient Greek belief that the

heavens—the realm of the Sun, Moon, planets, and stars—must be perfect and unchanging. The third objection was the ancient argument that stellar parallax ought to be detectable if the Earth orbits the Sun [Section 2.4]. Galileo (1564–1642), a contemporary and correspondent of Kepler, answered all three objections.

Galileo defused the first objection with experiments that almost single-handedly overturned the Aristotelian view of physics. His demonstration that gravity accelerates all objects by the same amount directly contradicted Aristotle's claim that heavier objects would fall to the ground faster. Aristotle also had taught that the natural tendency of any object was to come to rest. Galileo demonstrated that a moving object remains in motion *unless* a force acts to stop it, an idea now codified in Newton's first law of motion. Thus, he concluded that objects such as birds, stones, and clouds that are moving with the Earth should *stay* with the Earth unless some force knocks them away. This same idea explains why passengers in an airplane stay with the moving airplane even when they leave their seats.

Tycho's supernova and comet observations already had shown that the heavens could change, and Galileo shattered the idea of heavenly perfection after he built a telescope in late 1609. (Contrary to a common belief, he did *not* invent the telescope.) Through his telescope, he saw sunspots on the Sun, which were considered "imperfections" at the time. He also used his telescope to prove that the Moon has mountains and valleys like the "imperfect" Earth

THINKING ABOUT . . .

Aristotle

Aristotle (384–322 B.C.) is among the best-known philosophers of the ancient world. Both his parents died while he was a child, and he was raised by a family friend. In his 20s and 30s, he studied under Plato (427–347 B.C.) at Plato's Academy. He later founded his own school, called the Lyceum, where he studied and lectured on virtually every subject. Historical records tell us that his lectures were collected and published in 150 volumes. About 50 of these volumes survive to the present day.

Many of Aristotle's scientific discoveries involved the nature of plants and animals. He studied more than 500 animal species in detail, including dissecting specimens of nearly 50 species, and came up with a strikingly modern classification system. For example, he was the first person to recognize that dolphins should be classified with land mammals

rather than with fish. In mathematics, he is known for laying the foundations of mathematical logic. Unfortunately, he was far less successful in physics and astronomy, areas in which many of his claims turned out to be wrong.

Interestingly, Aristotle's philosophies were not particularly influential until many centuries after his death. His books were preserved and valued by Islamic scholars, but they were unknown in Europe until they were translated into Latin in the twelfth and thirteenth centuries. Aristotle achieved his near-reverential status only after St. Thomas Aquinas (1225–1274) integrated Aristotle's philosophy into Christian theology. In the ancient world, Aristotle's greatest influence came indirectly, through his role as the tutor of Alexander the Great [Section 3.3].

FIGURE 5.10 The shadows cast by mountains and crater rims near the dividing line between the light and dark portions of the lunar face prove that the Moon's surface is not perfectly smooth.

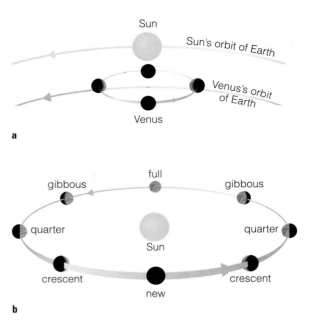

FIGURE 5.11 (**a**) In the Ptolemaic system, Venus follows a circle upon a circle that keeps it close to the Sun in our sky. Therefore, its phases would range only from new to crescent. (**b**) Galileo observed Venus go through a complete set of phases and therefore proved that it orbits the Sun.

by noticing the shadows cast near the dividing line between the light and dark portions of the lunar face (Figure 5.10). If the heavens were not perfect, then the idea of elliptical (rather than circular) orbits was not so objectionable.

The absence of observable stellar parallax had been of particular concern to Tycho. Based on his estimates of the distances of stars, Tycho believed that his naked-eye observations were sufficiently precise to detect stellar parallax if the Earth orbited the Sun. Refuting Tycho's argument required showing that the stars were more distant than Tycho had thought and therefore too distant for him to have observed stellar parallax. Although Galileo didn't actually prove this fact, he provided strong evidence in its favor. In particular, he saw with his telescope that the Milky Way resolved into countless individual stars, which helped him argue that the stars were far more numerous and more distant than Tycho had imagined.

The true death knell for an Earth-centered universe came with two of Galileo's earliest discoveries through the telescope. First, he observed four moons clearly orbiting Jupiter, *not* the Earth. Soon thereafter, he observed that Venus goes through phases like the Moon, proving that Venus must orbit the Sun and not the Earth (Figure 5.11).

At last the Copernican revolution was complete. Thanks to Galileo, by the mid-1600s the scientific community was near-unanimous in accepting Kepler's model of planetary motion. The debate turned to the question of *why* Kepler's laws hold true.

5.4 The Force of Gravity

The Aristotelian view of the world, held in Europe as near-gospel truth, was in tatters. The Earth was not the center of the universe, and the laws of physics were not what Aristotle had believed. But the final blow was still to come. Aristotle had maintained that the heavens were totally distinct from Earth and that the physical laws on Earth could not be applied to understanding heavenly motion.

In 1666, Isaac Newton saw an apple fall to the ground. (The story of the apple may or may not be true, but Newton himself told the story.) He suddenly realized that the force that brought the apple to the ground and the force that held the Moon in orbit were the same. That insight brought the Earth and the heavens together in one *universe*. It also heralded the birth of the modern science of *astrophysics,* in which physical laws discovered on Earth are applied to phenomena throughout the cosmos.

Newton and the Universal Law of Gravitation

Isaac Newton was born prematurely in Lincolnshire, England, on Christmas day in 1642. His father, a farmer who had never learned to read or write, died 3 months before he was born. Newton had a difficult childhood and showed few signs of unusual talent. He attended Trinity College at Cambridge, where he earned his keep by performing menial labor, such as cleaning the boots and bathrooms of wealthier students and waiting on their tables.

Shortly after he graduated, the plague hit Cambridge, and Newton returned home. It was there that he saw the apple fall and began his detailed investigations of gravity, light, optics, and mathematics. Over the next 20 years, Newton's work completely revolutionized mathematics and science. Besides his work on motion and gravity, he conducted crucial experiments regarding the nature of light, built the first reflecting telescopes, and invented the branch of mathematics called *calculus*.

In 1687, Newton published *Philosophiae Naturalis Principia Mathematica* ("Mathematical Principles of Natural Philosophy"), usually referred to as *Principia*. In this book, Newton stated his three laws of motion and another law, called the **universal law of gravitation,** that describes the force of gravity.

Three simple statements summarize the universal law of gravitation:

- Every mass attracts every other mass through the force called *gravity.*

- The force of attraction between any two objects is *directly proportional* to the product of their masses. For example, doubling the mass of *one* object doubles the force of gravity between the two objects.

- The force of attraction between two objects decreases with the *square* of the distance between their centers. That is, the force follows an **inverse square law** with distance. For example, doubling the distance between two objects weakens the force of gravity by a factor of 2^2, or 4.

TIME OUT TO THINK *How does the gravitational force between two objects change if the distance between them triples? If the distance between them drops in half?*

Mathematically, Newton's law of universal gravitation is written:

$$F_g = G\frac{M_1 M_2}{d^2}$$

where F_g is the force of gravitational attraction, M_1 and M_2 are the masses of the two objects, and d is the distance between their *centers* (Figure 5.12). The

FIGURE 5.12 The law of universal gravitation.

symbol G is a constant called the **gravitational constant;** its numerical value was not known to Newton but has since been measured by experiments.

The "Why" of Kepler's Laws, and More

For almost 70 years after Kepler published his laws of planetary motion, scientists debated *why* these laws hold true. Newton settled the debate. By doing a bit of mathematics with the law of universal gravitation and his laws of motion, Newton was able to derive all three of Kepler's laws. In doing so, he also extended Kepler's laws in several important ways:

- Newton found that Kepler's first two laws apply not only to planets, but to *any* object going around another object under the force of gravity. The orbits of a satellite around the Earth, of a moon around a planet, of an asteroid around the Sun, and of binary stars around each other all are ellipses (with the center of mass of the two objects at one focus) in which orbital speeds are faster when the two objects are closer together and slower when they are farther apart.

- He found that orbits do not have to be **bound orbits** as they are with ellipses (Figure 5.13). **Unbound orbits** are also possible (in the shape

FIGURE 5.13 Orbits allowed by the law of gravity.

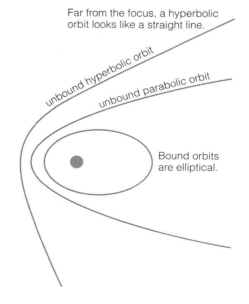

Far from the focus, a hyperbolic orbit looks like a straight line.

unbound hyperbolic orbit

unbound parabolic orbit

Bound orbits are elliptical.

of parabolas or hyperbolas). A comet with an unbound orbit comes in toward the Sun just once, looping around the Sun and never returning.

■ Perhaps most important, he found that Kepler's third law could be generalized in a way that often allows us to calculate the mass of one or both orbiting objects. This generalization, which we call **Newton's version of Kepler's third law,** provides the primary means by which we determine masses throughout the universe. (See problem 5 at the end of the chapter.)

5.5 Tides and Tidal Forces

If you've ever lived near an ocean, you're probably aware of the rising and falling of the tide twice each day. But what causes the tides, and why are there two each day? A complete answer to this question turns out to be fairly complex, but in a simple sense tides arise because the Moon's gravity tugs more strongly on the side of the Earth facing the Moon than it does on the opposite side (Figure 5.14). This creates two tidal bulges: one facing the Moon where the oceans are pulled out from the Earth, and one opposite the Moon where the Earth is being pulled out from under the oceans. As the Earth rotates, the two daily high tides occur when your location passes through these two tidal bulges, and the two low tides occur when your location is at the points halfway in between. More technically, tides affect the land as well as the oceans because the difference in the Moon's pull creates a "stretching force," or **tidal force,** that stretches the *entire Earth*. The ocean tides are much more noticeable than the land tides only because the liquid oceans stretch more readily than the solid land.

The Sun also exerts a tidal force on the Earth, causing the Earth to stretch along the Sun-Earth line. However, the greater distance to the Sun (than to the Moon) means that the *difference* in the Sun's pull on the near and far sides of the Earth is relatively small, and the overall tidal force caused by the Sun is only about one-third that caused by the Moon (Figure 5.15). When the tidal forces of the Sun and the Moon work together, as is the case at both new and full moons, we get the especially pronounced *spring tides* (so named because the water tends to "spring up" from the Earth). When the tidal forces of the Sun and the Moon oppose each other, as is the case at first- and third-quarter moons, we get the relatively small tides known as *neap tides.*

The tidal stretching of the Earth exerts internal friction, called **tidal friction,** that tends to fight against the Earth's rotation. As a result, the length of a day is gradually getting longer and (because angular momentum must be conserved) the Moon is moving gradually farther from the Earth. Although these changes are barely noticeable on human time scales, they add up over billions of years: Early in the Earth's history, a day probably was only 5 or 6 hours long, and the Moon probably was one-tenth or less of its current distance from Earth.

FIGURE 5.15 This diagram represents the gravitational pulls from the Sun (yellow arrows) and the Moon (black arrows) at four phases of the Moon. Tides are enhanced at both new and full moons (spring tides), when the tidal forces from the Sun and the Moon stretch the Earth along the same line. Tides are diminished at first- and third-quarter moons (neap tides), when the tidal forces from the Sun and the Moon stretch the Earth along different (and perpendicular) lines.

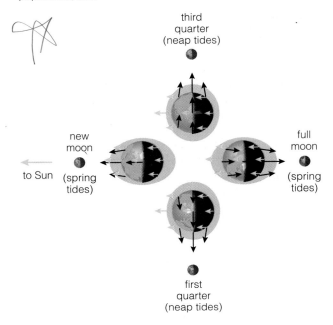

third quarter (neap tides)

new moon

to Sun (spring tides)

full moon (spring tides)

first quarter (neap tides)

FIGURE 5.14 Tidal bulges face toward and away from the Moon. Arrows represent the strength and direction of the gravitational attraction toward the Moon.

tidal bulge opposite Moon

tidal bulge toward Moon

Not to scale! The real tidal bulge raises the oceans by only about 2 meters.

Tidal friction has had even more dramatic effects on the Moon. The Moon probably once rotated much faster than it does today. But, just as the Moon exerts a tidal force on the Earth, the Earth exerts a tidal force on the Moon. In fact, the Earth exerts a greater tidal force on the Moon than vice versa because of its larger mass. Over time, this tidal force generated tidal friction within the Moon that gradually slowed the Moon's rotation until it reached its current state of always keeping the same face toward Earth [Section 2.3]. In later chapters, we will see many more examples of the important effects of tidal forces and tidal friction.

5.6 Orbital Energy and Escape Velocity

Consider a satellite in an elliptical orbit around the Earth. Its gravitational potential energy is greatest when it is farthest from the Earth, and smallest when it is nearest the Earth. Conversely, its kinetic energy is greatest when it is nearest the Earth and moving fastest in its orbit, and smallest when it is farthest from the Earth and moving slowest in its orbit. Throughout its orbit, its total *orbital energy*—the sum of its kinetic and gravitational potential energies—must be conserved.

Because any change in its orbit would mean a change in its total orbital energy, a satellite's orbit around the Earth cannot change if it is left completely undisturbed. If the satellite's orbit *does* change, it must somehow have gained or lost energy. For a satellite in low-Earth orbit, the Earth's thin upper atmosphere exerts a bit of drag that can cause it to lose energy and eventually plummet back to Earth. The satellite's lost orbital energy is converted to heat, which is why a falling satellite usually burns up in the atmosphere. Raising a satellite to a higher orbit requires that it gain energy by firing one of its rockets. The chemical potential energy of the rocket fuel is converted to gravitational potential energy as the satellite moves higher.

Generalizing from the satellite example shows that conservation of energy has a very important implication for understanding motion throughout the cosmos: *Orbits cannot change spontaneously*. For example, an asteroid or a comet passing near a planet cannot spontaneously be "sucked in" to crash on the planet. It can hit the planet only if its current orbit already intersects the planet's surface or if it somehow gains or loses orbital energy so that its new orbit

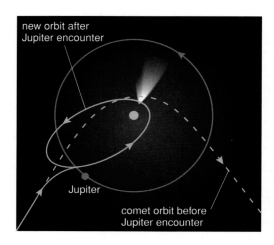

FIGURE 5.16 Depiction of a comet in an unbound orbit of the Sun that happens to pass near Jupiter. The comet loses orbital energy to Jupiter, thereby changing to a bound orbit.

intersects the planet's surface. Of course, if the asteroid or comet *gains* energy, something else must *lose* exactly the same amount of energy, and vice versa.

One way that objects can exchange orbital energy is through **gravitational encounters,** in which they pass near enough so that each can feel the effects of the other's gravity. For example, Figure 5.16 shows a gravitational encounter between Jupiter and a comet headed toward the Sun on an unbound orbit. The comet's close passage by Jupiter allows the comet and Jupiter to exchange energy: The comet loses orbital energy and changes to a bound, elliptical orbit; Jupiter must gain the energy the comet loses. However, because Jupiter is so much more massive than the comet, the effect on Jupiter is unnoticeable.

More generally, when two objects exchange orbital energy we expect one to lose energy and fall to a lower orbit while the other gains energy and is thrown to a higher orbit. If an object gains enough energy, it may end up in an unbound orbit that allows it to *escape* from the gravitational influence of the object it is orbiting. For example, if we want to send a space probe to Mars, we must use a large rocket that gives the probe enough energy to achieve an unbound orbit and ultimately escape the Earth's gravitational influence. Although it would probably make more sense to say that the probe achieved "escape energy," we say that it achieved *escape velocity* (see Figure 5.3).

If I have seen farther than others, it is because I have stood on the shoulders of giants.
ISAAC NEWTON

THE BIG PICTURE

We've covered a lot of ground in this chapter, from the scientific terminology of motion to the story of how universal motion finally was understood by Newton. Be sure you understand the following "big picture" ideas:

■ Understanding the universe requires understanding motion. Although the terminology of the laws of motion may be new to you, the *concepts* are familiar from everyday experience. Think about these experiences so that you'll better understand the less familiar astronomical applications of the laws of motion.

■ The Copernican revolution, which overthrew the ancient belief in an Earth-centered universe, did not occur instantaneously. It unfolded over a period of more than a century and involved careful observational, experimental, and theoretical work by many different people—especially Copernicus, Tycho Brahe, Kepler, and Galileo.

■ Newton's discovery of the universal law of gravitation allowed him to explain *how* gravity holds planets in their orbits. Perhaps even more important, it showed that the same physical laws we observe on Earth apply throughout the universe. This universality of physics opens up the entire universe as a possible realm of human study.

Review Questions

1. How does *speed* differ from *velocity*? Give an example in which you can be traveling at constant speed but not at constant velocity.

2. What do we mean by *acceleration*? Explain the units of acceleration (m/s^2).

3. What is the *acceleration of gravity* on Earth? If you drop a rock from very high, how fast will it be falling after 4 seconds (neglecting air resistance)?

4. What is *momentum*? How can momentum be affected by a *force*? What do we mean when we say that momentum will be changed only by a *net force*?

5. What is the difference between *mass* and *weight*?

6. What is *free-fall*, and why does it make you *weightless*? Have you ever been weightless? Explain.

7. Why does the Space Shuttle require a high speed to achieve orbit? What would happen if it were launched with a speed greater than the Earth's *escape velocity*?

8. State each of Newton's three laws of motion. For each law, give an example of its application.

9. What is the principle of *conservation of angular momentum*? How can a skater use this principle to vary his or her rate of spin?

10. Was Copernicus the first person to suggest a Sun-centered solar system? What advantages did his model have over the Ptolemaic model? In what ways did it fail to improve on the Ptolemaic model?

11. How were Tycho Brahe's observations important to the development of modern astronomy?

12. State Kepler's three laws of motion, and explain the meaning of each law.

13. Describe how Galileo helped spur acceptance of Kepler's Sun-centered model of the solar system.

14. What is the universal law of gravitation? Summarize what this law says in words, and then state the law mathematically. Define each variable in the formula.

15. What do we mean by *bound* and *unbound* orbits?

16. Why is Newton's version of Kepler's third law so important to our understanding of the universe?

17. Explain how the Moon creates tides on the Earth. How do tides vary with the phase of the Moon? Why?

18. What is *tidal friction*? Briefly describe how tidal friction has affected the Earth and the Moon.

19. Explain why orbits cannot change spontaneously. How can atmospheric drag cause an orbit to change? How can a *gravitational encounter* cause an orbit to change?

Discussion Question

1. *Kepler's Choice*. Casting aside the idea that orbits must be perfect circles meant going against deeply entrenched beliefs, and Kepler said that it shook his deep religious faith. Given that only two of Tycho's observations disagreed with a perfectly circular orbit—and only by 8 arcminutes—do you think most other people would have made the choice Kepler made to abandon perfect circles? Have you ever performed an experiment that disagreed with theory? Which did you question, the theory or your experiment? Why?

Problems

1. *Practice with Acceleration*.
   a. Some schools have an annual ritual that involves dropping a watermelon from a tall building. Suppose it takes 6 seconds for the watermelon to fall to the ground (which would mean it's about a 60-story building). If there were no air resistance so that the watermelon would fall with the acceleration of gravity, how fast would it be going when it hit the ground? Answer in m/s, km/hr, and mi/hr.

b. As you sled down a steep, slick street, you accelerate at a rate of 4 m/s². How fast will you be going after 5 seconds? Give your answer in m/s, km/hr, and mi/hr.

c. You are driving along the highway at a speed of 70 miles per hour when you slam on the brakes. If you decelerate at an average rate of −20 miles per hour per second, how long will it take to come to a stop?

2. *Gees.* Acceleration is sometimes measured in *gees,* or multiples of the acceleration of gravity: 1 gee means $1 \times g$, or 9.8 m/s²; 2 gees means $2 \times g$, or 2×9.8 m/s² = 19.6 m/s²; and so on. Suppose you experience 6 gees of acceleration in a rocket.

a. What is your acceleration in meters per second squared?

b. You will feel a compression force from the acceleration. How does this force compare to your normal weight?

c. Do you think you could survive this acceleration for long? Explain.

3. *Eclipse Frequency in the Past.* Over billions of years, the Moon has been gradually moving farther from the Earth. Thus, the Moon used to be substantially nearer the Earth than it is today.

a. How would the Moon's past angular size in our sky compare to its present angular size? Why?

b. How would the length of a lunar month in the past compare to the length of a lunar month today? Why? (*Hint:* Think about Kepler's third law as it would apply to the Moon orbiting the Earth.)

c. Based on your answers to parts (a) and (b), would eclipses (both solar and lunar) have been more or less common in the past? Why?

4. *The Gravitational Law.* Use the universal law of gravitation to answer each of the following questions.

a. How does tripling the distance between two objects affect the gravitational force between them?

b. Compare the gravitational force between the Earth and the Sun to that between Jupiter and the Sun. The mass of Jupiter is about 318 times the mass of the Earth.

c. Suppose the Sun were magically replaced by a star with twice as much mass. What would happen to the gravitational force between the Earth and the Sun?

5. *Measuring Masses.* Mathematically, Newton's version of Kepler's third law reads:

$$p^2 = \frac{4\pi^2}{G(M_1 + M_2)} a^3$$

where *p* is the orbital period in *seconds,* *a* is the average distance in *meters,* M_1 and M_2 are the masses of the two objects in kilograms, and the gravitational constant is

$$G = 6.67 \times 10^{-11} \frac{m^3}{kg \times s^2}$$

a. The Moon orbits the Earth in an average of 27.3 days at an average distance of 384,000 kilometers. Use these facts to determine the mass of the Earth. You may neglect the mass of the Moon and assume $M_{Earth} + M_{Moon} \approx M_{Earth}$.

b. Jupiter's moon Io orbits Jupiter every 42.5 hours at an average distance of 422,000 kilometers from the center of Jupiter. Calculate the mass of Jupiter. (*Hint:* Assume $M_{Jupiter} + M_{Io} \approx M_{Jupiter}$.)

c. Calculate the orbital period of the Space Shuttle in an orbit 300 kilometers above the Earth's surface. (*Hint:* Assume $M_{Earth} + M_{satellite} \approx M_{Earth}$.)

d. Pluto's moon Charon orbits Pluto every 6.4 days with a semimajor axis of 19,700 kilometers. Calculate the *combined* mass of Pluto and Charon. Compare this combined mass to the mass of the Earth.

6. *Measuring Masses in Other Star Systems.* Use Newton's version of Kepler's third law to answer the following questions. (*Hint:* The calculations for this problem are so simple that you will not need a calculator.)

a. Imagine another solar system, with a star of the same mass as the Sun. Suppose there is a planet in that solar system with a mass twice that of Earth orbiting at a distance of 1 AU from the star. What is the orbital period of this planet? Explain.

b. Suppose a solar system has a star that is four times as massive as our Sun. If that solar system has a planet the same size as Earth orbiting at a distance of 1 AU, what is the orbital period of the planet? Explain.

7. *Head-to-Foot Tides.* You and the Earth attract each other gravitationally, so you should also be subject to a tidal force resulting from the difference between the gravitational attraction felt by your feet and that felt by your head (at least when standing). Explain why you can't feel this tidal force.

8. *Research: Players in the Copernican Revolution.* A number of interesting personalities played important roles in the Copernican revolution besides Copernicus, Tycho, Kepler, and Galileo. Research one or more of these people and write a short biography of their scientific lives and their contributions to the Copernican revolution. Among the people you might consider: Nicholas of Cusa (1401–1464), who argued for a Sun-centered solar system and believed that other stars might be circled by their own planets; Leonardo da Vinci (1452–1519), who made many contributions to science, engineering, and art and also believed the Earth *not* to be the center of the universe; Rheticus (1514–1574), a student of Copernicus who persuaded him to publish his work; William Gilbert (1540–1603), who studied the Earth's magnetism and influenced the thinking of Kepler; Giordano Bruno (1548–1600), who was burned at the stake for his beliefs in the Copernican system and the atomism of Democritus; and Francis Bacon (1561–1626), whose writings contributed to the acceptance of experimental methods in science.

Light, Telescopes, and Spacecraft

6

ANCIENT OBSERVERS COULD DISCERN ONLY THE MOST BASIC features—such as color and brightness—of the light that they saw. Over the past several hundred years, we have discovered that light contains far more information. Remarkably, analysis of light with proper instruments can reveal the chemical composition of distant objects, their temperature, how fast they rotate, and much more. Light, the cosmic messenger, brings the stories of distant objects to our home here on Earth.

In this chapter, we'll explore the basic properties of light. We'll learn why the light visible to our eyes is only the tip of the iceberg when it comes to the full spectrum of light, and we'll see how we can extract a treasury of information by examining light spectra in detail. We'll also discuss the basic features of telescopes, the instruments with which we collect light from the far reaches of the universe and of spacecraft, which play an increasingly important role in astronomy.

FIGURE 6.1 A prism reveals that white light contains a spectrum of colors from red to violet.

6.1 Light in Everyday Life

Even without opening your eyes, it's clear that light is a form of energy, called *radiative energy* [Section 4.1]. Outside on a hot, sunny day, you can feel the radiative energy of sunlight being converted to thermal energy as it strikes your skin. On an economic level, you know that light is a form of energy because you have to pay for it. The rate at which a light bulb uses energy is usually printed on it—something like "100 watts." A watt is a unit of **power,** which describes the rate of energy use; each watt of power means that 1 joule of energy is used each second. Thus, for every second that you leave a 100-watt light bulb turned on, you will have to pay the utility company for 100 joules of energy. Interestingly, the power requirement of an average human—about 10 million joules per day—is about the same as that of a 100-watt light bulb.

Another basic property of light is what our eyes perceive as *color.* You've probably seen a prism split light into a **spectrum** (plural, spectra), or rainbow of colors (Figure 6.1). You can also produce a spectrum with a **diffraction grating**—a piece of plastic or glass etched with many closely spaced lines. The colors in a spectrum are unusually pure forms of the basic colors red, orange, yellow, green, blue, and violet. The wide variety of all possible colors comes from mixtures of these basic colors in varying proportions; *white* is simply what we see when the basic colors are mixed in roughly equal proportions. Your television takes advantage of this fact to simulate a huge range of colors by combining only three specific colors of red, green, and blue light.

TIME OUT TO THINK *If you have a magnifying glass handy, hold it close to your TV set to see the individual red, blue, and green dots. If you don't have a magnifying glass, try splashing a few droplets of water onto your TV screen. What do you see? What are the drops of water doing?*

Energy carried by light can interact with matter in four general ways:

- **Emission:** When you turn on a lamp, electricity flowing through the filament of the light bulb heats it to a point at which it *emits* visible light.

- **Absorption:** If you place your hand near a lit light bulb, your hand *absorbs* some of the light, and this absorbed energy makes your hand warmer.

- **Transmission:** Some forms of matter, such as glass or air, *transmit* light; that is, they allow light to pass through them.

- **Reflection:** A mirror *reflects* light in a very specific way, similar to the way a rubber ball bounces off a hard surface, so that the direction of a reflected beam of light depends on the direction of the incident (incoming) beam of light (Figure 6.2a). Sometimes, reflection is more random, so that an incident beam of light is **scattered** into many different directions. The screen in a movie theater scatters a narrow beam of light from the projector into an array of beams that reach every member of the audience (Figure 6.2b).

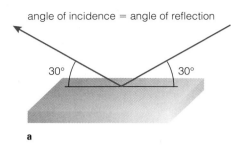

angle of incidence = angle of reflection

30°   30°

a

screen

projection booth

b

FIGURE 6.2 (**a**) A mirror reflects light along a path determined by the angle at which the light strikes the mirror. (**b**) A movie screen scatters light into an array of beams that reach every member of the audience.

Particular materials affect different colors of light differently. Red glass transmits red light but absorbs other colors. A healthy lawn looks green because it reflects green light but absorbs all other colors.

Now let's put all these ideas together and think about what happens when you walk into a dark room and turn on the light switch. The light bulb begins to emit white light, which is a mix of all the colors in the spectrum. Some of this light exits the room, transmitted through the windows. The rest of the light strikes the surfaces of objects inside the room, where each object's material properties determine the colors absorbed or reflected. The light coming from each object therefore carries an enormous amount of information about the object's location, shape and structure, and material makeup. You acquire this information when light enters your eyes, where it is absorbed by special cells (called *cones* and *rods*) that use the energy of the absorbed light to send signals to your brain. Your brain interprets the messages carried by the light, recognizing materials and objects in the room in the process we call *vision*.

6.2 Properties of Light

In fact, light carries even more information than your ordinary vision can recognize. Just as a microscope can reveal structure that is invisible to the naked eye, modern instruments can reveal otherwise invisible details in the spectrum of light. Learning to interpret these details is the key to unlocking the vast amount of information carried by light. Let's begin by examining fundamental properties of light.

Particle or Wave?

In our everyday lives, waves and particles appear to be very different. After all, no one would confuse the ripples on a pond with a baseball. However, light behaves as *both* a particle *and* a wave. Like particles, light comes in individual "pieces," called **photons,** that can hit a wall one at a time. Like ripples on a pond, light can make particles bob up and down.

First, let's ignore the particle properties of light and look at its wave nature. Light is an **electromagnetic wave**—a wave in which electric and magnetic fields vibrate. (The concept of a *field* is a bit abstract,

a

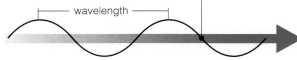

Wavelength is the distance between adjacent peaks of the electric field.

Frequency is the number of waves passing any point each second.

wavelength

All light travels with speed c = 300,000 km/s.

b

FIGURE 6.3 (**a**) A row of electrons would wriggle up and down as light passes by, showing that light carries a vibrating electric field. It also carries a magnetic field (not shown) that vibrates perpendicular to the direction of the electric field vibrations. (**b**) Characteristics of light waves. Because all light travels at the same speed, light of longer wavelength must have lower frequency.

but you can think of electric or magnetic fields as places where charged particles would feel electric or magnetic forces.) Like a leaf on a rippling pond, an electron will bob up and down when an electromagnetic wave passes by; if you could set up a row of electrons, they would wriggle like a snake (Figure 6.3a). The **wavelength** is the distance between adjacent peaks of the electric or magnetic field, and the **frequency** is the number of peaks that pass by any point each second (Figure 6.3b). All light travels at the same speed—about 300,000 kilometers per second, or 3×10^8 m/s—regardless of its wavelength or frequency. Therefore, light with a shorter wavelength must have a higher frequency, and vice versa.

Measuring the particle properties of light requires thinking of each photon as a distinct entity. Just as a baseball carries a specific amount of kinetic energy, each photon of light carries a specific amount of radiative energy. The shorter the wavelength of the light (or, equivalently, the higher its frequency), the higher the energy of the photons. For example, a photon with a wavelength of 100 nanometers (nm) has more energy than a photon with a 120-nm wavelength. (A nanometer is a billionth of a meter: 1 nm = 10^{-9} m.)

The Many Forms of Light

Because light consists of electromagnetic waves, the spectrum of light is called the **electromagnetic spectrum.** Photons of light can have *any* wavelength or frequency, so in principle the complete electro-

magnetic spectrum extends from a wavelength of zero to infinity. For convenience, we usually refer to different portions of the electromagnetic spectrum by different names (Figure 6.4).

The **visible light** that we see with our eyes has wavelengths ranging from about 400 nm at the blue end of the rainbow to about 700 nm at the red end. Light with wavelengths somewhat longer than red light is called **infrared,** because it lies beyond the red end of the rainbow. Light with very long wavelengths is called **radio** (or *radio waves*)—thus, radio is a form of light, *not* a form of sound.

On the other end of the spectrum, light with wavelengths somewhat shorter than blue light is called **ultraviolet,** because it lies beyond the blue (or violet) end of the rainbow. Light with even shorter wavelengths is called **X rays,** and the shortest-wavelength light is called **gamma rays.** Note that visible light is an extremely small part of the entire electromagnetic spectrum: The reddest red that our eyes can see has only about twice the wavelength of the bluest blue, but the radio waves from your favorite radio station are a billion times longer than the X rays used in a doctor's office.

As we move from the radio end toward the gamma-ray end of the spectrum, the wavelengths become shorter, and therefore the frequencies and energies

---

Common Misconceptions: You Can't Hear Radio Waves and You Can't See X Rays

Most people associate the term *radio* with sound, but radio waves are a form of *light* with long wavelengths—too long for our eyes to see. Radio stations encode sounds (e.g., voices, music) as electrical signals, which they broadcast as radio waves. What we call "a radio" in daily life is an electronic device that receives these radio waves and decodes them to re-create the sounds played at the radio station. Television is also broadcast by encoding information (both sound and pictures) in the form of light called radio waves.

X rays are also a form of light, with wavelengths far too short for our eyes to see. In a doctor's or dentist's office, a special machine works rather like the flash on an ordinary camera but emits X rays instead of visible light. This machine flashes the X rays at you, and a piece of photographic film records the X rays that are transmitted through your body. Note that you never *see* the X rays; you see only an image left on film by the transmitted X rays.

low energy   700 nm   400 nm   high Energy

FIGURE 6.4 The electromagnetic spectrum. The unit of frequency, hertz, is equivalent to waves per second. For example, 10^3 hertz means that 10^3 wave peaks pass by a point each second.

of the photons increase. Visible photons happen to have enough energy to activate the molecular receptors in our eyes. Ultraviolet photons, with shorter wavelength than visible light, carry more energy—enough to harm our skin cells, causing sunburn or skin cancer. X-ray photons have enough energy to transmit easily through skin and muscle but not so easily through bones or teeth. That is why photographs taken with X-ray light allow doctors and dentists to see our underlying bone structures.

6.3 Light and Matter

Whenever matter and light interact, matter leaves its fingerprints. Examining the color of an object is a crude way of studying the clues left by the matter it contains. If we take light and disperse it into a spectrum, we can see the spectral fingerprints in more detail.

Figure 6.5 shows a schematic spectrum of light from a celestial body such as a planet. The spectrum is a graph that shows the amount of radiation, or **intensity,** at different wavelengths. At wavelengths where a lot of light is coming from the celestial body, the intensity is high; at wavelengths where there is little light, the intensity is low. Our goal is to see how the bumps and wiggles in this graph convey a wealth of information about the celestial body in question. Let's begin by going through a short list of ways that matter interacts with light and showing the spectra that result.

Absorption and Emission by Thin Gases

You know that gases can absorb light; for example, ozone in the Earth's atmosphere absorbs ultraviolet light from space, preventing it from reaching the ground. You also know that gases can emit light, which is what makes interstellar clouds glow so beautifully. But *how* do gases absorb or emit light?

Recall that the electrons in atoms can have only specific energies, somewhat like the specific heights of the rungs on a ladder [Section 4.4]. If an electron in an

FIGURE 6.5 A schematic spectrum obtained from the light of a distant object. The "rainbow" at bottom shows how the light would appear when viewed through a prism or diffraction grating. The graph shows the corresponding intensity of the light at each wavelength. Note that the intensity is high where the "rainbow" is bright and low where it is dim (such as in places where the "rainbow" shows dark lines).

FIGURE 6.6 (**a**) Photons emitted by various energy level transitions in hydrogen. (**b**) The visible emission line spectrum from heated hydrogen gas. These lines come from transitions in which electrons fall from higher energy levels to level 2. (**c**) If we pass white light through a cloud of cool hydrogen gas, we get this absorption line spectrum. These lines come from transitions in which electrons jump from energy level 2 to higher levels.

atom is bumped from a lower energy level to a higher one—by a collision with another atom, for example— it will eventually fall back to the lower level. The energy that the atom loses when the electron falls back down must go somewhere, and often it goes to *emitting* a photon of light. The emitted photon must have exactly the same amount of energy that the electron loses, which means that it has a specific wavelength (and frequency).

Figure 6.6a shows the allowed energy levels in hydrogen, along with the wavelengths of the photons emitted by various downward *transitions* of an electron from a higher energy level to a lower one. For example, transitions from level 2 to level 1 emit an ultraviolet photon of wavelength 121.6 nm, and transitions from level 3 to level 2 emit a red visible-light photon of wavelength 656.3 nm. If you heat some hydrogen gas so that collisions are continually bumping electrons to higher energy levels, you'll get an **emission line spectrum** consisting of the photons emitted as each electron falls back to lower levels (Figure 6.6b).

TIME OUT TO THINK *If nothing continues to heat the hydrogen gas, all the electrons eventually will end up in the lowest energy level (the ground state, or level 1). Use this fact to explain why we should* not *expect to see an emission line spectrum from a very cold cloud of hydrogen gas.*

Photons of light can also be absorbed, causing electrons to jump *up* in energy—but only if an incoming photon happens to have precisely the right amount of energy. For example, just as an electron moving downward from level 2 to level 1 in hydrogen emits a photon of wavelength 121.6 nm, absorbing a photon with this wavelength will cause an electron in level 1 to jump up to level 2.

Suppose a lamp emitting white light illuminates a cloud of hydrogen gas from behind. The cloud will absorb photons with the precise energies needed to bump electrons from a low energy level to a higher one, while all other photons pass right through the cloud. The result is an **absorption line spectrum** that looks like a rainbow with light missing at particular wavelengths (Figure 6.6c).

helium

sodium

neon

FIGURE 6.7 Emission line spectra for helium, sodium, and neon. The patterns and wavelengths of lines are different for each element, giving each a unique spectral fingerprint.

FIGURE 6.8 This spectrum of molecular hydrogen (H_2) shows that molecular spectra consist of lines bunched into broad *molecular bands*.

If you compare the bright emission lines in Figure 6.6b to the dark absorption lines in Figure 6.6c, you will see that the lines occur at the same wavelengths regardless of whether the hydrogen is absorbing or emitting light. Absorption lines simply correspond to upward jumps of the electron between energy levels, while emission lines correspond to downward jumps.

The energy levels of electrons in each chemical element, and in each ion of each chemical element, are unique [Section 4.4]. As a result, each element and ion produces its own distinct set of spectral lines, giving it a unique "spectral fingerprint." For example, Figure 6.7 shows emission line spectra for (neutral) helium, sodium, and neon. Each different type of molecule (made from two or more atoms) also produces its own unique set of spectral lines; these lines tend to come in bunches called **molecular bands** (Figure 6.8).

Thermal Radiation: Every Body Does It

While thin gases tend to produce simple emission line or absorption line spectra, more "complex" objects—such as planets, stars, and you—produce what we call **thermal radiation** (sometimes called "blackbody radiation"). This term arises because, if an object emitted *only* thermal radiation, its spectrum would depend only on its *temperature*. No real object is a perfect "thermal emitter," but many objects make close approximations. Therefore, it is useful to understand the two rules that describe how the spectrum of a thermal emitter depends on its temperature:

FIGURE 6.9 A fireplace poker gets brighter as it is heated, demonstrating rule 1 for thermal radiation: hotter objects emit more total radiation per unit surface area. In addition, its "color" moves from infrared to red to white as it is heated, demonstrating rule 2: hotter objects emit photons with higher average energy.

1. *Hotter objects emit more total radiation per unit surface area.* The radiated energy is proportional to the *fourth* power of the temperature expressed in Kelvin (*not* in Celsius or Fahrenheit). For example, a 600 K object has twice the temperature of a 300 K object and therefore radiates $2^4 = 16$ times as much total energy per unit surface area.

2. *Hotter objects emit photons with a higher average energy,* which means a shorter average wavelength.

You can see the first rule in action by playing with a light that has a dimmer switch: When you turn the switch up, the filament in the light bulb gets hotter and the light brightens; when you turn it down, the filament gets cooler and the light dims. (You can verify the changing temperature by placing your hand near the bulb.) You can see the second rule in action with a fireplace poker (Figure 6.9). When the poker is relatively cool, it emits only infrared radiation,

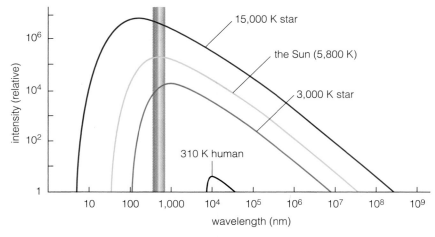

FIGURE 6.10 Graphs of idealized thermal radiation spectra. Note that, per unit surface area, hotter objects emit more radiation at every wavelength, demonstrating rule 1 for thermal radiation. The peaks of the spectra occur at shorter wavelengths (higher energies) for hotter objects, demonstrating rule 2 for thermal radiation.

which we cannot see. As it gets hot, it begins to glow red ("red hot"). If the poker continues to heat up, the average wavelength of the emitted photons gets shorter, moving toward the blue end of the visible spectrum. By the time it gets very hot, the mix of colors emitted by the poker looks white ("white hot").

Figure 6.10 shows how these two rules affect the spectra of idealized thermal emitters. The spectra of hotter objects show bigger "humps" because they emit more total radiation per unit area (rule 1). Hotter objects also have the peaks of their humps at shorter wavelengths because of the higher average energy of their photons (rule 2). Note that hotter objects emit more light at *all* wavelengths but the biggest difference appears at the shortest wavelengths. An object with a temperature of 310 K, which is about human body temperature, emits mostly in the infrared and emits no visible light at all—that explains why we don't glow in the dark! A relatively cool star, with a 3,000 K surface temperature, emits mostly red light; that is why some bright stars in our sky appear red, such as Betelgeuse (in Orion) and Antares (in Scorpio). The Sun's 5,800 K surface emits most strongly in green light (around 500 nm), but the Sun looks yellow or white because it also emits other colors throughout the visible spectrum. Hotter stars emit mostly in the ultraviolet, but because our eyes cannot see ultraviolet they appear blue or blue-white in color. If an object were heated to a temperature of millions of degrees, it would radiate mostly X rays. Some astronomical objects are indeed hot enough to emit X rays, such as disks of gas encircling exotic objects like neutron stars and black holes (see Chapter 15).

Summary of Spectral Formation

We can now summarize the circumstances under which objects produce thermal, absorption line, or emission line spectra. (These rules are often called *Kirchhoff's laws*.)

■ Any "complex" object produces thermal radiation over a broad range of wavelengths. If the object is hot enough to produce visible light, as is the case with the filament of a light bulb, we see a smooth, *continuous* rainbow when we disperse the light through a prism or a diffraction grating (Figure 6.11a). On a graph of intensity versus wavelength, the rainbow becomes the characteristic hump of a thermal radiation spectrum.

■ When thermal radiation passes through a thin cloud of gas, the form of the fingerprints the cloud leaves depends on its temperature. If the background source of thermal emission is hotter than the cloud, the balance between emission and absorption in the cloud's spectral lines tips toward absorption. We then see absorption lines cutting into the thermal spectrum. This is the case when the light from the hot light bulb passes through a cool gas cloud (Figure 6.11b). On the graph of intensity versus wavelength, these absorption lines create dips in the thermal radiation spectrum; the width and depth of each dip depend on how much light is absorbed by the chemical responsible for the line.

■ If the background source is colder than the cloud, or if there is no background source at all, the spectrum is dominated by bright emission

FIGURE 6.11 (**a**) A "complex" object, such as a light bulb filament, produces a continuous spectrum of thermal radiation. (**b**) If thermal radiation passes through a thin gas that is cooler than the thermal emitter, dark absorption lines are superimposed on the continuous spectrum. (**c**) Viewed against a cold, dark background, the same gas produces an emission line spectrum.

lines produced by the cloud's atoms and molecules (Figure 6.11c). These lines create narrow peaks on a graph of intensity versus wavelength.

Reflected Light

We've now covered enough material to understand spectra emitted by objects generating their own light, such as stars and clouds of interstellar gas. But most of our daily experience involves *reflected* (or *scattered*) light. The source of the light is thermal radiation from the Sun or a lamp. After this light strikes the ground, clouds, people, or other objects, we see only the wavelengths of light that are reflected. For example, a red sweatshirt absorbs blue light and reflects red light, so its visible spectrum looks like the thermal radiation spectrum of its light source—the Sun—but with blue light missing.

In the same way in which we distinguish lemons from limes, we can use color information in reflected light to learn about celestial objects. Different fruits, different rocks, and even different atmospheric gases reflect and absorb light at different wavelengths. Although the absorption features that show up in spectra of reflected light are not as distinct as the emission and absorption lines for thin gases, they still provide useful information. For example, the surface materials of a planet determine how much light of different colors is reflected or absorbed. The reflected light gives the planet its color, while the absorbed light heats the surface and helps determine its temperature.

Putting It All Together

Figure 6.12 again shows the complicated spectrum that we began with in Figure 6.5, but this time with labels indicating the processes responsible for its various features. What can we say about this object from its spectrum? The hump of thermal emission

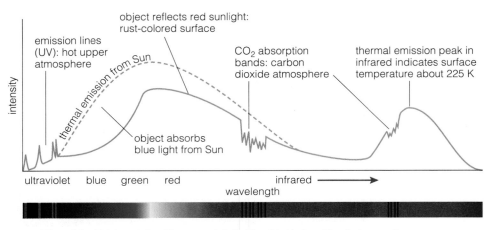

FIGURE 6.12 The spectrum of Figure 6.5, with interpretation. We can conclude that the object looks red in color because it absorbs more blue light than red light from the Sun. The absorption lines tell us that the object has a carbon dioxide atmosphere, and the emission lines tell us that its upper atmosphere is hot. The hump in the infrared (near the right of the diagram) tells us that the object has a surface temperature of about 225 K. It is a spectrum of the planet Mars.

shows that this object has a surface temperature of about 225 K, well below the freezing point of water. The absorption bands in the infrared come mainly from carbon dioxide, which tells us that the object has a carbon dioxide atmosphere. The emission lines in the ultraviolet come from hot gas in a high, thin layer of the object's atmosphere. The reflected light looks like the Sun's 5,800 K thermal radiation except that the blue light is missing, so the object must be reflecting sunlight and must look red in color. Perhaps by now you have guessed that this figure represents the spectrum of the planet Mars.

6.4 The Doppler Shift

The volume of information about temperature and composition that light contains is truly amazing, but we can learn even more once we identify the various lines in a spectrum. Among the most important pieces of information contained in light is information about motion. In particular, we can determine the radial motion (toward or away from us) of a distant object from changes in its spectrum caused by the **Doppler effect.**

You've probably noticed the Doppler effect on *sound;* it is especially easy to notice when you stand near train tracks and listen to the whistle of a train (Figure 6.13). As the train approaches, its whistle is relatively high pitched; as it recedes, the sound is relatively low pitched. Just as the train passes by, you hear the dramatic change from high to low pitch— a sort of "weeeeeeee–ooooooooooh" sound. You can visualize the Doppler effect by imagining that the

train's sound waves are bunched up ahead of it, resulting in shorter wavelengths and thus the high pitch you hear as the train approaches. Behind the train, the sound waves are stretched out to longer wavelengths, resulting in the low pitch you hear as the train recedes.

The Doppler effect causes similar shifts in the wavelengths of light. If an object is moving toward us, then its entire spectrum is shifted to shorter wavelengths. Because shorter wavelengths are bluer when we are dealing with visible light, the Doppler shift of an object coming toward us is called a **blue-shift.** If an object is moving away from us, its light is shifted to longer wavelengths; we call this a **redshift** because longer wavelengths are redder when we are dealing with visible light. Note that the terms *blueshift* and *redshift* are used even when we are not dealing with visible light.

Spectral lines provide the reference points we use to identify and measure Doppler shifts (Figure 6.14). For example, suppose we recognize the pattern of hydrogen lines in the spectrum of a distant object. We know the **rest wavelengths** of the hydrogen lines—that is, their wavelengths in stationary clouds of hydrogen gas—from laboratory experiments in which a tube of hydrogen gas is heated so that the wavelengths of the spectral lines can be measured. If the hydrogen lines from the object appear at longer wavelengths, then we know that they are redshifted and the object is moving away from us; the larger the shift, the faster the object is moving. If the lines appear at shorter wavelengths, then we know that they are blueshifted and the object is moving toward us.

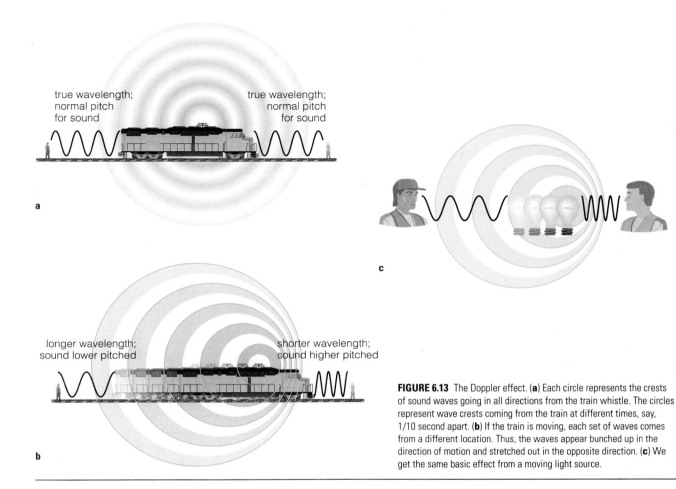

true wavelength; normal pitch for sound

true wavelength; normal pitch for sound

a

c

longer wavelength; sound lower pitched

shorter wavelength; sound higher pitched

b

FIGURE 6.13 The Doppler effect. (**a**) Each circle represents the crests of sound waves going in all directions from the train whistle. The circles represent wave crests coming from the train at different times, say, 1/10 second apart. (**b**) If the train is moving, each set of waves comes from a different location. Thus, the waves appear bunched up in the direction of motion and stretched out in the opposite direction. (**c**) We get the same basic effect from a moving light source.

Laboratory spectrum
Lines at rest wavelengths.

FIGURE 6.14 Spectral lines provide the crucial reference points for measuring Doppler shifts.

Object 1
Lines redshifted: Object is moving away from us.

Object 2
Greater redshift: Object is moving away faster than Object 1.

Object 3
Lines blueshifted: Object is moving toward us.

Object 4
Greater blueshift: Object is moving toward us faster than Object 3.

6.5 Collecting Light with Telescopes

Now that we understand what we can learn from light, it's time to turn our attention to how we collect light. Telescopes are essentially giant eyes that allow us to collect far more light than we could with our naked eyes alone. Telescopes are characterized by two key properties:

- **Light-collecting area** measures the ability to collect light. The larger the light-collecting area, the more light is collected. The light-collecting area of the world's largest visible-light telescopes is more than a million times that of the human eye.

- **Angular resolution** tells us how much detail we can see. For example, an angular resolution of 1 arcminute (about that of the human eye) means that two stars will appear distinct only if they appear greater than 1 arcminute ($\frac{1}{60}°$) apart in the sky. (Otherwise, they will look like a single star.) The smaller the angular resolution, the more powerful the telescope. The Hubble Space Telescope has an angular resolution of 0.05 arcsecond, which would allow you to read this book from a distance of about 800 meters ($\frac{1}{2}$ mile).

In principle, larger telescopes have both a greater light-collecting area and a better angular resolution. However, the angular resolution may be degraded if a telescope is poorly made, and the Earth's atmosphere also can limit the angular resolution of a telescope (unless it is in space). Therefore, astronomers seek ways to maximize both light-collecting area and angular resolution when designing new telescopes.

Common Misconceptions: Magnification and Telescopes

You are probably familiar with how binoculars, magnifying glasses, and telephoto lenses for cameras make objects appear larger in size—the phenomenon we call magnification. Many people assume that astronomical telescopes are also characterized by their magnification. In fact, astronomers are more interested in a telescope's light-gathering power and resolution. The magnification of most telescopes is set to achieve the maximum amount of detail given the telescope's resolution—no more, no less. After all, there's little point in magnifying an image if it doesn't show any more detail.

Basic Telescope Design

Telescopes come in two basic designs: *refracting* and *reflecting*. A **refracting telescope** operates much like an eye, using transparent glass lenses to focus the light from distant objects (Figure 6.15a). The earliest telescopes, including those built by Galileo, were refracting telescopes. The world's largest refracting telescope, completed in 1897, has a lens that is 1 meter (40 inches) in diameter and a telescope tube that is 19.5 meters (64 feet) long (Figure 6.15b).

A **reflecting telescope** uses a precisely curved **primary mirror** to gather and focus light (Figure 6.16a). Today, most astronomical research uses reflecting telescopes. Note that the *prime focus,* the point at which rays that come straight into the telescope converge, lies *in front of* the mirror. Thus, if we place a camera at the prime focus, it will block some of the light entering the telescope. This does not create a serious problem as long as the camera is small compared to the telescope mirror. If the telescope is large enough, a "cage" in which an astronomer can work can be suspended over the prime focus (Figure 6.16b).

An alternative and more common arrangement for reflecting telescopes is to place a **secondary mirror** just below the prime focus. The secondary mirror reflects the light to a location that is more convenient for viewing or for attaching instruments (Figure 6.17): through a hole in the primary mirror (a *Cassegrain focus*), to a hole in the side (a *Newtonian focus,* particularly common in telescopes made for amateurs), or to a third mirror that deflects light toward instruments that are not attached to the telescope (a *coudé focus*).

For most of the latter half of the twentieth century, telescopes hardly changed at all. The 5-meter Hale telescope on Mount Palomar outside San Diego remained the world's most powerful for more than 40 years after its opening in 1947. Building larger telescopes posed a difficult technological challenge because large mirrors tend to sag under their own weight. A 6-meter telescope built in Russia in 1976 is larger than the 5-meter Hale, but it suffers several problems of optical quality that make it far less useful. Recent technological innovations have made it possible to build very large, low-weight mirrors, fueling a boom in large telescope construction. For example, each of the twin 10-meter Keck telescopes in Hawaii has a primary mirror consisting of 36 smaller mirrors that function together (Figure 6.18). Indeed, the Hale telescope is no longer even among the top 10 in size (including telescopes currently under construction).

FIGURE 6.15 (**a**) The basic design of a refracting telescope. (**b**) The 1-meter refractor at the University of Chicago's Yerkes Observatory is the world's largest refracting telescope.

lens

starlight

focus

eyepiece

a to eye

b

FIGURE 6.16 (**a**) The basic design of a reflecting telescope. (**b**) An astronomer in the cage at the focal plane of the Mount Palomar telescope. The primary mirror lies several meters below.

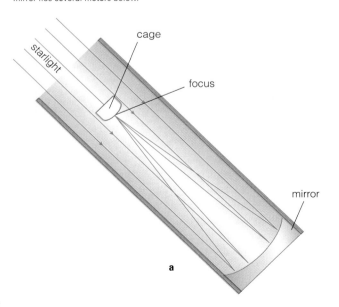

cage

starlight

focus

mirror

a

b

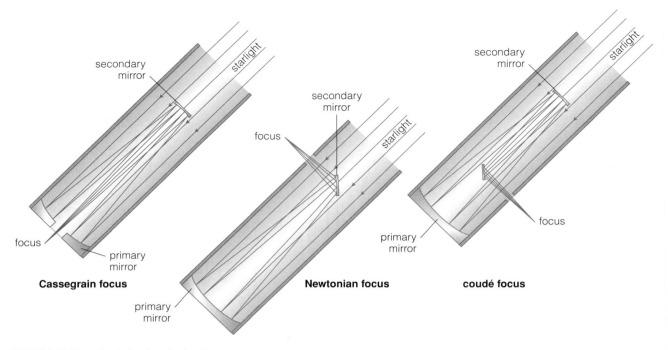

FIGURE 6.17 Alternative designs for reflecting telescopes.

FIGURE 6.18 The Keck telescopes use 36 hexagonal mirrors to function as a single 10-meter primary mirror. These mirrors make the honeycomb pattern that surrounds the hole with the man in this photo.

Uses of Telescopes

Astronomers use many different kinds of instruments and detectors to extract the information contained in the light collected by a telescope. Every astronomical observation is unique, but most observations fall into one of three basic categories:

■ **Imaging** yields pictures of astronomical objects. At its most basic, an imaging instrument is simply a camera that puts film or an electronic detector such as a CCD (charge-coupled device) in a position to record light focused by the telescope. Sometimes astronomers place *filters* in front of

focused starlight

1. Slit: Allows only the light from the object of interest to pass through.

2. Collimating mirror: Makes all the reflected rays parallel.

3. Diffraction grating: Disperses reflected light into a spectrum.

5. CCD: Records an image of the spectrum.

4. Camera mirror: Focuses the spectrum onto the CCD.

FIGURE 6.19 The basic design of a spectrograph.

the camera that allow only particular colors or wavelengths of light to pass through.

- **Spectroscopy** involves dispersing light into a spectrum. Instruments called *spectrographs* use diffraction gratings or other techniques to disperse light into spectra, which are then recorded with a detector such as a CCD (Figure 6.19).

- **Timing** tracks how the light intensity hitting the detector varies with time. If an object varies slowly enough, a timing experiment may be as simple as comparing a set of images or spectra obtained on different nights. For more rapidly varying sources, some instruments can actually record the arrival time of every individual photon striking a detector.

Major observatories typically have several different instruments capable of each of these tasks, and some instruments can perform all three basic tasks.

Atmospheric Effects on Observations

A telescope on the ground does not look directly into space but rather looks through the Earth's sometimes murky atmosphere. In fact, the Earth's atmosphere creates several problems for astronomical observations. The most obvious problem is weather—an optical telescope is useless under cloudy skies. Another problem is that our atmosphere scatters the bright lights of cities, causing what astronomers call **light pollution.** Somewhat less obvious is the fact that the atmosphere distorts light. The ever-changing motion, or **turbulence,** of air in the atmosphere bends light in constantly shifting patterns. This turbulence causes the familiar twinkling of stars and blurs astronomical images.

Common Misconceptions: Twinkle, Twinkle Little Star

Twinkling, or apparent variation in the brightness and color of stars, is *not* intrinsic to the stars. Instead, just as light is bent by water in a swimming pool, starlight is bent by the Earth's atmosphere. Air turbulence causes twinkling because it constantly changes how the starlight is bent. Hence, stars tend to twinkle more on windy nights and when they are near the horizon (and therefore viewed through a thicker layer of atmosphere). Planets also twinkle, but not nearly as much as stars: Because planets have a measurable angular size, the effects of turbulence on any one ray of light are compensated for by the effects of turbulence on others, reducing the twinkling seen by the naked eye (but making planets shimmer noticeably in telescopes). While twinkling may be beautiful, it blurs telescopic images. Avoiding the effects of twinkling is one of the primary reasons for putting telescopes in space. There, above the atmosphere, astronauts and telescopes do not see any twinkling.

TIME OUT TO THINK *If you look down a long, paved street on a hot day, you'll notice the images of distant cars and buildings rippling and distorting. How are these distortions similar to the twinkling of stars? Why do you think these distortions are more noticeable on hot days than on cooler days?*

Many of these atmospheric problems can be mitigated by choosing appropriate sites for observatories. The key criteria are that the sites be dark (limiting light pollution), dry (limiting rain and clouds), calm (limiting turbulence), and high (placing them above at least part of the atmosphere). Islands are often ideal, and the 4,300-meter (14,000-foot) summit of Mauna Kea on the big island of Hawaii is home to many of the world's best observatories (Figure 6.20).

The ultimate solution to atmospheric problems is to put telescopes in space. That is the primary reason why the Hubble Space Telescope was built and why it is so successful even though its 2.4-meter primary mirror is much smaller than that of many telescopes on the ground (Figure 6.21). However, a

FIGURE 6.20 Observatories on the summit of Mauna Kea in Hawaii. The twin domes at the far right house the two Keck telescopes.

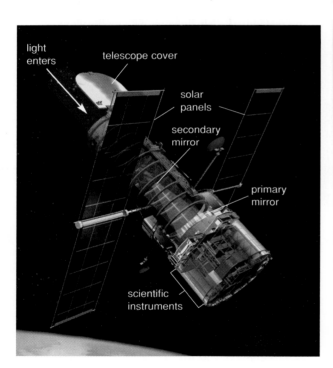

FIGURE 6.21 This diagram shows basic components of the Hubble Space Telescope, which orbits the Earth. The entire observatory is roughly the size of a school bus.

new technology called **adaptive optics** can overcome atmospheric turbulence for ground-based telescopes, essentially by changing the shape of a telescope's mirrors many times per second in just the right way to overcome the distortions caused by the atmosphere (Figure 6.22).

a   b

FIGURE 6.22 (a) Atmospheric distortion makes this ground-based image of a double star look like a single star. (b) Using the same telescope, but with adaptive optics, the two stars are clearly distinguishable.

Telescopes Across the Spectrum

If we studied only visible light, we'd be missing much of the picture. Planets are relatively cool and emit primarily infrared light. The hot upper layers of stars like the Sun emit ultraviolet and X-ray light. Many objects emit radio waves, including some of the most exotic objects in the cosmos—objects that may contain *black holes* buried in their centers [Section 15.4]. Some violent events even produce gamma rays that travel through space to the Earth. Indeed, most objects emit light over a broad range of wavelengths.

Unfortunately, the atmosphere poses a major problem that no Earth-bound technology can overcome: It prevents most forms of light from reaching the ground at all. Figure 6.23 shows the depth to

FIGURE 6.23 Different wavelengths of light penetrate to different depths in the Earth's atmosphere.

FIGURE 6.24 This diagram shows how a set of grazing incidence mirrors can focus X rays.

which different forms of light penetrate into the Earth's atmosphere. Only radio waves, visible light, parts of the infrared spectrum, and the longest wavelengths of ultraviolet light reach the ground. Observing other wavelengths requires placing telescopes in space (or, in a few cases, very high in the atmosphere on airplanes or balloons.).

Today, astronomers study light across the entire spectrum. With the exception of telescopes for radio, some infrared, and visible-light observations, most of these telescopes are in space. Indeed, while the Hubble Space Telescope is the only major visible-light observatory in space (it is also used for infrared and ultraviolet observations), many less famous observatories are in Earth orbit for the purpose of making observations in nonvisible wavelengths.

Telescopes for nonvisible wavelengths generally require the use of somewhat different telescope or detector technologies. For example, ordinary telescopes are warm enough to emit significant amounts of infrared light, and this emission can interfere with attempts to observe infrared light from the cosmos. As a result, some infrared tele-

scopes are equipped with sophisticated cooling systems. At the other end of the spectrum, X rays and gamma rays have sufficient energy to penetrate ordinary telescope mirrors, making them useless for these forms of light. X rays can be focused by slightly bending their paths with what astronomers call *grazing incidence* mirrors (Figure 6.24). (Other technologies are used for gamma rays.)

Radio telescopes use large metal dishes as "mirrors" to reflect radio waves. However, the long wavelengths of radio waves mean that very large telescopes are necessary in order to achieve reasonable angular resolution. The largest radio dish in the world, the Arecibo Telescope in Puerto Rico, stretches 305 meters (1,000 feet) across a natural valley (Figure 6.25). Despite its large size, Arecibo has an angular resolution of only about 1 arcminute at commonly observed radio wavelengths (e.g., 21 cm)—a few hundred times worse than the visible-light resolution of the Hubble Space Telescope.

In the 1950s, radio astronomers developed an ingenious technique for improving the angular resolution of radio telescopes. This technique, called

FIGURE 6.25 The Arecibo radio telescope in Puerto Rico is the world's largest single radio dish.

FIGURE 6.26 The Very Large Array (VLA) in New Mexico consists of 27 telescopes that work together through interferometry.

interferometry, links several separate radio dishes together in an array that resolves details as if it were one enormous radio telescope. One famous example, the Very Large Array (VLA) near Socorro, New Mexico, consists of 27 individual radio dishes that can be moved along railroad tracks laid down in the shape of a Y (Figure 6.26). The light-gathering capability of the VLA's 27 dishes is simply equal to their combined area, equivalent to a single telescope 130 meters across. But the VLA's best angular resolution, achieved by spacing the individual dishes as widely as possible, is like that of a single radio telescope with a diameter of almost 40 kilometers (25 miles). Today, astronomers can achieve even higher angular resolution by linking radio telescopes around the world.

Common Misconceptions: Closer to the Stars?

Many people mistakenly believe that space telescopes are advantageous because their locations above the Earth make them closer to the stars. You can quickly realize the error of this belief by thinking about scale. On the 1-to-10-billion scale discussed in Section 1.2, the Hubble Space Telescope is so close to the surface of the millimeter-diameter Earth that you would need a microscope to resolve its altitude. Thus, the distances to the stars are effectively the same no matter whether a telescope is on the ground or in space. The real advantages of space telescopes all arise from being above the many observational problems presented by the Earth's atmosphere.

In principle, interferometry can improve angular resolution not only for radio waves but also for any other form of light. In practice, interferometry becomes increasingly difficult for light with shorter wavelengths, and astronomers are only beginning to succeed at infrared and visible interferometry. Nevertheless, one reason why *two* Keck telescopes were built on Mauna Kea is so they can be used for infrared interferometry and someday for optical interferometry. The potential value of such interferometers is enormous. In the future, infrared interferometers may be able to obtain spectra from individual planets around other stars, allowing spectroscopy that could determine the compositions of their atmospheres and help determine whether they harbor life.

6.6 Spacecraft

Most people probably don't equate spacecraft with light and telescopes, but nearly all astronomical spacecraft use light in some way, and many carry sophisticated telescopes and detectors (such as cameras and spectrographs). Thus, we conclude this chapter with a brief discussion of spacecraft.

With a few exceptions, such as the Space Shuttle and on the Apollo missions to the Moon, nearly all spacecraft are *robotic,* meaning that no humans are aboard. Their operations are partly automatic and partly controlled by people who send instructions via radio signals from Earth. Broadly speaking, most robotic spacecraft fall into four main categories:

1. **Earth-orbiters:** Simply being in space is enough to overcome the observational problems presented by the Earth's atmosphere, so most space-based astronomical observatories orbit the Earth.

2. **Flybys:** Close-up study of other planets requires sending spacecraft to them. Flybys follow unbound orbits past their destinations—they fly past a world just once and then continue on their way. For example, *Voyager 2*'s orbit allowed it to make flybys of Jupiter, Saturn, Uranus, and Neptune (Figure 6.27).

3. **Orbiters** (of other worlds): A flyby can study a planet or moon close up—but only once. For more detailed or longer-term studies, we need a spacecraft that goes into orbit around another world.

4. **Probes and landers:** The most "up close and personal" study of other worlds comes from spacecraft that send probes into their atmospheres or landers to their surfaces.

Table 6.1 lists a few of the most significant robotic missions to other worlds. The primary constraint on building more spacecraft is cost: Besides the costs of developing, building, and operating a

FIGURE 6.27 The trajectory of *Voyager 2*, which made flybys of each of the four jovian planets in our solar system.

Table 6.1 Selected Robotic Missions to Other Worlds

Name	Type	Destination	Arrival Date	Mission Highlights
Cassini	Orbiter	Saturn	2004*	Includes lander for Titan
Near Earth Asteroid Rendezvous (NEAR)	Orbiter	Eros (asteroid)	2000*	First spacecraft dedicated to in-depth study of an asteroid
Mars Polar Lander	Lander	Mars	1999*	Land near Martian South Pole
Lunar Prospector	Orbiter	Moon	1998	Detailed mapping of lunar surface; search for ice in craters near lunar poles
Mars Global Surveyor	Orbiter	Mars	1997	Detailed imaging of surface from orbit
Mars Pathfinder	Lander	Mars	1997	Carried *Sojourner*, the first robotic rover on Mars
Galileo	Orbiter	Jupiter	1995	Dropped probe into Jupiter; close-up study of moons
Magellan	Orbiter	Venus	1990	Detailed radar mapping of surface of Venus
Voyager 1	Flyby	Jupiter, Saturn	1979, 1980	Unprecedented views of Jupiter and Saturn; continuing out of solar system
Voyager 2	Flyby	Jupiter, Saturn, Uranus, Neptune	1979, 1981, 1986, 1989	Only mission to Uranus and Neptune; continuing out of solar system
Viking 1 and 2	Orbiter/lander	Mars	1976	First landers on Mars

*Scheduled arrival date.

FIGURE 6.28 An artist's conception of a lunar observatory.

spacecraft, it can cost tens to hundreds of millions of dollars just to launch it into space. Fortunately, new technologies (particularly miniaturization, which reduces launch weight) are making spacecraft cheaper, and you're likely to read about more and more new space missions as time goes on. Someday astronomers hope to build an observatory on the Moon (Figure 6.28). Because the Moon has no atmosphere, it offers all the advantages of telescopes in space, but it also offers the ease of operating telescopes on a solid surface.

> *May the warp be the white light of morning,*
> *May the weft be the red light of evening,*
> *May the fringes be the falling rain,*
> *May the border be the standing rainbow.*
> *Thus weave for us a garment of brightness.*
>
> SONG OF THE SKY LOOM (NATIVE AMERICAN)

THE BIG PICTURE

We've discussed a lot about light, telescopes, and spacecraft in this chapter. "Big picture" ideas that will help you keep your understanding in perspective include the following:

- There is far more to light than meets the eye. By dispersing light into a spectrum with a prism or a diffraction grating, we discover a wealth of information about the object from which the light has come. Most of what we know about the universe comes from information that we receive in the form of light.

- The visible light that our eyes can see is only a small portion of the complete electromagnetic spectrum. Different portions of the spectrum may contain different pieces of the story of a distant object, so it is important to study spectra at many wavelengths.

- By studying the spectra of a distant object, we can determine its composition, surface temperature, motion toward or away from us, and more.

- Telescopes work much like giant eyes, enabling us to see the universe in great detail. New technologies for making larger telescopes, along with advances in adaptive optics and interferometry, are making ground-based telescopes more powerful than ever.

- For the ultimate in observing the universe, space is the place! Telescopes in space allow us to detect light from across the entire spectrum while also avoiding any distortion caused by Earth's atmosphere. Flybys, orbiters, and landers allow us to make detailed studies of other worlds in our solar system that are not possible from Earth.

Review Questions

1. What is the difference between *energy* and *power,* and what units do we use to measure each?

2. What do we mean when we speak of a *spectrum* produced by a prism or *diffraction grating?*

3. Describe each of the four basic ways that light can interact with matter: *emission, absorption, transmission,* and *reflection.* How is *scattering* related to reflection?

4. What is a *photon?* In what way is a photon like a particle? In what way is it like a wave? Explain why we say that light is an *electromagnetic wave.*

5. Explain why light with a shorter wavelength must have a higher frequency, and vice versa.

6. Describe the *electromagnetic spectrum.* In terms of frequency, wavelength, and energy, distinguish among *radio, infrared, visible light, ultraviolet, X rays,* and *gamma rays.*

7. Explain how spectral lines can be used to determine the composition and temperature of their source.

8. Describe the two laws of *thermal radiation* and give a few examples of the use of each law. Explain how the effects of both laws can be seen in Figure 6.10. Why is "white hot" hotter than "red hot"?

9. Summarize the circumstances under which objects produce thermal, emission line, or absorption line spectra as shown in Figure 6.11.

10. Study Figure 6.12 carefully, and explain each feature in the spectrum.

11. Describe the *Doppler effect* for light. Explain why light from objects moving toward us shows a *blueshift* while light from objects moving away from us shows a *redshift.* What can we say about an object with a large Doppler shift, compared to an object with a small Doppler shift?

12. Briefly distinguish between a *refracting telescope* and a *reflecting telescope.*

13. What do we mean by a telescope's *light-collecting area* and *angular resolution?*

14. Define and differentiate between the three basic categories of astronomical observation: *imaging, spectroscopy,* and *timing.*

15. Explain how the Earth's atmosphere affects astronomical observations made from the ground.

16. Study Figure 6.23 in detail, and describe how deeply each portion of the electromagnetic spectrum penetrates the Earth's atmosphere.

17. Briefly describe the advantages of putting telescopes in space.

18. What is *interferometry,* and why is it useful?

19. Describe each of the four main categories of robotic spacecraft.

Discussion Questions

1. *The Changing Limitations of Science.* In 1835, French philosopher Auguste Comte stated that the composition of stars could never be known by science. Although spectral lines had been seen in the Sun's spectrum by that time, not until the mid-1800s did scientists recognize that spectral lines give clear information about chemical composition (primarily through the work of Foucault and Kirchhoff). Why might our present knowledge have seemed unattainable in 1835? Discuss how new discoveries can

change the apparent limitations of science. Today, other questions seem beyond the reach of science, such as the question of how life began on Earth. Do you think that such questions will ever be answerable by science? Defend your opinion.

2. *A Lunar Observatory.* Do the potential benefits of building an astronomical observatory on the Moon justify its costs at the present time? If it were up to you, would you recommend that Congress begin funding such an observatory? Defend your opinions.

Problems

1. *Spectral Summary.* Clearly explain how studying an object's spectrum can allow us to determine each of the following properties of the object.

   a. The object's surface chemical composition.

   b. The object's surface temperature.

   c. Whether the object is a thin cloud of gas or something more substantial.

   d. Whether the object has a hot upper atmosphere.

   e. The speed at which the object is moving toward or away from us.

2. *Planetary Spectrum.* Suppose you take a spectrum of light coming from a planet that looks blue to the eye. Do you expect to see any visible light in the planet's spectrum? Is the visible light emitted by the planet, reflected by the planet, or both? Which (if any) portions of the visible spectrum do you expect to find "missing" in the planet's spectrum? Explain your answers clearly.

3. *Hotter Sun.* Suppose that the surface temperature of the Sun were about 12,000 K, rather than 6,000 K.

   a. How much more thermal radiation would the Sun emit?

   b. How would the thermal radiation spectrum of the Sun be different?

   c. Do you think it would still be possible to have life on Earth? Explain.

4. *The Doppler Effect.* In hydrogen, the transition from level 2 to level 1 has a rest wavelength of 121.6 nm. Suppose you see this line at a wavelength of 120.5 nm in Star A, at 121.2 nm in Star B, at 121.9 nm in Star C, and at 122.9 nm in Star D. Which stars are coming toward us? Which are moving away? Which star is moving fastest relative to us (either toward or away from)?

5. *The Expanding Universe.* Recall from Chapter 1 that we know the universe is expanding because (1) all galaxies outside our Local Group are moving away from us and (2) more distant galaxies are moving faster. Explain how Doppler shift measurements allow us to know these two facts.

6. *Project: Twinkling Stars.* Using a star chart, identify 5 to 10 bright stars that should be visible in the early evening. On a clear night, observe each of these stars for a few minutes. Note the date and time, and for each star record the following information: approximate altitude and direction in your sky, brightness compared to other stars, color, how much the star twinkles compared to other stars. Study your record. Can you draw any conclusions about how brightness and position in your sky affect twinkling? Explain.

7. *Research: Planetary Mission.* Choose a current mission to a planet. Using the Internet, find as much information as you can about the mission. Write a two- to three-page summary of the mission, including discussion of its purpose, design, and cost.

PART III

LEARNING FROM OTHER WORLDS

$\mathcal{F}$ormation of the $\mathcal{S}$olar $\mathcal{S}$ystem

7

How old is the Earth, and how did it come to be? Is it unique? Our ancestors could do little more than guess, but today we can answer with reasonable confidence: The Earth and the rest of our solar system formed from a great cloud of gas and dust about 4.6 billion years ago. Whereas ancient astronomers believed the Earth to be fundamentally different from objects in the heavens, we now see the Earth as just one of many worlds.

As we examine the evidence that helped shape the modern scientific view of the formation of our solar system, we'll highlight one common technique of scientific research: sifting through seemingly unrelated facts to find the most important trends, patterns, and characteristics—and then attempting to explain them. The explanation must be based on known physical laws, and the success or failure of a theory depends on how well it matches the facts. As we discover more and more planets around other stars, we hope to test our theories further and to modify them as necessary.

7.1 Comparative Planetology

Galileo's telescopic observations began a new era in astronomy in which the Sun, the Moon, and the planets could be studied for the first time as *worlds,* rather than merely as lights in the sky. During the early part of this new era, astronomers studied each world independently, but in the twentieth century we discovered that the similarities among worlds are often more profound than the differences.

Today, we know that the Sun, planets, moons, and other bodies in our solar system all formed at about the same time from the same cloud of interstellar gas and dust (*interstellar* means "between the stars") and in accord with the same physical laws. Thus, the differences between these worlds must be attributable to physical processes that we can study and understand by comparing the worlds to one another. This approach is called **comparative planetology;** the term *planetology* is used broadly to include moons, asteroids, and comets as well as planets. The idea is that we can learn more about the Earth, or any other world, by studying it in the context of other objects in our solar system—rather like learning about a person by studying his or her family, friends, and culture.

Before beginning the comparative study of worlds, we must be sure that the comparison will provide valid lessons. The basic premise of comparative planetology—that the similarities and differences among worlds in our solar system can be traced to common physical processes—assumes that the planets share a common origin. Thus, we begin our study of comparative planetology by seeking clues to our origins from the current state of our solar system; then we examine the modern theory of solar system formation that explains much of what we see.

7.2 The Origin of the Solar System: Four Challenges

Learning about the origin of the solar system is easiest if we begin by looking at our solar system with a "big picture" view, rather than focusing on individual worlds. Imagine that we have the perspective of an alien spacecraft making its first scientific survey of our solar system. What would we see?

First, as we learned from looking at a scale model of our solar system in Chapter 1, we'd find that our solar system is mostly empty—interplanetary space is a near-vacuum containing only very low density gas and tiny, solid grains of dust. But we'd soon focus on the widely spaced planets, mapping their orbits, measuring their sizes, compositions, and densities, and taking inventory of their moons and ring systems. Figure 7.1 shows the resulting schematic map, and Table 7.1 summarizes the planetary properties. Asteroids and comets are also listed in the table because these objects orbit the Sun, though with orbital properties somewhat different from those of the planets.

Studying the data in the map and the table reveals several general features of our solar system. Let's investigate by grouping our observations into four major challenges that must be met by any theory claiming to explain how our solar system developed.

FIGURE 7.1 Side view of the solar system. Arrows indicate the orientation of the rotation axes of the planets and their orbital motion. (Planetary tilts in this diagram are aligned in the same plane for easier comparison. Planets not to scale.) Seen from above, all orbits except those of Mercury and Pluto are nearly circular. Most moons orbit in the same direction as the planets orbit and rotate—counterclockwise when seen from above Earth's North Pole.

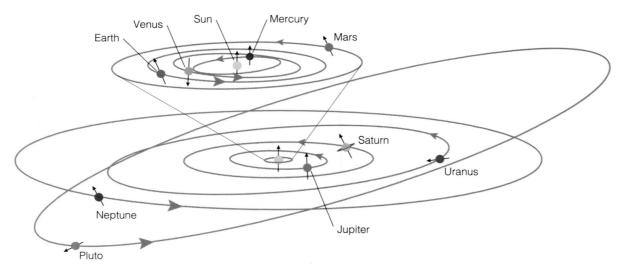

Table 7.1 Planetary Facts*

Photo	Planet	Average Distance from Sun (AU)	Temperature[†]	Relative Size	Average Equatorial Radius (km)	Average Density (g/cm³)	Composition	Moons	Rings?
	Mercury	0.387	700 K	·	2,440	5.43	Rocks, metals	0	No
	Venus	0.723	740 K	•	6,051	5.24	Rocks, metals	0	No
	Earth	1.00	290 K	•	6,378	5.52	Rocks, metals	1	No
	Mars	1.52	240 K	•	3,397	3.93	Rocks, metals	2 (tiny)	No
	Most asteroids	2–3	170 K	·	≤500	1.5–3	Rocks, metals	?	No
	Jupiter	5.20	125 K	⬤	71,492	1.33	H, He, hydrogen compounds[‡]	16	Yes
	Saturn	9.53	95 K	⬤	60,268	0.70	H, He, hydrogen compounds[‡]	18	Yes
	Uranus	19.2	60 K	●	25,559	1.32	H, He, hydrogen compounds[‡]	21	Yes
	Neptune	30.1	60 K	●	24,764	1.64	H, He, hydrogen compounds[‡]	8	Yes
	Pluto	39.5	40 K	·	1,160	2.0	Ices, rock	1	No
	Most comets	0–50,000	a few K[§]	·	a few km?	<1?	Ices, dust	?	No

*Appendix C gives a more complete list of planetary properties.　[†]Surface temperatures for all objects except Jupiter, Saturn, Uranus, and Neptune, for which cloud-top temperatures are listed.　[‡]Includes water (H_2O), methane (CH_4), and ammonia (NH_3).
[§]Comets passing close to the Sun warm considerably, especially their outer layers.

Challenge 1: Patterns of Motion

If we look closely at the map of solar system orbits in Figure 7.1, we'll see several striking patterns:

- All planets orbit the Sun in the same direction—counterclockwise when seen from high above the Earth's North Pole.

- All planetary orbits lie nearly in the same plane.

- Almost all planets travel on nearly circular orbits, and the spacing between planetary orbits increases with distance from the Sun according to a fairly regular trend. The most notable exception to this trend is an extra-wide gap between Mars and Jupiter that is populated with asteroids.

- Most planets rotate in the same direction in which they orbit—counterclockwise when seen from above the Earth's North Pole—with fairly small axis tilts (i.e., less than about 25°).

- Almost all moons orbit their planet in the same direction as the planet's rotation and near the planet's equatorial plane.

- The Sun rotates in the same direction in which the planets orbit.

These patterns show that motion in the solar system is generally quite organized. If each planet had come into existence independently, we might expect planetary motions to be much more random. (Note that the motions of asteroids are slightly more random, and those of comets even more so, as we will discuss shortly.)

The first challenge for any theory of solar system formation is to explain why motions in the solar system are generally so orderly.

Challenge 2: Categorizing Planets

Before reading any further, study the properties of the planets in Table 7.1. Can you categorize the planets into groups? How many categories do you need? What characteristics do planets within a category share?

Astronomers classify most of the planets in two distinct groups: the rocky **terrestrial planets** and the gas-rich **jovian planets.**

The word *terrestrial* means "Earth-like" (*terra* is the Greek word for Earth), and the terrestrial planets are the four planets of the inner solar system—Mercury, Venus, Earth, and Mars. These are the worlds most like our own. They are relatively small, close to the Sun, and close together. They

have solid, rocky surfaces and an abundance of metals deep in their interiors. They have few moons, if any, and none have rings. Note that many scientists count our Moon as a fifth terrestrial world because it shares these general characteristics, although it's not technically a planet.

The word *jovian* means "Jupiter-like," and the jovian planets are the four outer solar system planets—Jupiter, Saturn, Uranus, and Neptune. The jovian planets are much larger than the terrestrial planets, farther from the Sun, and widely separated from each other. These huge planets have little in common with Earth. They are made mostly of hydrogen, helium, and **hydrogen compounds** such as water, ammonia, and methane. They contain relatively small amounts of rocky material only deep in their cores. Jovian planets do not have solid surfaces; if you plunged into the atmosphere of a jovian planet, you would sink deeper and deeper until you were crushed by overwhelming pressure. Each jovian planet has rings and an extensive system of moons. These solid satellites are made mostly of low-density ices and rocks. Table 7.2 contrasts the general traits of the terrestrial and jovian planets.

The second challenge for any theory of solar system formation is to explain why the inner and outer solar system planets divide so neatly into two classes.

TIME OUT TO THINK *Are the distinctions in Table 7.2 clear-cut? Examine Table 7.1 to compare, for example, the radii of the largest terrestrial planet and the smallest jovian planet.*

Note that Pluto is left out in the cold, both literally and figuratively. On one hand, it is small and solid like the terrestrial planets. On the other hand, Pluto is far from the Sun, cold, and made of low-density ices. Pluto also has an unusual orbit, far more eccentric than the orbit of any other planet and substantially inclined to the plane in which the other planets orbit the Sun. For a long time, scientists considered Pluto to be a lone misfit. However, some scientists now classify Pluto with other icy bodies in the outer solar system: comets.

Table 7.2 Comparison of Terrestrial and Jovian Planets

Terrestrial Planets	Jovian Planets
Smaller size and mass	Larger size and mass
Higher density (rocks, metals)	Lower density (light gases, hydrogen compounds)
Solid surface	No solid surface
Closer to the Sun (and closer together)	Farther from the Sun (and farther apart)
Warmer	Cooler
Few (if any) moons and no rings	Rings and many moons

Challenge 3: Asteroids and Comets

No formation theory is complete without an explanation of the most numerous objects in the solar system: asteroids and comets. **Asteroids** are small, rocky bodies that orbit the Sun primarily in the **asteroid belt** between the orbits of Mars and Jupiter. Asteroids orbit the Sun in the same direction as the planets. Their orbits generally lie close to the plane of planetary orbits, although they are usually a bit tilted. Some asteroids have elliptical orbits that are quite eccentric compared to the nearly circular orbits of the planets. Almost 9,000 asteroids have been identified and cataloged, but these are probably only the largest among a much greater number of small asteroids. The largest asteroids are a few hundred kilometers in radius—much less than half the Moon's radius.

Comets are small, icy bodies that spend most of their lives well beyond the orbit of Pluto; we generally recognize them only on the rare occasions when one dives into the inner solar system and grows a spectacular tail. Based on the orbits of the relatively few comets that reach the inner solar system, astronomers have determined that many billions of comets must be orbiting the Sun, primarily in two broad regions. The first region, called the **Kuiper belt** (pronounced koy-per), begins somewhere in the vicinity of the orbit of Neptune (about 30 AU from the Sun) and extends to perhaps 100 AU from the Sun. Comets in the Kuiper belt have orbits that lie fairly close to the plane of planetary orbits, and they travel around the Sun in the same direction as the planets. The second region populated by comets, called the **Oort cloud,** is a huge, spherical region centered on the Sun and extending perhaps halfway to the nearest stars. The orbits of comets in the Oort cloud are completely random, with no pattern to their inclinations, orbital directions, or eccentricities.

The third challenge for any theory of solar system formation is to explain the existence and general properties of the large numbers of asteroids and comets.

Challenge 4: Exceptions to the Rules

The first three challenges involve explaining general patterns in our solar system. But some objects don't fit these patterns. For example, Mercury and Pluto have larger orbital eccentricities and inclinations than the other planets. The rotational axes of Uranus and Pluto are substantially tilted, and Venus rotates backward—clockwise, rather than counterclockwise, as viewed from high above Earth's North Pole. Unlike the other terrestrial planets, Earth has a large moon. Pluto has a moon almost as big as itself. While most moons of the jovian planets orbit with the same orientation as their planet's rotation, a few orbit in the opposite direction.

Table 7.3 Four Major Characteristics of the Solar System

> **Large bodies in the solar system have orderly motions.** All planets and most satellites have nearly circular orbits going in the same direction in nearly the same plane. The Sun and most of the planets rotate in this same direction as well.
>
> **Planets fall into two main categories:** small, rocky terrestrial planets near the Sun and large, hydrogen-rich jovian planets farther out. The jovian planets have many moons and rings made of rock and ice.
>
> **Swarms of asteroids and comets populate the solar system.** Asteroids are concentrated in the asteroid belt, and comets populate the regions known as the Kuiper belt and the Oort cloud.
>
> **Several notable exceptions to these general trends stand out,** such as planets with unusual axis tilts or surprisingly large moons, and moons with unusual orbits.

Allowing for these and other exceptions to the general patterns is the fourth challenge for any theory of solar system formation.

Summarizing the Challenges

Table 7.3 summarizes the major characteristics of our solar system that we have discussed as four challenges to be explained by our solar system formation theory. In the remainder of this chapter, we will examine the leading theory of solar system formation and put it to the test to see whether it can meet the four challenges. In this sense, we will be following the idealized scientific method by starting with the facts, putting forward a theory, and putting the theory to the test. In doing so, we have the benefit of hindsight: The historical path to our modern theory of the solar system was actually quite complex, with many competing theories being proposed and eventually discarded because they were unable to meet the challenges we have laid out.

7.3 The Nebular Theory of Solar System Formation

Over the past three centuries, several models have been put forth in attempts to meet the four challenges for a solar system formation theory. In the past few decades, a tremendous amount of evidence has accumulated in support of one model. This model, which we will call the **nebular theory,** holds that our solar system formed from a giant, swirling **interstellar cloud** of gas and dust (Figure 7.2). Such a cloud is also called a **nebula** (the Latin word for "cloud"). More generally, we now have evidence that all **star systems**—systems like our solar system that contain at least one star and, in some cases,

FIGURE 7.2 This photograph shows the central region of the Orion Nebula, an interstellar cloud in which star systems—possibly including planets—are forming. The photo is actually a composite of more than a dozen separate images taken with the Hubble Space Telescope. (See Figure 16.12 for a complete view of the Orion Nebula.)

planets or other orbiting material—form similarly from interstellar clouds.

Our galaxy, like the entire universe, originally contained only hydrogen and helium (and trace amounts of lithium). By the time our solar system formed, 4.6 billion years ago, about 2% of the original hydrogen and helium in the galaxy had been converted into heavier elements by fusion inside stars. This material was mixed into the interstellar clouds by **stellar winds** and stellar explosions. Although 2% sounds like a small amount, it was more than enough to form the rocky terrestrial planets—and us. Nevertheless, hydrogen and helium remain the most abundant elements by far and make up the bulk of the Sun and the jovian planets.

TIME OUT TO THINK *Could a solar system like ours have formed with the first generation of stars after the Big Bang? Explain.*

Collapse of the Solar Nebula

An individual star system forms from just a small part of a giant interstellar cloud. We refer to the collapsed piece of cloud that formed our own solar system as the **solar nebula.** The solar nebula collapsed under its own gravity; the collapse may have been triggered by a cataclysmic event such as the impact of a shock wave from a nearby exploding star. Before the collapse, the low-density gas of the interstellar cloud may have been

a few light-years in diameter. As it collapsed to a diameter of about 200 AU—roughly twice the present-day diameter of Pluto's orbit—three important processes gave form to our solar system (Figure 7.3).

First, the temperature of the solar nebula increased as it collapsed. Such heating represents energy conservation in action [Section 4.2]. As the cloud shrank, its gravitational potential energy was converted to the kinetic energy of individual gas particles falling inward. These particles crashed into one another, converting the kinetic energy of their inward fall into the random motions of thermal energy. Some of this energy was radiated away as thermal radiation. The solar nebula became hottest near its center, where much of the mass collected to form the **protosun** (the prefix *proto* comes from a Greek word meaning "earliest form of"). The protosun eventually became so hot that nuclear fusion ignited in its core—at which point our Sun became a full-fledged star.

Second, like an ice skater pulling in her arms as she spins, the solar nebula rotated faster and faster as it shrank in radius. This increase in rotation rate represents the conservation of angular momentum in action [Section 5.2]. The rotation helped ensure that not all of the material in the solar nebula collapsed onto the protosun: The greater the angular momentum of a rotating cloud, the more spread out it will be.

Third, the solar nebula flattened into a disk—the **protoplanetary disk** from which the planets eventually formed. This flattening is a natural consequence of collisions between particles, which explains why flat disks are so common in the universe (e.g., the disks of spiral galaxies like the Milky Way, ring systems around planets, and *accretion disks* around neutron stars or black holes [Section 15.3]). A cloud may start with any size or shape, and different clumps of gas within the cloud may be moving in random directions at random speeds. When the cloud collapses, these different clumps collide and merge, giving the new clumps the average of their differing velocities. The result is that the random motions of the clumps in the original cloud become more orderly as the cloud collapses, changing the cloud's original lumpy shape into a rotating, flattened disk. Similarly, collisions between clumps of material in highly elliptical orbits reduce their ellipticities, making their orbits more circular. You can see a similar effect if you shake some pepper into a bowl of water and quickly stir it in a random way. Because the water molecules are always colliding with one another, the motion of the pepper grains will settle down into a slow rotation representing the average of the original, random velocities.

These three processes—heating, spinning, and flattening—explain the tidy layout of our solar system. The flattening of the protoplanetary disk explains why all planets orbit in nearly the same plane. The spinning explains why all planets orbit in the same direction and also plays a role in making most planets

a The original cloud is large and diffuse, and its rotation is almost imperceptibly slow.

b The cloud heats up and spins faster as it contracts.

c The result is a spinning, flattened disk, with mass concentrated near the center.

FIGURE 7.3 This sequence of paintings shows the collapse of an interstellar cloud. In our solar nebula, the hot, dense central bulge became the Sun, and the planets formed in the disk.

Central star blocked out

200 AU

a

a (Left) A dust disk around the star Beta Pictoris taken by the Hubble Space Telescope at optical wavelengths. (Right) Twin dust disks around binary stars, taken by the VLA at radio wavelengths.

500 AU

b

b Protoplanetary disks around stars in the constellation Auriga (left) and in the Orion Nebula (right).

FIGURE 7.4 Evidence for disks around other stars.

rotate in this same direction. The fact that collisions in the protoplanetary disk tend to make highly elliptical orbits more circular explains why most planets in our solar system have nearly circular orbits today.

Evidence Concerning Nebular Collapse

The theory that our solar system formed from interstellar gas may sound reasonable, but we cannot ac- cept it without hard evidence. Fortunately, we have strong observational evidence that the same processes are occurring elsewhere in our galaxy. A collapsing nebula emits thermal radiation [Section 6.3], primarily in the infrared. We've detected such infrared radiation from many other nebulae where star systems appear to be forming. We've even seen several structures around other stars that look similar to protoplanetary disks (Figure 7.4). Other support for the nebular theory comes from sophisticated computer models that simulate the formation processes it describes.

All in all, the nebular theory meets the challenge of explaining most of the orderly motions of the planets and satellites. Now let's turn to our second challenge: explaining the strikingly different characteristics of the terrestrial and jovian planets.

7.4 Building the Planets

The churning and mixing of the gas in the solar nebula ensured that its composition was about the same throughout: roughly 98% hydrogen and helium, and 2% heavier elements such as carbon, nitrogen, oxygen, silicon, and iron. How did the planets and other bodies in our solar system end up with such a wide variety of compositions when they came from such uniform material? To answer this question, we must investigate how the material in the protoplanetary disk came together to form the planets.

Condensation: Sowing the Seeds of Planets

In the center of the collapsing solar nebula, gravity drew much of the material together into the proto-sun. In the rest of the spinning protoplanetary disk, however, the gaseous material was so spread out that, at first, gravity was not strong enough to pull it together to form planets. The formation of planets therefore required "seeds"—solid chunks of matter that gravity could eventually draw together. Understanding these seeds is the key to explaining the differing compositions of the planets.

The vast majority of the material in the solar nebula was gaseous: High temperatures kept virtually all the ingredients of the solar nebula vaporized near the protosun. Farther out the nebula was still primarily gaseous because the hydrogen and helium that made up 98% of the solar nebula remain gaseous even at extremely low temperatures. But the 2% of

material consisting of heavier elements could form solid seeds where temperatures were low enough. The formation of solid or liquid particles from a cloud of gas is called **condensation.** Pressures in the solar nebula were so low that liquid droplets rarely formed, but solid particles could condense in the same way that snowflakes condense from water vapor in our atmosphere. We refer to such solid particles as **condensates.** The different kinds of planets and satellites formed out of the different kinds of condensates present at different locations in the solar nebula.

The ingredients of the solar nebula fell into four categories based on their condensation temperatures.

- **Metals** include iron, nickel, aluminum, and other materials that are familiar on Earth but less common on the surface than ordinary rock. Some of these metals can remain solid at temperatures as high as 1,600 K. These materials made up less than 0.2% of the solar nebula's mass.

- **Rocks** are materials common on the surface of the Earth, primarily silicon-based minerals. Rocks are solid at temperatures and pressures found on Earth but typically melt or vaporize at temperatures of 500–1,300 K depending on their type. Rocky materials made up about 0.4% of the nebula by mass.

- **Hydrogen compounds** are molecules such as methane (CH_4), ammonia (NH_3), and water (H_2O) that solidify into **ices** below about 150 K. These compounds were significantly more abundant than rocks and metals, making up 1.4% of the nebula's mass.

- **Light gases** (hydrogen and helium) never condense under solar nebula conditions. These gases made up the remaining 98% of the nebula's mass.

The great temperature differences between the hot inner regions and the cool outer regions of the nebula determined what kinds of condensates were available to form planets (Figure 7.5). Let's examine

FIGURE 7.5 Temperature differences in the solar nebula led to different kinds of condensed materials, sowing the seeds for two different kinds of planets.

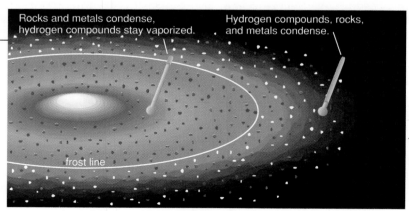

Rocks and metals condense, hydrogen compounds stay vaporized.

Hydrogen compounds, rocks, and metals condense.

frost line

these regions, beginning near the protosun and moving outward to cooler regions. Very near the protosun, where the nebula temperature was above 1,600 K, there were no condensates—everything remained gaseous. Farther out, where the temperature was slightly lower, metal flakes appeared. Near the distance of Mercury's orbit, flakes of rock joined the mix. Moving outward past the orbits of Venus and Earth, more varieties of rock minerals condensed. Near the location of the future asteroid belt, temperatures were low enough to allow minerals containing small amounts of water to condense as well. Dark, carbon-rich materials also condensed here.

Only beyond the **frost line,** which lay between the present-day orbits of Mars and Jupiter, were temperatures low enough (150 K ≈ −123°C) for hydrogen compounds to condense into ices. (Water ice did not form at the familiar 0°C because the pressures in the nebula were 10,000 times lower than on Earth.) Thus, the outer solar system contained condensates of all kinds: rocks, metals, and ices. However, ice flakes were nearly three times more abundant than flakes of rock and metal because of the greater abundance of hydrogen compounds in the solar nebula.

Accretion: Assembling the Planetesimals

The first solid flakes that condensed from the solar nebula were microscopic in size. They orbited the protosun in the same orderly, circular paths as the gas from which they condensed. Individual flakes therefore moved with nearly the same speed as their neighboring flakes, allowing them to collide very gently. At this point, the flakes were far too small to attract one another by gravity, but they were able to stick together through electrostatic forces—the same "static electricity" that makes hair stick to a comb. Thus, the flakes grew slowly into larger particles. As the particles grew in mass, gravity began to aid the process of their sticking together, accelerating their growth. This process of growing by colliding and sticking is called **accretion.** The growing objects formed by accretion are called **planetesimals,** which essentially means "pieces of planets." Small planetesimals probably came in a variety of shapes, still reflected in many small asteroids today. Larger planetesimals (i.e., those several hundred kilometers across) became spherical due to the force of gravity pulling everything toward the center.

The sizes and compositions of the planetesimals depended on the temperature of the surrounding solar nebula. In the inner solar system, where only rocky and metallic flakes condensed, planetesimals were made of rock and metal. That is why the terrestrial planets ended up being composed of rocks and

metals. Moreover, because rocky and metallic elements made up only about 0.6% of the material in the solar nebula, the planetesimals in the inner solar system could not grow very large, which explains why the terrestrial planets are relatively small.

Beyond the frost line between the orbits of Mars and Jupiter, where temperatures were cold enough for ices to condense, planetesimals were built from ice flakes in addition to flakes of rock and metal. Because ice flakes were much more abundant, these planetesimals were made mostly of ices and could grow to much larger sizes than could planetesimals in the inner solar system. The largest icy planetesimals of the outer solar system became the cores of the jovian planets.

The densities of the various objects in the solar system support this theory. Objects near the Sun are composed of higher-density materials like metals and rocks. In the outer solar system, solid objects such as moons and comets are made mostly of low-density ices with a smaller proportion of rocks and metals.

Nebular Capture: Making the Jovian Planets

Accretion proceeded rapidly in the outer solar system, since the presence of ices meant much more material in solid form. Some of the icy planetesimals of the outer solar system quickly grew to sizes many times larger than the Earth. At these large sizes, their gravity was strong enough to capture the far more abundant hydrogen and helium gas from the surrounding nebula. As they accumulated substantial amounts of gas, the gravity of these growing planets grew larger still—allowing them to capture even more gas. The process by which icy planetesimals act as seeds for capturing far larger amounts of hydrogen and helium gas, called **nebular capture,** led directly to

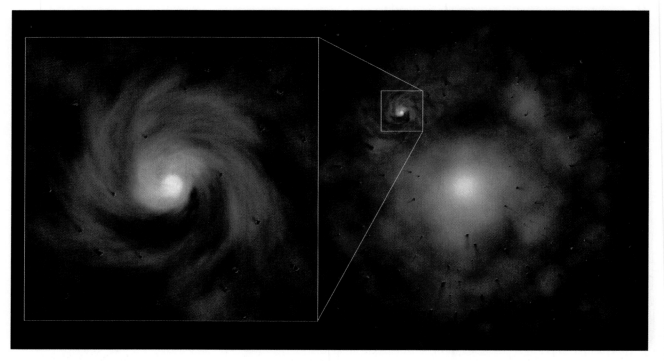

FIGURE 7.6 Large icy planetesimals in the cold outer regions of the solar nebula captured significant amounts of hydrogen and helium gas. This process of nebular capture created jovian nebulae, resembling the solar nebula in miniature, in which the jovian planets and satellites formed. This painting shows a jovian nebula (enlarged at left) located within the entire solar nebula.

the formation of the jovian worlds (Figure 7.6). It explains their huge sizes and the large abundance of hydrogen and helium reflected in their low average densities.

Nebular capture also explains the formation of the diverse satellite systems of the jovian planets. As the early jovian planets captured large amounts of gas from the solar nebula, the same processes that formed the protoplanetary disk—heating, spinning, and flattening—formed similar but smaller disks of material around these planets. Condensation (of metals, rocks, and a lot of ice) and accretion took place within these **jovian nebulae,** essentially creating a miniature solar system around each jovian planet. The spinning disks of the jovian nebulae explain why most of the jovian planet satellites orbit in nearly circular paths lying close to the equatorial plane of their parent planet and also why they orbit in the same direction in which their planet rotates. The composition of the jovian nebulae explains why jovian planets possess systems of large, icy satellites. Their densities of 1–3 g/cm³ reflect their mixtures of icy and rocky condensates. Temperature differences within the jovian nebulae may have led to density differences between satellites analogous to the density differences between planets.

Our theory of condensation, accretion, and nebular capture meets the second challenge for our solar system formation theory: It explains the general differences between terrestrial and jovian planets. (Our theory can also explain the presence of planetary rings, but we postpone this interesting discussion until Chapter 9.) However, our theory does not yet explain why the spacing between the planets increases with distance from the Sun, although we can come up with some reasonable hypotheses. For example, the rapid accretion in the outer solar system may have allowed a few protoplanets to gobble up all their neighbors, leaving the jovian planets more widely spaced.

The Solar Wind: Clearing Away the Nebula

What happened to the remaining gas of the solar nebula? Apparently, this gas was blown into interstellar space by the **solar wind,** a flow of low-density plasma [Section 4.3] ejected by the Sun in all directions. Although the solar wind is fairly weak today, we have evidence that it was much stronger when the Sun was young—strong enough to have swept huge quantities of gas out of our solar system. The clearing of the gas interrupted the cooling process in the nebula. Had the cooling continued longer, ices might have condensed in the inner solar system. Instead, the solar wind swept the still-vaporized hydrogen compounds away from the inner solar system, along with the remaining hydrogen and helium throughout the solar system. When the gas cleared, the compositions of objects in the early solar system were essentially set.

7.5 Leftover Planetesimals

We are now ready to turn to our third challenge: explaining the existence of asteroids and comets. The strong wind from the young Sun cleared excess gas from the solar nebula, but many planetesimals remained scattered between the newly formed planets. These "leftovers" became comets and asteroids. Like the planetesimals that formed the planets, they formed from condensation and accretion in the solar nebula. Thus, their compositions followed the pattern determined by condensation: planetesimals of rock and metal in the inner solar system, and icy planetesimals in the outer solar system.

Asteroids and comets originally must have had nearly circular orbits in the same plane as the orbits of the planets. But the strong gravity of the jovian planets tugged and nudged the orbits of planetesimals even at great distances. The result was that the remaining planetesimals ended up with more highly elliptical orbits than the planets, and sometimes with large tilts relative to the plane of planetary orbits. Most of these asteroids and comets eventually either crashed into one of the planets or were flung out of the solar system, but many others still survive today.

Asteroids are the rocky leftover planetesimals of the inner solar system. Some asteroids are scattered throughout the inner solar system, but most are concentrated in the "extra-wide" gap between Mars and Jupiter that contains the *asteroid belt*. This region probably once contained enough rocky planetesimals to form another terrestrial planet. However, the gravity of Jupiter (the largest and closest jovian planet) tended to nudge the orbits of these planetesimals, sending most of them on collision courses with the planets or with one another. The present-day asteroid belt contains the remaining planetesimals from this "frustrated planet formation." Although thousands of asteroids remain in the asteroid belt, their *combined mass is less than 1/1,000 of Earth's mass.* Jupiter's gravity continues to nudge these asteroids, changing their orbits and sometimes leading to violent, shattering collisions. Debris from these collisions often crashes to Earth in the form of meteorites.

Comets are the icy leftover planetesimals of the outer solar system. Depending on where they formed, most comets ended up in one of two main groups.

The icy planetesimals that cruised the space between Jupiter and Neptune couldn't grow to more than a few kilometers in size before suffering either a collision or a gravitational encounter with one of the jovian planets. Those that escaped being swallowed up by the jovian planets tended to be flung off at high speeds in random directions. Some may have been cast away at such high speed that they completely escaped the solar system and now drift through interstellar space. But billions of these small, icy planetesimals ended up in orbits with very large average distances from the Sun. These are the comets of the *Oort cloud.*

Beyond the orbit of Neptune, the icy planetesimals were much less likely to be destroyed by collisions or cast off by gravitational encounters. Instead, they remained in orbits going in the same direction as planetary orbits and concentrated near the plane of planetary orbits (but with somewhat more randomness than the orbits of the planets). They were also able to continue their accretion, and many may have grown to hundreds or even thousands of kilometers in diameter. These are the comets of the *Kuiper belt.* Pluto is probably the largest member of this class.

The nebular theory has thus met our third challenge: explaining the existence of asteroids and comets. Moreover, it has suggested an explanation for the seemingly anomalous planet Pluto. Now we turn to our fourth and final challenge: explaining the exceptions to the general trends in our solar system.

The Early Bombardment: A Rain of Rock and Ice

The collision of a leftover planetesimal with a planet is called an **impact,** and the responsible planetesimal is called the **impactor.** On planets with solid surfaces, impacts leave the scars we call **impact craters.** Impacts were extremely common in the young solar system; in fact, impacts were part of the accretion process in the late stages of planetary formation. The vast majority of impacts occurred in the first few hundred million years of our solar system's history.

The heavy bombardment of planetary surfaces in the early solar system would have resembled a rain of rock and ice from space (Figure 7.7). The surface of the Earth was once scarred like the Moon's surface, but most impact craters on Earth were erased long ago by erosion and other geological processes. Only the outlines of a few large craters are recognizable on Earth. But vast numbers of craters remain on worlds that experience less erosion or other geological activity, such as the Moon and Mercury. Indeed, one way

FIGURE 7.7 Around 4 billion years ago, Earth, its Moon, and the other planets were heavily bombarded by leftover planetesimals. This painting shows the young Earth and Moon glowing with the heat of accretion, and with an impact in progress on the Earth.

of estimating the age of a planetary surface (the time since the surface last changed in a substantial way) is to count the number of craters: If there are many craters, the surface must still look much as it did at the time of the early bombardment, about 4 billion years ago. Although most impacts occurred long ago, asteroids and comets still occasionally crash into planets.

The early rain of rock and ice did more than just scar planetary surfaces. It also brought the materials from which atmospheres, oceans, and polar caps eventually formed. Remember that the terrestrial planets were built from planetesimals of metal and rock. But Earth's oceans, the polar caps of Earth and Mars, and the atmospheres of Venus, Earth, and Mars are all made from hydrogen compounds that remained gaseous in the inner solar nebula. These materials must have arrived on the terrestrial planets after their initial formation, most likely brought by impacts of planetesimals formed farther out in the solar system. We don't yet know whether the impactors came from the asteroid belt, where rocky planetesimals contained small amounts of water and other hydrogen compounds that had condensed as ices, or whether they were comets containing huge amounts of ice. Either way, the water we drink and the air we breathe probably once were part of planetesimals floating beyond the orbit of Mars.

Captured Moons

We can easily explain the orbits of most jovian planet satellites by their formation in a jovian nebula that swirled around the forming planet. But some moons have unusual orbits—orbits in the "wrong" direction (opposite the rotation of their planet) or with large inclinations to the planet's equator. These unusual moons are probably leftover planetesimals that were *captured* into orbit around a planet. Because of the random nature of the capture process, the captured moons would not necessarily orbit in the same direction as their planet or in its equatorial plane.

The two small moons of Mars—Phobos and Deimos—probably were asteroids captured by this process (Figure 7.8). They resemble asteroids seen in the asteroid belt and are much darker and lower in density than Mars. Jupiter probably also captured several of its moons. Other jovian planets may have captured moons, including one that is particularly large—Triton, the largest moon of Neptune. Triton is considerably larger than Pluto and orbits Neptune in a direction opposite to Neptune's rotation. It may be a captured object from the Kuiper belt [Section 9.5]. The one unusual moon that cannot be explained by this process is our own: Our Moon is much too large to have been captured by Earth.

a Phobos

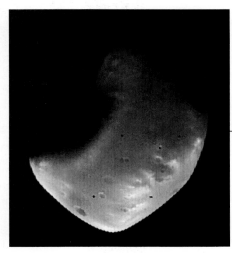

b Deimos

FIGURE 7.8 The two moons of Mars, shown here in photos taken by the Viking spacecraft, are probably captured asteroids. Phobos is only about 13 km across and Deimos is only about 8 km across—making each of these two moons small enough to fit within the boundaries of a typical large city.

Giant Impacts and the Formation of Our Moon

The largest planetesimals remaining as the planets formed may have been huge—some may have been the size of Mars. When one of these planet-size planetesimals collided with a planet, the spectacle would have been awesome. Such a **giant impact** could have significantly altered the planet's fate.

What if an object the size of Mars had collided with Earth? Computer simulations address this interesting question. The Earth would have shattered from the impact, and material from the outer layers would have ended up in orbit around the Earth (Figure 7.9). There this material could have reaccreted to form a large satellite. Depending on exactly where and how fast the giant impactor struck the Earth, the blow might also have tilted the Earth's axis, changed its rotation rate, or completely torn it apart. Today, such a giant impact is the leading hypothesis for explaining the origin of our Moon. The Moon's composition is similar to that of the Earth's outer layers, exactly what we would predict for the kind of collision shown in Figure 7.9. Moreover, we can rule out the idea of the Moon forming simultaneously with Earth. In that case, Earth and the Moon would have formed from the same material and therefore should have the same density, but the Moon's density is considerably lower than the Earth's.

In fact, many of the unusual properties of specific planets—properties that defy the general trends expected by the nebular theory—may be the results of giant impacts. Mercury may have lost much of its outer, rocky layer in a giant impact, leaving it with a huge metallic core. A giant impact might even have contributed to the slow, backward rotation of Venus, which may have had a "normal" rotation until the arrival of the giant impactor. Giant impacts probably were also responsible for the axis tilts of many planets (including Earth) and for tipping Uranus on its

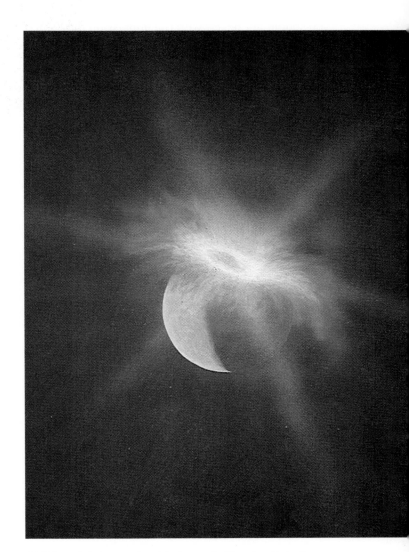

FIGURE 7.9 Artist's conception of the impact of a Mars-size object with Earth, as may have occurred soon after Earth's formation. The ejected material comes mostly from the outer rocky layers and accretes to form the Moon, which is poor in metal.

side. Pluto's moon Charon may have formed in a giant-impact process similar to the one that formed our Moon.

Unfortunately, we can do little to test whether a particular giant impact really occurred billions of years ago. But no other idea so effectively explains the formation of our Moon and other "oddities" that we've discussed. Moreover, giant impacts certainly should have occurred, given the number of large leftover planetesimals predicted by the nebular theory. Random giant impacts are the most promising explanation for the many exceptional circumstances noted in the fourth challenge.

Summary: Meeting the Challenges

The nebular theory explains the great majority of important facts contained in our four challenges. But you should not be left with the impression that competing theories were never put forth or that solar system formation is now a "solved problem." Theories have evolved hand-in-hand with the discovery of the nature of our solar system, and what we have presented here is the culmination of that great endeavor to date. Planetary scientists are still struggling with more quantitative aspects, seeking reasons for the exact sizes, locations, and compositions of the planets. Perhaps as we discover other planetary systems and examine their properties, we will be able to improve our understanding of solar system formation.

Assuming that the nebular theory is correct, it is interesting to ask whether our solar nebula was "destined" to form the solar system we see today. The first stages of planet formation were orderly and inevitable according to the nebular theory. Nebular collapse, condensation, and the first stages of accretion were relatively gradual processes that probably would happen all over again if we turned back the clock. But the final stages of accretion, and giant impacts in particular, are inherently random in nature and probably would not happen again in just the same way. A larger or smaller planet might form at Earth's location and might suffer from a larger giant impact or from none at all. We don't yet know whether these differences would fundamentally alter the solar system or simply change a few "minor" details—such as the possibility of life on Earth.

As we conclude the discussion of *how* our solar system formed, it's natural to ask *when*. We have a surprisingly good answer: The solar system began forming about 4.6 billion years ago. The answer comes from a meticulous examination of meteorites that are leftovers from that period. All rocks, including meteorites, contain minute fractions of various radioactive elements that decay or break apart and become new elements. The rates of decay for the various elements can be measured to very high accu-

racy in the laboratory, so scientists can determine how long those elements have been decaying within the rock. Virtually all meteorites, and all the various elements within the meteorites, give the same answer of about 4.6 billion years as the age of our solar system.

7.6 Planetary Systems Beyond Our Solar System

How common are planetary systems? What are the odds of another Earth-like planet? Are habitable planets as common in the real universe as they are in science fiction? We cannot definitively answer these questions—yet. But our observational and theoretical tools are helping us guess at possible answers.

Our theory of solar system formation suggests that other forming stars should also be surrounded by spinning, flattened disks in which planets might form, and numerous protoplanetary disks have been observed around stars. However, about half the "stars" in our sky are actually binary or multiple star systems. Our expectations for planets in multiple star systems have recently been raised. Protoplanetary disks have been seen in binary star systems (Figure 7.4a, right), and new calculations show that planets can safely orbit binary stars, either close to either star or in a large orbit around both stars. Orbits in between might not be stable; perhaps this is the explanation behind Figure 7.10, in which a protoplanet appears to have been ejected from a binary star system.

FIGURE 7.10 This infrared image shows a possible protoplanet (indicated by arrow) ejected from a binary star system (bright spot in image), leaving behind a bright, curved trail where the protoplanet plowed through a dust cloud surrounding the stars. (Some astronomers, however, think it's nothing more than a background star coincidentally aligned with a jet.) (Hubble Space Telescope.)

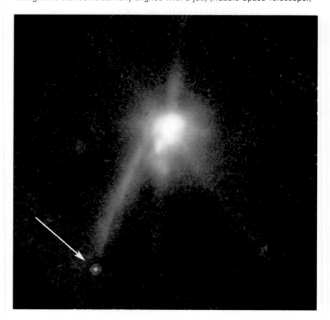

We also need to think about what the term *planet* really means. Does a newly discovered object around another star have to be like one of our solar system's planets to qualify? Where do we draw the line between large planets and small stars? For the time being, let's be generous with our definition: Detect first and classify later.

At the beginning of the 1990s, we had no conclusive proof that planets existed around any star besides our Sun. By 1999, however, astronomers had detected about twenty planetlike objects around other stars—including three planets around one star—providing the first definitive answer to the age-old question of whether other solar systems exist.

There are two basic ways to detect "extrasolar planets" (planets beyond our solar system). The most direct method involves detecting light emitted or reflected by a planet. The difficulty with this technique is that any light coming from a planet is likely to be overwhelmed by light coming from the star it orbits. For example, a Sun-like star would be a *billion times* brighter than the reflected light from an Earth-like planet. Because even the best telescopes blur the light from stars at least a little [Section 6.5], finding the small blip of planetary light amid the glare of scattered starlight would be very difficult. Direct detection should be easiest for large planets orbiting at relatively large distances around their stars. As of 1998, we had yet to directly detect a true planet orbiting a star, but we have found planetlike objects. For example, direct detection revealed an object called Gliese 229B that orbits about 40 AU from the star Gliese 229A. With a mass between 20 and 50 times that of Jupiter, Gliese 229B is too big to be

considered a planet but too small to have the nuclear fusion in its core that would qualify it as a star. Astronomers categorize Gliese 229B as a type of "failed star" called a *brown dwarf* [Section 14.2]. Within a couple of decades, high-resolution interferometers [Section 6.5]—particularly ones that can detect the infrared thermal radiation from planets—may allow us to obtain images of true planets in other solar systems. NASA is evaluating an interferometer in space as the best method of studying extrasolar planets.

The second method is indirect: detecting a planet by virtue of the small gravitational "tug" it exerts on its star. This tug should cause the star's position to shift very slightly back and forth against the sky with each orbit of the planet. Such shifts are easiest to detect when they are caused by a large planet far from its star—though not too far, since observers must wait for one or more complete orbits to be sure of what they've seen. This technique has purportedly found two Jupiter-size planets around the nearby star Lalande 21185. In some cases, it is easier to detect the gravitational effect of an orbiting planet by looking for Doppler shifts in a star's spectrum [Section 6.4]. An orbiting planet causes its star to alternately move slightly toward and away from us, which makes the star's spectral lines alternately shift toward the blue and toward the red (Figure 7.11a). This method works best for massive planets very near their stars and has been the most successful method so far. One early discovery was a 0.6-Jupiter-mass planet that tugs on the star 51 Pegasi (Figure 7.11b). This planet lies so close to its star that its year lasts only four of our days, and its temperature is probably over 1,000 K.

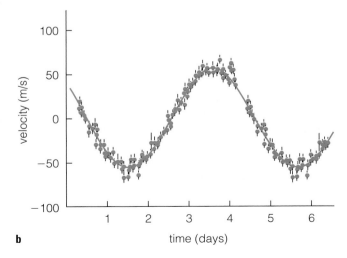

FIGURE 7.11 (**a**) Doppler shifts allow us to detect the slight motion of a star caused by an orbiting planet. (**b**) A periodic Doppler shift in the spectrum of 51 Pegasi shows the presence of a large planet with an orbital period of about 4 days. Dots are actual data points; bars through dots represent measurement uncertainty.

Planets Around Sun-like Stars					
inner solar system		· Mercury · Venus · Earth	· Mars		
HD 187123		· 0.52M_{Jup}			
HD 75289		· 0.4M_{Jup}			
Tau Boo		· 3.64M_{Jup}			
51 Peg		· 0.44M_{Jup}			
Ups And		· 0.63M_{Jup}	● 2.0M_{Jup}		● 4.5M_{Jup}
HD 217107		● 1.28M_{Jup}			
Gliese 86		● 3.6M_{Jup}			
Rho1 55 Cancri		· 0.85M_{Jup}			
HD 195019		● 3.43M_{Jup}			
Gliese 876		● 2.1M_{Jup}			
Rho Cr B		● 1.1M_{Jup}			
HD 168443		● 5.04M_{Jup}			
HD 114762		● 11M_{Jup}			
70 Vir		● 7.4M_{Jup}			
HD 210277			● 1.36M_{Jup}		
16 Cyg B			● 1.74M_{Jup}		
47 UMa				● 2.42M_{Jup}	
14 Her					● 4M_{Jup}

```
0          1          2          3
       orbital semimajor axis (AU)
```

FIGURE 7.12 Most of the planets found around other stars are closer to their stars and more massive than the planets in our solar system. (Planet sizes are not to scale with their distances shown.)

Many of the recently discovered planets seem quite different from those of our solar system (Figure 7.12). For example, several of the objects are much more massive than Jupiter yet lie quite close to their stars—a situation very different from that in our solar system, where all the jovian planets are found in the outer solar system. Recent theoretical work has shown that the nebula itself can help drag a jovian planet inward toward its star. Apparently, the time at which the solar wind clears out the nebula may determine the final location of the jovian planets. In our solar system, the nebula must have been lost relatively early.

Do these unusual planets require us to throw out our theories of solar system formation? Not necessarily, because we may be seeing only the rare exceptions to general rules. Large planets in fast orbits are

much easier to find than planets like Earth or even Jupiter. Systems like our own may be quite common but hard to detect. Our early discoveries of extrasolar planets are like glimpsing animals in the rain forest: The jungle appears full of brightly colored parrots and frogs, but far more animals are present that don't catch our eye. However, if future surveys show that planets like Jupiter and Earth are rare, planetary scientists will have to go back to the drawing board.

The future looks bright for the expansion of planetary science to other solar systems. New technologies are rapidly improving our ability to find and study extrasolar planets, and we already know of more planets around other stars than around our own Sun. For the latest news in planet discoveries, check the book web site.

The evolution of the world may be compared to a display of fireworks that has just ended: some few red wisps, ashes and smoke. Standing on a cooled cinder, we see the slow fading of the suns, and we try to recall the vanished brilliance of the origin of the worlds.

G. LEMAÎTRE (1894–1966), Astronomer and Catholic Priest

THE BIG PICTURE

We've seen that the nebular theory accounts for the major characteristics of our solar system. As you continue your study of the solar system, keep in mind the following "big picture" ideas.

- Close examination of the solar system unveils a wealth of patterns, trends, and groupings, leading us to the conclusion that all the planets formed from the same cloud of gas at about the same time.

- Chance events may have played a large role in determining how individual planets turned out. No one knows how different the solar system might be if it started over.

- Planet-forming processes are apparently universal. The discovery of protoplanetary disks and full-fledged planets around other stars brings planetary science to the brink of an exciting new era.

Review Questions

1. What is *comparative planetology?* What is its basic premise? Does comparative planetology apply only to the nine planets of our solar system? Explain.

2. Briefly summarize the observed patterns of motion in our solar system.

3. Summarize the differences between terrestrial planets and jovian planets. Why is the Moon grouped with the terrestrial planets? Where does Pluto fit in?

4. What are *asteroids?* Where are they found? What are *comets,* and how do those of the *Oort cloud* and the *Kuiper belt* differ in terms of their orbits?

5. Summarize the four challenges that any theory of the solar system must explain. List a few of the key exceptions to the rules that must be explained.

6. What is the *nebular theory?* How does it get its name?

7. Describe the processes that led the solar nebula to collapse into a spinning disk. Also describe the evidence supporting the idea that our solar system had a *protosun* and a *protoplanetary disk* early in its history.

8. Distinguish between *metals, rocks, hydrogen compounds,* and *light gases* in terms of condensation temperatures and relative abundance. Give a few examples of substances that fall into each category.

9. Explain how temperature differences in the solar nebula led to the condensation of different materials at different distances from the protosun. What do we mean by the *frost line* in the solar nebula?

10. What is *accretion?* What are *planetesimals?*

11. Describe the process of *nebular capture* that allowed the jovian planets to grow to very large sizes (compared to terrestrial planets). How does this process explain why jovian planets have extensive satellite systems?

12. What is the *solar wind?* How did its strength in the past differ from its strength today? What role did it play in ending the growth of the planets?

13. Briefly describe and explain the present distribution of asteroids and comets in our solar system.

14. Why were impacts so much more common in the past than today? How can we use this fact to estimate the age of a planetary surface by counting its impact craters?

15. What are the clues that a planetary satellite was captured instead of forming with the planet?

16. Describe the process by which a giant impact may have led to the formation of the Moon. What other oddities of our solar system might be explained by giant impacts?

17. Summarize how the nebular theory meets the four challenges laid out in this chapter. Describe a few of the remaining unanswered questions about the origin of our solar system.

18. Why is direct detection of extrasolar planets so difficult? Describe how we can indirectly detect extrasolar planets today, and summarize the current evidence concerning planets in other solar systems.

Discussion Question

1. *Lucky to Be Here?* In considering the overall process of solar system formation, do you think it was very likely for a planet like Earth to have formed? Could random events in the early history of the solar system have prevented us from even being here today? Defend your opinion. Do your opinions on these questions have any implications for your belief in the possibility of Earth-like planets around other stars?

Problems

1. *Two Classes of Planets.* Explain in terms a friend or roommate would understand why the jovian planets are lower in density than the terrestrial planets even though they all formed from the same cloud.

2. *A Cold Solar Nebula.* Suppose the entire solar nebula had cooled to 50 K before the solar wind cleared it away. How would the composition and sizes of the terrestrial planets be different from what we see today? Explain your answer in a few sentences.

3. *No Nebular Capture.* Suppose the solar wind had cleared away the solar nebula before the process of nebular capture was completed in the outer solar system. How would the jovian planets be different? Would they still have satellites? Explain your answer in a few sentences.

4. *Angular Momentum.* Suppose our solar nebula had begun with much more angular momentum than it did. Do you think planets could still have formed? Why or why not? What if the solar nebula had started with zero angular momentum? Explain your answers in one or two paragraphs.

5. *Jupiter's Action.* Suppose that, for some reason, the planet Jupiter had never formed. How do you think the distribution of asteroids and comets in our solar system would be different? How would these differences have affected Earth? Explain your answer in a few sentences.

6. *A Backward Planet.* Imagine a solar system in which all planets except one orbit the Sun in the same direction. Can the nebular theory explain the origin of this planet? How could you alter the theory of solar system formation to explain the "backward planet"? Explain your answer in a few sentences. (*Hint:* Think of "backward" planetary satellites.)

7. *51 Pegasi.* The star 51 Pegasi has about the same mass as our Sun, and the planet discovered around it has an orbital period of 4.23 days. The mass of the planet is estimated to be 0.6 times the mass of Jupiter.

   a. Use Kepler's third law to find the planet's average distance (semimajor axis) from its star. (*Hint:* Because the mass of 51 Pegasi is about the same as the mass of our Sun, you can use Kepler's third law in its original form, $p^2 = a^3$ [see Chapter 5]; be sure to convert the period into years.)

   b. Briefly explain why, according to our theory of solar system formation, it is surprising to find a planet the size of the 51 Pegasi planet orbiting at this distance.

   c. Hypothesize as to how the 51 Pegasi planet might have come to exist. Explain your hypothesis in a few sentences.

8. *Web Project: New Planets.* Starting from the book web site, search the Web for up-to-date information about discoveries of planets in other solar systems. Create a personal "Web journal," complete with pictures from the Web, describing at least three recent discoveries of new planets. For each case, write one or two paragraphs in your journal describing the method used in the discovery, comparing the discovered planet to the planets of our own solar system, and summarizing whether the discovery poses any new challenges to our theory of solar system formation.

The Terrestrial Worlds

8

As we live our daily lives, we can't help but take Earth's pleasant conditions for granted: a temperature neither boiling nor freezing, abundant water, a protective atmosphere, and a relatively stable environment thanks to a moderate level of geological activity. We need look only as far as our neighboring terrestrial worlds to see how fortunate we are.

Since the beginning of the space age, we've documented the widely different surfaces and atmospheres of the planets, and we can now make detailed comparisons among them. We've learned that, even though all the terrestrial worlds are similar in composition and formed at about the same time, their subsequent histories diverged dramatically because of a few basic properties of each world. Thus, by comparing the geological and atmospheric evolution of the terrestrial worlds, we can gain a better appreciation for how the Earth works and why it has turned out to be advantageous for us.

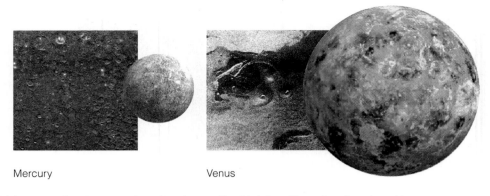

Mercury   Venus

FIGURE 8.1 Global views of the terrestrial planets to scale and representative surface close-ups. The global view of Venus shows its surface without its atmosphere, based on radar data from the Magellan spacecraft; all other images are photos (or composite photos) taken from spacecraft.

8.1 Principles of Comparative Planetology

The five terrestrial worlds—Mercury, Venus, Earth, the Moon, and Mars—share a common ancestry in their birth from the solar nebula, but their present-day surfaces and atmospheres show vast differences (Figure 8.1). Mercury and the Moon are ancient, battered worlds densely covered by craters, with some areas of volcanic plains. Venus is shrouded in a thick, broiling atmosphere, and its surface has been flooded with lava and twisted by internal stresses. Mars, despite its middling size, has the solar system's largest volcanoes and is the only planet other than Earth where running water played a major role in shaping the surface. Earth has surface features similar to all those on other terrestrial worlds and more—including a unique layer of living organisms that covers almost the entire surface of the planet. Our purpose here is to understand how the profound differences among the terrestrial worlds came to be.

Geology and atmospheric science began as studies just of the Earth, but the space age has broadened these endeavors to include all the planets and their satellites. *Comparative planetology* is the study of the solar system through examining and understanding the similarities and differences among the planets. It is based on the idea that the planets have a common origin, are subject to the same physical laws, and undergo the same physical processes to varying degrees. Differences among the planets are rarely accidents but rather can be traced to a planet's fundamental properties, such as size, location in the solar system, composition, or other important factors.

Comparative planetology allows a far better appreciation of our world—and our place in the universe—than could possibly be achieved by even the most advanced study of the Earth alone.

In this chapter, we will begin by examining the governing principles of comparative planetology, focusing first on geological processes and then on atmospheres. We'll then examine our neighboring terrestrial worlds in detail, using the tools developed in the beginning of the chapter.

8.2 Planetary Geology

The study of surface features and the processes that create them is called **geology.** The root *geo* means "Earth," and *geology* originally referred only to the study of the Earth. Today we speak of *planetary geology,* the extension of geology to include all the solid bodies in the solar system, whether rocky or icy.

Geology is relatively easy to study on Earth, where we can examine the surface in great detail, but is clearly much more challenging on the other planets. The Moon is the only alien world from which we've collected rocks, some gathered by the Apollo astronauts and others by Russian robotic landers in the 1970s. We can also study the dozen or so meteorites from Mars that have landed on Earth. Aside from these few rocks, we have little more than images taken from spacecraft from which to decode the history of the other planets over the past 4.6 billion years. It's like studying people's facial expressions to understand what they're feeling inside—and what their childhood was like! Fortunately, this decoding is considerably easier for planets than for people.

Earth

Earth's Moon

Mars

Viewing the Terrestrial Worlds

Spacecraft have visited and photographed all the terrestrial worlds. Besides ordinary, visible-light photographs, we also have spacecraft images taken with infrared and ultraviolet cameras, spectroscopic data, and in some cases three-dimensional data compiled with the aid of radar. Finally, we have close-up photographs of selected locations on all the terrestrial worlds but Mercury, taken by spacecraft that have landed on their surfaces (Figure 8.2). The result is that we now understand the geology of the terrestrial worlds well enough to make detailed and meaningful comparisons among them.

Comparative planetary geology (one part of comparative planetology) hinges on the principle that a planet's surface features can be traced back to its fundamental properties. For example, we saw in Chapter 7 that a planet's composition depends primarily on how far from the Sun it formed. In the rest of this section, we look for similar relationships between surface features and the fundamental properties of the planets. But because surface geology depends largely on a planet's interior, we must first look inside the terrestrial worlds.

Inside the Terrestrial Worlds

Planets are all approximately spherical and quite smooth relative to their size. For example, compared to the Earth's radius of 6,378 km, the tallest mountains could be represented by grains of sand on a typical globe. It's easy to understand why the gaseous jovian planets ended up nearly spherical: Gravity always acts to pull material together, and a sphere is the most compact shape possible. It's a bit subtler when we deal with the rocky terrestrial worlds, because rock can resist the pull of gravity. Indeed, many small moons and asteroids are "potato-shaped" precisely because their weak gravity is unable to overcome the rigidity of their rocky material. We need to take a closer look at the behavior of rock.

We often think of rocks as the very definition of strength ("solid as a rock"), but rocks are not always as solid as they may seem. When subjected to sustained stress over millions or billions of years, rocky material slowly deforms and flows. In fact, rock acts more like Silly Putty™, which stretches when you pull it slowly but breaks if you pull it sharply (Figure 8.3). And just as Silly Putty becomes more pliable if you heat it, warm rocks are weaker and more deformable than cooler rocks.

The rocky terrestrial worlds became spherical because of rock's ability to flow. When objects exceed about 500 km in diameter, gravity can overcome the strength of solid rock and make a world spherical in less than a billion years. If the object is molten, as was the case with the terrestrial worlds in their early histories, it can become spherical much more quickly. (For similar reasons, Earth's ocean surface is smoother and more spherical than its rocky surface.)

Gravity also gives the terrestrial worlds similar internal structures. You know that in a mixture of oil and water the less dense oil rises to the top while the denser water sinks to the bottom. The process by which gravity separates materials according to density is called **differentiation** (because it results in layers made of *different* materials). The terrestrial worlds underwent differentiation early in their histories, at the time when they were molten throughout

FIGURE 8.2 Surface views of the terrestrial worlds. No spacecraft have landed on Mercury, so an artist's conception is shown; all other images are photos.

a Mercury

b Venus

c Earth

d Moon

e Mars

their interiors. The result was the formation of three layers of differing composition within each terrestrial world (Figure 8.4). Dense metals such as iron and nickel sank through molten rocky material to form a **core.** Rocky material came to rest above the core, forming a thick **mantle.** A "scum" of the lowest-density rocks rose to form a **crust.**

The terms *core, mantle,* and *crust* are defined by the composition within each layer. But characterizing the layers by *rock strength* rather than composition turns out to be more useful for understanding geological activity. From this view, we speak of an outer layer of relatively rigid rock called the **lithosphere** (*lithos* means "stone" in Greek). The lithosphere generally encompasses the crust and the uppermost portion of the mantle. Beneath the lithosphere, the higher temperatures allow rock to deform and flow much more easily even though it is not actually molten. Thus, the lithosphere is essentially a layer of rigid rock that "floats" on the softer rock below. (On the Earth, the lithosphere is broken into *plates* that move as the

FIGURE 8.3 Silly Putty stretches when pulled slowly but breaks cleanly when pulled rapidly. Rock behaves just the same, but on a longer time scale.

FIGURE 8.4 Interior structures of the terrestrial worlds, in order of size. The thickness of outer layers is exaggerated for clarity.

underlying rock flows gradually, creating the phenomenon called _continental drift_ [Section 11.2].)

The strength of the lithosphere determines the level of geological activity on the planet's surface. Interior heat and motion will lead to volcanoes and other geological activity on a planet with a relatively thin lithosphere. In contrast, geological activity is inhibited on a planet with a thick, strong lithosphere. The most important factor determining lithospheric thickness is internal temperature: Higher internal temperature makes rocks softer, leading to a thinner rigid lithosphere. Figure 8.4 compares the interior structures of the terrestrial worlds. The greatest difference is that smaller worlds have thicker lithospheres, indicating that they have cooler interiors.

The secret to understanding the differences between planetary lithospheres lies in understanding how planets are heated and how they cool off. Planetary _surfaces_ are all warmed by sunlight, but the high temperatures _inside_ the planets today are due to radioactive heating. Rocks contain small amounts of radioactive elements such as uranium, thorium, and potassium. When the nuclei of radioactive elements decay, subatomic particles fly off at high speeds, colliding with neighboring atoms and heating them up. In essence, this transfers some of the mass-energy of the radioactive elements ($E = mc^2$ [Section 4.2]) to the heat of the planetary interior.

Some of a planet's internal heat is always escaping the interior and being radiated away to space. Heat flows outward from the hot interior toward the cooler surface through _conduction_ and _convection_. **Conduction** transfers heat between solid objects in contact with each other; it makes heat flow from your hand to a glass full of ice water. Conduction is the main process by which heat flows upward through the lithosphere. Planets with warm interiors also lose heat through **convection,** in which hot material expands and rises while cooler material contracts and falls. Convection can occur any time a substance is strongly heated from underneath; you can see convection whenever you heat a pot of soup on the stove, and you're probably also familiar with it in the context of weather, in which warm air rises while cool air falls in our atmosphere. Each small region of rising and falling material is called a **convection cell.**

The interiors of the terrestrial planets are slowly cooling as their heat escapes. By now, 4.6 billion years after their formation, much of the planets' original heat from accretion and differentiation (Figure 7.7) has leaked away. Even the rate of radioactive decay declines as the planets age (because a particular nucleus can decay only once). The interior cooling gradually changes a planet's structure, making the lithosphere thicker and leaving any regions of molten rock lying deeper inside.

Ultimately, the single most important factor in determining how long a planet stays hot is its size: Larger planets stay hot longer, just as larger baked potatoes stay hot longer than small ones. You can see why size matters most by picturing a large planet as a smaller planet wrapped in extra layers of rock. The extra rock acts as an insulating blanket, so heat from the center takes longer to reach the surface on the larger planet than it would on the smaller one. The same principle explains many everyday phenomena. For example, crushing a cube of ice into smaller pieces increases the total amount of surface area from which heat is lost, while the total amount of ice remains the same. Thus, the crushed ice will cool a drink more quickly than would the original ice cube.

TIME OUT TO THINK _Can you think of other examples in everyday life of a small object cooling off faster than a large one? What about a smaller object warming up more quickly than a large one? How do these examples relate to the issue of geological activity on the terrestrial worlds?_

The overall result is that the large terrestrial worlds—Earth and Venus—have cooled the least over the life of the solar system. These worlds therefore have thin lithospheres and substantial geological activity. The smaller worlds—Mercury and the Moon—have mostly cooled off and therefore have very thick lithospheres that essentially make them geologically "dead." Mars, intermediate in size, has cooled significantly, resulting in an intermediate interior temperature.

A planet's interior is also where its magnetic field, if any, is generated. The metallic cores of all the terrestrial worlds are electrically conducting; if the core materials are also in motion, they can create a magnetic field that surrounds the planet. Although our understanding of planetary magnetic fields is still rudimentary, we know that the combined motions of planetary rotation and convection (in a molten core) are necessary to create a magnetic field. Earth, with its convecting core and rapid rotation, has the strongest magnetic field, while the Moon, with its small, frozen core and slow rotation, has none. Mercury lies in between. Mars lacks a significant magnetic field today, but the latest measurements from the Mars Global Surveyor spacecraft reveal the imprint of a stronger field in Mars's first billion years. Apparently, as the core cooled and solidified, the magnetic field faded away. Venus lacks any magnetic field, due to either its slow rotation or an as-yet-unexplained lack of core convection.

The presence or absence of a magnetic field may not seem important to the massive rocky worlds, but it actually matters a great deal to a planet's atmosphere. Magnetic fields are responsible for forming a protective bubble around a planet called a *magnetosphere*; as we'll see, this can have significant effects on the evolution of a planet's atmosphere.

8.3 Shaping Planetary Surfaces

When we look around the Earth, we find an apparently endless variety of geological surface features. The diversity increases when we survey the other planets. But on closer examination, geologists have found that almost all the features observed result from just four major **geological processes** that affect planetary surfaces:

- **Impact cratering:** the excavation of bowl-shaped depressions (*impact craters*) by asteroids or comets striking a planet's surface.

- **Volcanism:** the eruption of molten rock, or *lava*, from a planet's interior onto its surface.

- **Tectonics:** the disruption of a planet's surface by internal stresses.

Common Misconceptions: Pressure and Temperature

You might think that Earth's interior is hot just because the pressures are high. After all, if we compress a gas from low pressure to high pressure, it heats up. But the same is not necessarily true of rock. High pressure hardly compresses rock, so the compression causes little increase in temperature. Thus, while high pressures and temperatures sometimes go together in planets, they don't have to. In fact, after all the radioactive elements decay (billions of years from now), Earth's deep interior will become quite cool even though the pressure will be the same as it is today. The temperatures inside Earth and the other planets can remain high only if there is a source of heat.

- **Erosion:** the wearing down or building up of geological features by wind, water, ice, and other phenomena of planetary weather.

Planetary geology once consisted largely of cataloging the number and kinds of geological features found on the planets. The field has advanced remarkably in recent decades, however, and it is now possible to describe how planets work in general and what kinds of geological features we expect to find on different planets. This progress has been made possible by an examination of the geological processes in detail and the determination of what factors affect and control them.

Impact Cratering

Impact cratering occurs when a leftover planetesimal (such as a comet or an asteroid) crashes into the surface of a terrestrial world. Impacts can have devastating effects on planetary surfaces (Figure 8.5). Impactors typically hit planets at speeds of between 30,000 and 250,000 km/hr (10–70 km/s) and thus pack enough energy to vaporize solid rock and excavate a **crater.** Craters are generally circular because the impact blasts out material in all directions, no matter which direction the impactor came from. Debris from the blast shoots high into the atmosphere and then rains down over an area much larger than that of the impact crater itself. In large impacts, some of the atmosphere may be blasted away into space, and some of the rocky ejecta may completely escape from the planet.

Craters come in all sizes and a few different shapes, but most are small and bowl-shaped. Small craters far outnumber large ones because there are

FIGURE 8.5 Artist's conception of the impact process. The last frame shows how, in a larger crater, the center can rebound just as water does after you drop a pebble in.

Lunar highlands are ancient and heavily cratered.

Lunar maria are huge impact basins that were flooded by lava. Only a few small craters appear on the maria.

a

b

FIGURE 8.6 (**a**) The Moon's surface shows both heavily and lightly cratered areas. (**b**) This map of the Moon shows the differences between the lightly-cratered maria (dark areas) and the heavily-cratered highlands. At the lower left is an impact basin known as Mare Orientale.

far more small objects orbiting the Sun than large ones. When the largest (and rarest) impactors strike a planet, they form **impact basins.** Looking at the Moon with binoculars, you can see examples of lava-filled impact basins up to 1,100 kilometers across called **lunar maria** (Figure 8.6). These large impacts violently fractured the Moon's lithosphere, making cracks through which lava escaped to flood the impact basins.

At the small extreme, the most common impactors are sand-size particles called *micrometeorites*. Such tiny particles burn up as meteors in the atmospheres of Venus, Earth, and Mars. But on worlds that lack a significant atmosphere, such as Mercury and the Moon, the countless impacts of micrometeorites gradually pulverize the surface rock to create a layer of powdery "soil." On the Moon, the Apollo astronauts and their rovers left marks in this

FIGURE 8.7 Reminders of the Apollo missions to the Moon will last for millions of years.

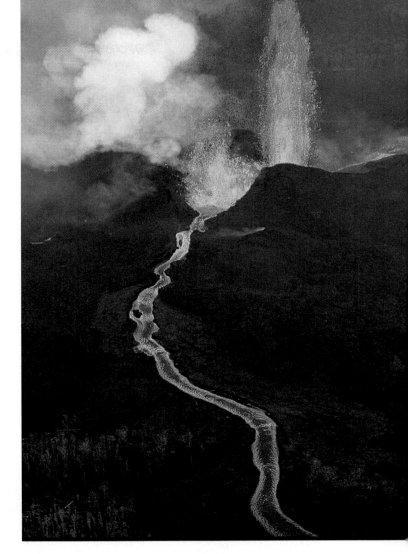

FIGURE 8.8 Eruption of an active volcano on the flanks of Kilauea on the Big Island in Hawaii.

powdery surface (Figure 8.7). Because of the lack of wind and rain, these footprints and tire tracks will last millions of years, but they will eventually be erased by micrometeorite impacts.

Of the four geological processes, impact cratering is the most universal. Craters formed on all the terrestrial worlds, regardless of their size or location in the solar system. The "rain of rock and ice" that ended around 3.8 billion years ago saturated (completely covered) all solid surfaces with impact craters. We can therefore use the present-day abundance of craters on a planet's surface to learn about its subsequent geological history. A surface region that is still saturated with craters, such as the *lunar highlands,* must have remained essentially undisturbed. (See Figure 8.6.) In contrast, in regions like the *lunar maria* that now have few craters, the original craters must have been somehow "erased." The flood of lava that formed the lunar maria covered any craters that had formed within the impact basin, and the few craters that exist today on the maria must have formed from impacts occurring after the lava flows solidified. These craters tell us that the impact rate since the end of heavy bombardment has been quite small: The lunar maria have only 3% as many craters as the lunar highlands, but radioactive dating of moon rocks shows that the maria are still 3–3.5 billion years old.

TIME OUT TO THINK *Earth must also have been saturated with impact craters early in its history, but we see relatively few impact craters on Earth today. What processes erase impact craters on Earth?*

Volcanism

The second major geological process is volcanism. We find evidence of volcanoes and lava flows on all the terrestrial worlds, as well as on a few moons of the outer solar system. Volcanoes erupt when underground lava finds a path through the lithosphere to the surface (Figure 8.8). Molten rock is generally less dense than solid rock, so it has a natural tendency to rise. The lava source may be squeezed by tectonic forces that drive the magma upward under pressure. Any trapped gases expand as the molten rock rises, further pushing lava upward.

The appearance of volcanic features depends on the consistency of the lava that erupts onto the surface ("runny" vs. "sticky"). The common rock type known as **basalt** makes relatively runny lava. The runniest basalt lavas flow far and flatten out before solidifying, creating vast volcanic plains such as the

FIGURE 8.9 Volcanoes produce different types of features depending primarily on the consistency of the lava erupted.

a "Runny" lava makes flat lava plains. (Lava plains [maria] on the Moon.)

b Intermediate lava makes shallow-sloped shield volcanoes. (Olympus Mons [Mars].)

c "Sticky" lava makes steep-sloped stratovolcanoes. (Mount St. Helens.)

lunar maria (Figure 8.9a). Somewhat less runny basalt lavas solidify before they can completely spread out, resulting in **shield volcanoes** (so-named because they are shield-shaped). Shield volcanoes can be very tall, but they are not very steep; most have slopes of only 5°–10° (Figure 8.9b). The mountains of the Hawaiian Islands are shield volcanoes; measured from the ocean floor to their summits, the Hawaiian mountains are the tallest (and widest) on Earth. Tall, steep volcanoes such as Mount St. Helens are made from very "sticky" lavas that can't flow very far before solidifying (Figure 8.9c).

Volcanism occurs on planets with high internal temperatures and thin lithospheres, where molten rock lies relatively close to the surface. While all the terrestrial worlds had relatively thin lithospheres early in their histories, the smaller worlds have cooled off and their lithospheres have thickened. Only the largest terrestrial planets (Earth and Venus) have warm enough interiors and thin enough lithospheres for ongoing volcanic activity. On these worlds, volcanoes have obliterated most of the ancient craters that once covered the surface.

Volcanism plays an additional key role in planetary evolution through **outgassing,** the release of gases from a planet's interior. Most of the ingredients that make up today's planetary atmospheres were locked in the interior after formation but are released when lava reaches the surface. The gases that help drive volcanic eruptions accumulate to form the atmosphere (and even oceans), as we'll discuss further later in this chapter.

Tectonics

Tectonics, our third geological process, refers to the processes by which the internal forces and stresses acting on the lithosphere create surface features. Tectonic features take a wide variety of forms (Figure 8.10). Mountains may rise where the crust is compressed, as in the case of the Appalachian Mountains of the eastern United States. Huge valleys and cliffs may form where the crust is pulled apart; examples include the Guinevere Plains on Venus and

Appalachian Mountains in eastern United States

Guinevere Plains on Venus

Compression in crust can make mountains.

Extension can make cracks and valleys.

convection cells

FIGURE 8.10 Tectonic forces can produce a wide variety of features. Mountains and fractured plains are among the most common.

New Mexico's Rio Grande Valley here on Earth. Other forms of internal stress come from plumes of hot mantle material that push up on the lithosphere or from the expansion or contraction that comes with the warming or cooling of a planet.

Like volcanoes, tectonic features are more common on larger worlds that remain hot inside, and the two kinds of features are often found together. The thick lithospheres of smaller worlds are generally too strong to allow widespread deformation.

Erosion

The last of the four major geological processes is erosion, which encompasses a variety of processes connected by a single theme: the breakdown and transport of rocks by an atmosphere. Wind, rain, rivers, flash floods, and glaciers are just a few examples of processes that contribute to erosion on Earth. Erosion not only breaks down existing geological features (wearing down mountains and forming gullies, riverbeds, and deep valleys), but also builds new ones (such as sand dunes, river deltas, and glacial deposits).

Erosion is significant only on planets with substantial atmospheres, which means the larger planets. Larger planets are better able to generate atmospheres by outgassing; their stronger gravity also tends to prevent their atmospheres from escaping to space. In general, a thick atmosphere causes more erosion than a thin atmosphere, but a planet's rotation rate is also important. Slowly rotating planets have correspondingly slow winds and therefore weak or nonexistent wind erosion. Surface temperature also matters: Under the right conditions, some atmospheric ingredients can form rain and snow, which are very effective at eroding rock. We will expand on all these issues in the next section.

8.4 Planetary Atmospheres

Planetary atmospheres play a controlling role in climate and planetary habitability. We have also seen how a planet's geology and atmosphere can be closely intertwined, as in the cases of volcanism and erosion. In the pages that follow, we'll examine what atmospheres are like, how they behave, and how they change over time.

All the planets of our solar system have atmospheres to varying degrees, as do several of the solar system's larger moons. The jovian planets are essentially atmosphere throughout—they are largely made of gaseous material. In contrast, the atmospheres of solid bodies—terrestrial worlds, jovian moons with atmospheres, and Pluto—are relatively thin layers of gas that contain only a minuscule fraction of a world's mass. Although we will concentrate on understanding the atmospheres of the terrestrial worlds, similar principles apply to the atmospheres of all the planets and satellites.

Comparing Planetary Atmospheres

The terrestrial atmospheres are even more varied than the terrestrial geologies (Table 8.1). The Moon and

Mercury have very little atmosphere at all—you would seem to be facing the blackness of space even when standing on their surfaces. If you could condense the entire atmosphere of the Moon into solid form, you would have so little material that you could store it in a basement. At the other extreme, Venus is shrouded by a thick, cloudy atmosphere made of carbon dioxide. The pressure is 90 times greater than that on Earth, with a searing temperature hotter than your oven set on high. The atmosphere of Mars is also made mostly of carbon dioxide, but much less of it than on Venus. The result is very thin air with a pressure so low that your body tissues would bulge painfully if you stood on the surface without wearing a full spacesuit. In fact, the pressure and temperature are both so low that liquid water would rapidly disappear, with some evaporating and some freezing into ice. Thus, the fact that we see dried-up riverbeds on Mars tells us that its atmosphere must have been very different in the past. And Earth's unusual atmosphere of nitrogen and oxygen is the only one under which human life is possible, at least without the creation of an artificial environment.

Why did the atmospheres of the planets turn out to be so different, and what makes Earth's atmosphere so special? The best way to answer these questions is to examine and compare planetary atmospheres. Each atmosphere is unique, but, as we saw with geology in the previous sections, similar properties and processes determine the characteristics of all atmospheres. First we will explore the general behavior of atmospheres, and then we will learn how and why the terrestrial atmospheres came to differ so profoundly.

Table 8.1 Atmospheres of the Terrestrial Worlds

World	Composition	Surface Pressure*	Winds, Weather Patterns	Clouds, Hazes
Mercury	Atoms of helium, sodium, oxygen, potassium	10^{-15} atmosphere	None: too little atmosphere	None
Venus	96% CO_2, 3.5% N_2	90 atmospheres	Slow winds, no violent storms	Sulfuric acid clouds
Earth	77% N_2, 21% O_2, 1% argon, H_2O (variable)	1 atmosphere	Winds, hurricanes	H_2O clouds, pollution
Moon	Atoms of sodium, potassium	10^{-15} atmosphere	None: too little atmosphere	None
Mars	95% CO_2, 2.7% N_2, 1.6% argon	0.007 atmosphere	Winds, dust storms	H_2O and CO_2 clouds, dust

*Compared to the pressure at sea level on Earth.

a

FIGURE 8.11 (**a**) Most of Earth's atmosphere (gray layer at top) is confined to a layer less than 100 km thick. (**b**) The atmosphere does not end suddenly; some gas is present even at altitudes of hundreds of kilometers, causing the glow visible around the tail of the Space Shuttle.

b

Atmospheric Structure

If you've ever driven up a mountain, you know that the atmosphere changes with altitude. Your popping ears and shortness of breath are proof that the pressure is lower as you go higher, and the chill tells you that temperature also decreases with altitude. Let's investigate the origin of atmospheric structure—that is, the variation in pressure and temperature with altitude.

Planetary atmospheres exist in a perpetual balance between the downward weight of their gases and the upward push of their gas pressure. The higher you go in an atmosphere, the less the weight of the gas above you. Thus, the pressure must also become less as you go upward, which explains why the pressure decreases as you climb a mountain or ascend in an airplane. You can visualize what happens by imagining the atmosphere as a very big stack of pillows. The pillows at the bottom are very compressed because of the weight of all the pillows above. As you go upward, the pillows are less and less compressed because less weight lies on top of them. But atmospheres do not have clear upper boundaries; instead, they gradually fade away with increasing altitude (Figure 8.11).

The temperature variations in an atmosphere arise from the way different wavelengths of sunlight

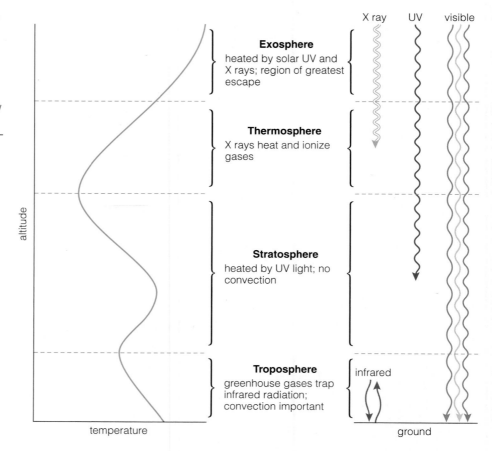

FIGURE 8.12 The structure of a generic planetary atmosphere: Solar X rays are absorbed in the thermosphere, ultraviolet light is absorbed in the stratosphere, and visible light reaches the ground. Planets that lack ultraviolet-absorbing molecules will lack a stratosphere, and planets with very little gas will have only an exosphere.

are absorbed in different layers of the atmosphere. Figure 8.12 shows the structure of a "generic" planetary atmosphere; it shows how and why the temperature varies with altitude. We will now investigate the processes that create these basic atmospheric layers in detail, from the ground up.

Visible Light: Warming the Surface and Coloring the Sky

The Sun emits most of its energy in the form of visible light, and atmospheric gases are generally transparent in this wavelength range. Thus, it's primarily visible light that reaches a planet's surface and warms it. In the absence of an atmosphere (or at least in the absence of any greenhouse effect), the surface temperature of any planet is determined by three basic properties: its distance from the Sun (which determines the intensity of sunlight), the fraction of sunlight absorbed by the surface (i.e., not reflected away), and the planet's rotation rate (which determines whether the planet is warmed evenly or bakes on one side while freezing on the other). The planet absorbs a certain amount of sunlight and reaches the temperature at which it emits exactly the same amount of energy in the form of thermal radia-

tion. Most planetary surfaces are in the temperature range at which they emit their thermal radiation in the infrared [Section 6.3].

Although most visible light passes directly through the atmosphere, some is *scattered* (redirected) by molecules in the atmosphere. This scattering of sunlight makes the sky bright and prevents us from seeing stars in the daytime. If our atmosphere did not scatter light, you'd be able to block out the light of the Sun simply by holding your hand in front of it, and then you'd have no trouble seeing other stars.

The scattering of light by the atmosphere also explains why Earth's sky is blue. It turns out that molecules scatter blue light much more effectively than red light. Thus, although the Sun illuminates the atmosphere with all colors of light, effectively only the blue light gets scattered. When the Sun is overhead, this scattered blue light reaches your eyes from all directions, which explains why the sky appears blue (Figure 8.13). At sunset (or sunrise), the sunlight must pass through a greater amount of atmosphere on its way to you. Thus, most of the blue light is scattered away from you altogether (making someone else's sky blue), leaving only red light to color your sunset.

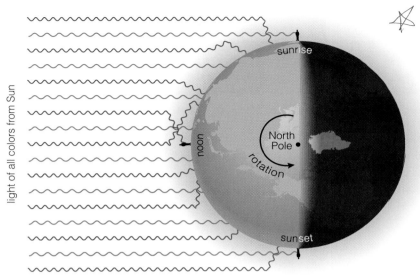

FIGURE 8.13 Atmospheric gases scatter blue light more than they scatter red light. During most of the day, you therefore see blue photons coming from most directions in the sky, making the sky look blue. But only the red photons reach your eyes at sunrise or sunset, when the light must travel a longer path through the atmosphere to reach you.

light of all colors from Sun

Infrared Light, the Greenhouse Effect, and the Troposphere

Figure 8.14 shows how the **greenhouse effect** works. Water vapor, carbon dioxide, and other *greenhouse gases* in the atmosphere absorb some of the infrared radiation emitted upward from the planet's surface. These gases therefore warm up and emit infrared thermal radiation themselves—but in all directions. Some of this radiation is directed back down toward the surface, making the surface warmer than it would be from absorbing visible sunlight alone. The more greenhouse gases present, the greater the degree of greenhouse warming. (The name *greenhouse effect* comes from botanical greenhouses, usually built mostly of glass, which keep the air inside them warmer than it would be otherwise. But it turns out that greenhouses trap heat simply by not letting hot air rise— a very different way of trapping heat than that taking place in planetary atmospheres.)

TIME OUT TO THINK *Clouds on Earth are made of water, and H₂O is a very effective greenhouse gas—even when the water vapor condenses into droplets in clouds. Use this fact to explain why clear nights tend to be colder than cloudy nights. (Hint: Sunlight warms the surface only in the daytime, but the surface radiates its heat away both day and night.)*

Table 8.2 lists the temperatures of the terrestrial worlds with and without the greenhouse effect. The greenhouse effect is responsible for warming Mars (where it adds 5 K), Earth (adding 40 K), and Venus

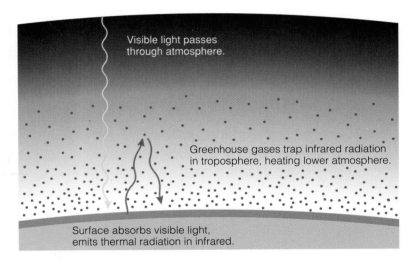

Visible light passes through atmosphere.

Greenhouse gases trap infrared radiation in troposphere, heating lower atmosphere.

Surface absorbs visible light, emits thermal radiation in infrared.

FIGURE 8.14 The greenhouse effect: The troposphere becomes warmer than it would be if it had no greenhouse gases.

(adding nearly 500 K!). Ironically, the greenhouse effect on Earth is almost entirely due to gases that make up less than 2% of our atmosphere: mainly water vapor, carbon dioxide, and methane. The major constituents of Earth's atmosphere: N_2 and O_2, have no effect on infrared light and do not contribute to the greenhouse effect. If they did, Earth might be too warm for life.

Most of what we call weather occurs in the troposphere. The fact that the greenhouse effect warms air near the ground drives *convection* in the troposphere. The warm air near the ground expands and rises, while cooler air near the top of the troposphere contracts and falls. Convection continually churns

Table 8.2 Temperatures of the Terrestrial Worlds

World	Distance to Sun (AU)	Reflectivity† (0 = black, 1 = white)	Length of Day	"No Atmosphere" Temperature*	Observed Average Surface Temperature
Mercury	0.38	0.11	176 days	440 K	700 K (day), 100 K (night)
Venus	0.72	0.72	117 days	230 K	740 K
Earth	1.00	0.36	1 day	250 K	288 K
Moon	1.00	0.07	28 days	273 K	400 K (day), 100 K (night)
Mars	1.52	0.25	≈ 1 day	218 K	223 K

*Assumes rapid rotation, in which case day and night temperatures would be the same.
†Fraction of light reflected by planets.

FIGURE 8.15 Earth's winds and storms are driven in part by the planet's relatively rapid rotation.

the air in the troposphere and raises air laden with water vapor to cooler altitudes, where it condenses to form clouds and rain. Winds in the troposphere are driven in part by the planet's rotation; fast winds and hurricane-like storms are found only on planets with rapid enough rotation (Figure 8.15).

Ultraviolet Light and the Stratosphere

As we move upward through the stratosphere, the primary source of atmospheric heating is absorption of ultraviolet light from the Sun. This heating is stronger at higher altitudes because the ultraviolet light gets absorbed before it reaches lower altitudes. As a result, the stratosphere gets warmer with increasing altitude—the opposite of the situation in the troposphere. Convection therefore cannot occur in the stratosphere: Heat cannot rise if the air is even hotter higher up. The stratosphere gets its name because the lack of convection makes its air relatively stagnant and *stratified* (layered), rather like a sitting jar of oil and water. The lack of convection

Common Misconceptions: Higher Altitudes Are Always Colder

Many people think that the low temperatures in the mountains are just the result of lower pressures, but Figure 8.12 shows that it's not that simple. The higher temperatures near sea level on Earth are a result of the greenhouse effect trapping more heat at lower altitudes. If Earth had no greenhouse gases, mountaintops wouldn't be so cold—or, more accurately, sea level wouldn't be so warm.

Common Misconceptions: The Greenhouse Effect Is Bad

The greenhouse effect is often in the news, usually in discussions about environmental problems. But the greenhouse effect itself is not a bad thing. In fact, we could not exist without it: Without the greenhouse effect, Earth's average temperature would be 250 K (−23°C, −9°F). Thus, the greenhouse effect is the only reason why our planet is not frozen over. Why, then, is the greenhouse effect discussed as an environmental problem? It is because human activity is adding more greenhouse gases to the atmosphere—which might change the Earth's climate [Section 11.5]. After all, while the greenhouse effect makes the Earth livable, it is also responsible for the searing 740 K temperature of Venus—proving that it's possible to have too much of a good thing.

also means that the stratosphere essentially has no weather and no rain.

Note that a planet can have a stratosphere *only* if its atmosphere contains molecules that are particularly good at absorbing ultraviolet photons. In fact, Earth is the only terrestrial planet with such an ultraviolet-absorbing layer and hence the only terrestrial planet with a stratosphere—at least in our solar system. We'll discuss why Earth is unique in this way in Chapter 11.

X Rays and the Thermosphere and Exosphere

A planet's upper atmosphere is strongly heated by solar X rays in the daytime, usually to much higher temperatures than the gases at the surface. That is why the upper region of the atmosphere is called the *thermo*sphere. Despite its high temperatures, the gas in the thermosphere would not feel hot to your skin because its density and pressure are so low.

If a planet has very little atmosphere, it will consist of just an exosphere. The exosphere marks the fuzzy boundary between the atmosphere and space. (The prefix *exo* means "outermost" or "outside.") The high temperatures in the exosphere give the individual atoms or molecules high speeds, allowing some of them to escape from the planet. The Space Shuttle and many satellites actually orbit the Earth within its exosphere, and the small amount of gas in the exosphere can exert a bit of atmospheric drag. (See Figure 8.11b.) That is why satellites in low-Earth orbit eventually spiral deeper into the atmosphere and burn up.

Magnetospheres and the Solar Wind

The Sun emits one more type of energy that we haven't yet considered: the low-density breeze of charged particles that we call the solar wind [Section 7.4]. Although the solar wind does not significantly affect atmospheric structure, it can have other important influences. Among the terrestrial planets, only Earth has a strong enough magnetic field to divert the charged particles of the solar wind. Earth's strong magnetic field creates a **magne-**

tosphere that acts like a protective bubble surrounding the planet (Figure 8.16a). Charged particles cannot easily pass through a magnetic field, so a magnetosphere diverts most of the solar wind around a planet. In contrast, solar wind particles can impact the exospheres of Venus and Mars—and the surfaces of Mercury and the Moon—because they lack strong magnetic fields and magnetospheres.

Some solar wind particles manage to infiltrate Earth's magnetosphere at its most vulnerable points, near the magnetic poles. Once inside, they have a difficult time escaping. The ions and electrons accumulate to make **charged particle belts** encircling the planet. (The charged particle belts around the Earth are called the *Van Allen belts* after their

FIGURE 8.16 A planet's magnetosphere acts like a protective bubble that shields the surface from charged particles coming from the solar wind. Among the terrestrial planets, only the Earth has a strong enough magnetic field to create a magnetosphere. The Earth's magnetosphere allows charged particles to strike the atmosphere only near the poles, thereby creating the phenomena of the aurora borealis and aurora australis.

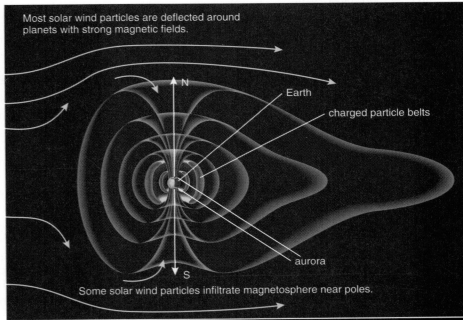

Most solar wind particles are deflected around planets with strong magnetic fields.

N

Earth

charged particle belts

aurora

S

Some solar wind particles infiltrate magnetosphere near poles.

a This diagram shows how Earth's magnetosphere interacts with the solar wind.

b Earth's aurora as seen from the Space Shuttle.

discoverer.) The high energies of the particles in charged particle belts can be very hazardous to spacecraft and astronauts passing through them.

Particles trapped in magnetospheres also cause the beautiful spectacle of light called *auroras*. If a trapped charged particle has enough energy, it can follow the magnetic field all the way down to a planet's atmosphere in the north and south polar regions. There the charged particles collide with atmospheric atoms and molecules, causing them to radiate and produce auroras (Figure 8.16b). On Earth, auroras are best viewed from Canada, Alaska, and Russia in the Northern Hemisphere (the *aurora borealis,* or northern lights) and from Australia, Chile, and Argentina in the Southern Hemisphere (the *aurora australis,* or southern lights).

8.5 Atmospheric Origins and Evolution

Today's headlines are full of stories about the human effects on Earth's atmosphere, such as ozone depletion, global warming, and air pollution. But the atmospheres of Earth and the other terrestrial planets have seen much larger changes in their billions of years of history. What causes such large changes? To answer this question, we must examine both how a planet gets its atmospheric gases and how it can lose them. As we will see, atmospheric changes have been just as important as surface changes, and the two are usually connected.

Where did the terrestrial worlds get their atmospheres? They were too small to capture significant amounts of gas from the solar nebula before the solar wind cleared it away [Section 7.4]. Any gas that the terrestrial worlds did capture from the solar nebula consisted primarily of hydrogen and helium (just like the solar nebula as a whole). But these light gases escape easily from the terrestrial worlds and by now are long gone. The atmospheres of the terrestrial worlds therefore must have formed after the worlds themselves.

The most important source of atmospheric gases is *outgassing,* the release of gases by volcanic eruptions. Recall that these gases were originally brought to the terrestrial planets by comets and asteroids during accretion and the early bombardment that ended 3.8 billion years ago [Section 7.6]. The high pressures inside the planets trapped these gases in rocks in much the same way that a pressurized bottle traps the bubbles of carbonated beverages. When the rocks melt and erupt onto the surface as lava, the release of pressure expels the gases. The gases may be released either with the lava itself or from nearby vents (Figure 8.17). The most common gases expelled are water (H_2O), carbon dioxide (CO_2), nitrogen (N_2), and sulfur-bearing gases (H_2S or SO_2).

FIGURE 8.17 Volcanoes give off H_2O, CO_2, N_2, and sulfur-bearing gases. Some gas is given off from the lava itself, and some leaks out from nearby gas vents. (Volcanoes National Park, Hawaii.)

Because outgassing depends on volcanism, it is most important on larger planets that retain their internal heat longer. Outgassing supplied most of the gas that became the atmospheres of Venus, Earth, and Mars. The Moon and Mercury, however, are not completely airless: Bombardment by the solar wind, micrometeorites, and even larger impactors supplies a small amount of material to their atmospheres.

An atmosphere can lose gas in several important ways: *thermal escape, bombardment, atmospheric cratering, condensation,* and *chemical reactions* (Figure 8.18). The terrestrial planets lost any original hydrogen and helium they once captured from the solar nebula through the process of **thermal escape,** in which an atom or a molecule in the exosphere moves fast enough to escape the pull of gravity. Three factors determine whether an atmospheric gas can be lost to space by thermal escape:

1. The planet's *escape velocity,* or the speed at which a particle must travel to escape the pull of gravity—assuming it is traveling in the right direction and doesn't collide with anything else. The more massive the planet, the higher its escape velocity [Section 5.6].

2. The temperature. Individual particles in a gas move with a wide range of random speeds [Section 4.2], but the higher the temperature, the faster their *average* speed.

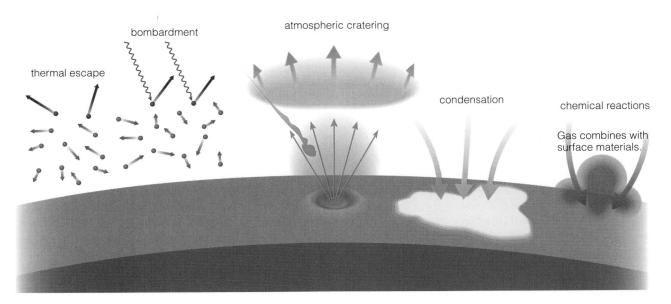

thermal escape

bombardment

atmospheric cratering

condensation

chemical reactions

Gas combines with surface materials.

FIGURE 8.18 The five major processes by which atmospheres lose gas.

3. The mass of the gas particles. At a particular temperature, lighter particles (such as H or He) move around at faster average speeds than heavier ones (such as O_2 or CO_2) and therefore are more likely to achieve escape velocity.

Escape can be driven from the outside by impacts, both large and small. *Bombardment* by the particles in the solar wind can lead to slow and steady atmospheric loss, ejecting even atoms that are too heavy for thermal escape. In general, planets with protective magnetic fields—such as Earth—are less susceptible to this type of atmospheric loss. Larger impacts can also cause major atmospheric escape. The same impacts that create craters on the surface may blast away a significant amount of the atmosphere as well in a process sometimes called *atmospheric cratering*. For the most part, atmospheric cratering happened only during the early bombardment stage of the solar system. We expect it to have been most important on smaller worlds, where gravity's hold on the atmosphere is weaker. Mars may have lost a significant amount of atmospheric gas through atmospheric cratering.

Two other atmospheric loss processes allow atmospheric gas to be absorbed back into the solid planet. Gases may simply *condense* into liquid or solid form if a planet cools down, especially in the cold polar regions. Finally, gases may become locked up in the surface through *chemical reactions* with rocks or liquids—yet another example of the strong links between planetary atmospheres and surfaces. We'll encounter examples of all five of these atmospheric loss processes when we tour the terrestrial worlds in the next section.

8.6 A Tour of the Terrestrial Worlds

Now that we've examined the processes that shape the geology and atmospheres of the terrestrial worlds, we're ready to return to the issue that lies at the heart of this chapter: why the terrestrial worlds ended up so different from one another. We'll take a brief "tour" of the terrestrial worlds. We could organize the tour by distance from the Sun, by density, or even by alphabetical order. But we'll choose the planetary property that has the strongest effect on geology: size, which controls a planet's internal heat. We'll start with our Moon, the smallest terrestrial world.

The Moon (1,738-km radius, 1.0 AU from Sun)

When you look at the Moon on a clear night, you can see much of its global geological history with your naked eye. The Moon is unique in this respect; other planets are too far away for us to see any surface details, and Earth is so close that we see only the local geological history. Twelve Apollo astronauts visited the lunar surface between 1969 and 1972, taking photographs, making measurements, and collecting rocks. Thanks to these visits and more recent observations of the Moon from other spacecraft, we know more about the Moon's geology than that of any other object, with the possible exception of the Earth.

The Moon's composition differs somewhat from the Earth's. According to the leading theory, our Moon formed very early in the history of our solar system when a giant impact blasted away some of Earth's rocky outer layers without dredging up metallic core material [Section 7.6]. The Moon therefore has a lower metal content than Earth and a smaller (or possibly nonexistent) metal core. In addition, lunar rocks collected by astronauts contain a very low amount of trapped gases compared to Earth rocks. Presumably, heat from the giant impact vaporized the gases and allowed them to escape into space before the Moon formed from the debris.

As the Moon accreted following the giant impact, the heat of accretion melted its outer layers. The lowest-density molten rock rose upward through the process of differentiation, forming an ocean of lava. This ocean cooled and solidified during the heavy bombardment that took place early in the history of the solar system, leaving the Moon's surface crowded with craters. We still see this ancient, heavily cratered landscape in the lunar highlands (Figure 8.19a). Lunar samples confirm that the highlands are composed of low-density rocks. As the bombardment tailed off, around 3.8 billion years ago, a few larger impactors struck the surface and formed impact basins.

Even as the heat of accretion leaked away, radioactive decay kept the Moon's interior molten long enough for volcanism to reshape parts of its surface. Between about 3 and 4 billion years ago, molten rock welled up through cracks in the deepest impact basins, forming the lunar maria. Long, winding channels (Figure 8.19b, a close-up of Figure 8.9a) must once have been rivers of molten rock that helped fill the lunar maria. Because the lava that filled the maria rose up from the Moon's mantle, the maria contain dense, iron-rich rock that is darker in color than the rock of the lunar highlands. The contrasts between the light-colored rock of the highlands and the dark rock of the maria make the "man-in-the-Moon" pattern that some people imagine when they look at the full moon.

Almost all the geological features of the Moon were formed by impacts and volcanism. Tectonic activity was inhibited by the Moon's thick lithosphere, and erosion never occurred due to the lack of a significant atmosphere. Over time, the Moon's small size allowed its interior to cool, thickening the lithosphere. Today, the Moon's lithosphere probably extends to a depth of 1,000 km, making it far too thick to allow further volcanic or tectonic activity. The Moon has probably been in this geologically "dead" state for 3 billion years.

Although geological activity and outgassing ceased billions of years ago, the Moon still holds some surprises. The *Lunar Prospector* orbiter recently discovered evidence of ice locked up in the Moon's polar regions. The water probably came from comet impacts that momentarily created a thin lunar atmosphere. The water molecules from these impacts bounced around the surface at random until they condensed into ice in the polar craters. The bottoms of these craters lie in nearly perpetual shadow, keeping them so cold that the water remains perpetually frozen. A thin atmosphere supplied by bombardment also surrounds the Moon, but it is continually lost by thermal escape and the impact of the solar wind.

The Moon is of prime interest to those hoping to build colonies in space. The *Lunar Prospector* arrived

FIGURE 8.19 Impact cratering and volcanism are the most important geological processes on the Moon.

a Astronaut explores a small crater.

b Site of an ancient lava river.

at the Moon in early 1998 and began mapping the surface in search of promising locations for a permanent human base. Although no concrete plans were in place as of 1999, several nations are exploring the possibility of building a human outpost on the Moon within the next couple of decades. The possibility of a water supply in polar craters may become an important issue.

Mercury (2,440-km radius, 0.39 AU from Sun)

Mercury is the least studied of the terrestrial worlds. Its proximity to the Sun makes it difficult to study through telescopes, and it has been visited by only one spacecraft: *Mariner 10,* which collected data during three rapid flybys of Mercury in 1974–75. *Mariner 10* obtained images of only one hemisphere of Mercury (Figure 8.20b). The influence of Mercury's gravity on *Mariner 10*'s orbit allowed an accurate determination of Mercury's mass and high average density. Mercury is about 61% iron by mass, with a core that extends to perhaps 75% of its radius. Mercury's high metal content is due to its formation close to the Sun, possibly enhanced by a giant impact that blasted away its outer, rocky layers [Section 7.6].

Craters are visible almost everywhere on Mercury, indicating an ancient surface (Figure 8.20c). A huge impact basin called the *Caloris Basin* (Figure 8.20a) covers a large portion of one hemisphere. Smooth patches between many of the craters tell us that volcanic lava once flowed and covered up many of the smallest craters. Mercury has no large volcanic maria, but smaller lava plains are found almost everywhere, suggesting that Mercury went through at least as much volcanic activity as the Moon. Much like the Moon, Mercury passed through phases of accretion, differentiation, and heavy cratering. Radioactive heat accumulated enough to melt the interior again, leading to the volcanism that created the lava plains.

Tectonic processes played a more significant role on Mercury than on the Moon, as evidenced by tremendous cliffs. The cliff shown in Figure 8.20c is several hundred kilometers long and up to 3 kilometers high in places. The crater in the center of the image apparently crumpled during the formation of the cliff. This cliff, and many others on Mercury, probably formed when tectonic forces compressed the crust. However, nowhere on Mercury do we find evidence of extension (stretching) to match this crustal compression. Can it be that the whole planet simply shrank? Apparently so. Early in its history, Mercury gained much more internal heat from accretion and differentiation than did the Moon, owing to its larger proportion of iron and its stronger gravity. Then, as the core cooled, it contracted by perhaps as much as 20 kilometers in radius. This contraction crumpled and compressed the crust, forming the many compressional cliffs. The contraction probably also closed off volcanic vents, ending Mercury's volcanic activity.

Mercury lost its internal heat relatively quickly because of its small size. Its days of volcanism and

FIGURE 8.20 Mercury, like the Moon, is a heavily cratered object with evidence of volcanism.

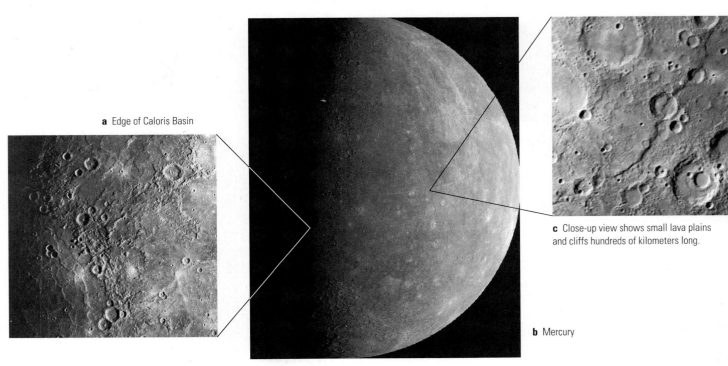

a Edge of Caloris Basin

b Mercury

c Close-up view shows small lava plains and cliffs hundreds of kilometers long.

tectonic shrinking probably ended within its first billion years. Mercury, like the Moon, has been geologically dead for most of its existence. But recent radar observations suggest that Mercury, like the Moon, may have water ice in its polar craters, probably for the same reason as on the Moon. Mercury also has a tenuous atmosphere created by bombardment, although its exact origin remains unclear.

Mercury may soon lose its unenviable status of being least explored. Another of NASA's "better, cheaper, faster" Discovery missions may be launched to Mercury in 2004 to map its entire surface, explore its unexpected magnetic fields, and study its atmosphere.

Mars (3,397-km radius, 1.52 AU from Sun)

Mars has captured the human imagination, from the first astronomers to science fiction writers to today's scientists and aerospace engineers. Much of the appeal comes from the fact that it has long seemed the most promising place to look for life beyond Earth. Early telescopic observations of Mars revealed several uncanny resemblances to Earth. A Martian day is just over 24 hours, and the Martian rotation axis is tilted about the same amount as Earth's. Mars

also has polar caps, which we now know to be composed primarily of frozen carbon dioxide, with smaller amounts of water ice. Telescopic observations also showed seasonal variations in surface coloration over the course of the Martian year (about 1.9 Earth years). All these discoveries led to the perception that Mars and Earth were at least cousins, if not twins. By the early 1900s, many astronomers—as well as the public—envisioned Mars as nearly Earthlike, possessing water, vegetation that changed with the seasons, and possibly intelligent life.

In 1877, Italian astronomer Giovanni Schiaparelli reported seeing linear features across the surface of Mars through his telescope. He named these features *canali,* the Italian word for "channels." This report inspired the American astronomer Percival Lowell to build an observatory in Flagstaff, Arizona, for the study of Mars. In 1895, Lowell claimed that canali were *canals* being used by intelligent Martians to route water around their planet. He suggested that Mars was falling victim to unfavorable climate changes and that water was a precious resource that had to be managed carefully. Lowell's studies drove rampant speculation about the nature of the Martians, fueling science fiction fantasies. The public mania drowned out the skepticism of astronomers who saw no canals through their telescopes or in their photographs.

The debate about Martian canals and cities was finally put to rest in 1965 with photos of the barren, cratered surface taken during the *Mariner 4* flyby and by many other spacecraft since. Schiaparelli's *canali* turned out to be nothing more than dark dust deposits redistributed by seasonal dust storms (Figure 8.21b). Dust also settles onto the polar caps, making alternating layers of darker dust and brighter frost (Figure 8.21a and c). The dust storms leave Mars with a perpetually dusty sky, giving it a pale pink color. Several other

FIGURE 8.21 (**a**) This photo, taken by the *Viking Orbiter,* shows the ice cap at a Martian pole during the Martian winter. (**b**) Mars Global Surveyor photo of a dust storm (curving clouds) brewing over Mars's northern polar cap. (**c**) Close-up of the edge of a polar cap, showing alternating layers of ice and dust. Each layer is a few dozen meters thick and may take thousands of years to form.

a Mars's southern hemisphere is heavily cratered.

b Valles Marineris is a huge valley created in part by tectonic stresses.

c A mantle plume was probably responsible for the volcanoes in the bottom half of the image and the networks of valleys near the top and bottom.

d Olympus Mons—the largest shield volcano in the solar system.

FIGURE 8.22 These photos of the Martian surface, taken from orbiting spacecraft, show that impact cratering has been an important process on Mars, but volcanism and tectonics have been even more important in shaping the current Martian surface.

spacecraft studied Mars in the 1960s and 1970s, culminating with the impressive Viking missions in 1976. Each of the two Viking spacecraft deployed an orbiter that mapped the Martian surface from above and landers that returned surface images and regular weather reports for almost 5 years. More than 20 years passed before the next successful missions to Mars, the Mars Pathfinder and Mars Global Surveyor missions in 1997. More missions to Mars are planned for every favorable launch window (roughly every 2 years) in the next decade and beyond.

The Viking and Mars Global Surveyor images show a much wider variety of geological features than seen on the Moon or Mercury. Some areas of the southern hemisphere are heavily cratered (Figure 8.22a), while the northern hemisphere is covered by huge volcanoes surrounded by extensive volcanic plains. Volcanism was expected on Mars—40% larger than Mercury, it should have retained a hot interior much longer.

Mars has a long, deep system of valleys called *Valles Marineris* running along its equator (Figure 8.22b). Named for the Mariner 9 spacecraft that first imaged it, Valles Marineris is as long as the United States is wide and almost four times deeper than the Grand Canyon. No one knows exactly how it formed; in fact, parts of the canyon are completely enclosed by high cliffs on all sides, so neither flowing lava nor water could have been responsible. Nevertheless, the linear "stretch marks" around the valley are evidence of tectonic stresses on a very large scale.

a This Viking photo (from orbit) shows ancient riverbeds that were probably created billions of years ago.

b This photo shows winding channels and sandbars that were probably created by catastrophic floods. The arrow shows the *Pathfinder* landing site.

c View from the floodplain (see **b** above) from the *Mars Pathfinder;* the *Pathfinder* landing site is now known as Carl Sagan Memorial Station.

FIGURE 8.23 Unlike the Moon and Mercury, Mars shows extensive evidence of erosion.

More cracks are visible on the nearby *Tharsis Bulge,* a continent-size region that rises well above the surrounding Martian surface (Figure 8.22c). Tharsis was probably created by a long-lived plume of rising mantle material that bulged the surface upward while stretching and cracking the crust. The plume was also responsible for releasing vast amounts of basaltic lava that built up several gigantic shield volcanoes, including *Olympus Mons,* the largest shield volcano in the solar system (Figure 8.22d). Olympus Mons has roughly the same shallow slope as the volcanic island of Hawaii, but it is three times larger in every dimension. Its base is some 600 kilometers wide, and it stands 26 kilometers high—some three times higher than Mount Everest. Planetary geologists estimate that the volcanoes went dormant hundreds of millions of years ago. Geologically speaking, this wasn't so long ago, and it remains possible that the Martian volcanoes will someday come back to life.

The Martian robotic explorers have also revealed geological features unlike any seen on the Moon or Mercury—features caused by erosion. Figure 8.23a no doubt reminds you of a dry riverbed on Earth seen from above. The eroded craters visible on some of the oldest terrain suggest that rain fell on the Martian surface billions of years ago. In some places, erosion has even acted below the surface, where

underground water and ice have broken up or dissolved the rocks. When the water and ice disappeared, a wide variety of odd pits and troughs were left behind. This process may have contributed to the formation of parts of Valles Marineris.

Underground ice may be the key to one of the biggest floods in the history of the solar system. Figure 8.23b shows a landscape hundreds of kilometers across that was apparently scoured by rushing water. Geologists theorize that heat from a volcano melted great quantities of underground ice, unleashing a catastrophic flood that created winding channels and large sandbars along its way.

The Mars Pathfinder spacecraft went to this interesting region to see the flood damage close up. The lander released the *Sojourner* rover, named after Sojourner Truth, an African-American heroine of the Civil War era who traveled the nation advocating equal rights for women and blacks. The six-wheeled rover, no larger than a microwave oven, carried cameras and instruments to measure the chemical composition of nearby rocks. Together the lander and rover confirmed the flood hypothesis: Gentle, dry, winding channels were visible in stereo images; rocks of many different types had been jumbled together in the flood; and the departing waters had left rocks stacked against each other in just the same manner that floods do here on Earth (Figure 8.23c).

Despite evidence that water likely flowed in the past, Mars is dry today. The stark contrast between Mars's past and present is captured in a dramatic Mars Global Surveyor image (Figure 8.24a) in which a channel that once carried water is now bone-dry and filled with sand dunes. The strong Martian winds still stir up planet-wide dust storms, redistributing dust on the surface and slowly grinding away at ancient geological features. Thus, Mars is still geologically active, though only a shadow of its former self. Ironically, even though he lacked our modern geological evidence, Percival Lowell's supposition that the planet was drying up was basically correct.

An explanation of Mars's geological evolution requires a close look at its atmospheric evolution. The abundant geological signs of water erosion on Mars present a major enigma in our understanding of the planet's climate. The evidence includes not only obvious features such as dried-up streambeds, but also subtler features such as eroded crater rims and floors. Further evidence comes from the study of meteorites believed to have come to Earth from Mars [Section 11.6]. These *Martian meteorites* contain types of minerals (clays and carbonates) that can form only in the presence of liquid water. Very recently, Mars Global Surveyor measurements may have identified the shoreline of an ocean that once filled much of the northern hemisphere (Figure 8.24b), although this conclusion is controversial.

The fact that liquid water once flowed on Mars means that its past must have been very different from its current "freeze-dried" state. The surface temperature must have been much warmer, and therefore Mars must have had a much stronger greenhouse effect and much higher atmospheric pressure than it does now. Computer simulations show that these requirements can be met by a carbon dioxide atmosphere about 400 times denser (i.e., having greater pressure) than the current Martian atmosphere—or about twice as dense as Earth's atmosphere.

The idea that Mars had a much denser atmosphere in the past is reasonable. Mars has plenty of ancient volcanoes, and calculations show that these volcanoes could have supplied all of the carbon dioxide needed to warm the planet. Moreover, if Martian volcanoes outgas carbon dioxide and water in the same proportions as volcanoes on Earth, Mars would have had enough water to fill oceans tens or even hundreds of meters deep.

The bigger question is not whether Mars once had a denser atmosphere, but where those atmospheric gases are today. Mars must somehow have lost most of its carbon dioxide gas. This loss reduced the strength of the greenhouse effect until the planet essentially froze over. The fate of the lost carbon dioxide is not completely clear. Some is locked up in the polar caps and some in carbonate minerals like those identified in the Martian meteorites. Perhaps

---

FIGURE 8.24 Envisioning a watery Mars.

a Mars Global Surveyor image highlighting Mars's change from a wet planet to a dry one. The horizontal channel extending across the lower part of the image was probably once filled with water, but it is now filled with sand dunes.

b Topographic map of Mars, showing the lowest regions in blue and the highest in red and white. Billions of years ago, an ocean may have filled the dark blue region covering much of Mars's northern hemisphere. The map is based on laser measurements by the Mars Global Surveyor spacecraft.

these carbonates formed in the same way as carbonates on Earth—through chemical reactions on seafloors after carbon dioxide dissolved in liquid oceans [Section 11.3]. Some of the gas in the early Martian atmosphere may also have been blasted away by large impacts (atmospheric cratering). Bombardment by the solar wind may have been important, especially after Mars's protective magnetosphere faded away as the core froze. We hope to learn more about the relative importance of these loss processes on Mars when future robotic spacecraft make better measurements of the distribution of carbonate rocks and the properties of Mars's current atmosphere.

The vast amounts of water once present on Mars are also gone. Some water is tied up in the polar caps and underground ice, but most has probably been lost forever. Mars lacks an ultraviolet-absorbing stratosphere, and water molecules are easily broken apart by ultraviolet photons. Once water molecules were broken apart, the hydrogen atoms would have been rapidly lost to space through thermal escape. Some of the remaining oxygen was probably lost by solar wind bombardment, and the rest was probably drawn out of the atmosphere through chemical reactions with surface rock. This oxygen literally rusted the Martian rocks, giving the "red planet" its distinctive tint.

The history of the Martian atmosphere holds important lessons for us on Earth. Mars was apparently once a world with pleasant temperatures and streams, rain, glaciers, lakes, and possibly oceans. It had all the necessities for "life as we know it." But the once hospitable planet turned into a frozen and barren desert at least 3 billion years ago, and it is unlikely that Mars will ever be warm enough for its frozen water to flow again. If life once existed on Mars, it is either extinct or hidden away in a few choice locations, such as hot springs around not-quite-dormant volcanoes. As we contemplate the possibility of future climate change on Earth, Mars presents us with an ominous example of how much things can change.

The Mars Global Surveyor and Pathfinder missions are paving the way for more adventurous missions. Two were launched in 1998: The *Mars Climate Orbiter* will thoroughly map out Mars's atmosphere and study climate change, while the *Mars Polar Lander* will actually dig a trench in the polar soil and examine it microscopically. Meanwhile, the *Mars Global Surveyor* continues to search for optimal landing sites for future rovers—one rover will search for evidence of life past and present, and another will gather rock samples for return to Earth. For the latest news on Mars missions, check the book web site.

Venus (6,051-km radius, 0.72 AU from Sun)

Venus is our closest neighbor in distance and in size, but it could hardly be more different from Earth. Its thick carbon dioxide atmosphere creates an extreme greenhouse effect that bakes the surface to unbearable temperatures. Venus's larger size allowed it to retain more interior heat than Mars, leading to greater volcanic activity. The associated outgassing produced the vast quantities of carbon dioxide in Venus's atmosphere, as well as the much smaller amounts of nitrogen and sulfur dioxide.

Venus is perpetually shrouded by some of the thickest and most unusual clouds in the solar system (Figure 8.25). The clouds are made of sulfuric acid dissolved in water droplets. These droplets form high in Venus's troposphere, where temperatures are 400 K cooler than on the surface. The droplets grow in the clouds until they form a sulfuric acid rain that falls toward the surface. But the droplets evaporate completely long before they reach the ground. The sulfuric acid is formed by chemical reactions involving sulfur dioxide and water, both of which come from volcanic eruptions. The atmosphere continuously loses sulfur dioxide through chemical reactions with surface rocks, so the presence of sulfuric acid clouds suggests that volcanoes on Venus must still be active (at least occasionally) in order to replenish the cloud-making ingredients.

FIGURE 8.25 Clouds on Venus are all that can be seen in this ultraviolet image taken by the *Pioneer Venus Orbiter.*

Venus's thick cloud cover prevents us from seeing through to its surface, but we can study its geological features with radar. *Radar mapping* involves bouncing radio waves off the surface from a spacecraft and using the reflections to create three-dimensional images of the surface. From 1990 to 1993, the Magellan spacecraft used radar to map the surface of Venus, discerning features as small as 100 meters across. During its 4 years of observations, *Magellan* returned more data than all previous planetary missions combined. Like the planet itself, almost all the geological features are named for female goddesses and famous women in history.

Venus is a large step up in size from Mars—it is almost as big as Earth. Thus, we would expect it to have a history of early impact cratering, followed by even more volcanism and tectonics than we have seen on Mars. *Magellan's* observations have borne

FIGURE 8.26 The surface of Venus is covered with abundant lava flows and tectonic features, along with a few large impact craters. Because these images were taken by the Magellan spacecraft radar, dark and light areas correspond to how well radio waves are reflected, not visible light. Nonetheless, geological features stand out well. In (**b**), the data have been converted to a three-dimensional perspective view; heights are magnified by a factor of more than 20, so the volcano is not actually as steep-sided as it appears.

a Two of Venus's relatively rare impact craters.

b Shield volcanoes like this one are common on Venus. (Height exaggerated to show detail.)

out this expectation. Most of Venus is covered by relatively smooth, rolling plains with few mountain ranges. The surface has some impact craters (Figure 8.26a), but very few compared to the Moon or even Mars. Venus lacks very small craters, because small impactors burn up in its dense atmosphere. But even large craters are rare, indicating that they must have been erased by other geological processes. Volcanism is clearly at work on Venus, as evidenced by an abundance of volcanoes and lava flows. Many are shield volcanoes, suggesting familiar basaltic eruptions (Figure 8.26b).

The most remarkable features on Venus are tectonic in origin. Its crust is quite contorted; in some regions, the surface appears to be fractured in a regular pattern (Figure 8.26c; see also Figure 8.10). Many of the tectonic features are associated with volcanic features, suggesting a particularly strong linkage between volcanism and tectonics on Venus. One striking example of this linkage is a type of roughly circular feature called a *corona* (Latin for "crown") that probably resulted from a hot, rising plume in the mantle (Figure 8.26d). The plume pushed up on the crust, forming concentric tectonic stretch marks on the surface. The plume also forced lava to the surface, dotting the area with volcanoes.

Does Venus, like Earth, have plate tectonics that moves pieces of its lithosphere around? Convincing evidence for plate tectonics might include deep trenches and linear mountain ranges like those we see on Earth, but *Magellan* found no such evidence on Venus. The apparent lack of plate tectonics suggests that Venus's lithosphere is quite different from Earth's. Some geologists theorize that Venus's lithosphere is thicker and stronger, preventing its surface from fracturing into plates. Another possibility is that plate tectonics happens only occasionally on Venus. The uniform but low abundance of craters suggests that most of the surface formed around a billion years ago. Plate tectonics is one candidate for the process that wiped out older surface features and created a fresh surface at that time. For the last billion years, only volcanism, a bit of impact cratering, and small-scale tectonics have occurred. Changing levels of volcanism may have affected the atmosphere and even the strength of the greenhouse effect.

c Tectonic forces have fractured and twisted the crust in this region.

d The circular cracks and volcanic bumps make up a corona, probably caused by the pressure of a mantle plume below.

One might expect Venus's thick atmosphere to drive strong erosional processes, but the view both from orbit and from the surface suggests that erosion has only a minimal effect. The Soviet Union landed two probes on the surface of Venus in 1975. Before the 700 K heat destroyed them, they returned images of a bleak, volcanic landscape with little evidence of erosional activity (see Figure 8.2b). Venus's slow rotation is apparently insufficient to drive significant surface winds, and its high surface temperature evaporates the corrosive rain before it reaches the surface. The lack of strong erosion on Venus leaves the tectonic contortions of its surface exposed to view, even though many features probably approach a billion years in age. Earth's terrain might look equally stark and rugged from space if not for the softening touches of wind, rain, and life.

As is the case for Mars, our understanding of Venus is incomplete until we consider its atmospheric evolution and its history of water. Venus is incredibly dry, with only minuscule amounts of water vapor in its atmosphere and not a drop on the surface, even though outgassing should have supplied a substantial amount of water. Water vapor is the most abundant gas from volcanic outgassing on Earth, and we expect similar gases to come out of volcanoes on Venus since both planets have similar overall compositions. Did Venus once have huge amounts of water? If so, what happened to it?

The leading theory suggests that Venus did indeed outgas plenty of water. But Venus's proximity to the Sun kept surface temperatures too high for oceans to form as they did on Earth. Outgassed water vapor accumulated in the atmosphere along with carbon dioxide and other gases. Because both water and carbon dioxide are strong greenhouse gases, Venus rapidly grew warmer still. Before the water disappeared, the surface may have had double its current high temperature. Solar ultraviolet photons broke apart water molecules, and the high temperature allowed rapid escape of hydrogen (just as on Mars). The remaining oxygen was lost by bombardment by the solar wind (made easier by Venus's lack of a protective magnetosphere) and by chemical reactions with surface rocks. Thus, the water that once was on Venus is gone forever.

It's difficult to prove that Venus lost an ocean's worth of water billions of years ago, but some evidence comes from the gases that didn't escape. Recall that most hydrogen nuclei contain just a single proton. But a tiny fraction of all hydrogen atoms in the universe contain a neutron in addition to the proton. This isotope of hydrogen is called *deuterium*, or sometimes "heavy hydrogen." Deuterium can be present in water molecules in place of one of the hydrogen atoms (making such molecules HDO, or "heavy water," rather than H_2O). A deuterium atom is therefore released into the atmosphere when an ultraviolet photon breaks apart such a water molecule. Both hydrogen and deuterium escape, but less deuterium escapes due to its higher mass. If Venus once had huge amounts of water molecules that broke apart, the more rapid escape of ordinary hydrogen should have left its atmosphere enriched in deuterium. Indeed, measurements show that deuterium is at least a hundred times more abundant (relative to ordinary hydrogen) on Venus than anywhere else in the solar system.

The next time you see Venus shining brightly as the morning or evening "star," consider the radically different path it has taken from that of Earth—and thank your lucky star. If Earth had formed a bit closer to the Sun or if the Sun had been a shade hotter, our planet might have suffered the same greenhouse-baked fate as Venus.

Looking Forward to Earth (6,378-km radius, 1.0 AU from Sun)

We've toured our neighboring terrestrial worlds and found a clear trend in the relationship of their geological activity to size. Soon after the intense cratering ended about 3.8 billion years ago, the larger planets underwent extensive reworking of their surfaces through volcanic and tectonic activity, which removed the evidence of earlier craters. The largest terrestrial worlds may still have ongoing tectonic and volcanic activity.

The picture is more complex when we consider the current states of the atmospheres and climates of the terrestrial worlds and their corresponding effects on erosional activity. Planetary size determines the amount of volcanic outgassing, but a planet's distance from the Sun plays a major role in atmospheric evolution. All in all, we'd expect Earth to be most like Venus in terms of volcanism and tectonics because of their similar sizes. But in terms of erosion, Earth appears to be most like its smaller cousin Mars, with a surface sculpted by running water. Yet even this analysis doesn't come close to explaining Earth's unique geology, its unusual atmosphere, or even its suitability for life.

It's ironic that astronomers began their studies of Mars and Venus expecting these planets to be variants of Earth. Instead, we've discovered that they are fundamentally different from Earth, particularly in the areas of atmospheric composition and climate. Rather than being a "role model" for understanding other planets, Earth looks more like the odd one out. The lesson of planetary exploration is that in order to understand Earth we must understand the solar system as a whole: its formation, its sister terrestrial worlds, and even the jovian planets and the smallest bodies orbiting the Sun. We are halfway through this list already, and after two more chapters we will come full circle back to Earth, ready to explore and explain our unique circumstances in Chapter 11.

Nothing is rich but the inexhaustible wealth of nature. She shows us only surfaces, but she is a million fathoms deep.

RALPH WALDO EMERSON

THE BIG PICTURE

With what we have learned in this chapter, we now have a complete "big picture" view of how the terrestrial planets started out so similar yet ended up so different. As you continue your study of the solar system, keep the following important ideas in mind:

- Much of planetary geology can be distilled down to a few basic geological processes that depend on a handful of basic planetary properties.

- Every terrestrial world was once as heavily cratered as the Moon is today, but craters have been erased on the other worlds to varying degrees—depending mainly on each planet's size.

- Planetary atmospheres are not static. Complete atmospheric transformation over the age of the solar system appears to be the rule for large planets, not the exception.

- Understanding the differences and similarities among Venus, Earth, and Mars offers some enlightening insights into how planets can change.

- Our understanding of planetary geology and atmospheres explains much of what we see on terrestrial worlds in our solar system and might help us learn what happens on planets in other solar systems.

Review Questions

1. Briefly explain how the terrestrial planets ended up spherical in shape. Would you also expect small asteroids to be spherical? Why or why not?

2. How does *differentiation* occur, and under what circumstances? Explain how differentiation led to the *core–mantle–crust* structure of the terrestrial worlds.

3. What is the *lithosphere* of a planet, and what determines its thickness? What primary factor determines how long a planetary interior remains hot? Explain.

4. What conditions appear to give a planet a *magnetic field*? Contrast the magnetic fields of the terrestrial planets. What is a *magnetosphere*? What are *charged particle belts*? Describe how the solar wind affects magnetospheres and how auroras are produced.

5. Define each of the four major geological processes: *impact cratering, volcanism, tectonics,* and *erosion.*

6. Explain how we can estimate the geological age of a planetary surface from its number of impact craters. How do we know that the *lunar maria* are younger than the *lunar highlands*?

7. Give two reasons why lava may be forced to a planetary surface. What kinds of volcanic features are made by "runny" versus "sticky" lava?

8. Describe several types of internal stresses that can drive tectonic activity on a planet. Briefly explain how interior temperature affects tectonics.

9. What factors lead to erosional activity on a planet? Which terrestrial planets have erosional features?

10. In a few words, summarize the basic characteristics of the atmospheres of each of the five terrestrial worlds.

11. Explain how to interpret the generic atmospheric structure profile shown in Figure 8.12. Define the *exosphere, thermosphere, stratosphere,* and *troposphere.* Briefly describe how the *greenhouse effect* works and on what planets it is most important.

12. Briefly explain why our sky is blue in the daytime and why sunrises and sunsets are red.

13. What is *convection*? Why is it important in a planet's interior? Why is it important in a planet's atmosphere?

14. What is *outgassing*? Which planets undergo a lot of outgassing, and why? Briefly describe the five processes by which a planetary atmosphere can lose gas.

15. What factors determine whether thermal escape will be important for an atmospheric gas on a planet? Use this idea to explain why Mercury and the Moon retain no substantial atmospheres at all and why the other terrestrial planets retain most gases but not hydrogen or helium.

16. Briefly summarize the geological history of the Moon. How and when did the maria form? Why is the Moon geologically "dead" today?

17. Briefly summarize the geological history of Mercury. What is the *Caloris Basin*? How did Mercury get its many huge cliffs? How do we know that Mercury has a huge iron core (relative to its size)?

18. Describe the *Tharsis Bulge, Valles Marineris,* and *Olympus Mons.* What geological process or processes were responsible for making them?

19. What geological processes are most important on Venus's surface? Does Venus have plate tectonics?

20. Briefly describe the evidence for climate change on Mars. Where did all the atmospheric gas go? Is it likely that Mars will ever again have a much thicker and warmer atmosphere? Why or why not?

21. Why should we expect Venus and Earth to have had very similar early atmospheres? How did the two planetary atmospheres become so different? In particular, what happened to the water on Venus? Explain the evidence supporting the idea that Venus has lost enormous amounts of water.

Discussion Questions

1. *Geological Connections.* Consider the four geological processes: impact cratering, volcanism, tectonics, and erosion. Which two do you think are most closely connected to each other? Explain your answer, giving several ways in which the processes are connected.

2. *Mars: Past and Future.* Summarize the evolution of the Martian atmosphere since planetary formation, emphasizing the important source and loss processes. What is your prediction for the future of the Martian atmosphere?

Problems

1. *Earth vs. Moon.* Why is the Moon heavily cratered, but not the Earth? Is the difference due mostly to a difference in the size of the two objects, their distance from the Sun, their composition, or some other factor? Explain in a paragraph or two.

2. *Miniature Mars.* Suppose Mars had turned out to be significantly smaller than its current size—say, the size of our Moon. How would this have affected the number of geological features due to each of the four major geological processes (impact cratering, volcanism, tectonics, and erosion)? Do you think Mars would still be a good candidate for harboring extraterrestrial life? Why or why not? Summarize your answers in two or three paragraphs.

3. *Mystery Planet.* It's the year 2098, and you are designing a robotic mission to a newly discovered planet around a star that is nearly identical to our Sun. The planet is as large in radius as Venus, rotates with the same daily period as Mars, and lies 1.2 AU from its star. Your spacecraft will orbit but not land.

   a. Some of your colleagues believe that the planet has no metallic core. How could you support or refute their hypothesis? (*Hint:* Remember that metal is dense and electrically conducting.)

   b. Other colleagues believe that the planet formed with few if any radioactive elements. How could images of the surface support or refute their hypothesis?

   c. Others suspect it has no atmosphere, but the instruments designed to study the planet's atmosphere fail due to a software error. How could you use the spacecraft's photos of geological features to determine whether a significant atmosphere is (or was) present on this planet?

4. *Atmospheric Structure.* Study Figure 8.12, which shows the atmospheric structure for a generic terrestrial planet. For each of the following cases, make a sketch similar to Figure 8.12 showing how the atmospheric structure would be different, and explain the differences in words.

   a. Suppose the planet had no greenhouse gases.

   b. Suppose the Sun emitted no solar ultraviolet light.

   c. Suppose the Sun had a higher output of X rays.

5. *Project: Geological Properties of Silly Putty.*

   a. Roll room-temperature Silly Putty into a ball and measure its diameter. Place the ball on a table and gently place one end of a heavy book on it. After 5 seconds, measure the height of the squashed ball.

   b. Warm the Silly Putty in hot water and repeat the experiment in problem 5a. Then chill the Silly Putty in ice water and repeat. How did the different temperatures affect the rate of "squashing"?

   c. Did heating and cooling have a small or large effect on the rate of "squashing"? How does this experiment relate to planetary geology?

6. *Project: Planetary Cooling in a Freezer.* To simulate the cooling of planetary bodies of different sizes, use a freezer and two small *plastic* containers of similar shape but different sizes. Fill them with cold water and place them in the freezer. Checking every hour or so, record the time and your estimate of the thickness of the "lithosphere" (the frozen layer) in the two containers.

   a. Make a graph in which you plot the lithospheric thickness versus time for each container. In which container does the lithosphere thicken fastest?

   b. How long does it take the water in each container to freeze completely? What is the ratio of the two freezing times?

   c. Describe in a few sentences the relevance of your experiment to planetary geology.

7. *Terraforming Mars.* Some people have proposed that we might someday *terraform* Mars, making it more Earth-like so that we might live there more easily. One of the first steps in terraforming Mars would be to somehow warm it up so that its ice sublimes into the atmosphere and increases the atmospheric pressure. Suggest at least one way you could cause Mars to warm up if you had unlimited resources. Describe your suggestion(s) in one or two paragraphs, and also discuss whether you think your idea might ever be practical. (*Hint:* Think about what factors control temperature and climate change.)

8. *Web Project: Martian Weather.* Follow the links from the book web site to find the latest weather report for Mars from spacecraft and other satellites. What season is it in the northern hemisphere? When was the most recent dust storm? What surface temperature was most recently reported from Mars's surface, and where was the lander located? Summarize your findings by writing a 1-minute script for a television news update on Martian weather.

9. *Web Project: Mars Exploration.* Follow the links from the book web site to find the latest results from current Mars missions.

   a. In a few sentences, describe the main scientific goals of one of the missions.

   b. Print out (or describe) an image of geological interest to you, and analyze it using the techniques developed in this chapter.

   c. Explain how the latest results have changed or improved our understanding of Mars.

10. *Amateur Astronomy: Observing the Moon.* Any amateur telescope is adequate to identify geological features on the Moon. The light highlands and dark maria should be evident, and the shading of topography will be seen in the region near the terminator (the line between night and day). Try to observe near first- or third-quarter phase. Sketch the Moon at low magnification, and then zoom in on a region of interest, noting its location on your sketch. Sketch your field of view, label its features, and identify the geological process that created them. Look for craters, volcanic plains, and tectonic features. Estimate the horizontal size of the features by comparing them to the size of the whole Moon (radius = 1,738 km).

Jovian Planet Systems

9

In Roman mythology, the namesakes of the jovian planets are rulers among gods: Jupiter is the king of the gods, Saturn is Jupiter's father, Uranus is the lord of the sky, and Neptune rules the sea. But even the most imaginative of our ancestors did not foresee the true majesty of the four jovian planets. The smallest, Neptune, is still large enough to contain the volume of more than 50 Earths. The largest, Jupiter, has a volume some 1,400 times that of Earth. These worlds are totally unlike the terrestrial planets. They are essentially giant balls of gas, with no solid surface on which to stand.

Why should we care about a set of worlds so different from our own? Apart from satisfying natural curiosity about the diversity of the solar system, the jovian planet systems serve as a testing ground for our general theories of comparative planetology. Are our theories good enough to predict the nature of these planets and their elaborate satellite systems, or are different processes at work? We'll find out in this chapter as we investigate the nature of the jovian planets, along with their intriguing moons and beautifully complex rings.

FIGURE 9.1 Jupiter, Saturn, Uranus, and Neptune, shown to scale with Earth for comparison.

9.1 The Jovian Worlds: A Different Kind of Planet

The jovian planets have long held an important place in many human cultures. Jupiter is the third-brightest object that appears regularly in our night skies, after the Moon and Venus, and ancient astronomers carefully charted its 13-year trek through the constellations of the zodiac. Saturn is fainter and slower-moving, but people of many cultures also mapped its 27-year circuit of the sky. Ancient skywatchers did not know of the planets Uranus and Neptune, but astronomers charted their orbits soon after their discoveries in 1781 and 1846, respectively.

Scientific study of the jovian worlds dates back to Galileo [Section 5.3]; however, scientists did not fully appreciate the differences between terrestrial and jovian planets until they established the absolute scale of the solar system (that is, in units such as kilometers instead of AU). Once they had established distances, they could calculate the truly immense sizes of the jovian planets from their *angular* sizes (as measured through telescopes) [Section 6.5]. Knowing the distance scale also allowed them to measure the masses of the jovian planets; recall that Newton's version of Kepler's third law allows the calculation of a planet's mass from the orbital period and average distance of one of its moons [Section 5.4]. Together, the measurements of size and mass revealed the low densities of the jovian planets, proving that these worlds are very different from the Earth.

The slow pace of discovery about the jovian planets gave way to revolutionary advances with the era of spacecraft exploration. The first spacecraft to the outer planets, *Pioneer 10* and *Pioneer 11,* flew past Jupiter and Saturn in the early 1970s. The Voyager missions followed less than a decade later. Both *Voyager 1* and *Voyager 2* visited Jupiter in 1979 and Saturn in 1981. *Voyager 2* continued on a "grand tour" of the jovian planets, flying past Uranus in 1986 and Neptune in 1989. The results were spectacular. Figure 9.1 shows a montage of the jovian planets compiled by the Voyager spacecraft. Exploration of the jovian planet systems continues with the Galileo spacecraft, orbiting Jupiter since 1995, and the Cassini mission, scheduled for arrival at Saturn in 2004.

Jupiter, Saturn, Uranus, and Neptune are so different from the terrestrial planets that we must create an entirely new mental image of the term *planet.*

Table 9.1 Comparison of Bulk Properties of the Jovian Planets

Planet	Average Distance from Sun (AU)	Mass (Earth masses)	Radius (Earth radii)	Average Density (g/cm³)	Bulk Composition
Jupiter	5.20	317	11.2	1.33	Mostly H, He
Saturn	9.53	90	9.4	0.70	Mostly H, He
Uranus	19.2	14	4.11	1.32	Hydrogen compounds and rocks, H and He
Neptune	30.1	17	3.92	1.64	Hydrogen compounds and rocks, H and He

Table 9.1 summarizes the bulk properties of these giant worlds. The most obvious difference between the jovian and terrestrial planets is size. Jupiter's mass in comparison to that of Earth is like the mass of a full-grown adult in comparison to that of a squirrel. By volume, Earth in comparison to Jupiter is like a pea in a cup of soup.

The overall composition of the jovian planets—particularly Jupiter and Saturn—is more similar to that of the Sun than to any of the terrestrial worlds. Their primary chemical constituents are the light gases hydrogen and helium, although their atmospheres also contain many hydrogen compounds and their cores consist of a mixture of rocks, metals, and hydrogen compounds. Although these cores are small in proportion to the jovian planet volumes, each is still more massive than any terrestrial planet. Moreover, the pressures and temperatures near the cores are so extreme that the "surfaces" of the cores do not resemble terrestrial surfaces at all. In a sense, the general structures of the jovian planets are opposite those of the terrestrial planets: Whereas the terrestrial planets have thin atmospheres around rocky bodies, the jovian planets have proportionally small, rocky cores surrounded by massive layers of gas (Figure 9.2). A hypothetical spacecraft descending into Jupiter, for example, would have to travel *tens of thousands* of kilometers before reaching the core.

The jovian planets are Sun-like in composition, but their masses are far too low to provide the interior temperatures and densities needed for nuclear fusion.

Calculations show that nuclear fusion is possible only with a mass at least 80 times that of Jupiter.

The jovian planets also rotate much more rapidly than any of the terrestrial worlds, but precisely defining their rotation rates can be difficult because they are not solid. In fact, different latitude ranges on the jovian planets appear to rotate at different rates. Despite these difficulties, measurements show that the jovian "days" range from only about 10 hours on Jupiter and Saturn to 16–17 hours on Uranus and Neptune.

Even the shapes of the jovian planets are somewhat different from the shapes of the terrestrial planets. Gravity makes the jovian planets approximately spherical, but their rapid rotation rates make them slightly squashed. Material near the equator, where speeds around the rotation axis are highest, is flung outward in the same way that you feel yourself flung outward when you ride a merry-go-round. The result is that jovian planets bulge around the equator.

We can trace almost all the major characteristics of the jovian planets back to their formation [Section 7.4]. The jovian planets formed far from the Sun where icy grains condensed in the solar nebula, yielding far more solid matter to accrete into planetesimals. As the massive ice/rock cores of the future jovian planets accreted, their strong gravity captured hydrogen and helium gas from the surrounding nebula. Some of this gas formed flattened "miniature solar nebulae" (or *jovian nebulae*) around the jovian planets. Just as solid grains condensed and combined into planets in the full solar nebula,

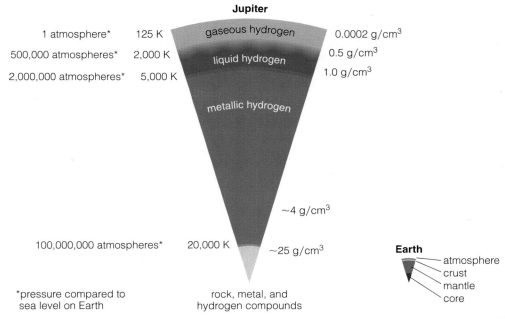

FIGURE 9.2 Jupiter's interior structure, labeled with the pressure, temperature, and density at various depths. Earth's interior structure is shown to scale for comparison. Note that Jupiter's core is only slightly larger than Earth but is about 10 times more massive.

solid grains in the nebulae surrounding the jovian planets condensed to form satellite systems. The deviations from the general patterns, such as the large 98° axis tilt of Uranus, probably arose from giant impacts as the planets were forming. In the rest of this chapter, we will explore the features of the jovian planets and their satellites in greater depth, putting our theories of planetary development to the test.

9.2 Jovian Planet Interiors

As was the case with the terrestrial planets, developing a true understanding of the jovian planets requires looking deep inside them. Probing jovian planet interiors is even more difficult than probing terrestrial interiors, but several techniques allow us to learn what lies below the clouds. We've already discussed how Earth-based observations yielded the sizes and average densities of the jovian planets and how spacecraft measurements of magnetic and gravitational fields provide additional clues to their interior structure.

Spectroscopy from Earth and from spacecraft reveals the chemical compositions of the jovian upper atmospheres. Their deeper atmospheres are generally more difficult to probe, but two recent events provided rare opportunities. In July 1994, fragments of comet Shoemaker–Levy 9 slammed into Jupiter,

blasting material from deeper in the atmosphere out into the open, where astronomers could study it through telescopes [Section 10.6]. Then, in December 1995, NASA's Galileo spacecraft dropped a scientific probe into Jupiter's atmosphere. The Galileo probe survived to a depth of about 300 kilometers—a significant distance, but still much less than 1% of Jupiter's 70,000-kilometer radius—providing our first direct data from within a jovian world.

We'll begin our discussion of the jovian interiors by using Jupiter as a prototype; then we'll apply the principles of comparative planetology to understand the other jovian planet interiors.

Inside Jupiter

Imagine plunging head-on into Jupiter in a futuristic spacesuit that allows you to survive the incredible interior conditions. Near the cloud tops, you'll find the temperature to be a brisk 125 K (−148°C), the density to be a low 0.0002 g/cm³, and the atmospheric pressure to be about 1 "atmosphere"—the same as the pressure at sea level on Earth. The deeper you go, the higher the temperature, density, and pressure become (see Figure 9.2). By a depth of 7,000 km—about 10% of the planet's radius—you'll find that the temperature has increased to a scorching 2,000 K, the density has reached 0.5 g/cm³ (half that of water), and the pressure is about 500,000 atmospheres. Under these conditions, hydrogen acts more like a liquid than a gas. By the time you reach a depth of 14,000 km, the density is about the same as that of water and the temperature is near 5,000 K. The pressure of 2 million atmospheres is so great that it forces hydrogen into an even more compact form: *metallic hydrogen,* which is so compact that all the molecules share electrons, just as happens in everyday metals. The metallic and hence electrically conducting nature of this interior hydrogen is important in generating Jupiter's strong magnetic field. Continuing your descent, you'll reach Jupiter's core at a depth of 60,000 km, about 10,000 km from the center. The core is a mix of hydrogen compounds, rocks, and metals, but these materials bear little resemblance to familiar solids or liquids because of the extreme 20,000-K temperature and 100-million-atmospheres pressure. The core materials probably remain mixed together, with a density of about 25 g/cm³, rather than separating into layers of different composition as in the terrestrial planets. The total mass of Jupiter's core is about 10 times that of Earth, so the core alone would make an impressive planet.

Our study of the terrestrial planets taught us that internal heat sources can have a strong effect on the behavior of planets, so it is natural to ask if Jupiter and the other jovian planets are also affected by internal heat sources. Jupiter apparently has a tremendous amount of internal heat—so much that it emits almost twice as much energy as it receives from the Sun. (For comparison, Earth's internal heat adds only 0.005% as much energy to the surface as does sunlight.) The magnitude of this internal heat source is much larger than can be explained by radioactive

decay, so we must search for other explanations. The most probable explanation is that Jupiter is still slowly contracting, as if it has not quite finished forming from a jovian nebula. This contraction is so gradual that we cannot measure it directly, but it converts gravitational potential energy to thermal energy, keeping Jupiter hot inside.

Comparing Jovian Planet Interiors

Now that we've discussed the interior of Jupiter, what can we say about the other jovian planets? Look back at the summary of the bulk properties of the jovian planets in Table 9.1. These properties may seem difficult to explain at first, because no clear correlation exists between size, density, and composition. But further examination of the behavior of gaseous materials removes the mystery.

Building a planet of hydrogen and helium is a bit like making one out of very fluffy pillows (Figure 9.3a). Imagine assembling a planet pillow by pillow. As each new pillow is added, those on the bottom are more compressed by those above. As the lower layers are forced closer together, their mutual gravitational attraction increases, compressing them even further. At first the stack grows substantially with each additional pillow, but eventually the growth slows until adding pillows hardly increases the height of the stack. This example explains why Jupiter is only slightly larger than Saturn in radius even though it is more than three times more massive. The extra mass of Jupiter compresses its interior to a much greater extent, making Jupiter's average density almost twice

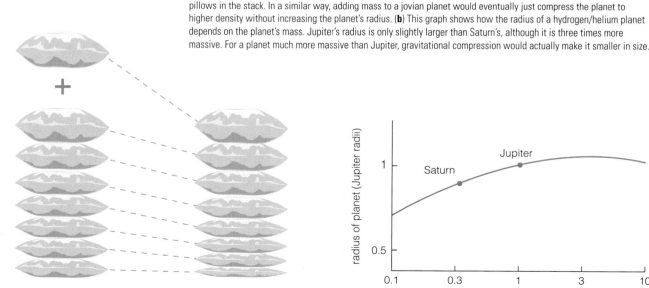

FIGURE 9.3 (**a**) Adding pillows to a stack may increase its height at first, but eventually it just compresses all the pillows in the stack. In a similar way, adding mass to a jovian planet would eventually just compress the planet to higher density without increasing the planet's radius. (**b**) This graph shows how the radius of a hydrogen/helium planet depends on the planet's mass. Jupiter's radius is only slightly larger than Saturn's, although it is three times more massive. For a planet much more massive than Jupiter, gravitational compression would actually make it smaller in size.

a

b

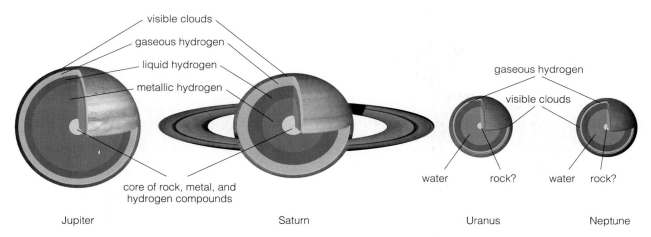

Jupiter   Saturn   Uranus   Neptune

FIGURE 9.4 These diagrams compare the interior structures of the jovian planets (shown approximately to scale). All four planets have cores equal to about 10 Earth masses of rock, metal, and hydrogen compounds, and they differ primarily in the hydrogen/helium layers that surround the cores.

that of Saturn. In fact, Saturn's average density of 0.7 g/cm^3 is less than that of water. More precise calculations show that Jupiter's radius is almost the maximum possible for a jovian planet: If more hydrogen and helium were added to Jupiter, they would actually compress the interior enough to make the planet *smaller* instead of larger (Figure 9.3b). In fact, the smallest stars are significantly smaller in radius than Jupiter, even though they are 80 times more massive.

Using the same logic that explains why Saturn's density is lower than Jupiter's, we would predict that a hydrogen/helium planet smaller than Saturn should have an even lower density. The higher densities of Uranus and Neptune therefore tell us that they cannot have the same composition as Jupiter and Saturn. Instead, Uranus and Neptune must have a much larger fraction of higher-density material, such as hydrogen compounds and rocks, and a smaller fraction of "plain" hydrogen and helium (Figure 9.4).

Saturn is large enough for interior densities, temperatures, and pressures to eventually become high enough to force gaseous hydrogen into liquid and then metallic form, just as in Jupiter. But these conditions occur relatively deeper in Saturn than in Jupiter because of its smaller size. Uranus and Neptune have even smaller layers of hydrogen and helium, so pressures are not high enough to form liquid or metallic hydrogen. However, their cores of rock, metal, and hydrogen compounds may be liquid, making for very odd "oceans" buried deep inside them.

Like Jupiter, Saturn and Neptune emit nearly twice as much energy as they receive from the Sun. But Uranus has no detectable excess internal heat. The reasons for these differences are not yet clear, but they probably have an effect on the weather occurring on these planets.

9.3 Jovian Planet Atmospheres

The jovian planets lack the geology of the terrestrial planets because they do not have solid surfaces. But whatever they lack in geology they more than make up for with their atmospheres. Fortunately, their atmospheric processes are quite similar to those we've already discussed for the terrestrial atmospheres, despite their much greater extent and very different compositions. Let's again begin by studying Jupiter as a prototype.

Jupiter's Atmosphere

Jupiter's atmosphere is mostly hydrogen and helium, but it also contains trace amounts of an interesting and important assortment of other hydrogen compounds, including methane (CH_4), ammonia (NH_3), and water (H_2O). Although these hydrogen compounds make up only a minuscule fraction of Jupiter's atmosphere, they are responsible for virtually all aspects of its appearance. Some of these compounds condense to form the clouds so prominent in telescope and spacecraft images, and others are responsible for Jupiter's great variety of colors. Without these compounds, Jupiter would be a uniform, colorless ball of gas.

TIME OUT TO THINK *Jupiter possesses substantial amounts of gases like methane, propane, and acetylene, highly flammable fuels used here on Earth. There is plenty of lightning on Jupiter to provide sparks, so why aren't we concerned about Jupiter exploding? (Hint: What's missing from Jupiter's atmosphere that's necessary for ordinary fire?)*

FIGURE 9.5 Artist's conception of the Galileo probe entering Jupiter's atmosphere. This view, looking outward from beneath the clouds, shows the suitcase-size probe falling with its parachute extended above it.

Jupiter is the only jovian planet that we've directly sampled. On December 7, 1995, following a 6-year trip from Earth, the Galileo spacecraft released a scientific probe (the size of a large suitcase) designed to study the atmosphere of Jupiter (Figure 9.5). The probe collected temperature, pressure, composition, and radiation measurements for about an hour as it descended into Jupiter's atmosphere, until it was finally destroyed by the ever-increasing pressures and temperatures. The probe sent its data via radio signals back to the Galileo spacecraft in orbit around Jupiter, which relayed the data back to Earth. When combined with observations by other spacecraft and telescopes, the probe data give astronomers a good understanding of Jupiter's atmosphere.

As on the terrestrial planets, Jupiter's weather occurs mostly in its troposphere, where higher temperatures at lower altitudes drive vigorous convection. This convection is responsible for the thick clouds that enshroud Jupiter. As a parcel of gas rises upward through the troposphere, it encounters grad-

ually lower temperatures. Soon the surrounding temperature is low enough for water vapor to condense into liquid droplets or flakes of ice, forming water clouds similar to those on Earth. As the gas continues to rise, temperatures become low enough for another minor atmospheric ingredient, ammonium hydrosulfide (NH_4SH), to condense and form clouds. Still higher, the temperatures fall to the point at which ammonia (NH_3) condenses to form clouds of ammonia crystals—the prominent white clouds visible on Jupiter. Thus, Jupiter has several layers of clouds of different compositions. The relentless motion of Jupiter's atmosphere continuously regenerates the extensive clouds.

As on Earth, Jupiter's wind patterns are driven by the planet's rotation. But Jupiter's rotation is so much more rapid than Earth's that the atmosphere breaks up into many swirling bands encircling the entire planet (Figure 9.6a). The bands of rising air are called **zones,** and they appear white in color because ammonia clouds form as the air rises to high, cool

Belts are warm, red, low-altitude clouds.

Zones are cool, white, high-altitude clouds.

a

b

FIGURE 9.6 Jupiter's belts and zones correspond to clouds of different composition at different altitudes. (**a**) The color difference between belts and zones is evident in this Hubble Space Telescope image taken at visible wavelengths. (**b**) In this infrared image taken nearly simultaneously with (**a**), brightness indicates higher temperature deeper in the atmosphere.

altitudes. The adjacent **belts** of falling air are depleted in the cloud-forming ingredients and do not contain any white ammonia clouds. Instead, they appear dark in color because we can see down to the red or tan ammonium-hydrosulfide clouds that form at lower altitudes. Infrared images confirm the temperature differences between Jupiter's warm belts and cool zones (Figure 9.6b).

Jupiter's **Great Red Spot** is perhaps the most dramatic weather pattern in the solar system (Figure 9.7). We know that it has been prominent in Jupiter's southern hemisphere for at least three centuries, because it has been seen ever since telescopes became powerful enough to detect it. The Great Red Spot is huge—more than twice as wide as the Earth. Could it be a tremendous storm like a hurricane on Earth? Not quite. The Great Red Spot actually spins in the opposite direction that a hurricane would. Astronomers speculate that the storm is kept spinning by the influence of nearby belts and zones.

Jupiter's vibrant colors remain perplexing despite decades of intense study. Observations tell us that the high clouds of ammonia in the zones are usually white and that the lower clouds of ammonium hydrosulfide in the belts are brown or red. But pure ammonium-hydrosulfide crystals are as white as ammonia crystals, so the colors of the belt clouds must come from ingredients besides the ammonium-hydrosulfide crystals themselves. Sulfur compounds are a plausible suspect, because they form a variety

of reds and tans (also seen on the surface of Jupiter's moon Io). Phosphorus compounds (including phosphine, PH_3) are another possibility. Whatever their origin, the red and tan compounds are found primarily at lower (and therefore warmer) altitudes. The chemical reactions that form them evidently require the extra energy of Jupiter's internal heat, or possibly lightning deep below the clouds.

As far as we know, Jupiter's climate is steady and unchanging. Jupiter has no appreciable axis tilt and therefore has no seasons. In fact, Jupiter's polar temperatures are quite similar to its equatorial temperatures. Solar heating alone might leave the poles relatively cool, but Jupiter's internal heat source keeps the planet uniformly warm.

Comparing Jovian Planet Atmospheres

The most striking difference among the jovian atmospheres is their colors. Starting with Jupiter's distinct red colors, the jovian planets turn to Saturn's more subdued yellows, to Uranus's faint blue-green tinge, and finally to Neptune's pronounced blue hue (see Figure 9.1). What's responsible for this regular progression of colors? The primary constituents in all the jovian atmospheres (hydrogen and helium) are colorless, so the planetary colors must come from trace gases or chemical reactions that create colored compounds.

Jupiter

Uranus

Saturn

Neptune

FIGURE 9.7 Close-ups of cloud patterns on the four jovian planets. Scientists combine series of images like these to make weather movies similar to those for Earth shown on the evening news. Jupiter's Great Red Spot is a huge high-pressure storm that is large enough to swallow two or three Earths. The Great Dark Spot is visible in the Neptune image. Jupiter, Saturn, and Neptune images by the Voyager spacecraft at visible wavelengths, Uranus image by the Hubble Space Telescope at infrared wavelengths.

Saturn's reds and tans almost certainly come from the same compounds that produce these colors on Jupiter—whatever those compounds may be. The blue colors of Uranus and Neptune have a completely different explanation: abundant methane gas, which absorbs red light but transmits blue light. The clouds then reflect (scatter) the blue light upward, where it is again transmitted through the gas above.

The Voyager spacecraft cameras monitored the meteorology of the four jovian planets with excellent resolution (Figure 9.7). Jupiter's weather phenomena (clouds, belts, zones, and other storms) are by far the strongest and most active. Saturn also possesses belts and zones (in more subdued colors), along with some small storms and an occasional large storm, but it lacks any weather feature as prominent as Jupiter's Great Red Spot. (For unknown reasons, Saturn's winds are even stronger than Jupiter's.) Neptune's atmosphere is also banded and has a high-pressure storm, called the Great Dark Spot, that acts much like Jupiter's Great Red Spot. The greatest surprise in

the weather patterns on the jovian planets is the weather on Uranus, the planet tipped on its side. When the Voyager spacecraft flew past in 1986, it was midsummer in Uranus's northern hemisphere, with the summer pole pointed almost directly at the Sun. Photographs revealed virtually no clouds, belts, or zones, which scientists attributed to the lack of a significant internal heat source on Uranus. But recently the Hubble Space Telescope has detected storms raging on the planet (Figure 9.7). The seasons are changing, and the winter hemisphere is seeing sunlight for the first time in decades.

9.4 Jovian Planet Magnetospheres

Each jovian planet is surrounded by a bubble-like magnetosphere consisting of the planet's magnetic field and the particles trapped within it. The jovian planets themselves dwarf the terrestrial planets, and their magnetospheres are even more impressive.

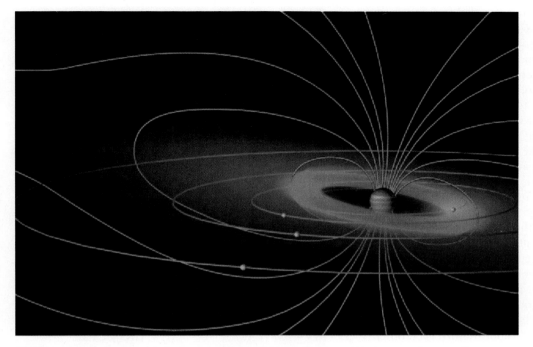

FIGURE 9.8 Artists' conception of Jupiter's magnetosphere, with Io embedded deep within. An orange glow around Io represents its escaping atmosphere, and the red donut represents the charged particle belt formed by Io. The blue lines show Jupiter's strong magnetic field.

Jupiter's magnetic field is awesome—about 20,000 times stronger than Earth's. This strong magnetic field deflects the solar wind some 3 million km (about 40 Jupiter radii) before it even reaches Jupiter. If our eyes could see this part of Jupiter's magnetosphere, it would be larger than the full moon in our sky.

Jupiter's magnetosphere contains a great number of charged particles, some captured from the solar wind but most originating from Jupiter's volcanically active moon, Io. The charged particles bombard the surfaces of Jupiter's icy moons and can generate thin bombardment atmospheres—just as bombardment creates the thin atmospheres of Mercury and the Moon [Section 8.5]. On Io, the bombardment leads to the escape of the atmospheric gases continuously released by volcanic outgassing. As a result, Io loses atmospheric gases faster than any other object in the solar system. The escaping gases (sulfur, oxygen, and a hint of sodium) are ionized and feed a donut-shaped charged particle belt, called the *Io torus*, that approximately traces Io's orbit (Figure 9.8).

None of the other jovian planets has anything like the Io torus (because none has satellites quite like Io), but their magnetospheres do hold some big surprises. Both Uranus's and Neptune's magnetic fields are tipped more than 45°. Planetary scientists are still at a loss to explain these large tilts.

9.5 A Wealth of Worlds: Satellites of Ice and Rock

More than 50 known moons (*satellites*) orbit the jovian planets. Figure 9.9 shows the larger ones, and Appendix C lists fundamental data for all the known moons. It's worth taking a few minutes to study the figure and appendix, looking for patterns, before you read on.

Broadly speaking, the jovian moons can be divided into three groups: small moons less than about 300 km across, medium-size moons ranging from about 300 to 1,500 km in diameter, and large moons more than 1,500 km in diameter. More than half the known moons fall into the small category, and additional small moons may yet be discovered. The small moons generally look more like potatoes than spheres, because their gravities are too weak to force their rigid material into spheres.

Nearly all the medium-size and large moons follow the orbital patterns we expect from formation in "miniature solar nebulae": They have approximately circular orbits that lie close to the equatorial plane of their parent planet, and they orbit in the same direction in which their planet rotates. These moons are planetlike in almost all ways. They are approximately spherical, each has a solid surface with its own unique geology, and some possess atmospheres, hot interiors, and even magnetic fields. A few are planetlike in size as well. The two largest moons—Jupiter's moon Ganymede and Saturn's moon Titan—are larger than the planet Mercury. Four others are larger than Pluto: Jupiter's moons Io, Europa, and Callisto and Neptune's moon Triton.

Nearly all moons share an uncanny trait: They always keep the same face turned toward their planet, just as our Moon always shows the same face to Earth. This trait is called *synchronous rotation,* because it means that a moon's rotation period and orbital period are synchronized, or the same (see Figure 2.19), and it is a result of tidal forces acting on the moon. [Section 5.5].

Incidentally, similar tidal processes can also affect planets. Pluto and its moon Charon are relatively close in size, so their mutual tidal forces have forced them to rotate synchronously with each other (Figure 9.10). Like two dancers, each always keeps the same face toward the other. Tidal forces from the Moon also cause the Earth's rotation to gradually slow down [Section 5.5]. In a few hundred billion years (if the Earth and the Moon stay together that long), the Earth will always show the same face to the Moon.

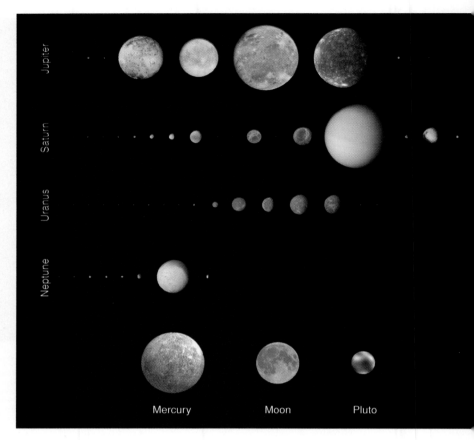

FIGURE 9.9 The larger satellites of the jovian planets, with sizes (but not distances) shown to scale. Mercury, the Moon, and Pluto are included for comparison.

Ice Geology

Prior to the Voyager missions, most scientists expected the jovian moons to be cold and geologically dead. After all, only two of the moons are even as large as Mercury, and Mercury has been geologically dead for a long time. Instead, Voyager provided many surprises, revealing worlds with spectacular past and present geological activity. We've since learned even more—about Jupiter's moons in particular—thanks primarily to data from the Galileo spacecraft.

How can moons have so much more geological activity than similar-size (or larger) planets? In a few cases, the answer involves unforeseen heat sources that we'll discuss shortly. But another important factor is the icy composition of most jovian moons, because ice melts and deforms at much lower temperatures than rock.

All in all, the *ice geology* of the jovian moons bears many similarities to the rock geology of the terrestrial planets [Section 8.3]. *Impact cratering,* occurring mostly during the time of the early bombardment more than 4 billion years ago, probably left all the satellites with battered surfaces. *Volcanism* is certainly present on some jovian moons, and *tectonics* seems likely

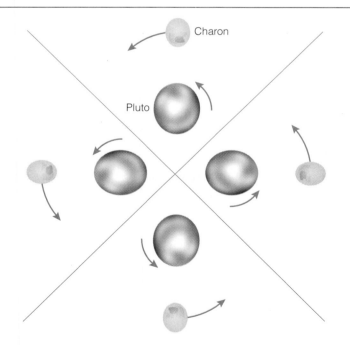

FIGURE 9.10 Pluto and Charon rotate synchronously with each other, so that each always shows the same face to the other. If you stood on Pluto, Charon would remain stationary in your sky, always showing the same face (but going through phases like the phases of our Moon). Similarly, if you stood on Charon, Pluto would remain stationary in your sky, always showing the same face (and going through phases). Tidal bulges are exaggerated for clarity.

as well. Both of these processes are driven by internal heat and hence take place more easily on icy moons than on rocky planets. We know very little about *erosion* in the outer solar system, although we expect it to be relatively rare. Most satellites possess either no atmosphere or a very thin one, so in most cases wind or rain erosion is unlikely. Nevertheless, many of the jovian moons have interesting features that go beyond simple generalities. There are far too many moons to study all of them in depth here, but as we undertake our brief tour of the jovian moon systems we will stop to investigate a few of the more intriguing moons in detail.

The Galilean Satellites of Jupiter: Io, Europa, Ganymede, and Callisto

The first stops on our tour are the Galilean satellites of Jupiter—the four moons that Galileo saw through his telescope and that disproved the idea that all celestial bodies circle the Earth [Section 5.3]. These four moons are large enough that they would count as planets if they orbited the Sun. They bear the names of four mythological lovers of the Roman god Jupiter: Io, Europa, Ganymede, and Callisto.

Io Just a few months before the Voyager spacecraft reached Jupiter, a group of scientists made the astonishing prediction that we would find active volcanism on Io. This prediction went against the common belief that only much larger bodies could have substantial geological activity. But the prediction proved correct. Direct proof came from the discovery of towering volcanic plumes reaching hundreds of kilometers above the surface and spreading fallout over almost a million square kilometers of Io's surface (Figure 9.11a). The indirect proof was just as mindboggling: the complete lack of impact craters on the surface—not a single one, to the resolution of Voyager's cameras. Io's eruptions have buried *all* its impact craters.

Io's volcanoes take many forms. The plumes photographed by the Voyager and Galileo spacecraft are not eruptions of molten rock like those from volcanoes on terrestrial planets. Instead, they resemble *geysers* on Earth, in which underground water is heated to the boiling point and shoots up in plumes of hot water and steam. If Yellowstone's Old Faithful geyser erupted under Io's low gravity and thin atmosphere, its plume would shoot up about 30 kilome-

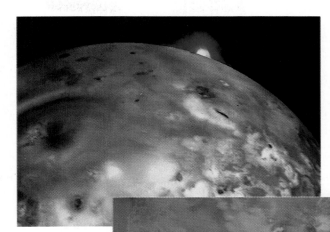

a An erupting volcanic plume rising hundreds of kilometers above Io's surface.

b The reddish color of the now-cooled lava flows extending from this volcano on Io (center black dot) suggests they were once molten sulfur.

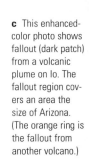

c This enhanced-color photo shows fallout (dark patch) from a volcanic plume on Io. The fallout region covers an area the size of Arizona. (The orange ring is the fallout from another volcano.)

FIGURE 9.11 Io is the most volcanically active body in the solar system.

ters. Io's plumes shoot up hundreds of kilometers because they are driven by boiling sulfur and sulfur dioxide, which are even more explosive when boiling than water. Not all volcanoes on Io erupt so violently. Liquid sulfur flows out of some vents, creating distinctly red lava flows on the surface (Figure 9.11b). Some of the higher mountains found in some regions of Io were probably built by basaltic lava flows similar to those of terrestrial volcanoes. Many of Io's volcanic vents glow red-hot. Galileo and Voyager images confirmed that fallout from volcanic plumes can cover areas the size of Arizona in a matter of months (Figure 9.11c). Io's volcanoes also produce a

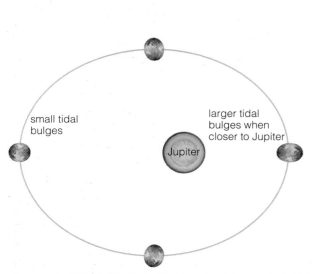

a Because Io's orbit is slightly elliptical, the strength and direction of Io's tidal bulges change. The bulges and orbital eccentricity are exaggerated.

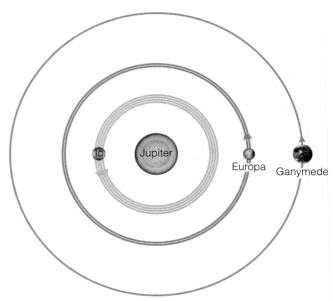

b About every seven Earth days (one Ganymede orbit, two Europa orbits, and four Io orbits), the three moons line up as shown. The small gravitational tugs repeat and make all three orbits slightly elliptical.

FIGURE 9.12 Tidal heating explained.

very thin sulfur-dioxide atmosphere—about a billion times less dense than our own atmosphere.

Why is Io so geologically active? After all, it is about the size of our own geologically dead Moon, and it is not as icy in composition as most other jovian satellites. Io must have an additional internal heat source besides the usual combination of accretion, differentiation, and radioactivity that heats the interiors of the terrestrial worlds [Section 8.2]. This fourth internal heat source is **tidal heating,** and its theory was the basis for the pre-Voyager prediction of volcanic activity on Io.

How does tidal heating work? Tidal forces lock Io in synchronous rotation so that it keeps the same face toward Jupiter. But Io's orbit is slightly elliptical, so its orbital speed and distance from Jupiter vary. As a result, the strength and even the direction of the tidal force change very slightly over its orbit: Io is continuously flexed by Jupiter (Figure 9.12a). The flexing of Io's rocky crust and interior releases heat, just as flexing warms Silly Putty. In fact, Io's tidal heating releases more than 200 times more heat (per gram of mass) than the radioactive heat driving Earth's geology, which explains why Io is the most volcanically active body in the solar system.

But why is Io's orbit slightly elliptical, when almost all other large satellites' orbits are virtually circular? Io's neighbors are responsible: Gravitational nudges between Io, Europa, and Ganymede maintain **orbital resonances** between them (Figure 9.12b). During the time in which Ganymede completes one orbit of Jupiter, Europa completes exactly two orbits and Io completes exactly four orbits. The satellites therefore are periodically lined up, and the gravitational tugs these satellites exert on one another add up over time. The satellites are always tugged in exactly the same direction, and the satellite orbits become very slightly elliptical as a result. (If the satellites didn't line up periodically, the tugs would be exerted in random directions at random times—as ineffective as pushing a child on a swing at random times.) The resulting tidal forces are strongest on Io because it is closest to Jupiter, but they are also important on Europa and Ganymede. The case of Europa is particularly interesting (Figure 9.13).

Europa Europa's bizarre, fractured crust is proof enough that tidal heating has taken hold there (Figure 9.14a). The icy surface is nearly devoid of impact craters and may be only a few million years old. Jumbled icebergs (Figure 9.14b) suggest that an ocean of liquid water lies just a few kilometers below the surface, extending down to perhaps 100 km deep. The icy shell overlying the ocean is repeatedly flexed by tidal forces, resulting in a global network of cracks (Figure 9.14c).

In some ways, Europa resembles a rocky terrestrial planet wrapped in an icy crust. The rocky object that makes up most of Europa is undoubtedly geologically active below the layer of ocean and ice. Perhaps basaltic lavas erupt on Europa's seafloors, sometimes violently enough to jumble up the icy crust

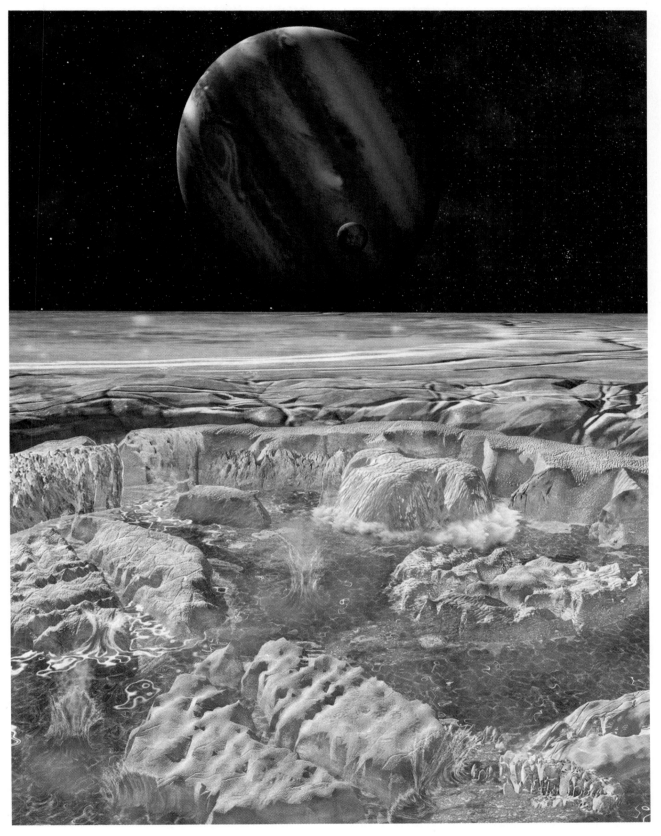

FIGURE 9.13 Tidal heating may give Europa a subsurface ocean beneath its icy crust. This artist's conception imagines a region where the crust has been disrupted by an undersea volcano.

b Some regions show jumbled crust with icebergs, apparently frozen in slush. This figure combines low-resolution images and high-resolution close-ups from the Galileo spacecraft.

a Europa's icy crust is criss-crossed with cracks.

FIGURE 9.14 Europa is one of the most intriguing moons in the solar system.

c Close-up photos show that many surface cracks have a double-ridged pattern.

above (see Figure 9.14b). And, just possibly, tidal heating makes Europa's oceans as hospitable for life as our own oceans [Section 11.6].

Ganymede Photographs of Ganymede tell the story of an interesting geological history. Parts of the surface have many impact craters, indicating that these regions are billions of years old (Figure 9.15a). But other areas show unusual "grooved terrain" unlike anything we've seen in terrestrial geology (Figure 9.15b). The grooves may have formed as Ganymede's crust expanded and fractured early in its history. Ganymede also shows some evidence of tectonics similar to plate tectonics on Earth—some chunks of the surface appear to have shifted relative to one another. Surprisingly, Ganymede has its own magnetic field—perhaps indicating a molten, convecting core.

Callisto The outermost Galilean satellite, Callisto, is about what we might expect for outer solar system satellites: a heavily cratered iceball (Figure 9.16a). The bright patches on its surface are caused by cratering: Large impactors dig up cleaner ice from deep down and spread it over the surface. The odd rings in the upper left of Callisto's image also come from impacts: The largest impactors shatter the icy sphere in a way that produces concentric rings around the impact site.

Despite its relatively large size (the third-largest moon in the solar system), Callisto lacks volcanic and tectonic features. Internal heat sources must be very small. Callisto has no tidal heating because it shares no orbital resonances with other satellites. Nonetheless, the surface holds some surprises. Close-up images show the surface to be covered by a dark, powdery substance concentrated in low-lying areas, leaving ridges and crests bright white (Figure 9.16b). No one knows the nature of this material or how it got there.

FIGURE 9.15 Ganymede, the largest moon in the solar system.

a Ganymede's numerous craters (bright spots) show that its surface is older than Europa's.

b The brighter, ridged regions of Ganymede's surface, called grooved terrain, have few craters and indicate relatively recent geological activity.

FIGURE 9.16 Callisto shows no evidence of volcanic or tectonic activity.

a Heavy cratering indicates an ancient surface.

b Close-up photos show a dark powder overlying the low areas of the surface.

Titan and the Medium-Size Moons of Saturn

Leaving Jupiter's moons behind, our satellite tour takes us next to Saturn. Here we find an amazing moon shrouded in a thick atmosphere: Titan. It is Saturn's only large moon and is the second-largest moon in the solar system after Ganymede (see Figure 9.9).

Titan's hazy and cloudy atmosphere hides its surface (Figure 9.17a). The atmosphere is about 90% nitrogen—making it the only world besides Earth where nitrogen is the dominant atmospheric constituent. The rest of Titan's atmosphere consists of argon, methane, and other hydrogen compounds, including ethane (C_2H_6). The methane and ethane actually give Titan an appreciable greenhouse effect, but the surface temperature is still a frigid 93 K ($-180°C$). The surface pressure on Titan is only somewhat higher than on Earth—about 1.5 bars at the surface, which would be fairly comfortable if not for the lack of oxygen and the cold temperatures.

How did Titan end up with such an unusual atmosphere? Titan is composed mostly of ices, including methane and ammonia ice. Some of this ice sublimed long ago to form an atmosphere on Titan. Over billions of years, solar ultraviolet light broke apart the ammonia molecules (NH_3) into hydrogen and nitrogen. The light hydrogen escaped from the atmosphere (through thermal escape [Section 8.5]), but the heavier nitrogen molecules remained and accumulated. Similarly, methane (CH_4) may have been converted to ethane, which could make clouds, rain, and even an ocean on Titan (Figure 9.17b).

Because we cannot see its surface, we do not know whether Titan is geologically active. However, given its relatively large size and icy composition, it

FIGURE 9.17 Saturn's moon Titan.

a Titan is enshrouded by a hazy, cloudy atmosphere.

b Artist's conception of the surface of Titan, showing the possible ethane oceans.

c A recent image from the Keck Telescope taken at infrared wavelengths can see through Titan's clouds to the surface. The dark areas may be oceans.

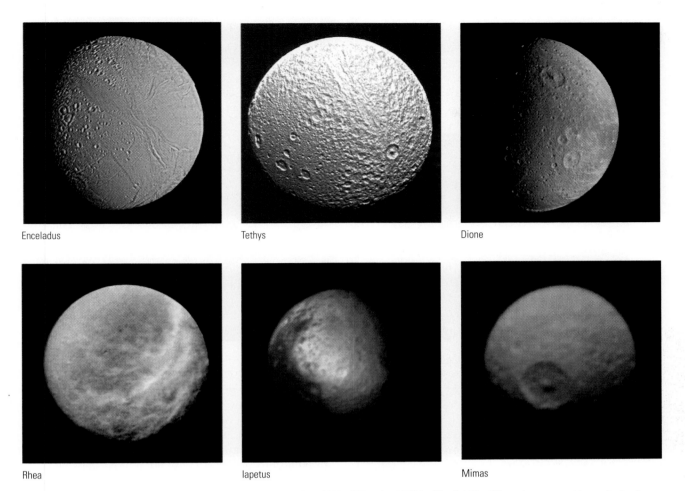

FIGURE 9.18 Saturn's six medium-size moons, which range in diameter from 390 km (Mimas) to 1,530 km (Rhea). All but Mimas show some evidence of volcanism and/or tectonics.

Enceladus · Tethys · Dione · Rhea · Iapetus · Mimas

seems likely that Titan has a rich geological history. Our understanding of Titan should improve dramatically when the Cassini spacecraft reaches Saturn in 2004 [Section 6.6]. In addition to cameras and spectrographs, *Cassini* carries a radar mapper similar to the *Magellan* radar that mapped the surface of Venus [Section 8.6]. Thus, we will see the surface of Titan at high resolution and at last will learn about its geology—and whether the surface is covered by an ethane ocean. *Cassini* will also drop a probe (named *Huygens,* for a seventeenth-century pioneer in astronomy and optics) into Titan's atmosphere that will take pictures and measure atmospheric properties on the way down. Just in case, the probe is designed to float in liquid ethane.

TIME OUT TO THINK *Based on your understanding of the four geological processes on icy satellites, what kind of surface features do you think might be present on Titan?*

Saturn's medium-size moons (Rhea, Iapetus, Dione, Tethys, Enceladus, and Mimas) show evi-

dence of substantial geological activity (Figure 9.18). While all show some cratered surfaces, Enceladus has grooved terrain similar to Ganymede's, and the others show evidence of flows of "ice lava." But these satellites are considerably smaller than Ganymede: Enceladus is barely 500 kilometers across—small enough to fit inside the boundaries of Colorado. How can such small moons support so much geological activity? The answer probably lies with "ice geology," in which icy combinations of water, ammonia, and methane can melt, deform, and flow at remarkably low temperatures. Thus, even small objects with relatively little internal heat can sustain fascinating geological activity. Iapetus is particularly bizarre, with some regions distinctly bright and others distinctly dark. The dark regions appear to be covered by a thin veneer of dark material, but we still don't know what it is or how it got there.

Mimas (radius = 195 km), the smallest of the "medium-size" moons of Saturn, is essentially a heavily cratered iceball (Figure 9.18). The impact that created its largest crater probably came close to breaking Mimas apart.

FIGURE 9.19 The surface of Miranda shows astonishing tectonic activity despite its small size. The cliff walls seen in the inset are higher than those of the Grand Canyon on Earth.

The Medium-Size Moons of Uranus

Uranus is orbited by 5 medium-size moons (see Figure 9.9) and 10 smaller moons discovered by Voyager. Astronomers have recently discovered 3 more moons, including 2 that orbit the planet backward. Miranda, the smallest of Uranus's medium-size moons (radius = 235 km), is the most surprising (Figure 9.19). Prior to the Voyager flyby, scientists expected Miranda to be a cratered iceball like Saturn's similar-size moon Mimas (see Figure 9.18). Instead, Voyager images of Miranda show tremendous tectonic features and relatively few craters. Why should Miranda be so much more geologically active than Mimas? Our best guess is tidal heating from a temporary orbital resonance with another satellite of Uranus billions of years ago. Evidently, Mimas never underwent a similar period of tidal heating.

Triton, the Backward Moon of Neptune

Last stop on our satellite tour is Neptune, so distant that only two of its moons (Triton and Nereid) were known to exist before the Voyager visit. Triton is the only large moon in the Neptune system, and only one other moon (Proteus) even makes it into our "medium-size" category.

Triton may appear to be a typical satellite, but it is not. Its orbit is retrograde (it travels in a direction opposite to Neptune's rotation) and highly inclined to Neptune's equator. These are telltale signs of a captured satellite. But Triton is not small and potato-shaped like most captured asteroids. Instead it is large, spherical, and icy. In fact, it is larger than the planet Pluto. Nonetheless, it is almost certain that Triton once orbited the Sun instead of orbiting Neptune.

A quick study of the Triton photos shown in Figure 9.20 reveals a surface unlike any other in the solar system, and one that poses many unanswered questions. Are the flat surfaces like the lunar maria? Are the wrinkly ridges (nicknamed "cantaloupe terrain") tectonic in nature? Are the bright polar caps similar to those on Mars? Clearly, Triton is more than a cratered iceball. It appears to have undergone some sort of icy volcanism in the past. Such geological activity is astonishing in light of Triton's extremely low (40 K) surface temperature, but it probably involved low-melting-point mixtures of water, methane, and ammonia ices. The internal heating source for Triton's geological activity is not known, but it may have involved tidal heating during its capture. To cap off Triton's odd geology, the sublimation of surface ices creates a thin atmosphere. Thin as it is, the atmosphere creates wind streaks on the surface, and unknown processes pump unusual plumes of gas and particles into the atmosphere.

cantaloupe terrain

frost deposits?

wind streaks

a Triton's southern hemisphere as seen by *Voyager 2.*

FIGURE 9.20 Neptune's moon Triton.

b The flat regions in this close-up photo of Triton's surface may be lava-filled impact basins similar to the lunar maria.

If Triton really is a captured satellite, where did it come from? Could other objects like it still be orbiting the Sun? Intriguingly, we know of at least one object that appears to be quite similar to Triton: the planet Pluto. But that's a story for the next chapter.

Satellite Summary: The Active Outer Solar System

Our brief tour took us to most of the medium-size and large moons in the solar system. The major lesson of the tour is that icy-satellite geology is harder to predict than terrestrial-planet geology. Size isn't everything when it comes to geological activity. Ices form lavas more easily than rocks, so any icy object can have more geological activity than a rocky object of the same size. Satellites made of ices that include methane and ammonia may have even greater potential geological activity. In addition, we saw many cases that demonstrate how tidal heating can supply added heat even when other internal heat sources are no longer important. The jovian planet satellites have turned out to be far more interesting and instructive than anyone could have imagined before the Voyager expeditions.

9.6 Jovian Planet Rings

We have completed our comparative study of the jovian planets themselves and of their major moons, but we have one more topic left to cover: their amazing rings. Saturn's rings have dazzled and puzzled astronomers since Galileo first saw them through his small telescope and suggested that they resembled "ears" on Saturn. For a long time, Saturn's rings were thought to be unique in the solar system, but we now know that all four jovian planets have rings. As we did for the planets themselves, we'll look at one ring system in detail and then examine the others for important similarities and differences. Saturn's rings are the clear choice as the standard for comparison.

FIGURE 9.21 (**a**) Earth-based telescopic view of Saturn. (**b**) Voyager image of Saturn's rings against the disk. (**c**) Artist's conception of particles in a ring system. All the particles are moving slowly relative to one another and occasionally collide.

Saturn's Rings

You can see Saturn's rings through a backyard telescope, but learning about their nature requires higher resolution. Even through large telescopes on Earth, the rings appear to be continuous, concentric sheets of material separated by gaps (Figure 9.21a). Spacecraft images reveal these "sheets" to be made of many more individual rings, each broken up into even smaller concentric *ringlets,* and different regions of a ring to range from transparent to opaque (Figure 9.21b). But even these appearances are somewhat deceiving. If we could wander into the rings, we'd see that they are made of countless individual particles orbiting Saturn together, each obeying Kepler's laws and occasionally colliding with nearby particles (Figure 9.21c). The particles range in size from large boulders to dust grains—far too small to be photographed, even when spacecraft like *Voyager* or *Cassini* pass nearby.

Spectroscopy reveals that Saturn's ring particles are made of relatively reflective (high-albedo) water ice. The rings look bright where there are enough particles to intercept most of the Sun's light and scatter it back toward us and appear more transparent where there are fewer particles. In regions where light cannot easily pass through, neither can another ring particle. In the densest parts of the rings, each particle collides with another every few hours.

TIME OUT TO THINK *Which ring particles travel faster: those at the inner edge of Saturn's rings, or those at the outer edge? (Hint: Think about Kepler's third law.) Can you think of a way to confirm your answer with telescopic observations?*

Saturn's rings lie in a thin plane directly above the planet's bulging equator, so they share Saturn's 27° axis tilt. Over the course of Saturn's year, the rings present a changing appearance to the Earth and the Sun, appearing wide open at Saturn's solstices and edge-on at its equinoxes. The rings are perhaps the thinnest known astronomical structure: They are over 270,000 kilometers across but only a few tens of *meters* thick.

Where do the rings come from? An important clue is that the rings lie close to the planet in a region where tidal forces are very influential. Within two to three planetary radii (of any planet), the tidal forces tugging an object apart become comparable to the gravitational forces holding it together. This region is called the **Roche zone.** Only relatively small objects held together by nongravitational forces (such as the electrostatic forces that hold solid rock—or spacecraft or human beings—together) can survive within the Roche zone. Thus, there are two basic scenarios for the origin of the rings. One possibility is that a wandering moon strayed too close to Saturn and was torn apart. A more likely scenario is that tidal forces prevented the material in Saturn's rings from accreting into a single large moon, instead forming many smaller moons.

Rings and Gaps

By the time the Voyager spacecraft were approaching Saturn, astronomers knew of at least six distinct rings around Saturn. Voyager scientists had hoped to find reasons for the six rings, but instead they found that the total number of individual rings and gaps may be as high as 100,000. Theorists are still struggling to explain all the rings and gaps, but some general ideas are now clear.

Rings and gaps are caused by particles bunching up at some orbital distances and being forced out at others. This bunching does not happen spontaneously but happens when gravity nudges the orbits of ring particles in some way. One source of nudging is tiny moons—called *gap moons*—located within the rings themselves. The gravity of a gap moon tugs gently on nearby particles, effectively nudging them farther away to other orbits. This clears a gap in the rings around the moon's orbit. In some cases, Voyager images show two gap moons forcing particles trapped between them into line. The gap moons are said to act as *shepherd moons* in such cases, because they shepherd the ring particles.

In some cases, ring particles orbiting Saturn may also be nudged by the gravity from Saturn's larger and more distant moons. For example, a ring particle orbiting about 120,000 kilometers from Saturn's center will circle the planet in exactly half the time it takes the moon Mimas to orbit. Every time Mimas returns to a certain location, the ring particle will also return to its original location and will experience the same gravitational nudge from Mimas. The nudges reinforce one another and eventually clear a gap in the rings (Figure 9.22a). This "Mimas 2:1 resonance" is essentially the same type of *orbital resonance* that affects the orbits of Io, Europa, and Ganymede. It is responsible for the large gap in Saturn's rings easily visible from Earth (called the *Cassini division*). Many similar resonances are responsible for other gaps, and resonances can even create beautiful ripples within the rings (Figure 9.22b).

FIGURE 9.22 (**a**) The largest gap in Saturn's rings, called the Cassini division, is caused by an orbital resonance with the moon Mimas. (**b**) Another Mimas resonance creates remarkable ripples in Saturn's rings. The dark spots in the image are calibration marks for the camera.

Cassini division

a

b

Jupiter

Uranus

Saturn

Neptune

FIGURE 9.23 Four ring systems. The planets are not shown to scale. Uranus's rings were photographed by the Hubble Space Telescope, the others by Voyager. The Neptune frame is made of two images taken on either side of the bright planet.

Although we now know three causes of rings and gaps—gap moons, shepherd moons, and orbital resonances—we have not yet identified specific causes for the vast majority of the features within Saturn's rings. Perhaps the Cassini spacecraft will identify more gap moons and shepherd moons, or perhaps it will reveal even more puzzling structures.

Comparing Planetary Rings

The ring systems of Jupiter, Uranus, and Neptune are so much fainter than Saturn's ring system that it took almost four centuries longer to discover them. Ring particles in these three systems are far less numerous, generally smaller, and much darker. Saturn's rings were the only ones known until 1977, when the rings of Uranus were discovered during observations of a *stellar occultation*—a star passing behind Uranus as seen from the Earth. During the occultation, the star "blinked" on and off nine times before it disappeared behind Uranus, and nine more times as it emerged. Scientists concluded that these nine "blinks" were caused by nine thin rings encircling Uranus. Similar observations of stars passing behind Neptune gave more confounding results: Rings appeared to be present some of the time, but not at other times. Could Neptune's rings be incomplete or transient?

The Voyager spacecraft provided some definitive answers. Voyager cameras first discovered thin rings around Jupiter in 1979. Then, after providing incredible images of Saturn's rings, *Voyager 2* confirmed the existence of rings around Uranus as it flew past in 1986. In 1989, *Voyager 2* passed by Neptune and found that it does, in fact, have partial rings—at least when seen from Earth. The space between the ring segments is filled with dust that was not detectable from Earth.

A family portrait of the jovian ring systems shows many similarities (Figure 9.23). All rings lie in their planet's equatorial plane within the Roche zone. Particle orbits are fairly circular, with small orbital tilts relative to the equator. Gaps and ringlets are probably due to gap moons, shepherd moons, and orbital resonances. The differences also offer important insights: The larger size, higher reflectivity, and much greater number of particles in Saturn's rings compel us to wonder if different processes are at work there.

Origin of the Rings

Where do rings come from? How old are they? Did they form with their planet out of the nebula, or are they a more recent phenomenon? These questions long puzzled astronomers when they had only one case (Saturn) to study, but the discovery that all jovian planets have rings has made the answers clearer.

Part of the answer lies in determining how long rings will last. Unlike planets and large satellites, ring particles may not last very long. Ring particles orbiting Saturn bump into one another every few hours, and even with low-velocity collisions the particles are chipped away over time. Basketball-size ring particles around Saturn are ground away to dust in a few million years. Other processes dismantle rings around the other jovian planets. In summary, particles in the four ring systems cannot last as long as the 4.6-billion-year age of our solar system.

If ring particles are rapidly disappearing, some source of new particles must keep the rings supplied. The most likely source is collisions. As ring particles are ground away by collisions, small moons (perhaps the gap moons) occasionally collide and create more ring particles. Meteorites may also strike these moons, with the impacts releasing even more ring particles. We now believe that the "miniature solar nebulae" that surrounded the jovian planets formed many small satellites and that the gradual dismantling of these satellites creates the dramatic ring systems we see today. The appearance of any particular ring system probably changes dramatically over millions and billions of years. The spectacle of Saturn's brilliant rings may be a special treat of our epoch, one that could not have been seen a billion years ago and that may not last long on the time scale of our solar system.

Do there exist many worlds, or is there but a single world? This is one of the most noble and exalted questions in the study of Nature.

ST. ALBERTUS MAGNUS (1206–1280)

THE BIG PICTURE

In this chapter, we saw that the jovian planets really are a "different kind of planet"—and, indeed, a different kind of planetary system. The jovian planets dwarf the terrestrial planets, and even some of their moons are as large as terrestrial worlds. As you continue your study of the solar system, keep the following "big picture" ideas in mind:

■ The jovian planets may lack solid surfaces on which geology can work, but they are interesting and dynamic worlds with rapid winds, huge storms, strong magnetic fields, and interiors in which common materials behave in unfamiliar ways.

■ Despite their relatively small sizes and frigid temperatures, many jovian satellites are geologically active by virtue of their icy compositions. Ironically, it was cold temperatures in the solar nebula that led to icy compositions and hence geological activity.

■ Ring systems owe their existence to small satellites formed from the "miniature solar nebulae" that produced the jovian planets billions of years ago. The rings we see today are composed of particles liberated from those satellites surprisingly recently.

■ Understanding the jovian planets forced us to modify many of our earlier ideas about the solar system, in particular by adding the concepts of ice geology, tidal heating, and orbital resonances. Each new set of circumstances offers further opportunities to learn how the universe works.

Review Questions

1. Briefly describe the past and present space exploration of the jovian worlds. Summarize what we've learned about the jovian planets in terms of composition, rotation rate, and shape. How did the jovian planets form?

2. Describe the interior structure of Jupiter. Contrast Jupiter's interior structure with the structures of the other jovian planets. Why is Jupiter denser than Saturn? Why is Neptune denser than Saturn?

3. How do we think Jupiter generates its internal heat? Does a similar process operate on any of the terrestrial planets?

4. What is the *Great Red Spot?* Is it the same type of storm as a hurricane on Earth? Explain.

5. Describe the possible origins of Jupiter's vibrant colors. Contrast these with the origins of the colors of the other jovian planets.

6. Does Jupiter have seasons? Why or why not?

7. What is the *Io torus?* Why don't Saturn, Uranus, or Neptune have a similar "torus"? What is unusual about the magnetospheres of the other jovian planets?

8. Briefly describe some of the general characteristics of the jovian moons.

9. What is *synchronous rotation?* What causes it? Explain why our Moon, most of the jovian moons, and Pluto and Charon rotate synchronously.

10. How is the *ice geology* of the jovian moons similar to the rock geology of the terrestrial planets? How is it different?

11. Describe the volcanoes of Io and the mechanism of *tidal heating* that generates the internal heat needed for these volcanoes.

12. What are *orbital resonances?* Explain how the resonance between Io, Europa, and Ganymede makes their orbits slightly elliptical. How do orbital resonances relate to tidal heating?

13. Describe the evidence suggesting that Europa may have a subsurface ocean of liquid water.

14. Describe the general surface characteristics of Ganymede and Callisto. What do these characteristics tell us about internal heating on these worlds?

15. Describe the atmosphere of Titan. What evidence suggests that Titan may have oceans of liquid ethane? Why do we expect to learn much more about Titan in 2004?

16. Describe a few general features of Saturn's medium-size moons. Why is it surprising to find evidence of geological activity on some of these moons, and how do we explain it? How do we explain the odd appearance of Uranus's moon Miranda?

17. How do we know that Triton is *not* a typical large jovian satellite? Describe Triton's general characteristics and give possible ideas of where Triton came from.

18. Describe the appearance and structure of Saturn's rings. Where do the rings lie in relation to Saturn's *Roche zone?*

19. Briefly describe how *gap moons* and orbital resonances can lead to gaps within ring systems.

20. Describe two possible scenarios for the origin of planetary rings. What makes us think that ring systems must be continually replenished?

Discussion Questions

1. *Jovian Mission.* We can study terrestrial planets up close by landing on them, but jovian planets have no surfaces to land on. Suppose you are in charge of planning a long-term mission to "float" in the atmosphere of a jovian planet. Describe the technology you will use and how you will ensure survival for any people assigned to this mission.

2. *Extrasolar Planets.* Many of the newly discovered planets orbiting other stars are more massive than Jupiter and much closer to their stars. Assuming that they are in fact jovian (as opposed to terrestrial or something else), how would you expect these new planets to differ from the jovian planets of our solar system? How would any magnetospheres, satellites, or rings be different? Explain.

Problems

1. *The Importance of Rotation.* Suppose the material that formed Jupiter came together without any rotation, so that no "jovian nebula" formed and the planet today wasn't spinning. How else would the jovian system be different? Think of as many effects as you can, and explain each in a sentence.

2. *Minor Ingredients Matter.* Suppose the jovian planet atmospheres were composed 100% of hydrogen and helium rather than 98% of hydrogen and helium. How would the atmospheres be different in terms of color and weather? Explain.

3. *Disappearing Satellite.* Io loses about a ton of sulfur dioxide per second to Jupiter's magnetosphere.

   a. At this rate, what fraction of its mass would Io lose in 5 billion years?

   b. Suppose sulfur dioxide currently makes up 1% of Io's mass. When will Io run out of this gas at the current loss rate?

4. *Project: Jupiter's Moons.* Using binoculars or a small telescope, view the moons of Jupiter. Make a sketch of what you see, or take a photograph. Repeat your observations several times (nightly, if possible) over a period of a couple of weeks. Can you determine which moon is which? Can you measure their orbital periods? Can you determine their approximate distances from Jupiter? Explain.

5. *Project: Saturn and Its Rings.* Using binoculars or a small telescope, view the rings of Saturn. Make a sketch of what you see, or take a photograph. What season is it in Saturn's northern hemisphere? How far do the rings extend above Saturn's atmosphere? Can you identify any gaps in the rings? Describe any other features you notice.

6. *Web Project: Galileo and Cassini Missions.* Follow the links from the book web site to find the latest news on the Galileo and Cassini missions.

   a. Print out (or describe) an image from the Galileo mission of interest to you, and interpret it using the techniques developed in this chapter.

   b. In a few sentences, describe the main scientific goals of the Cassini mission. What is the current health of the spacecraft, and how is the mission progressing?

7. *Web Project: Oceans of Europa.* The possibility of subsurface oceans on Europa holds great scientific interest and may even mean that life could exist on Europa. Use the links from the book web site to find some of the latest research concerning Europa. Write a one- to two-page summary of what we have learned about Europa since the time this book was published.

Remnants of Rock and Ice:
Asteroids, Comets, and Pluto

ASTEROIDS AND COMETS MIGHT SEEM INSIGNIFICANT IN comparison to the planets and satellites we've been discussing. But there is strength in numbers, and the trillions of small bodies orbiting our Sun are far more important than their small size might suggest.

Few topics in astronomy have the potential to touch our lives more directly than asteroids and comets. The appearance of comets more than once altered the course of human history, when our ancestors acted upon superstitions related to comet sightings. More profoundly, asteroids or comets falling to the Earth have scarred our planet with impact craters and altered the course of biological evolution.

Asteroids and comets are also important for a very different reason: They are remnants from the birth of our solar system and therefore provide an important testing ground for our theories of how our solar system came to be. In this chapter, we will explore asteroids, comets, and those bodies that fall to Earth as meteorites. We will also examine the smallest planet, Pluto, and see that, while it may be a misfit among planets, it may be right at home among the smaller objects of our solar system.

10.1 Remnants from Birth

The objects we've studied so far—terrestrial planets, jovian planets, and large moons—have changed dramatically since their formation. Most planets have been transformed in almost every way: Their interiors have differentiated, their surfaces have been reshaped by geological processes, and their atmospheres have changed as some gases escape while others are added. But many small bodies remain virtually unchanged since their formation some 4.5 billion years ago. Thus, comets, asteroids, and meteorites carry the history of our solar system encoded in their compositions, locations, and numbers.

Using these small bodies to understand planetary formation is a bit like picking through the trash in a carpenter's shop to see how furniture is made. By studying the scraps, sawdust, paint chips, and glue, we can develop a rudimentary understanding of how the carpenter builds furniture. Similarly, the "scraps" left over from the formation of our solar system allow us to understand how the planets and larger moons came to exist. Indeed, much of our modern theory of solar system formation (presented in Chapter 7) was developed from studies of asteroids, comets, and meteorites.

In this chapter, we will investigate what we have learned from these remnants from the birth of our solar system. We'll also discuss the cosmic collisions that sometimes occur between leftover planetesimals and the existing planets. Before we continue, however, it's worth noting that the terminology used to describe these objects has a long history that can sometimes be confusing. The word *comet* comes from the Greek word for "hair," and comets get their name from the long, hairlike tails they display on the rare occasions when they come close enough to the Sun to be visible in our sky. *Asteroid* means "starlike," but asteroids are starlike only in their appearance through a telescope, not in any more fundamental properties. The term *meteor* refers to a bit of interplanetary dust burning up in our atmosphere; literally, it means "a thing in the air." Meteors are also sometimes called *shooting stars* or *falling stars* because some people once believed that they really were stars falling from the sky. Larger chunks of rock that survive the plunge through the atmosphere and hit the ground are called *fallen stars* or *meteorites,* which means "associated with meteors." To keep potential confusion to a minimum, we will use only the following terms and definitions in this book:

- *Asteroid:* a rocky leftover planetesimal orbiting the Sun.

- *Comet:* an icy leftover planetesimal orbiting the Sun—regardless of its size or whether or not it has a tail.

- *Meteor:* a flash of light in the sky caused by a particle entering the atmosphere, whether the particle comes from an asteroid or a comet.

- *Meteorite:* any piece of rock that falls from the sky, whether from an asteroid, a comet, or even another planet.

10.2 Asteroids

Asteroids are virtually undetectable to the naked eye and went unnoticed for almost two centuries after the invention of the telescope. The first asteroids were discovered about 200 years ago when astronomers were searching the "extra wide" gap between the orbits of Mars and Jupiter in hopes of discovering a previously unknown planet. It took 50 years to discover the first 10 asteroids (all large and bright), but today modern telescopes with advanced detectors discover that many on a typical night. Once an asteroid's orbit is calculated and verified, the asteroid is assigned a number (in order of verification), and the discoverer chooses a name (subject to the approval of the International Astronomical Union). The earliest discovered asteroids bear names of mythological figures; more recent discoveries carry names of scientists, cartoon heroes, pets, and rock stars.

The largest asteroid, Ceres, has a radius of about 500 kilometers—about half that of Pluto. About a dozen others are large enough that we would call them "medium-size" moons if they orbited a planet. Most asteroids are far smaller. There are probably more than 100,000 asteroids larger than 1 kilometer in diameter. However, even if they could all be put together, they'd make an object less than 1,000 kilometers in radius—far smaller than any terrestrial planet.

Painstaking observations have revealed the precise orbits of thousands of asteroids. The main **asteroid belt** lies between 2.2 and 3.3 AU from the Sun, though some interesting asteroids lie outside the belt (Figure 10.1). Asteroid orbits are distinctly elliptical and inclined relative to the ecliptic by as much as 20°–30°, but all orbit the Sun in the same direction as the planets. Science fiction movies often show the asteroid belt as a crowded and hazardous place, but the average distance between asteroids is millions of kilometers. Spacecraft have to be carefully targeted to fly close enough to an asteroid to photograph one.

Origin and Evolution of the Asteroid Belt

Why are asteroids concentrated in the asteroid belt, and why didn't a full-fledged planet form in this region? The answers probably lie with *orbital resonances.*

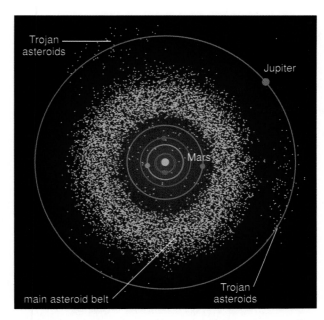

FIGURE 10.1 Positions of 8,777 asteroids for midnight, 1 January 2000. The asteroids themselves are much smaller than the dots on this scale.

Recall that an orbital resonance occurs whenever one object's orbital period is a simple ratio of another object's period, such as $\frac{1}{2}$, $\frac{1}{4}$, or $\frac{5}{3}$. In such cases, the two objects periodically line up with each other, and the extra gravitational attractions at these times can affect the objects' orbits. Consider, for example, the orbital resonances between Saturn's moons and particles in its rings: The moons are much larger than the ring particles and therefore force the ring particles out of resonant orbits, clearing *gaps* in the rings [Section 9.6]. In the asteroid belt, similar orbital resonances occur between asteroids and Jupiter, the most massive planet by far. For example, any asteroid in an orbit that takes 6 years to circle the Sun—half of Jupiter's 12-year orbital period—would receive the same gravitational nudge from Jupiter every 12 years and thus would soon be pushed out of this orbit. The same is true for asteroids with orbital periods of 4 years ($\frac{1}{3}$ of Jupiter's period) and 3 years ($\frac{1}{4}$ of Jupiter's period).

Today, we see the results of these orbital resonances

when we graph the number of asteroids with various orbital periods. Just as we would expect, there are virtually no asteroids with periods exactly $\frac{1}{2}$, $\frac{1}{3}$, or $\frac{1}{4}$ of Jupiter's. Thus, there are no asteroids with average distances from the Sun (semimajor axes) that correspond to these orbital periods. Figure 10.2 shows several other gaps in the asteroid belt, corresponding to orbital resonances that make other simple ratios with Jupiter's orbital period. (By a *gap* in the asteroid belt we mean a gap in *average* distance; because asteroid orbits are elliptical, some asteroids pass through the gaps on parts of their orbits.)

The gravitational tugs from Jupiter that created the gaps in the asteroid belt probably also explain why no planet ever formed in this region. Early in the history of the solar nebula, the region of the asteroid belt probably contained enough rocky material to form another planet as large as Earth or Mars. Just as in other parts of the solar nebula, collisions among planetesimals frequently disrupted their orbits. However, the asteroid belt had an additional source of disruption: gravitational resonances with the forming planet Jupiter. The result was that planetesimals could not collect together to form a planet. Over the past 4.5 billion years, the ongoing orbital disruptions

FIGURE 10.2 This graph shows the numbers of asteroids with different average distances (semimajor axes) from the Sun. Distances where we see few if any asteroids represent gaps. The ratio labeling each gap shows how the orbital period of objects at this distance compares to the 12-year orbital period of Jupiter. For example, the label $\frac{1}{3}$ means the orbital period of objects with this average distance is $\frac{1}{3}$ Jupiter's orbital period, or about 4 years.

have gradually kicked pieces of this "unformed planet" out of the asteroid belt altogether. Once booted from the main asteroid belt, these objects eventually either crash into a planet or a moon or are flung out of the solar system. Thus, the asteroid belt slowly loses mass, explaining why the total mass of all its asteroids is now less than that of any terrestrial planet.

Most of the asteroids found outside the asteroid belt—including the *Earth-approaching asteroids* that pass near Earth's orbit—are probably "impacts waiting to happen." But asteroids can safely congregate in two stable zones outside the main belt. These zones are found along Jupiter's orbit 60° ahead of and behind Jupiter (see Figure 10.1). The asteroids found in these two zones are called the *Trojan asteroids*. The Trojan asteroids are stable because of a different type of orbital resonance with Jupiter. In this case, any asteroid that wanders away from one of theses zones is nudged back *into* the zone by Jupiter's gravity.

Learning About Asteroids

The first step in learning about an asteroid is determining its precise orbit. Asteroids are recognizable in telescopic images because they move noticeably relative to the stars in just a short time (Figure 10.3). Once we detect an asteroid, we can calculate its orbit by applying Kepler's laws [Section 5.3]—even if we've observed only a small part of one complete orbit.

Most asteroids appear as little more than points of light even to our largest Earth-based telescopes, so the best way to study an asteroid in depth is to send

a Gaspra (16 km across)

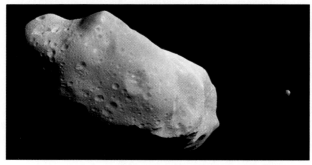

b Ida (53 km) and its tiny moon, Dactyl

c Mathilde (59 km)

FIGURE 10.4 Close-up views of three asteroids.

FIGURE 10.3 Because asteroids orbit the Sun, they move relative to the stars just as planets do. In this photograph, stars show up as white dots, and the motion of an asteroid makes it show up as a short streak.

a spacecraft out to meet it. Spacecraft have visited only a handful of asteroids close-up in their native environment of the asteroid belt and a few other presumed asteroids "in captivity" orbiting other planets (e.g., Phobos and Deimos orbiting Mars, and the backward outer satellites of Jupiter). The Galileo spacecraft on its way to Jupiter photographed the asteroid Gaspra (Figure 10.4a) and an asteroid pair known as Ida and Dactyl (Figure 10.4b). The Near-Earth Asteroid Rendezvous mission (NEAR) photographed the asteroid Mathilde (Figure 10.4c). NEAR's plan to match orbits with an asteroid called Eros went awry when the spacecraft's computers turned off the retrorockets too soon. The error turned the rendezvous into a flyby, giving us just a glimpse of the asteroid and postponing the rendezvous for more than a year. All these spacecraft images reveal battered worlds that are probably fragments of larger asteroids shattered by collisions.

10.3 Meteorites

No spacecraft has yet landed on an asteroid, but we have pieces of asteroids nonetheless—the rocks called *meteorites* that fall from the sky. The reality of rocks falling from the sky wasn't always accepted. In fact, Thomas Jefferson (who worked in science as well as politics), upon hearing of a meteorite fall in Connecticut, reportedly said, "It is easier to believe that Yankee professors would lie than that stones would fall from heaven." Today we know that rocks really do sometimes fall from the heavens. Some are large enough to do significant damage, blasting huge craters upon impact. Despite the forces of erosion, more than 100 impact craters can still be seen on Earth.

Rocks from space continue to rain down on Earth, and observers find a few meteorites every year by following the trajectories of very bright meteors called *fireballs*. Meteorites are often blasted apart in their fiery descent through our atmosphere, scattering fragments over an area several kilometers across. A direct hit on the head by a meteorite would be fatal, but there are no reliable accounts of human deaths from meteorites. However, meteorites have injured at least one human (by a ricochet, not a direct hit), damaged houses and cars, and killed animals. Most meteorites fall into the oceans, which cover three-quarters of the Earth's surface area.

Comparing Meteorites

Most meteorites are difficult to distinguish from terrestrial rocks without a detailed scientific analysis, but a few clues can help. Meteorites are usually covered with a dark, pitted crust resulting from their fiery passage through the atmosphere. Some can be readily distinguished from terrestrial rocks by their high metal content. The ultimate judge of extraterrestrial origin is laboratory analysis of the rock's composition. (Most types of meteorites contain enough metal to attract a magnet hanging on a string. If you suspect that you have found a meteorite, most science museums will analyze a small chip free of charge.) The presence of certain rare elements such as iridium is one major indicator of extraterrestrial origin; nearly all of Earth's iridium sank to the core long ago and hence is absent from surface rocks.

The tens of thousands of known meteorites fall into two basic categories. The vast majority of meteorites appear to be composed of a random mix of flakes from the solar nebula, and radioactive dating shows that they formed at the same time as the solar system itself—about 4.6 billion years ago [Section 7.5]. These **primitive meteorites** are our best source of information about conditions in the solar nebula (Figure 10.5a). Most primitive meteorites are com-

a

Stony primitive meteorite: metal flakes intermixed with rocky material.

Carbon-rich primitive meteorite: similar to meteorite at left, but with carbon compounds added in.

b

Differentiated iron meteorite: similar to a planetary core.

Differentiated stony meteorite: similar to volcanic rocks.

FIGURE 10.5 (**a**) Two examples of primitive meteorites. (**b**) Two examples of processed meteorites.

posed of rocky minerals with an important difference from Earth rocks: A small but noticeable fraction of pure metallic flakes is mixed in. (Iron in Earth rocks is chemically bound in minerals.) Some primitive meteorites also contain substantial amounts of carbon compounds, and some even contain a small amount of water.

A much smaller group of meteorites appears to have undergone substantial change since the formation of the solar system, and radioactive dating shows some of these meteorites to be younger than the primitive meteorites. We call this group **processed meteorites** because they apparently were once part of a larger object that "processed" the original material of the solar nebula into another form (Figure 10.5b). Some processed meteorites resemble the Earth's core in composition: a dense iron/nickel mixture with trace amounts of other metals. Other processed meteorites have much lower densities and are made of rocks more similar to the Earth's crust or mantle. A few even have a composition remarkably close to that of the lava erupted from terrestrial volcanoes [Section 8.3].

The Origin of Meteorites

We've seen that primitive meteorites may be either rocky or carbon-rich. Why are there two varieties? The answer lies in where they formed in the solar nebula. Carbon compounds condense only at the relatively low temperatures that were found in the solar nebula beyond about 3 AU from the Sun. Thus, we conclude that carbon-rich meteorites come from the outer portion of the asteroid belt lying beyond about 3 AU from the Sun. The remainder of the primitive meteorites lack carbon compounds and thus presumably formed closer to the Sun in what is now the inner part of the asteroid belt.

The processed meteorites tell a more complex story. Their compositions appear similar to the cores, mantles, or crusts of the terrestrial worlds. Thus, they must be fragments of worlds that underwent *differentiation* just like the terrestrial worlds [Section 8.2]. That is, they are fragments of worlds that must have been heated to high enough temperatures to melt inside, allowing metals to sink to the center and rocks to rise to the surface. The processed meteorites with basaltic compositions must come from lava flows that occurred on the surfaces of some of the larger asteroids during a time when they had active volcanism.

Meteorites resembling lava may simply have been chipped off the surface of a large asteroid by relatively small collisions. In contrast, the processed meteorites with corelike or mantlelike compositions must be fragments of worlds that completely shattered in collisions. These shattered worlds not only lost a chance to grow into planets, but they sent many fragments onto collision courses with other asteroids and other planets, including the Earth. Thus, processed meteorites essentially offer us an opportunity to study a "dissected planet."

If fragments such as the basaltic meteorites can be chipped off the surface of an asteroid, is it possible that they might also be chipped off a larger object such as a moon or a planet? In fact, a few processed meteorites have been found that don't appear to match the compositions of asteroids but instead appear to match the composition either of the Moon or of Mars. We now believe that some of these in fact were chipped off the Moon in impacts and that others were chipped off Mars. The analysis of these *lunar meteorites* and *Martian meteorites* is providing new insights into the conditions on the Moon and Mars. In at least one case, a Martian meteorite may be offering us clues about whether life once existed on Mars [Section 11.6].

10.4 Comets

Humans have watched comets in awe for millennia. The occasional presence of a bright comet in the night sky was hard to miss before the advent of big cities, inexpensive illumination, and television. In some cultures, these rare intrusions into the unchanging heavens foretold bad or good luck, and in most there was little attempt to interpret the event in astronomical terms. Nowadays, these leftover icy planetesimals can teach us about formation processes in the outer solar system, just as asteroids teach us about the inner solar system.

After Kepler developed his laws of planetary motion, other astronomers applied them to comets and found extremely eccentric orbits. Some comets visit the inner solar system only once before being ejected into interstellar space, others return on orbits spanning thousands of years, and a few come by more frequently. The most famous is Halley's comet, named for the English scientist Edmund Halley (1656–1742). Halley did not "discover" his comet but rather gained fame for recognizing that a comet seen in 1682 was the same one seen on a number of previous occasions. He used Newton's law of gravitation to calculate the comet's 76-year orbit and, in a book published in 1705, predicted that the comet would return in 1758. The comet was given his name when it reappeared as predicted, 16 years after his death. Halley's comet has returned on schedule ever since, including twice in the twentieth century—spectacularly in 1910 when it passed near the Earth, and unimpressively in 1986 when it passed by at a much greater distance. It will next visit in 2061. Several comets are discovered every year, and most have never been seen before by human eyes. In 1996 and 1997, we were treated to back-to-back brilliant comets: comet Hyakutake and comet Hale–Bopp (Figure 10.6).

The Flashy Lives of Comets

Comets are icy planetesimals from the outer solar system. Their composition has been described as "dirty snowballs": ices mixed with rocky dust. Comets are completely frozen when they are far from the Sun, and most are just a few kilometers across. But a comet takes on an entirely different appearance when it comes close to the Sun (Figure 10.7). The dirty snowball is the comet's **nucleus.** The sublimation of ices in the nucleus creates a rapidly escaping dusty atmosphere called the **coma,** along with a **tail** that points away from the Sun regardless of which way the comet is moving through its orbit. In most cases, we see two tails associated with comets: a

a

b

FIGURE 10.6 Brilliant comets can appear at almost any time, as demonstrated by the back-to-back appearances of (**a**) comet Hyakutake in 1996 and (**b**) comet Hale–Bopp in 1997, photographed at Mono Lake.

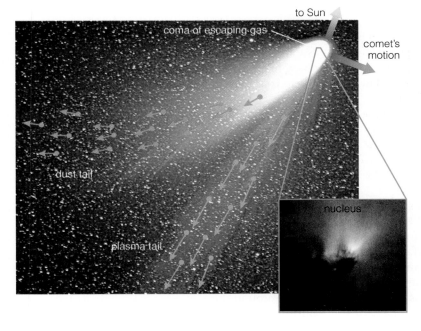

to Sun

coma of escaping gas

comet's motion

dust tail

plasma tail

nucleus

FIGURE 10.7 Anatomy of a comet. The inset photo is the nucleus of comet Halley photographed by the Giotto spacecraft; the coma and tails shown are those of comet Hale–Bopp.

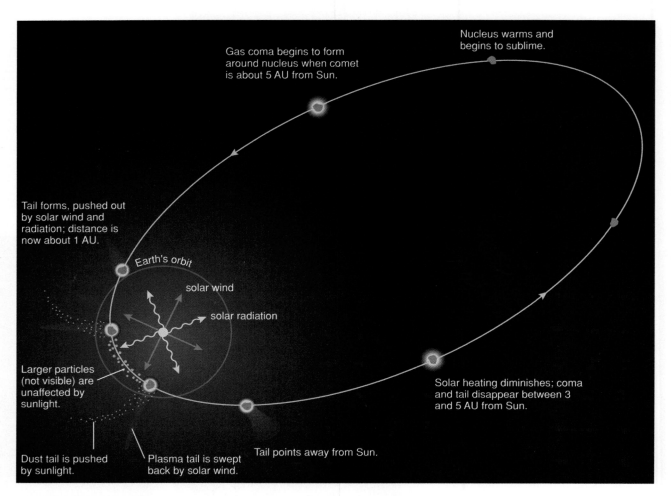

Gas coma begins to form around nucleus when comet is about 5 AU from Sun.

Nucleus warms and begins to sublime.

Tail forms, pushed out by solar wind and radiation; distance is now about 1 AU.

Earth's orbit

solar wind

solar radiation

Larger particles (not visible) are unaffected by sunlight.

Dust tail is pushed by sunlight.

Plasma tail is swept back by solar wind.

Tail points away from Sun.

Solar heating diminishes; coma and tail disappear between 3 and 5 AU from Sun.

FIGURE 10.8 Comets exist as bare nuclei over most of their orbits and grow a coma and tails only when they approach the Sun. (Diagram not to scale.)

plasma tail made of ionized gas, and a **dust tail** made of small solid particles. Although much of comet science focuses on the dramatic tail and coma, we begin our closer look at comets with the humbler nucleus.

Comet nuclei are rarely observable with telescopes, as they are quite small and are shrouded by the dusty coma. A spacecraft flyby of comet Halley revealed a dark, lumpy, potato-shaped nucleus about 16 kilometers long and 8 kilometers in width and depth. Despite its icy composition, comet Halley's nucleus is darker than charcoal, reflecting only 3% of the light that falls on it. It doesn't take much rocky soil or carbon-rich material to darken a comet.

How can such a small nucleus put on such a spectacular show? Comets spend most of their lives in the frigid outer limits of our solar system. As a comet accelerates toward the Sun, its surface temperature increases and ices begin to sublime into gaseous form. By the time the comet comes within about 5 AU of the Sun, sublimation begins to form a noticeable atmosphere that easily escapes the

comet's weak gravity. The escaping atmosphere drags away dust particles that were mixed with the ice and begins to create the dusty coma around the nucleus. Sublimation rates increase as the comet approaches the Sun, and gases jet away from patches on the nucleus at speeds of hundreds of meters per second (Figure 10.8).

As gas and dust move away from the comet, they are influenced by the Sun in different ways. The gases are ionized by ultraviolet photons from the Sun and then carried straight outward away from the Sun with the solar wind at speeds of hundreds of kilometers per second. These ionized gases form the *plasma tail,* which may extend hundreds of millions of kilometers away from the Sun. Dust-size particles experience a much weaker push caused by the pressure of sunlight itself (called *radiation pressure* [Section 14.4]), so the *dust tail* is swept away from the Sun in a slightly different direction (see Figure 10.8). The comet wheels around the Sun, always keeping its tails of dust and plasma pointed roughly away from the Sun.

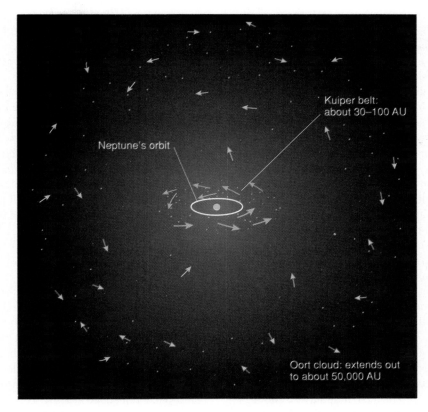

FIGURE 10.9 The Kuiper belt and the Oort cloud. Arrows indicate representative orbital motions of objects in the two regions. (Figure is not to scale.)

Comets also eject some larger, pebble-size particles that are not pushed away from the Sun. These form another, invisible, "tail" extending along the comet's orbit and are the particles responsible for meteor showers [Section 10.6].

After the comet loops around the Sun and begins to head back outward, sublimation declines, the coma dissipates, and some of the dust settles back to the surface. Nothing happens until the comet again comes sunward—in a century, a millennium, a million years, or perhaps never.

Active comets cannot last forever. In a close pass by the Sun, a comet may shed a layer 1 meter thick. Dust that is too heavy to escape accumulates on the surface. This thick dusty deposit helps make comets dark and may eventually block the escape of interior gas that makes comets so phenomenal. It is estimated that a comet loses about 0.1% of its volatiles on every pass around the Sun. No one is certain what happens after the volatiles stop escaping. Either the remaining dust disguises the dead comet as an asteroid until it crashes into a planet or is ejected from the solar system, or the comet comes "unglued" without the binding effects of ices and simply disintegrates along its orbit.

The Origin of Comets

Comets are a rarity in the inner solar system, with only a handful coming within the orbit of Jupiter at any one time. However, because comets cannot last for very many passes by the Sun, comets must come from enormous reservoirs at much greater distances: the Kuiper belt and the Oort cloud (Figure 10.9).

Most comet orbits fit no pattern—they do not orbit the Sun in the same direction as the planets, and their elliptical orbits can be pointed in any direction. Comets Hyakutake and Hale–Bopp both fall into this class. It is believed that such comets are visitors from the Oort cloud. Based on the number of comets that fall into the inner solar system, the Oort cloud must contain about a trillion (10^{12}) comets. However, Oort cloud comets normally orbit the Sun at such great distances—up to about 50,000 AU—that we have so far seen only the rare visitors to the inner solar system. The comets of the Oort cloud probably formed in the vicinity of the jovian planets and were flung into their large, random orbits by gravitational encounters with these planets.

In contrast, the comets of the Kuiper belt probably still lie in the same general region in which they formed, which is why their orbits match the pattern of the solar nebula as a whole. In the 1990s, astronomers began using the largest telescopes to search for Kuiper belt objects in their native environment rather than on their excursions into the inner solar system. The first discovery, given the name 1992QB1, orbits beyond Pluto with an orbital period of 296 years. It is a comet approximately 280 kilometers across. Subsequent searches have identified dozens of other objects that match the predicted orbital properties of Kuiper belt comets—fairly circular orbits in the same direction as the planets, with small orbital tilts. More than 160 Kuiper belt comets have been discovered so far, suggesting that at least 100,000 comets more than 100 km across must populate the Kuiper belt, along with a billion objects around 10 km across (the typical size of comets that enter the inner solar system). Thus, the total mass of the Kuiper belt is far greater than that of the asteroid belt.

What are the limits on Kuiper belt objects? An odd object called *Chiron* orbits the Sun between Saturn and Uranus. About 170 km across, Chiron once was classified as a surprisingly distant asteroid. However, Chiron recently developed a coma, suggesting that it is actually a comet from the Kuiper belt. An even larger comet, designated 1996T066, may be 600 km across. If a comet can be 600 km across, could one be larger still? It's time to consider the odd planet Pluto.

10.5 Pluto: Lone Dog or Part of a Pack?

Pluto was discovered in 1930 by American astronomer Clyde Tombaugh, culminating a search that began not long after the discovery of Neptune in 1846. Neptune's existence and location were predicted before its actual discovery by mathematical analysis of irregularities in the orbit of Uranus. Further analysis of Uranus's orbit suggested the existence of a "ninth planet," and Pluto was found only 6° from the predicted position of this planet. Initial estimates suggested that Pluto was much larger than Earth, but successively more accurate measurements derived ever-smaller sizes. We now know that Pluto has a radius of only 1,195 kilometers and a mass of just 0.0025 Earth mass—making it far too small to affect the orbit of Uranus. In fact, the discovery of Pluto near the predicted position of the "ninth planet" was coincidental; Uranus's supposed orbital irregularities were apparently just errors in measurement. Pluto might have escaped detection for decades without those errors.

Pluto has long seemed to be a misfit among the planets, fitting into neither the terrestrial nor the jovian category. Its 248-year orbit is also unusually elliptical and significantly tilted relative to the ecliptic. (See Figure 7.1.) Near perihelion, it actually comes closer to the Sun than Neptune—such was the case between 1979 and 1999. At aphelion, it is 50 AU from the Sun, putting it far beyond the realm of the jovian planets. Despite the fact that Pluto sometimes comes closer to the Sun than Neptune, there is no danger of the two planets colliding: For every two Pluto orbits, Neptune circles the Sun three times. Because of this stable *orbital resonance,* whenever Pluto is near Neptune's orbit, Neptune is always a safe distance away. The two will probably continue their dance of avoidance until the end of the solar system.

Pluto and Its Moon

Pluto's great distance and small size make it difficult to study, but the task became much easier with the discovery of Pluto's moon Charon in 1978. The original discovery was made when astronomers noticed a "bump" on Pluto's blurry image that moved from side to side over a 6.4-day period. Today we have much clearer pictures of Pluto and Charon from the Hubble Space Telescope (Figure 10.10). The discovery of the moon enabled astronomers to accurately determine the mass of Pluto by applying Newton's version of Kepler's third law. They also learned that Pluto's rotation axis is tipped 118° relative to its orbit, making it the third planet (in addition to Venus and Uranus) that rotates backward relative to the majority of planets.

Pluto's moon was discovered just in time, because astronomers soon learned that Charon's orbit was about to go edge-on as seen from the Earth—something that happens only every 124 years. From 1985 to 1990, astronomers carefully monitored the combined brightness of Pluto and Charon as they alternately eclipsed each other every few days. Detailed analysis of brightness variations allowed calculation of accurate sizes, masses, and densities for both Pluto and Charon, as well as the compilation of rough maps of their surface markings. Charon's diameter is more than half Pluto's, and its mass is about one-eighth the mass of Pluto. Furthermore, Charon orbits only 20,000 kilometers from Pluto. (For comparison, our Moon has a mass $\frac{1}{80}$ that of Earth and orbits 240,000 kilometers away.) Charon's large relative size has caused some scientists to speculate that it, like our Moon, formed in a giant impact. Some astronomers argue that Pluto and Charon qualify as a "double planet," but others argue that neither is large enough to qualify as a planet at all.

Pluto currently has a thin atmosphere of nitrogen and other gases formed by sublimation of surface

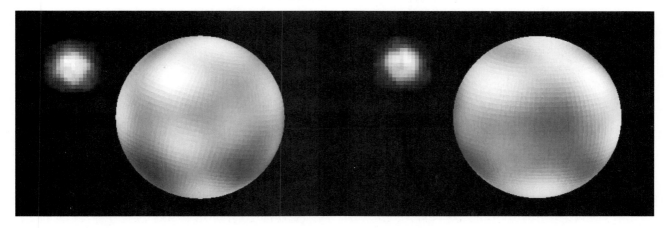

FIGURE 10.10 The surfaces of Pluto and Charon, as reconstructed from Hubble Space Telescope images. The pairs of spheres show views of opposite hemispheres.

ices. However, the atmosphere is steadily thinning because Pluto's elliptical orbit is now carrying it farther from the Sun. (Pluto will not reach aphelion until 2113.) As Pluto recedes, its atmospheric gases are refreezing onto its surface.

Despite the cold, the view from Pluto would be stunning. Charon would dominate the sky, appearing almost 10 times larger in angular size than our Moon. Pluto and Charon's mutual tidal pulls long ago made them rotate synchronously with each other (see Figure 9.10). Thus, Charon is visible from only one side of Pluto and would hang motionless in the sky, always showing the same face. Moreover, the synchronous rotation means that Pluto's "day" is the same length as Charon's "month" (orbital period) of 6.4 Earth days. Charon would neither rise nor set in Pluto's skies but instead would cycle through its phases in place. The Sun would appear more than a thousand times fainter than it appears here on Earth, and it would be no larger in angular size than Jupiter is in our skies.

Pluto: Planet or Kuiper Belt Object?

Pluto is an object in search of a category. It is a misfit among the planets, but it seems somewhat less out of place when considered in the context of the Kuiper belt. After all, it orbits in the same vicinity as Kuiper belt comets, it has a cometlike composition of icy and rocky materials, and it has a cometlike atmosphere that grows as it comes nearer the Sun in its orbit and fades as it recedes. Could Pluto be simply the largest known member of the Kuiper belt?

A closer look at the orbits of Kuiper belt objects uncovers an uncanny resemblance to the orbit of Pluto. Like Pluto, many Kuiper belt objects lie in stable orbital resonances with Neptune, and more

than a dozen Kuiper belt objects have the *same* period and semimajor axis as Pluto itself (and are nicknamed "Plutinos").

All in all, Pluto stands apart from Kuiper belt objects only in its size. But Pluto may not have been so unusual in the early solar system. Recall that Neptune's moon Triton is quite similar to (and larger than) Pluto, and its orbit indicates that it must be a captured object. Thus, Triton was once a "planet" orbiting the Sun. Giant impacts of other large icy objects may have created Charon and might even have given Uranus its large tilt. It remains possible that we will someday discover other objects with sizes similar to that of Pluto, though probably farther away or considerably darker. As of 1999, much less than 1% of the sky had been thoroughly searched for large Kuiper belt objects.

TIME OUT TO THINK *What is your definition of a planet? Be as rigorous as possible. Exactly how many planets are there, according to your criteria? Does your list include Pluto, Charon, Ceres or other asteroids, or Triton (a captured cousin of Pluto)? How big would a new Kuiper belt object have to be to meet your definition? Try to convince a classmate of your definition. Is your definition useful?*

Planet or not, Pluto remains the largest known object in the solar system never visited by spacecraft. Close-up observations of its surface, atmosphere, and unusual moon are sure to teach us as much about icy bodies as past missions did about rocky bodies. NASA's current goal is to send a mission to Pluto before its atmosphere completely refreezes onto the surface, but budget constraints may keep the mission from becoming a reality. (Check the text web site for updates on the status of a Pluto mission.)

FIGURE 10.11 The SL9 impacts allowed astronomers their most direct view ever of cosmic collisions.

a The Galileo spacecraft, on its way to Jupiter at the time, got a direct view of the impacts on Jupiter's night side.

b In under 20 minutes, the Hubble Space Telescope observed an impact plume rise thousands of kilometers above Jupiter's clouds and then collapse back down.

c This infrared photo shows the brilliant glow of a rising fireball from one of the impacts. Most of Jupiter appears dark because methane gas in its atmosphere absorbs infrared light.

10.6 Cosmic Collisions: Small Bodies Versus the Planets

The hordes of small bodies orbiting the solar system are slowly shrinking in number through collisions with the planets and ejection from the solar system. Many more must have roamed the solar system in the days of the early bombardment, when most impact craters were formed [Section 7.6]. But there are still plenty left, and cosmic collisions still occur on occasion, with important ramifications for Earth as well as for other planets.

Shoemaker–Levy 9 Impacts on Jupiter

We usually think about impacts in the context of solid bodies because such impacts leave long-lasting scars in the form of impact craters. But impacts must be even more common on the jovian planets than on the terrestrial planets because of their larger size and stronger gravity, although no impact craters can form on gaseous worlds. It's estimated that a major impact on Jupiter occurs about once every 1,000 years. We were privileged to witness one in 1994.

Comet-hunters Gene and Carolyn Shoemaker, along with their colleague David Levy, came across an unusual object that turned out to be a string of comet nuclei all in a row. The reason for the odd appearance of *Shoemaker–Levy 9*, or *SL9* for short,

became clear when astronomers calculated its orbit backward in time. The calculations showed that it had passed very close to Jupiter only a few months earlier (in July 1992). Because it had passed well within Jupiter's Roche zone [Section 9.6], it apparently was ripped apart by tidal forces. Prior to that, SL9 was probably a single comet nucleus orbiting Jupiter, unnoticed by human observers.

The orbital calculations also showed that, thanks to gravitational nudges from the Sun, SL9 was on a collision course with Jupiter. Astronomers had more than a year to plan for observations of the impacts due in July 1994. The impacts of the 22 identified nuclei would take almost a week, and virtually every telescope on Earth and in space would be ready and watching. The only uncertainty was whether or not anything would happen. Even the best Hubble Space Telescope images were unable to determine whether the comet nuclei were massive kilometer-size chunks or merely insubstantial puffs of dust.

The answer came within minutes of the first impact: Infrared cameras recorded an intense fireball of hot gas rising thousands of kilometers above the impact site, which lay just barely on Jupiter's far side (Figure 10.11). As Jupiter's rapid rotation carried the scene into view, the collapsing plume smashed down on Jupiter's atmosphere, leading to another episode of heating and more infrared glow. News of each

e In this visible-light photo from the Hubble Space Telescope, the high, dusty clouds left by the impacts appear as dark "scars" on Jupiter.

f This Hubble Space Telescope photo shows how the impact scars became smeared due to atmospheric winds. The scars disappeared completely within a few months.

d This infrared photo shows high clouds created by several impacts. These clouds reflect infrared light from the Sun because they lie high above the methane gas that absorbs infrared light.

g Artist's conception of the impacts viewed from the surface of Io.

subsequent impact spread around the world rapidly via the Internet, and everyone waited anxiously to hear each observatory report. The Hubble Space Telescope caught impact plumes in action as they rose into sunlight and collapsed back down, leaving dust clouds high in Jupiter's stratosphere. The impact scars lingered for months, but Jupiter has recovered—for now.

Comet Tails and Meteor Showers

Far smaller impacts happen on Earth all the time, lighting up the sky as *meteors*. If you watch the sky on a clear night, you'll typically see a few meteors each hour. Most meteors are created by single pieces of comet dust, each no larger than a pea, that enter our atmosphere at speeds of up to 250,000 km/hr (70 km/s). The particles and the surrounding gas are heated so much that we see a brief but brilliant flash as they are vaporized high in the atmosphere. An estimated 25 million meteors occur worldwide every day, adding hundreds of tons of comet dust to the Earth daily.

Earth's orbit intersects many comet orbits over the course of the year, and each produces a *meteor shower* during which we may see dozens of meteors per hour (Table 10.1). The most famous meteor shower, the *Perseids,* occurs every August when the Earth passes through the orbit of comet Swift–Tuttle. Although meteors all strike the Earth on parallel tracks, meteor showers appear to radiate from a particular direction

Table 10.1 Major Annual Meteor Showers

Shower Name	Approximate Date
Quadrantids	January 3
Lyrids	April 22
Eta Aquarids	May 5
Delta Aquarids	July 28
Perseids	August 12
Orionids	October 22
Taurids	November 3
Leonids	November 17
Geminids	December 14
Ursids	December 23

in the sky. You may have experienced a similar illusion while driving into a blizzard at night: The snowflakes seem to diverge from a single direction that depends on the combination of your speed and the wind speed (Figure 10.12a,b). (This illusion occurs even before the air flow around the car deflects the motion of the snowflakes.) Meteor showers get their name from the constellation from which they appear to radiate (Figure 10.12c). The Earth runs into more meteors on the side facing in the direction of Earth's orbital motion (just as more snow hits the front windshield while the car is moving), so the predawn sky is the best place to watch for meteors.

FIGURE 10.12 Snowflakes and meteor showers.

a Driving into a blizzard.

b Geometry of a meteor shower.

c Meteors photographed during the intense Leonid meteor shower in 1966. Note one meteor at top center headed straight for the camera.

a

b

FIGURE 10.13 Impacts on Earth. (**a**) Meteor Crater in Arizona was created about 50,000 years ago by an asteroid impact. The crater is more than a kilometer across and almost 200 meters deep, but the asteroid that made it was only about 50 meters across. (**b**) The arrow points to a layer of sediment laid down by the impact 65 million years ago. At the time, the rock layers above the arrow did not exist. Dust from the impact and soot from global wildfires settled down through the atmosphere onto the seafloor that once occupied this location in Colorado.

TIME OUT TO THINK *The associated "comet" for the Geminid meteor shower is an object called Phaethon, which is classified as an asteroid because we've never seen a coma or tail associated with it. How is it possible that Phaethon looks like an asteroid today but once shed the particles that create the Geminid meteor shower?*

Impacts and Mass Extinctions

Meteorites and impact craters bear witness to the fact that much larger impacts occasionally occur on Earth. Meteor Crater in Arizona (see Figure 10.13a) formed about 50,000 years ago when a metallic impactor roughly 50 meters across crashed to Earth with the explosive power of a 20-megaton hydrogen bomb. Although the crater is only a bit more than 1 kilometer across, an area covering hundreds of square kilometers was probably battered by the blast and ejecta. Far bigger impacts have occurred, sometimes with catastrophic consequences for life on Earth.

Collecting geological samples in Italy in 1978, the father–son team of Luis and Walter Alvarez discovered a thin layer of dark sediments that had ap-

parently been deposited 65 million years ago—about the same time that the dinosaurs and many other organisms suddenly became extinct. Subsequent studies found similar sediments deposited at the same time at many sites around the world (Figure 10.13b). Careful analysis showed this worldwide sediment layer to be rich in the element iridium, which is rare on Earth's surface. But iridium is common in primitive meteorites, which led the Alvarezes to a stunning conclusion: The extinction of the dinosaurs was caused by the impact of an asteroid or comet. This conclusion was not immediately accepted and still generates some controversy, but it now seems clear that a major impact coincided with the death of the dinosaurs. Moreover, while the dinosaurs were the most famous victims of this **mass extinction,** it seems that up to 99% of all living things were killed and that 75% of all *species* living on Earth were wiped out at that time.

How could an impact lead to mass extinction? The amount of iridium deposited worldwide suggests that the impactor must have been about 10 kilometers across. After a decade-long search, scientists identified what appears to be the impact crater from

a A 10-km impactor, hurtling toward Earth at perhaps 100,000 km/hr or more, is about to enter the atmosphere.

FIGURE 10.14 Instant geology—and biology: Artist's conception of the impact that led to a mass extinction 65 million years ago.

b One second to impact.

c Impact.

e Many years later, after the dust has settled, we see the 200-km-wide crater from the impact.

d One month later, debris from the impact fills our atmosphere with dust.

the event. Located off the coast of Mexico's Yucatán peninsula, it is 200 kilometers across, which is close to what we expect for a 10-kilometer impactor, and dates to 65 million years ago. The impact almost immediately sent a shower of debris raining across much of nearby North America and generated huge waves that may have sloshed more than 1,000 kilometers inland (Figure 10.14). Many North American species

thus may have been wiped out shortly after impact. For the rest of the world, death may have come more slowly. Heat from the impact and returning ejecta probably ignited wildfires in forests around the world. Evidence of such wildfires is found in the large amount of soot that is also present in the iridium-rich sediments from 65 million years ago. The impact also sent huge quantities of dust high into the stratosphere, where it remained for several years, blocking out sunlight, cooling the surface, and affecting atmospheric chemistry. Plants died for lack of sunlight, and effects propagated through the food chain.

Perhaps the most astonishing fact is not that 75% of all species died, but that 25% survived. Among the survivors were a few small, rodentlike mammals. These mammals may have survived because they lived in underground burrows and managed to store enough food to outlast the long spell of cold, dark days. Small mammals had first arisen at about the same time as the dinosaurs, more than 100 million years earlier. But the sudden disappearance of the dominant dinosaurs made these mammals dominant. With an evolutionary path swept wide open, they rapidly evolved into a large assortment of much larger mammals—including us.

The event 65 million years ago was but one of at least a dozen mass extinctions in Earth's distant past, and several of these have also been linked to impacts. Could a similar event occur in the future, wiping out all that we have accomplished?

FIGURE 10.15 Damage from the 1908 impact over Tunguska, Siberia.

The Asteroid Threat:
Real Danger or Media Hype?

The first step in analyzing the threat of future impacts is to examine the past evidence more clearly. Only one significant impact has clearly occurred in modern times. In 1908, an unusual explosion occurred in a sparsely inhabited region of Siberia. (The impact is known as the *Tunguska event.*) Entire forests were flattened and set on fire, and air blasts knocked over people, tents, and furniture up to 200 kilometers away (Figure 10.15). Seismic disturbances were recorded at distances up to 1,000 kilometers, and atmospheric pressure fluctuations were detected almost 4,000 kilometers away. Scientific studies were delayed for almost 20 years (due to the remoteness of the impact area and the politics of the time), but investigators eventually found meteoritic dust at the site—but no impact crater. These wide-reaching effects were apparently caused by a weak, stony meteorite only 30 meters across that exploded in the air before reaching the surface. Objects of this size probably strike our planet every century or so, usually over the ocean. But the death toll could be quite high if such an object struck a densely populated area.

Another way to gauge the threat of impacts is to look at asteroids that might strike Earth. The largest known *Earth-approaching asteroid* (Eros, the target of the NEAR mission) is about 40 kilometers long. Dozens of smaller Earth-approaching asteroids are known, down to sizes of tens of meters across. Orbital calculations show that none of these known objects will impact Earth in the foreseeable future.

However, the vast majority of Earth-approaching asteroids probably have not yet been detected. Astronomers estimate that some 2,000 undiscovered members of this class may be larger than 1 kilometer across, and another 300,000 may be larger than 100 meters across. Comets also pose a threat, particularly because their rapid plunge from the outer solar system gives little warning.

By combining observations of asteroids with information from past impact craters, we can estimate the frequency of impacts of various sizes (Figure 10.16). Larger impacts are obviously more devastating but thankfully are rare. Impactors a few meters across probably break up high in the Earth's atmosphere every few days, with each liberating the energy of an atomic bomb. However, the effects are dissipated before reaching the surface, so we do not notice them. Hundred-meter impactors that forge craters similar to Meteor Crater probably strike only about every 10,000 years or so. Impacts large enough to cause mass extinctions come tens of millions of years apart.

Thus, while it seems a virtual certainty that the Earth will be battered by many more large impacts, the chance of a major impact happening in our lifetime is quite small. Nevertheless, the chance is not zero, and until we know the orbits of *all* large Earth-approaching asteroids we cannot predict when the next impact might occur. Some people advocate a thorough search to find all potentially dangerous asteroids, a relatively cheap proposition. But others wonder whether the information would be useful: If

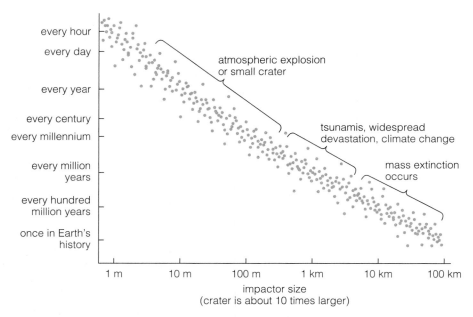

every hour
every day
every year
every century
every millennium
every million years
every hundred million years
once in Earth's history

1 m   10 m   100 m   1 km   10 km   100 km

impactor size
(crater is about 10 times larger)

atmospheric explosion or small crater

tsunamis, widespread devastation, climate change

mass extinction occurs

FIGURE 10.16 This graph shows how the frequency of impacts—and the magnitude of their effects—depends on the size of the impactor; note that smaller impacts are much more frequent than larger ones.

you knew an asteroid would hit next year, could you do anything about it? Schemes to use nuclear weapons to save Earth by demolishing or diverting an asteroid abound, but no one knows whether current technology is up to the task. For the time being, we can only hope that our number does not come up soon.

While some people see danger in Earth-approaching asteroids, others see opportunity because Earth-approaching asteroids bring valuable resources tantalizingly close to Earth. Iron-rich asteroids are particularly enticing, since they probably contain many precious metals that mostly sank to the core on Earth. In the not-too-distant future, it may prove technically feasible and financially profitable to mine metals from asteroids and return these resources to Earth. It may also be possible to gather fuel and water from asteroids for use in missions to the outer solar system.

As we look out into the Universe and identify the many accidents of physics and astronomy that have worked to our benefit, it almost seems as if the Universe must in some sense have known that we were coming.

FREEMAN DYSON

THE BIG PICTURE

In this chapter, we focused on the solar system's smallest objects and found that they can have big consequences. Asteroids, comets, and meteorites teach us much about the evolution of the solar system. And, in the case of the impact 65 million years ago, one such object may have made our existence possible. Keep in mind the following "big picture" ideas as you continue your studies:

- The smallest bodies in the solar system—asteroids and comets—are our best evidence of how the solar system formed.

- The small bodies are subjected to the gravitational whims of the planets, particularly the jovian planets. The subtleties of resonances play a major role in sculpting the outer solar system.

- The interplay of large and small bodies brings us meteorites to teach us our origins, comets and meteor showers to light up the sky, and impacts that can alter or obliterate life as we know it.

- Pluto is called the ninth planet, but it bears much more similarity to the thousands of Kuiper belt comets than to the other eight planets.

Review Questions

1. Briefly explain why comets, asteroids, and meteorites are so useful in terms of helping us understand the formation and development of our solar system.

2. Define and distinguish among each of the following: *asteroid, comet, meteor, meteorite.*

3. How does the largest asteroid compare in size to the terrestrial worlds? How many asteroids are larger than 1 km in radius? How does the total mass of all the asteroids compare to the mass of a terrestrial world?

4. What is the *asteroid belt,* and where is it located?

5. Briefly describe how orbital resonances with Jupiter explain the numerous *gaps* in the asteroid belt. How did these resonances prevent a planet from forming in the vicinity of the asteroid belt?

6. What happens to asteroids that are kicked out of the asteroid belt by a gravitational encounter or a collision? In this context, describe the origin of the Earth-approaching asteroids.

7. What are the Trojan asteroids, and where are they found? What is the orbital period of the Trojan asteroids? (*Hint:* They share an orbit with an object whose orbital period you know.)

8. Distinguish between *primitive meteorites* and *processed meteorites* in terms of composition. How do the origins of these two groups of meteorites differ?

9. Explain how the existence of processed meteorites tells us that some asteroids once had active volcanism. How is the study of a processed meteorite an opportunity to look at a piece of a "dissected planet"?

10. What is the origin of processed meteorites with basaltic composition? What evidence suggests that some of these come from the Moon or Mars?

11. How did Halley's comet get its name? Can Kepler's laws be applied to comets?

12. Describe the anatomy of a comet, including its *nucleus, coma, plasma tail,* and *dust tail.* Which of these components are present when a comet is close to the Sun? When it is far from the Sun? Explain.

13. Briefly describe the composition of the nucleus of Halley's comet. How does gas escape from the interior of the nucleus into space? Why do the tails point away from the Sun?

14. Why can't an active comet last forever? What happens to comets after many passes near the Sun?

15. Describe the Kuiper belt and the Oort cloud in terms of location and comet orbits.

16. Where did comets that are now in the Oort cloud originally form? How did they end up in the Oort cloud?

17. Describe the origin of the Kuiper belt comets. What do we know about the sizes of Kuiper belt comets?

18. Describe Pluto and Charon. How were they discovered? Why won't Pluto collide with Neptune even though their orbits cross? How did Charon probably form?

19. Briefly summarize the evidence suggesting that Pluto is a Kuiper belt comet. How does Neptune's moon Triton suggest that Pluto may not be unique in size among Kuiper belt comets? Is it possible that other Pluto-size objects remain to be discovered in our solar system? Explain.

20. Describe the impact of comet Shoemaker–Levy 9 on Jupiter. How was the impact studied?

21. Explain how meteor showers are linked to comets. Why do meteor showers seem to originate from a particular location in the sky?

22. Describe the evidence suggesting that the *mass extinction* that killed off the dinosaurs was caused by the impact of an asteroid or comet. How did the impact lead to the mass extinction?

23. How often should we expect impacts of various sizes on Earth? How could we alleviate the risk of major impacts?

Discussion Questions

1. *Impact Risk.* How can we compare the risk of death from an impact to the risk of death from other hazards? This question is very important if we are to assess how much money it is worth spending to search for potential impactors and ways of preventing them from hitting the Earth. One way of evaluating the risk involves multiplying the probability of an impact by the number of people it would kill. For example, the probability of a major impact that could kill half the world's population, or some 3 billion people, is estimated to be about 0.0000001 (10^{-7}) in any given year. If we multiply this low probability by the 3 billion people that would be killed, we find the "average" risk from an impact to be $10^{-7} \times (3 \times 10^9) = 300$ people killed per year—about the same as the average number of people killed in airline crashes. Do you think it is valid to say that the risk of being killed by a major impact is about the same as the risk of being killed in an airplane crash? Should we pay to limit this risk, as we pay to limit airline crashes? Defend your opinions.

2. *Rise of the Mammals.* Suppose the impact 65 million years ago had not occurred. How do you think our planet would be different? For example, do you think that mammals would still have eventually come to dominate the Earth? Would we be here? Defend your opinions.

Problems

1. *Orbital Resonances.* How would our solar system be different if orbital resonances had never been important? (For example, if Jupiter and asteroids continued

to orbit the Sun but did not affect each other through resonances.) Describe at least five ways in which our solar system would be different. Which differences would you consider merely superficial, and which differences would be more profound? Be sure to consider the effects of orbital resonances discussed both in this chapter and in Chapter 9.

2. *Life Story of an Iron Atom.* Imagine that you are an iron atom in an iron meteorite recently fallen to Earth. Tell the story of how you got here, beginning from the time when you were in a gaseous state in the solar nebula 4.6 billion years ago. Include as much detail as possible. Your story should be scientifically accurate, but also creative and interesting.

3. *The "Near Miss" of Toutatis.* The 5-km asteroid Toutatis passed a mere 3 million km from the Earth in 1992.

   a. In one or two paragraphs, describe what would have happened to the Earth if Toutatis had hit it.

   b. Suppose Toutatis was destined to pass *somewhere* within 3 million km of Earth. Calculate the probability that this "somewhere" would have meant that it slammed into Earth. Based on your result, do you think it is fair to say that the 1992 passage was a "near miss"? Explain. (*Hint:* You can calculate the probability by considering an imaginary dart-board of radius 3 million km on which the bulls-eye has the Earth's radius of 6,378 km.)

4. *Project: Tracking a Meteor Shower.* Armed with an expendable star chart and a flashlight covered in red plastic, set yourself up comfortably to watch a meteor shower listed in Table 10.1. Each time you see a meteor, record its path on your star chart. Record at least a dozen, and try to determine the *radiant* of the shower—the point in the sky from which the meteors appear to radiate. On your chart, also include the direction of the Earth's motion, which you can find by determining which sign of the zodiac rises at midnight.

5. *Project: Dirty Snowballs.* If there is snow where you live or study, make a filthy snowball. (The ice chunks that form behind tires work well.) How much dirt does it take to darken snow? Find out by allowing your dirty snowball to melt and measuring the approximate proportions of water and dirt afterward.

6. *Web Project: NEAR and Rosetta.* Starting from the text web site, search for recent discoveries from the Near-Earth Asteroid Rendezvous mission (NEAR) *or* the European Space Agency's Rosetta mission. Write a one- to two-page summary of what you learn, giving special attention to anything that affects (or will affect) our understanding of phenomena or processes described in this chapter.

Earth and Lessons on Life in the Universe

PERHAPS YOU'VE HEARD THE EARTH DESCRIBED AS THE "THIRD ROCK FROM THE SUN." At first glance, the perspective of comparative planetology may seem to support this dispassionate view of the Earth. However, a deeper comparison between Earth and its neighbors reveals a far more remarkable planet.

Earth is unique in the solar system in many ways. It is the only planet with surface oceans of liquid water and substantial amounts of oxygen in its atmosphere, and its geological activity is more diverse than that of any other terrestrial world. Most important, Earth is the only world known to harbor life, which has played a considerable role in shaping the Earth's surface and atmosphere.

So far, we have found that our theories of solar system formation and planetary evolution explain most of the general features of our solar system. In this chapter, we will apply and extend what we've learned in order to form a coherent picture of the combined geological, atmospheric, and biological evolution of our planet. We'll also discuss a few important lessons that the solar system teaches us about our future on Earth and that the Earth teaches us about the possibility of life elsewhere in the universe.

11.1 How Is Earth Different?

Take a look at Figure 11.1. The blue oceans and abundant white clouds show a planet unlike any other we have studied so far. It is, of course, our own planet Earth, and we are now ready to study it using the tools of comparative planetology we developed in the preceding four chapters.

We've already discussed many of the similarities and differences between Earth and other worlds. Earth's geology resembles that of the other terrestrial worlds, particularly Venus and Mars, in that it is characterized by the four processes of *impact cratering, volcanism, tectonics,* and *erosion* [Section 8.3]. But even here we find important differences. For example, Earth is the only planet on which the lithosphere is clearly broken into plates that move around in what we call *plate tectonics,* and erosion reshapes Earth's surface to a much greater extent than it does on any other world in the solar system.

Earth's atmosphere is also unlike those of its nearest neighbors, despite a common origin in *outgassing.* Venus and Mars have atmospheres composed primarily of carbon dioxide (CO_2), but Earth's is composed primarily of nitrogen (N_2) with substantial amounts of oxygen (O_2) mixed in. While one other world—Titan—has an atmosphere composed mostly of nitrogen, no other world has substantial quantities of atmospheric oxygen. Furthermore, while the atmospheres of Venus, Earth, and Mars all possess warm tropospheres near the surface and hot thermospheres at high altitudes [Section 8.4], only Earth possesses an ultraviolet-absorbing *stratosphere.*

The greatest differences between Earth and other worlds lie in two features totally unique to Earth. First, the surface of the Earth is covered by huge amounts of water. Oceans cover nearly three-fourths of the Earth's surface, with an average depth of about 3 kilometers (1.8 miles). Water is also significant on land, where it flows through streams and rivers, fills lakes and underground water tables, and sometimes lies frozen in glaciers. Frozen water covers nearly the entire continent of Antarctica in the form of the southern ice cap. Another ice cap sits atop the Arctic Ocean in the north and also covers the large island of Greenland. Water plays such an important role on Earth that some scientists treat it as a distinct planetary layer, called the **hydrosphere,** between the lithosphere and the atmosphere.

The second totally unique feature of Earth is its diversity of life. We find life nearly everywhere on Earth's surface, throughout the oceans, and even underground. The layer of life on Earth is sometimes called the **biosphere.** As we will see shortly, the biosphere helps shape many of the Earth's physical characteristics. For example, the biosphere explains the presence of oxygen in Earth's atmosphere. Without life, Earth's atmosphere would be very different.

FIGURE 11.1 Dawn over the Atlantic Ocean, photographed from the Space Shuttle.

Comparative planetology compels us to explore these unique properties of Earth closely. If we find that the formation and evolution of our Earth violated the rules we've derived for other planets—if Earth is "irregular"—life might be unique to Earth. If Earth does follow the rules—if Earth is "regular" or even "standard"—then life may be common in the universe.

11.2 Our Unique Geology

It's not surprising that Earth shows much more geological activity than Mercury or the Moon, since those small worlds have cooled since their formation and are now geologically "dead." But why have Earth and Venus ended up so different, despite very similar sizes? And, although Earth is significantly larger than Mars, both planets probably once had similar surface conditions that allowed liquid water to flow. Why did Earth remain hospitable, while Mars is now dry and barren? These mysteries warrant a closer look.

Figure 11.2 shows shaded relief maps comparing the surfaces of Venus, Earth, and Mars. While the three planets have many superficial similarities, they also have important differences. We find fewer impact craters on Earth than on Venus or Mars. Most of Earth's volcanic mountains are steep-sided and therefore must have been made from a much more viscous lava than that found on the other planets. And Earth shows far more evidence of tectonic activity—surface reshaping driven by internal stresses. To our current knowledge, Earth is the only terrestrial planet on which the lithosphere is split into distinct *plates*.

Why do the surfaces of Earth, Venus, and Mars look so different today, when all three planets share similar compositions and internal structures and all three have experienced at least some tectonic stresses and volcanic activity? To answer this question, we must compare the effects of the four geological processes on Earth to their effects on our neighbors.

The Earth's Interior

As we discussed in Chapter 8, the first step in understanding geological processes is to learn as much as possible about a planet's interior structure. In the case of the Earth, earthquakes have provided a valuable tool through the vibrations, or seismic waves, that they create. As the different kinds of waves pass through different interior layers, their speeds and directions change depending on the local density, temperature, and phase (solid or liquid). Careful analysis of global seismological data therefore allows geologists to construct a three-dimensional image of Earth's interior (Figure 11.3). The Earth's thin crust and the uppermost portion of the mantle make up a relatively cool and rigid *lithosphere* about 100 km thick [Section 8.2]. Below the lithosphere, the mantle is warm and partially molten in places; it supplies the magma for volcanic eruptions. Deeper in the mantle, higher pressures force the rock into the solid phase despite even higher temperatures. But even in solid form the mantle slowly flows, and convection continually carries heat up from below. The Earth's metallic core underlies the lower mantle. The outer region of

Venus

Earth

Mars

FIGURE 11.2 These shaded relief maps compare the surfaces of Venus, Earth, and Mars; the map for Earth shows the seafloor as it would appear if there were no oceans. Despite superficial similarities, there are important differences. For example, there are fewer impact craters on Earth or Venus than on Mars. Also note the unique distinction between the two kinds of surface (seafloor and continent) on Earth.

Labels on figure:
lithosphere
crust
mantle
convection cells
3,470 km
1,220
molten outer core
solid inner core

FIGURE 11.3 Internal structure of the Earth.

the core is molten, but high pressures in the inner core force the metal into solid form. Thus, the Earth has all the ingredients needed for exciting geology: a hot interior to supply energy for volcanism, convection to help generate tectonic stresses, and a relatively thin lithosphere that does not inhibit geological activity.

The Four Geological Processes on Earth

The first of the four geological processes, *impact cratering,* should have occurred at roughly similar rates on all the terrestrial worlds. All were subject to the intense early bombardment during the solar system's first few hundred million years. Four billion years ago, the Earth's surface may have been as saturated with craters as the densely cratered lunar highlands are today [Section 8.3]. We find no 4-billion-year-old craters on Earth now, so these ancient impacts must have been erased by other geological processes. The impact rate fell substantially after the early bombardment, but occasional impacts still occur—sometimes with devastating consequences [Section 10.6]. The effects of erosion make impact craters surprisingly difficult to identify on Earth. Nevertheless, remnants of more than 100 impact craters have been found. Large impacts should occur more or less uniformly over the Earth's surface, including both continents

and seafloors, because large impactors are not stopped by the oceans. Thus, the fact that we find fewer large craters on the seafloor than on the continents suggests that seafloor crust is much younger than is continental crust.

The rampant *erosion* on Earth arises primarily from processes involving water, but atmospheric winds also contribute. Wind tears away at geological features created by other processes, blows dust and sand across the globe, and builds features such as sand dunes. Note that wind erosion is more effective on Earth than on either Venus or Mars: Venus has very little surface wind because of its slow rotation, and wind on Mars does little damage because of the low atmospheric pressure. On Earth, water contributes to erosion through a remarkable variety of processes. Perhaps the most obvious are rain and rivers breaking down mountains, carving canyons, and transporting sand and silt across continents to deposit them in the sea. But water erosion occurs on microscopic levels too: Water seeps into cracks and crevices in rocks and breaks them down from the inside, especially when this water freezes and expands. Further examples of water-erosion processes include the slow movement of glaciers, the flow of underground rivers, and the pounding of the ocean surf.

Volcanism and *tectonics* are also very important processes on Earth. The many active volcanoes prove that volcanism still reshapes the surface, and many small-scale cliffs and valleys offer evidence of tectonics similar to that found on other terrestrial worlds. But plate tectonics may be unique to Earth. So, to gain a deeper understanding of our planet's closely linked volcanism and tectonics, we must study plate tectonics in more detail.

Plate Tectonics

If you cut up a map and rearrange the continents, you'll find that they fit together in surprising ways. For example, the east coast of South America fits quite nicely into the west coast of Africa. You might chalk this up to coincidence, but we also find similar types of distinct rocks and rare fossils in eastern South America and western Africa—suggesting that these two regions were once near each other. Based on such evidence, in 1912 German meteorologist and geologist Alfred Wegener proposed the idea of *continental drift*: that the continents gradually drift across

FIGURE 11.4 The discovery of mid-ocean ridges and trenches provided important evidence for the idea of continental drift. Solid lines show plate boundaries. Important geological features discussed later in the text are identified.

the surface of the Earth, over time scales of tens of millions of years.

For decades after Wegener made his proposal, most geologists rejected the idea that continents could move, and many ridiculed the idea openly. However, the idea of continental drift gained favor in the 1950s as supporting evidence began to accumulate. Mapping of the seafloors revealed surprising structures (Figure 11.4): high *mid-ocean ridges* extending a total length of about 60,000 km along the ocean floors and *trenches* in which the ocean depth can reach more than 8 kilometers. Geologists soon recognized that these features represented boundaries between **plates**—pieces of the lithosphere that apparently float upon the denser mantle below. Convection cells in the upper mantle move the plates around the surface, forcing them together, apart, or sideways at their boundaries—with profound geological consequences. This discovery offered a model to explain how continents could drift about, and Wegener's idea finally gained acceptance under the new name **plate tectonics.** (Recall that tectonics is geological activity driven by internal stresses, so *plate tectonics* refers to the motion of plates driven by internal stresses.) Today we know that the Earth's lithosphere is broken into more than a dozen plates. (See figure 11.4.) Most major earthquakes and volcanic eruptions occur along plate boundaries (Figure 11.5).

FIGURE 11.5 Earthquakes occur when plates slip violently along a plate boundary. This photo shows damage from the 1995 earthquake in Kobe, Japan.

TIME OUT TO THINK *Study the plate boundaries in Figure 11.4. Based on this diagram, explain why the West Coast states of California, Oregon, and Washington are more prone to earthquakes and volcanoes than other parts of the United States.*

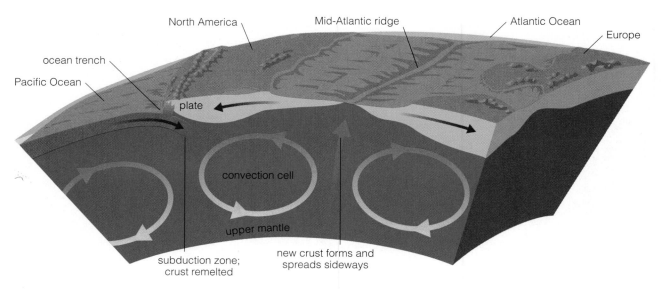

FIGURE 11.6 Plate tectonics acts like a giant conveyor belt, driven by mantle convection, that carries rock up from the mantle at mid-ocean ridges, transports it across the seafloor, and returns it down into the mantle at ocean trenches.

Over millions of years, plate tectonics acts like a giant conveyor belt for the Earth's crust, carrying rock up from the mantle, transporting it across the seafloor, and then returning it down into the mantle (Figure 11.6). The mid-ocean ridges mark **spreading centers** between plates—places where hot mantle material rises upward and then spreads sideways, pushing the plates apart. Worldwide along the mid-ocean ridges, new crust covers an area of about 2 km² every year—enough to replace the entire seafloor within a geologically short time of about 200 million years. Meanwhile, as new crust spreads over the surface, older crust must be returned to the mantle.

This return occurs at the locations of the deep ocean trenches, which mark places where one plate slides under another in a process called **subduction.** The conveyor-like process of plate tectonics is undoubtedly driven by convection, although the precise nature of the convection remains uncertain.

Overall, the plates move across the Earth's surface at speeds of a few centimeters per year—about the same speed at which your fingernails grow—and we can use this motion to project the locations of the continents millions of years into the past or future (Figure 11.7). Over the past billion years, the continents have slammed together, pulled apart, spun

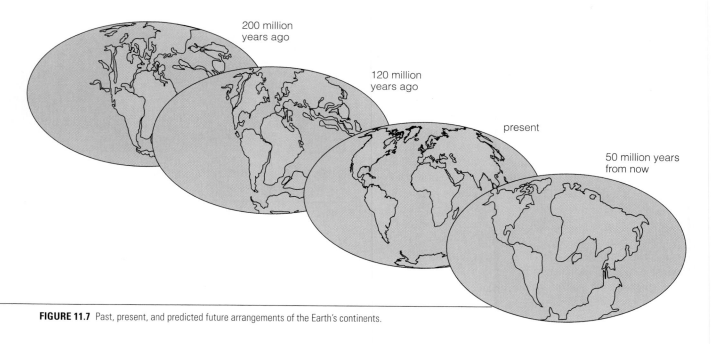

FIGURE 11.7 Past, present, and predicted future arrangements of the Earth's continents.

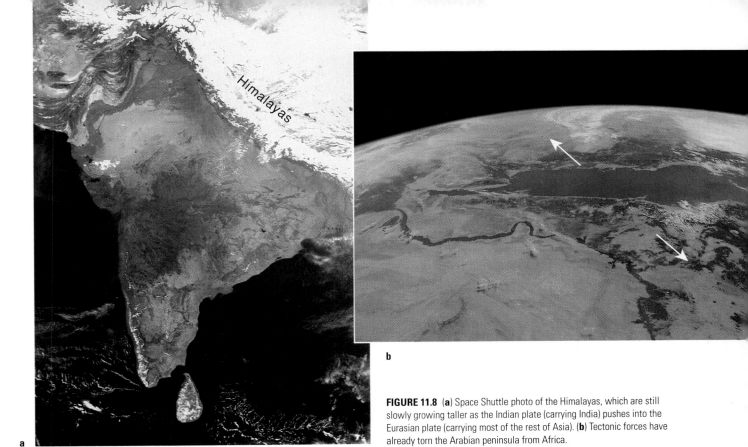

a

b

FIGURE 11.8 (a) Space Shuttle photo of the Himalayas, which are still slowly growing taller as the Indian plate (carrying India) pushes into the Eurasian plate (carrying most of the rest of Asia). (b) Tectonic forces have already torn the Arabian peninsula from Africa.

around, and changed places on the globe. Central Africa once lay at Earth's South Pole, and Antarctica was once nearer the equator. At various times, such as was the case about 200 million years ago, the continents were all merged into a single giant continent, often called *Pangaea* (which means "all lands"). The current arrangement of the continents is no more permanent than any other.

Most of Earth's interesting geology takes place near plate boundaries (see Figure 11.4). Where one plate subducts below another, the rock of the descending plate melts, rises, and fuels tremendous volcanoes. Alaska's Aleutian island chain and the islands of Japan and the Philippines are all located where one seafloor plate subducts below another. Earth's continents all originally formed this way. Volcanoes also occur where seafloor subducts below continental crust, such as Mount St. Helens in the Pacific Northwest, California's Sierra Nevada, and the Andes range of South America. (Some volcanoes, such as those that make up the Hawaiian Islands, occur in the middle of a plate, where molten material rises in the mantle.)

Tectonic forces continue to reshape the continents. When two continents collide, neither can subduct, and the resulting crunch makes huge mountains, as in the case of the Himalayas (Figure 11.8a). The Appalachians were formed by a similar process and were once as tall as the Himalayas. The forces of plate tectonics can also tear a continent apart, as is happening today in the East African rift zone and with the Arabian peninsula (Figure 11.8b).

Summary of Earth's Geology

Two fundamental differences between the geology of Earth and that of other terrestrial worlds stand out:

1. Rampant erosion driven by the action of plentiful water (related to Earth's atmospheric history, as we will see).

2. A different style of tectonics—plate tectonics—that recycles crust, drives high-viscosity volcanism to create continental crust, and causes frequent earthquakes.

Together, these two processes have reshaped the Earth's surface beyond recognition from its original state billions of years ago. They make the Earth's surface very young, perhaps the youngest in the solar system besides that of Io [Section 9.5].

11.3 Our Unique Atmosphere

In Chapter 8, we learned that the superficial differences in the atmospheres of the terrestrial worlds are fairly easy to understand. For example, the Moon and Mercury lack substantial atmospheres because of their small sizes, and a very strong greenhouse effect causes the high surface temperature on Venus. We also learned how solar heating, seasonal changes, and the Coriolis effect that comes from rotation lead to differing global circulation patterns on different worlds. But explaining how the compositions of the atmospheres came to be so different is more challenging, especially when we consider the cases of Venus, Earth, and Mars. Outgassing produced all three of these atmospheres, and the same volatiles—primarily water and carbon dioxide—were outgassed in all three cases. How, then, did Earth's atmosphere end up so different? In particular, any study of Earth's atmosphere must address four major questions:

1. Why did Earth retain most of its water—in the form of the oceans and other components of the hydrosphere—while Venus and Mars lost theirs?

2. Why does Earth have so little carbon dioxide (CO_2) in its atmosphere, when Earth should have outgassed about as much of it as Venus?

3. Why does Earth have so much more oxygen (O_2) than Venus or Mars? Where did it come from, and how does this highly reactive gas remain present in the atmosphere?

4. Why does Earth have an ultraviolet-absorbing stratosphere, while Venus and Mars do not?

To answer these questions, we must look into the history of Earth's atmosphere. The answers to all four questions turn out to be closely connected.

TIME OUT TO THINK *Recall that water ice could not condense in the region of the solar nebula where the terrestrial planets formed. How, then, did the water outgassed from volcanoes get here in the first place? (Hint: See Section 7.6.)*

Water and the Origin of the Hydrosphere

On a basic level, it's fairly easy to explain why the Earth retains so much water while Venus and Mars do not. As we learned in Chapter 8, Venus probably lost an "ocean-full" of water as solar ultraviolet photons split apart water molecules and the hydrogen atoms subsequently escaped to space. Mars probably lost some of its water in a similar way, and the rest is frozen at the polar caps and under the surface.

Let's look a little more closely at the sequence of events that led to these different histories. Picture a generic terrestrial planet billions of years ago: Erupting volcanoes were outgassing carbon dioxide, water vapor, and small amounts of nitrogen. As the water vapor increased, did this water vapor condense into oceans, freeze out as snow, or remain gaseous in the atmosphere? Even before Venus had a strong greenhouse effect, its proximity to the Sun made it warm enough to keep all its water gaseous in the atmosphere. The water vapor caused significant greenhouse warming and ensured that any additional water would also remain gaseous and warm the planet further. This tendency of the greenhouse effect to reinforce itself is called the **runaway greenhouse effect.** As volcanoes continued to belch out gases, Venus's atmosphere accumulated water vapor that later escaped into space. At the other extreme, Mars's temperature was low enough for the water vapor to freeze out of the atmosphere, resulting in thick polar caps.

On Earth, moderate temperatures allowed the water vapor to rain out of the atmosphere and liquid water to accumulate on the surface. A small amount of water vapor remained in the atmosphere and caused significant greenhouse warming, but additional water vapor released from volcanoes rained into the oceans and did not lead to a runaway greenhouse effect. The condensed water gradually formed the oceans and the other components of the hydrosphere. Thus, Venus lost its water because it was too hot, Mars lost its water because it was too cold, and Earth retained its water because conditions were "just right."

Where Is All the CO_2?

Measurements of outgassing by active volcanoes and estimates of past volcanic activity suggest that Earth must have outgassed nearly as much carbon dioxide throughout its history as Venus. But Venus today has an atmosphere with a CO_2 concentration of 96%, a surface pressure 90 times greater than that on Earth, and about 500°C of greenhouse warming. In contrast, the proportion of carbon dioxide in Earth's atmosphere is much less than 1%; in fact, we usually measure CO_2 concentration on Earth in units of *parts per million.* (For example, 300 parts per million means $300 \div 1,000,000 = 0.0003 = 0.03\%$.) Clearly, some atmospheric loss process must have removed nearly all of the outgassed CO_2 on Earth.

The secret of Earth's CO_2 is wrapped up in the history of our oceans. The primary process that removes atmospheric CO_2 on Earth involves liquid water. Carbon dioxide can dissolve in water, and the oceans actually contain about 60 times more carbon dioxide than the atmosphere (still a very small amount compared to the total amount that must have been outgassed). Most of the carbon dioxide is locked up in rocks on the seafloor through chemical reactions. Rainfall erodes silicate rocks on the Earth's

Rainfall erodes silicate
minerals on land.

CO_2 in the atmosphere

CO_2 dissolves in ocean.

Silicate minerals
react with dissolved
CO_2 to form
carbonate
rocks.

subduction of
carbonate rocks

release of
CO_2 by
volcanism

FIGURE 11.9 Recycling of CO_2
and the carbonate–silicate cycle.

surface and carries the eroded minerals to the oceans. There the minerals react with dissolved carbon dioxide to form **carbonate** minerals, which fall to the ocean floor, building up thick layers of carbonate rock such as *limestone*. This is part of the **carbonate–silicate cycle** (Figure 11.9). Because the process requires liquid water, it operates only on Earth, where it has removed about as much CO_2 from Earth's atmosphere as now remains in Venus's atmosphere. (If Venus's CO_2 could somehow be removed, the remaining atmosphere would have about as much nitrogen as Earth's.) Had Earth been slightly closer to the Sun, its warmer temperature might have evaporated the oceans, leaving the CO_2 in the atmosphere and causing a runaway greenhouse effect as on Venus. Conditions might not have been so conducive to "life as we know it."

The Earth today is just warm enough for liquid water and life thanks to a moderate greenhouse effect caused by the presence of water vapor and carbon dioxide in our atmosphere. Our good fortune might seem to be based on atmospheric properties alone, but in fact our luck lies deeper—with Earth's geological processes. Much of the CO_2 is injected into the atmosphere by volcanoes near subduction zones: Carbonate rocks carried beneath the seafloor by subduction melt and release their CO_2 (see Figure 11.9). The balance between the rate at which carbonate rocks form in the oceans and the rate at which carbonate rocks melt in subduction zones largely determines the amount of CO_2 in Earth's atmosphere. If plate tectonics—especially subduction—stopped on Earth, or if it occurred at a substantially different rate, the amount of atmospheric CO_2 would undoubtedly be different. The eventual effect on Earth's temperature might prove fatal to us.

The balance must have been different in Earth's past. Careful studies of ancient rocks show that liquid water was plentiful throughout Earth's history, so Earth's temperature has apparently been quite stable for billions of years. However, stars like the Sun increase in brightness as they age, and billions of years ago our Sun was probably about 30% fainter than it is today [Section 12.3]. With Earth's current abundance of greenhouse gases, our planet would have been frozen over. Thus, greenhouse gases must have been more abundant early in Earth's history. Venus, on the other hand, might have experienced hospitable conditions if it possessed "only" its current inventory of greenhouse gases—but it probably also had more greenhouse gases in its early history, leading to its runaway greenhouse effect. On Mars, extra greenhouse gases once provided the warm temperatures needed for liquid water to flow, but Mars lost so much of this gas that it became colder even as the Sun grew brighter.

The Origin of Oxygen, Ozone, and the Stratosphere

While both water and carbon dioxide are products of outgassing, molecular oxygen (O_2) is not. In fact, no geological process can explain the great abundance of oxygen (about 20%) in Earth's atmosphere. Moreover, oxygen is a highly reactive chemical that would rapidly disappear from the atmosphere if it were not continuously resupplied. Fire, rust, and the discoloration of freshly cut fruits and vegetables are everyday examples of **oxidation**—chemical reactions that remove oxygen from the atmosphere. Similar reactions between oxygen and surface materials (especially iron-bearing minerals) give rise to the reddish appearance of many of Earth's rock layers, such as those in Arizona's Grand Canyon. Thus, we must explain not only how oxygen got into the Earth's atmosphere in the first place, but also how the amount of oxygen remains relatively steady even while oxidation reactions tend to bind oxygen into rocks at a rapid rate.

The answer to the oxygen mystery is *life*. The process that supplies oxygen to the atmosphere is *photosynthesis,* which converts CO_2 to O_2. The carbon becomes incorporated into amino acids, proteins, and other components of living organisms. Today, plants and single-celled photosynthetic organisms return oxygen to the atmosphere in approximate balance with the rate at which animals and oxidation reactions consume oxygen, so the oxygen content of the atmosphere stays relatively steady. Earth originally developed its oxygen atmosphere when photosynthesis added oxygen at a rate greater than these processes could remove it from the atmosphere.

Life and oxygen also explain the presence of Earth's ultraviolet-absorbing stratosphere. In the upper atmosphere, chemical reactions involving solar ultraviolet light transform some of the O_2 into molecules of O_3, or *ozone.* The O_3 molecule is more weakly bound than O_2, which allows it to absorb solar ultraviolet energy even better, giving rise to the warm stratosphere and preventing harmful ultraviolet radiation from reaching the surface. Mars and Venus lack photosynthetic life and therefore have too little O_2 and too little ozone to form a stratosphere.

Summary of Earth's Atmosphere

Venus, Earth, and Mars all began on similar paths, releasing similar gases into their atmospheres by outgassing. But only Earth had conditions "just right" to support liquid oceans throughout its history. The oceans helped remove carbon dioxide from the atmosphere, ensuring that the greenhouse effect on Earth remained just strong enough to keep conditions hospitable for life, but not so strong as to create a runaway greenhouse effect like that on Venus.

But the physical connections between the Earth's geology and its atmosphere cannot fully explain the conditions on Earth today. In particular, the presence of oxygen and ozone can be explained only as products of life. Thus, to complete our understanding of the unique place of Earth in the solar system, we must now turn our attention to the role of life on Earth.

11.4 Life

Geologists and biologists have developed a fairly clear picture of the history of life on Earth, at least in broad outline. As we trace this history through time, you should ask yourself three key questions:

1. How did Earth's physical conditions affect the development of life?

2. How has life influenced the physical characteristics of Earth?

3. What can we learn about the prospects for finding life elsewhere in the solar system or universe?

FIGURE 11.10 (**a**) Microscopic fossil bacteria. (**b**) Stromatolite fossils (up to 3.5 billion years old). (**c**) Living stromatolites today. The biological origin of fossil stromatolites was questioned until living examples were found.

The Origin of Life (4.4–3.5 billion years ago)

Life arose on Earth at least 3.5 billion years ago, and possibly much earlier than that. We've found recognizable fossils in rocks as old as 3.5 billion years—microscopic fossils of single-celled bacteria and larger fossils of bacterial "colonies" called *stromatolites* (Figure 11.10). But indirect evidence from carbon isotopes suggests that life originated at least 350 million

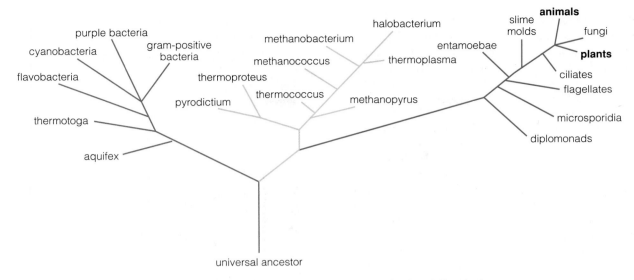

FIGURE 11.11 The tree of life, showing evolutionary relationships according to modern biology as of early 1998. Note that just two small branches represent *all* plant and animal species.

years earlier (3.85 billion years ago), even though such ancient rock samples are too contorted for fossils to remain recognizable. All life and all fossils tested to date show the same characteristic ratio of two carbon isotopes, carbon-12 and carbon-13. Rocks as old as 3.85 billion years show the same ratio, strongly suggesting that these rocks contain remnants of life. Beyond 3.85 billion years, we cannot yet determine whether life existed, since rocks more ancient than this no longer exist on Earth's surface. It is conceivable that life began as early as 4.4 billion years ago.

TIME OUT TO THINK *You may have noticed that, while we've been talking about life, we haven't actually defined the term. In fact, it's surprisingly difficult to draw a clear boundary between life and nonlife. How would you define life? Explain your reasoning.*

If life did originate much before 3.85 billion years ago, it may have been in for a rough time. The transition from intolerable conditions during planetary formation to more hospitable conditions was a gradual one. The early bombardment that continued after the end of accretion gradually died away over a period of a few hundred million years. But large impacts capable of sterilizing the planet by temporarily boiling all the oceans may have occurred as late as 4.0–3.8 billion years ago. Some biologists speculate that life may have arisen several times in the Earth's early history, only to be extinguished by the hostile conditions. If so, life might have turned out very different if one less—or one more—sterilizing impact had occurred.

An even deeper question concerns not just *when* life arose, but *how*. No one knows the precise answer to this question, but genetic testing makes us confident of one thing: Whether life arose once or multiple times, one particular type of organism came to dominate the entire Earth. That is, every organism living today apparently developed from a single ancestor (Figure 11.11).

Genetic testing also allows at least reasonable guesses as to which living organisms most resemble the common ancestor of all life. Surprisingly, the answer appears to be organisms living in the deep oceans around seafloor volcanic vents called *black smokers* (after the dark, mineral-rich water that flows out of them) and in hot springs in places like Yellowstone (Figure 11.12). These organisms thrive in temperatures as high as 125°C. (The high pressures at the seafloor prevent the water from boiling until it reaches 450°C.) Unlike most life at the Earth's surface, which depends on sunlight, the ultimate energy source for these organisms is chemical reactions in water volcanically heated by the internal heat of the Earth itself.

Of course, knowing that all life shares a common ancestor still does not tell us how that ancestor first arose. The step from simple chemical building blocks to our common DNA-bearing ancestor (from nonlife to life) is huge, to say the least. No fossils record the transition, nor has it ever been duplicated in the laboratory.

a

b

FIGURE 11.12 (**a**) Life around a black smoker deep beneath the ocean surface. (**b**) This aerial photo shows a hot spring in Yellowstone National Park that is filled with colorful bacterial life.

The bare necessities for "life as we know it"—chemicals, energy, and water—were undoubtedly available on the early Earth. The building blocks of life, including amino acids and nucleic acids, are composed primarily of carbon, oxygen, nitrogen, and hydrogen—chemicals that were readily available. Energy to fuel chemical reactions was present on the surface in the form of lightning and ultraviolet light from the Sun and was also present in the oceans in the heated water near undersea volcanoes. We know that oceans were present, because we find ancient carbonate rocks that must have formed in water.

Laboratory experiments demonstrate that the chemical building blocks of life could have formed spontaneously and rapidly in the conditions that prevailed early in Earth's history. In these experiments, researchers replicate the chemical conditions of the early Earth by, for example, including the types of ingredients outgassed by volcanoes and "sparking" the mixture with electricity to simulate lightning or other energy sources. Within just a few days, the mixture spontaneously becomes rich with amino acids and nucleic acids. Such chemical reactions were undoubtedly an important source of organic molecules on Earth, but they may not have been the only

source—some meteorites contain complex organic molecules.

With the building blocks for life plentiful, it may only have been a matter of time before chemical reactions created a molecule that could make a copy of itself. Such a *self-replicating molecule* might have spread quickly through regions with hospitable conditions. **Mutations**—errors in the copying process—could have created variations on the original molecule, some with increased complexity. If life really began in this way, then some ancient self-replicating molecule was the ancestor of modern DNA—and of all life on Earth.

Early Evolution in the Oceans (3.5–2.0 billion years ago)

Regardless of its origin, life soon thrived in the oceans in the form of single-celled organisms, tapping a variety of energy sources including sunlight (i.e., through photosynthesis). Individual organisms that survived and reproduced passed on copies of their DNA to the next generation. However, the transmission of DNA from one generation to the next is not always perfect. Mutations can change the arrangement of chemical bases in a strand of DNA, making the genetic information in a new cell slightly different from that of its parent. Mutations can be caused by many factors, including high-energy *cosmic rays* from space, ultraviolet light from the Sun, particles emitted by the decay of radioactive elements in the Earth, and various toxic chemicals. Most mutations are lethal, killing the cell in which the mutation occurs. However, some mutations may make a cell better able to survive in its surroundings, and the cell then passes on this improvement to its offspring, a process called **natural selection.**

Fossils show that evolution progressed remarkably slowly for most of Earth's history. For at least a billion years after life first arose, the most complex life-forms were still single-celled. Some 2 billion years ago, the land was still inhospitable because of the lack of a protective ozone layer. Continents much like those today were surrounded by oceans teeming with life, but the land itself was probably as barren as Mars is today, despite pleasant temperatures and plentiful rainfall.

Altering the Atmosphere
(beginning about 2 billion years ago)

The process of photosynthesis appears to have developed quite early in the history of life. Over billions of years, the abundant single-celled organisms in the ocean pulled carbon dioxide from the atmosphere and put back oxygen. For a long time, nearly all the highly reactive oxygen gas was pulled back out of the atmosphere by reactions with surface rocks. However, by about 2 billion years ago, the oxidation of surface rocks was fairly complete, and oxygen began to accumulate in the atmosphere. Today it constitutes about 20% of the atmosphere.

The buildup of oxygen had two far-reaching effects for life on Earth. First, it made possible the development of oxygen-breathing *animals.* No one knows precisely how or when the first oxygen-breathing organism appeared, but for that animal the world was filled with food. You can imagine how quickly these creatures must have spread around the world, and the fact that other organisms could now be eaten changed the "rules of the game" for evolution.

The second major effect of the oxygen buildup was the formation of the ozone layer, which made it safe for life to move onto the land. Although it took more than a billion years, natural selection eventually led to plants that could survive and thrive on land, and animals followed soon after, taking advantage of this new food source.

This epoch highlights the ability of life to fundamentally alter a planet. Not only did life change the Earth's atmosphere, but in doing so it made two other kinds of life possible: oxygen-breathing animals and land organisms.

An Explosion of Diversity
(beginning about 0.54 billion years ago)

As recently as 540 million years ago, most life-forms were still single-celled and tiny. Note that life had already been present on Earth for at least 3 billion years by that time, demonstrating the remarkable slowness with which evolution progressed for most of Earth's history. But the fossil record reveals a dramatic diversification of life beginning about 540 million years ago. We can trace the origin of most of today's plant and animal species—from insects to trees to vertebrates—back to this period. Apparently, once life began to diversify, the increased competition among the many more complex species accelerated the process of natural selection and the evolution of new species.

Some species have been more successful and more adaptable than others. Dinosaurs dominated the landscape of the Earth for more than 100 million years. But their sudden demise 65 million years ago paved the way for the evolution of large mammals—including us. The earliest humans appeared on the scene only a few million years ago (after 99.9% of Earth's history), and our few centuries of industry and technology have come after 99.99999% of Earth's history. Our population has been growing exponentially for the past few centuries, roughly doubling in just the past 40 years (Figure 11.13).

The tremendous growth of the human population suggests a chilling analogy: Exponential growth is also the mark of a cancerous tumor—which may seem very successful at surviving while it is growing but ultimately dies when it kills its host. Is it possible that our tremendous success is killing our host, the Earth? We can gain some perspective on this question by looking at a few lessons that the solar system teaches us about our relationship with the Earth.

FIGURE 11.13 This graph shows human population over the past 12,000 years. Note the tremendous population growth that has occurred in just the past few centuries.

TIME OUT TO THINK *The pace of change in human existence has accelerated since the advent of modern humans a few million years ago: Civilization arose about 10 thousand years ago, the industrial revolution began just a couple of centuries ago, and the computer era started just a few decades ago. Do you think the pace of change can continue to accelerate? What do you think the next stage will be?*

11.5 Lessons from the Solar System

The United States and other nations have carried out the exploration of the solar system for many reasons, including scientific curiosity, national pride, and technological advancement. Perhaps the most important return on this investment was one not expected: an improved understanding of our own planet and an enhanced ability to understand some very important global issues. In this section, we explore three particularly important issues about which comparative planetology offers important insights: global warming, ozone depletion, and mass extinctions.

Global Warming

Venus stands as a searing example of the effects of large amounts of atmospheric carbon dioxide. Earth remains habitable only because natural processes have locked up most of its carbon dioxide on the sea-floor. However, humans are now tinkering with this difference between the two planets by adding carbon dioxide and other greenhouse gases to the Earth's atmosphere. The primary way we release carbon dioxide is through the burning of fossil fuels (coal, natural gas, and oil)—the carbon-rich remains of plants

buried millions of years ago. But other human actions, such as deforestation and the subsequent burning of trees, also affect the carbon dioxide balance of our atmosphere. The inescapable result is that the amount of atmospheric carbon dioxide has risen steadily since the dawn of the industrial age—and continues to rise (Figure 11.14).

There is a growing scientific consensus that the average temperature of the Earth has warmed slightly (less than about 1°C) over the past 50 years—although there is much less agreement about why. It's likely that the cause is the human addition of greenhouse gases to the atmosphere, but it's also possible that the change is natural and would have occurred even without human activity. Some scientists even suggest that the Sun itself may be warming and thus contributing to the global temperature rise.

The question of whether human activity is inducing global warming is not merely academic: The consequences of a significant warming could be widespread and disastrous. Some simulations predict enough warming to melt polar ice and flood the Earth's densely populated coastal regions. Such warming would also increase evaporation from the oceans, tending to make storms of all types both more frequent and more destructive: Coastal regions would be hit by more hurricanes, intense thunderstorms and associated tornadoes would strike more frequently, and even winter blizzards would be more severe and more damaging. Rainfall patterns would also shift, and much of the world's current cropland might become unusable. The most ominous possibility is the collapse of entire ecosystems. If a regional climate changes more rapidly than local species can adapt or migrate, these species might become extinct. Given the complex interactions of the Earth's biosphere, hydrosphere, and atmosphere, the conse-

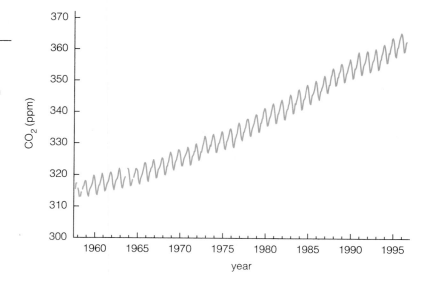

FIGURE 11.14 These data, collected for many years on Mauna Loa (Hawaii), show the increase in atmospheric carbon dioxide concentration. Yearly wiggles represent seasonal variations in the concentration, but the long-term trend is clearly upward. The concentration is measured in parts per million (ppm).

FIGURE 11.15 The extent of the ozone hole over Antarctica in 1979 (left) and 1998 (right). The ozone hole worsened between 1979 and 1998. Red indicates normal ozone concentration, and blue indicates severe depletion.

quences of such ecosystem changes are impossible to foresee.

Given the current uncertainties, it is impossible for anyone to know for sure whether global warming will even be a serious problem in the next century, let alone make specific predictions of its severity. Our studies of the planets show that surface temperatures depend on many factors and that seemingly minor changes can alter climates in surprising ways. Nevertheless, the fact that we cannot fully explain the current climate conditions on Earth suggests that we should be very careful about tampering with them.

TIME OUT TO THINK *The most obvious way to prevent global warming is to stop burning the fossil fuels that add greenhouse gases to the atmosphere. But much of the world economy depends on energy produced from fossil fuels. Given the uncertainties about the effects of global warming in the next century, what if any changes do you think are justified at present? What specific proposals would you make if you were a world leader?*

Ozone Depletion

For 2 billion years, the Earth's ozone layer has shielded the surface from hazardous ultraviolet radiation, but in the past two decades humans have begun to de-stroy the shield. The magnitude of the problem was not recognized until the discovery of an **ozone hole** over Antarctica in the mid-1980s (Figure 11.15). The "hole" is a place where the concentration of ozone in the stratosphere is dramatically lower than it would naturally be. The ozone hole appears for a few months during each Antarctic spring and has gradually worsened over the years. Since its discovery, satellite measurements have suggested that **ozone depletion** has caused ozone levels to fall by as much as a few percent worldwide. If ozone levels continue to decline, more solar ultraviolet radiation will reach the surface. Humans will face substantially greater risk from skin cancer, and plants and animals will suffer more genetic damage from mutations induced by ultraviolet radiation—with unknown consequences for the biosphere as a whole.

The apparent cause of the ozone depletion is human-made chemicals known as CFCs (chlorofluorocarbons), widely used in air conditioners and refrigerators and for many other purposes. Indeed, CFCs once seemed like an almost ideal chemical: They are useful in many industries, cheap and easy to produce, and chemically inert (i.e., they do not burn, break down, or react with anything on the Earth's surface). Ironically, their inertness is also their downfall; because CFCs are gases and are not destroyed by chemical reactions in the lower atmosphere, they eventually rise intact into the stratosphere. There they are broken down by the Sun's ultraviolet light, and one of the by-products is chlorine—a very reactive gas that catalytically destroys ozone without being consumed in the process.

Mars and Venus each played an important role in our understanding of ozone and chlorine. Mars lacks an ozone layer, so ultraviolet radiation reaches the surface. As a result, the surface is sterile: Terrestrial life could not survive on the surface, and the building blocks of life would be torn apart when exposed to the Sun. But it was the study of Venus that first alerted scientists to the dangers on Earth. Chlorine is a minor but important ingredient of Venus's atmosphere, and models of its chemical cycles showed that chlorine could rapidly destroy weak molecules such as ozone. When the models were altered to apply to Earth's atmosphere, CFCs and ozone depletion were connected.

This tale of three planets may have a happy ending, thanks to the awareness it brought of the dangers posed by ozone destruction. Today, international treaties ban the production of CFCs, although previously existing CFCs may still be used and recycled. If these treaties are obeyed, ozone destruction probably will not be a serious problem in the future, but the problem will remain at some level for decades to come.

TIME OUT TO THINK *You've just learned that your car air conditioner is broken and will cost $200 to fix. Then you hear about a shop that can fix it for only $100—undoubtedly by violating the laws concerning CFCs. Is the small benefit to our ozone layer worth the $100 difference?*

Mass Extinctions

When geology was a young science, its practitioners thought that entire mountain ranges and huge valleys were formed suddenly by cataclysmic events. Later geologists denounced this *catastrophism* and replaced it with *uniformitarianism,* which holds that geological change occurs gradually by processes acting over very long periods. Uniformitarianism certainly holds for three of our four geological processes, but impact cratering represents the "instant geology" of catastrophism. Many geologists initially rejected the importance of impact cratering as a geological process, but they were eventually convinced by the unambiguous evidence of impacts on the Moon, by astronomical studies of comets and asteroids, and by careful analysis of the few impact craters still present on Earth. Geology today is a combination of uniformitarianism and catastrophism.

Scientific ideas about evolution are undergoing a similar transformation. Biologists once assumed that evolution always proceeded gradually, with new species arising slowly and others occasionally becoming extinct. It now appears that the vast majority of all species in Earth's history died out in sudden *mass extinctions*—for which impacts are the prime suspect. The mass extinction that killed off the dinosaurs is only the most famous of these events [Section 10.6]; the fossil record shows evidence of at least a dozen other mass extinctions, some with even greater species loss than in the dinosaur event. While some species died out from direct effects of the impact, most probably died due to the disappearance from their ecosystem of species on which they depended. Furthermore, the survival or extinction of species appears to be random during mass extinctions: Surviving species show no discernible advantage over those that perish. In a sense, mass extinctions clear the blackboard, allowing evolution to get a fresh start with a new set of species. If we re-formed the Earth from scratch, life (and even intelligent life) might well arise again, but there is no scientific basis for thinking that evolution would follow the same course.

The story of mass extinctions teaches us about the danger of large impacts, but it also teaches a potentially more important lesson: Even without an impact, the Earth may be undergoing a mass extinction right now. In the normal course of evolution, the rate of species extinction is fairly low—perhaps one species lost per century. According to some estimates, human activity is now driving species to extinction so rapidly that half of today's species may be lost by the end of the twenty-first century. It is not just that humans are hunting and fishing many species to extinction. Far more are lost due to habitat destruction and pollution. And when a few species are lost by direct human influence, entire ecosystems that depend on those species may collapse. If global warming alters the climate significantly or if ozone depletion leads to more genetic damage in plants and animals, the rate of species extinction might increase even more. In any case, the "event" of losing half the world's species in just a few hundred years certainly qualifies as a mass extinction on a geological time scale, and the biological effects could be as dramatic as those of any impact. Are we unwittingly clearing the way for a new set of dominant species?

11.6 Lessons from Earth: Life in the Solar System and Beyond

Just as comparative planetology teaches us much about the Earth, the study of the Earth can teach us much about other worlds. In particular, it teaches us about the conditions for life and may help us answer what is surely one of the deepest philosophical questions of all time: Are we alone in the universe? It is fitting to close our study of the solar system by extending our understanding of biology to the planets and beyond. The study of the possibility of life elsewhere in the universe, called *astrobiology* (or *exobiology*), is a logical extension of comparative planetology.

In the time between the Copernican revolution [Section 5.3] and the space age, many people expected the other planets in our solar system to be Earth-like and to harbor intelligent life. In fact, a reward was supposedly once offered for the first evidence of intelligent life on another planet *other than Mars.* Venus, a bit closer to the Sun than Earth, was often pictured as a tropical paradise. Such expectations were dashed by the bleak images of Mars returned by spacecraft and the discovery of the runaway greenhouse effect on Venus. Many scientists began to believe that only Earth has the right conditions for life, intelligent or otherwise. Recently, the pendulum has begun to swing the other way, spurred primarily by two developments. First, biologists are learning that life thrives under a much wider range of conditions than once imagined. Second, planetary scientists are developing a much better understanding of conditions on other worlds. It now seems quite likely that conditions in at least some places on other worlds might be conducive to life.

The Hardiness, Diversity, and Probability of Life

Even on Earth, biologists long assumed that many environments were uninhabitable. But recent discoveries have found life surviving in a remarkable range of conditions. The teeming life surviving at temperatures as high as 125°C near underwater volcanic vents and in hot springs is only one of many surprises. Biologists have found microorganisms living deep inside rocks in the frozen deserts of Antarctica and inside basaltic rocks buried more than a kilometer underground. Some bacteria can even survive radiation levels once thought lethal—apparently, they have evolved cellular machinery that repairs mutations as fast as they occur. The newly discovered diversity of microscopic life has forced scientists to redraw the "tree of life" (see Figure 11.11), crowding familiar plants and animals into one corner. The majority of these microorganisms need neither sunlight, oxygen, nor "food" in the form of other organisms. Instead, they tap a variety of chemical reactions for their survival.

The new view of terrestrial biology forces us to rethink the possibility of life elsewhere. First, life on Earth thrives at extremes of temperature, pressure, and atmospheric conditions that overlap conditions found on other worlds. The Antarctic valleys, for example, are as dry and cold as certain parts of Mars, and the conditions found in terrestrial hot springs may have been duplicated on a number of planets and moons at certain times in the past. Second, life harnesses energy sources readily available on other planets. The basalt-dwelling bacteria, for example, would probably survive if they were transplanted to Mars. Third, life on Earth has evolved from a common ancestor into every imaginable ecological niche (and some unimaginable ones as well). The diversification of life on Earth has basically tested the limits of our planet, and there is no reason to doubt that it would do so on other planets. Thus, if life ever had a foothold on any other planet in the past, some organisms might still survive in surprising ecological niches today even if the planet has undergone substantial changes. The only real question is whether life ever got started elsewhere in the first place.

What is the probability of life arising from nonliving ingredients? This is probably the greatest unknown in exobiology. The fact that we exist and are asking the question does not tell us the probability; it merely tells us that it happened once. But the rapidity with which life arose on Earth may provide a clue. As we've discussed, we find fossil evidence for life dating almost all the way back to the end of the period of early bombardment in the solar system, suggesting that life arises easily and perhaps inevitably under the right conditions. Some people even speculate that primitive life arose many times during the heavy bombardment, only to be extinguished by violent impacts just as many times. If life indeed arises easily given the right conditions, we must search the solar system for those conditions, past or present.

TIME OUT TO THINK *The preceding discussion implies that the rapid appearance of life on Earth means that life is highly probable. Do you agree with this logic? What alternative conclusions could you reach?*

Life in the Solar System

Speculation about life in the solar system usually begins with Mars, for good reason. Before it dried out billions of years ago, its early atmosphere gave the surface hospitable conditions that rivaled those on Earth, with ample running water, the necessary raw chemical ingredients for life, and a variety of familiar energy sources. Many of Earth's organisms would have thrived under early Martian conditions, and some could even survive in places in today's Martian environment. Our first attempt to search for life on Mars came with the Viking missions to Mars in the 1970s, which included two landers equipped to search for the chemical signs of life. No life was found. But the landers sampled only two locations on the planet and tested soils only very near the surface. If life once existed on Mars, it either has become extinct or is hiding in other locations.

Today, a renewed debate about Martian life is under way, thanks in part to the study of a Martian meteorite found in Antarctica in 1984. The meteorite apparently landed in Antarctica 13,000 years ago, following a 16-million-year journey through space after being blasted from Mars by an impact. The rock itself dates to 4.5 billion years ago, indicating that it solidified shortly after Mars formed and therefore was present during the time when Mars was warmer and wetter. Painstaking analysis of the meteorite reveals indirect evidence of past life on Mars, including layered carbonate minerals and complex molecules (called polycyclic aromatic hydrocarbons), both of which are associated with life when they are found in Earth rocks. Even more intriguing, highly magnified images of the meteorite reveal eerily lifelike forms (Figure 11.16). These forms bear a superficial resemblance to terrestrial bacteria, although they are about a hundred times smaller—about the same size as recently discovered terrestrial "nanobacteria" and viruses. Nevertheless, many scientists dispute the conclusion that these features suggest the past existence of life on Mars, claiming that nonbiological causes can also explain many of the meteorite's unusual features.

Future missions to Mars will search for life using more sophisticated techniques than those used by the Viking missions. The best place to look for fossil remains of extinct life is probably in ancient valley bottoms or dried-up lake beds. If any hot springs surround Mars's not-quite-dormant volcanoes, they may be a good place to look for surviving life. A thorough search for Martian life will probably require the return of rock samples to Earth or human exploration of the planet.

Martian meteorites also remind us that the planets may occasionally exchange rocks dislodged by major impacts. The harsh conditions under which some life on Earth exists suggest that living organisms might survive such impacts and even survive the journey from one planet to another. Earth's basalt-dwelling bacteria, for example, could probably survive an impact-cratering event, the ensuing millions of years in space, and a violent impact on Mars. If a meteorite from Earth once landed in hospitable conditions on Mars, it might have introduced life to Mars or wiped out life already present. In a sense, Earth, Venus, and Mars have been "sneezing" on one another for billions of years. Life could conceivably have originated on any of these three planets and been transported to the others.

FIGURE 11.16 Microscopic view of seemingly lifelike structures in a Martian meteorite.

TIME OUT TO THINK *Suppose we someday discover living organisms on Mars. How will we be able to tell whether these organisms share a common ancestor with living organisms on Earth?*

Besides Mars, the best places to search for life probably are the satellites of the jovian planets. Several of them may meet the requirements of having liquid water, appropriate chemicals, and energy sources for life. In particular, Europa appears to have a planet-wide ocean beneath its icy crust [Section 9.5], the ice and rock from which Europa formed undoubtedly included the necessary chemicals, and its internal heating might lead to undersea volcanic vents. Thus, the real question about Europa may be whether it is possible that its ocean has existed for billions of years *without* developing life. NASA is already considering missions to search for signs of life on Europa. The most ambitious involves landing a robotic spacecraft on the surface that will melt its way through the icy crust to reach the ocean below.

Another enticing place to look for life past or present is Saturn's moon Titan. We've already found evidence of complex chemical reactions on Titan, many involving the same elements used by life on Earth. Liquid water and energy are in shorter supply, but both might have been supplied by impacts that heated and melted the icy surface early in Titan's history. Life might have arisen in a slushy pond during the brief period that it remained liquid and might have survived after the pond froze. Unlike the case on Earth, where early impacts probably sterilized the planet, impacts on Titan may have made life possible.

Despite our new awareness of the diversity of life on Earth, we may still be underestimating the range of conditions in which life can exist. Could life arise in liquids other than water, or possibly in an atmosphere? Some people have speculated that life could develop in the clouds of Jupiter or other jovian planets. Might life be based on different elements than those used on Earth? Are there other energy sources that we have not considered? Is our definition of life too narrow? Clearly, it will be a long time before we know all the places where life might exist even within our solar system.

Life Around Other Stars

Only a few places in our solar system seem hospitable to life, but many more hospitable worlds may be orbiting some of the hundred billion other stars in the Milky Way Galaxy or stars in some of the billions of other galaxies in the universe. Might some of the stars be orbited by planets that are as hospitable as our own Earth?

Interstellar clouds throughout the galaxy contain the basic chemical ingredients of life—carbon, oxygen, nitrogen, and hydrogen, as well as other elements. All stars are born from such interstellar clouds, so it seems likely that other solar nebulae should have given rise to planetary systems similar to our own [Section 7.6]. Terrestrial planets anywhere will receive light from their parent star, and if they are large enough, like Venus or Earth, they will have plenty of internal heat. Thus, the chemicals and energy sources for life should be present on many planets throughout the universe.

Planets with liquid water on their surfaces may be somewhat rarer, particularly if we consider only planets where oceans can endure for billions of years. Even if a planet is large enough to outgas substantial quantities of water and retain its atmosphere, the stories of Venus and Mars tell us that oceans are not guaranteed. To keep oceans for billions of years, the planet must lie within a range of distances from its star, sometimes called the **habitable zone,** that has temperatures just right for liquid water. How big is the habitable zone in our own solar system? We know that Venus is too close to the Sun, so this zone must begin outside the orbit of Venus. Mars is a borderline case: If it had been large enough to retain its atmosphere and sustain a stronger greenhouse effect, Mars might still be habitable today. Overall, the habitable zone around our star probably ranges from 0.8 to 1.5 AU. This zone may be broader around brighter stars and narrower around fainter stars. Computer models of solar system formation suggest that one or more terrestrial planets will usually form within a star's habitable zone, as long as the star is not part of a binary star system. The planets recently discovered around the star Upsilon Andromeda may lie within that star's habitable zone [Section 7.6].

The bottom line is that, according to our theories of solar system formation, planets with all the necessities for life should be quite common in the universe. The only major question is whether these ingredients combine to form life. The fact that life arose very early in Earth's history suggests that it may be very easy to produce life under Earth-like conditions, but we will not know for sure unless and until we find other life-bearing planets. NASA is currently developing plans for orbiting telescopes that may be able to detect ozone in the spectra of planets around other stars—and, at least in our solar system, substantial ozone implies life. In addition, radio astronomers are searching the skies in hopes of receiving a signal from some extraterrestrial civilization. Perhaps, in a decade or two, we will discover unmistakable evidence of life. On that day, if it comes, we will know that we are not alone.

TIME OUT TO THINK *Consider the following statements: (1) We are the only intelligent life in the entire universe. (2) Earth is one of many planets inhabited by intelligent life. Which do you think is true? Do you find either philosophically troubling?*

Looking outward to the blackness of space, sprinkled with the glory of a universe of lights, I saw majesty—but no welcome. Below was a welcoming planet. There, contained in the thin, moving, incredibly fragile shell of the biosphere is everything that is dear to you, all the human drama and comedy. That's where life is; that's where all the good stuff is.

LOREN ACTON, U.S. ASTRONAUT

THE BIG PICTURE

Our ancestors have observed the Sun, the Moon, and the planets for thousands of years, but only recently did we learn that these other worlds have much to teach us about our own Earth. Through our study of solar system formation and comparative planetology, we have learned to look at Earth from a very new perspective. Keep in mind the following "big pic-ture" ideas:

- Earth has been shaped by the same geological and atmospheric processes that shaped the other terrestrial worlds. Earth is not a special case from a planetary point of view but rather a place where natural processes led to conditions conducive to life.

- Most of Earth's unique features can be traced to the fact that abundant water has remained liquid throughout our planet's history—thanks to our distance from the Sun and the size of our planet.

- Life arose early on Earth and played a crucial role in shaping our planet's history. The abundant oxygen in our atmosphere is just one of many phenomena that demonstrate how life can transform a planet.

- Humans are ideally adapted to the Earth today, but there is no guarantee that the Earth will remain as hospitable in the future. The study of our solar system teaches us how planets can change.

- The conditions necessary for "life as we know it" are probably common in the universe and may even be found in our own solar system. So far, we have no proof that life exists elsewhere.

Review Questions

1. Briefly summarize how Earth is different from the other terrestrial planets. In particular, define *hydrosphere* and *biosphere*.

2. Briefly describe the interior structure of the Earth, including its molten outer core and its lithosphere.

3. Do we find evidence of impact cratering on Earth? Why do we find fewer craters on the seafloor than on the continents?

4. Why is erosion more important on Earth than on Venus or Mars? Describe how erosion affects the Earth's geology.

5. What do we mean by *plates* and *plate tectonics?* How fast do plates move? How has plate movement changed the arrangement of the continents over long time scales?

6. Briefly describe the conveyor-like process of plate tectonics, with emphasis on the creation of new crust at *spreading centers* and the destruction of old crust at *subduction zones.*

7. Describe how subduction has created the continents over billions of years. Give examples of how this process has affected Japan, the Philippines, and the western United States.

8. Briefly describe how tectonic stresses can create rift valleys and tall mountain ranges. Give examples of places that have been affected in each of these ways.

9. From the standpoint of comparative planetology, list four major questions about the Earth's atmosphere. Briefly answer each of them.

10. What is a *runaway greenhouse effect?* Why did it occur on Venus but not on Earth?

11. Describe how the *carbonate–silicate cycle* helps maintain the relatively small CO_2 content of our atmosphere.

12. What are *oxidation* reactions? If there were no life on Earth, would Earth's atmosphere still contain significant amounts of oxygen or ozone?

13. Approximately when did life arise on Earth? How do we know?

14. Briefly describe how life gradually altered the Earth's atmosphere until it reached its current state.

15. What is global warming? Briefly describe some of the potential dangers of global warming.

16. Briefly describe the phenomena and causes of the Antarctic *ozone hole* and worldwide *ozone depletion.*

17. Is the Earth currently undergoing a mass extinction? Explain.

18. Summarize why we now think that life can survive in a much broader range of conditions than we did just a few decades ago. Why is this important to the prospect of finding life elsewhere?

19. Describe the evidence from Martian meteorites suggesting that life may once have existed on Mars. Explain how life might have originated on one terrestrial planet and been transferred to others.

20. Why are Europa and Titan considered prospects for harboring life?

21. What is a *habitable zone?* Using this idea, briefly describe the prospect of finding life on planets around other stars.

Discussion Questions

1. *Cancer of the Earth?* In the text, we discussed how the spread of humans over the Earth resembles, at least in some ways, the spread of cancer in a human body. Cancers end up killing themselves because of the damage they do to their hosts. Do you think we are in danger of killing ourselves through our actions on the Earth? If so, what should we do to alleviate this danger? Overall, do you think the cancer analogy is valid or invalid? Defend your opinions.

2. *Contact.* Suppose we discover microbial life on another planet in our solar system. Would this discovery alter your view of our place in the universe? If so, how? What if we made contact with an intelligent species from another world? Do you think it is likely that either kind of life exists elsewhere in the universe? Do you think either kind will be discovered in your lifetime? Explain.

Problems

1. *Change in Formation Properties.* Consider Earth's four formation properties of size, distance, composition, and rotation rate. Choose one property, and suppose it had been different (e.g., smaller size, greater distance). Describe how this change might have affected Earth's subsequent history and the possibility of life on Earth.

2. *Defining Life.* Write a definition of life and explain the basis of your definition in a few sentences. Then evaluate whether each of the following three cases meets your definition. Explain why or why not in a paragraph for each case. (i) The first self-replicating molecule on Earth lies near the boundary of life and nonlife. Does it meet your definition of life? (ii) Modern-day viruses are essentially packets of DNA encased in a microscopic shell of protein. Viruses cannot replicate themselves; instead, they reproduce by infecting living cells and "hijacking" the cell's reproduction machinery to make copies of the viral DNA and proteins. Are viruses alive? (iii) Imagine that humans someday travel to other stars and discover a planet populated by what appear to be robots programmed to mine metal, refine it, and assemble copies of themselves. Examination of fossil "robots" shows that they have improved, perhaps because cosmic rays have caused errors in their programs.

Is this race of robots alive? Would your answer depend on whether another race had built the first robots?

3. *Ozone Signature.* Suppose a powerful future telescope is able to take a spectrum of a terrestrial planet around another star. The spectrum reveals the presence of significant amounts of ozone. Why would this discovery strongly suggest the presence of life on this planet? Would it tell us whether the life is microscopic or more advanced? Summarize your answers in one or two paragraphs.

4. *Essay: Explaining Ourselves to the Aliens.* Imagine that someday we make contact with intelligent aliens and even learn to communicate. Even so, many facets of life we take for granted may be utterly incomprehensible to them: music (perhaps they have no sense of hearing), money (perhaps there has never been a need), love (perhaps they don't feel this emotion), meals (perhaps they photosynthesize). Write a page attempting to explain one of these concepts, or another of your choosing. Remember that even the words in your explanation require definition; start at the lowest possible level.

5. *Research: Local Geology.* Write a one- to two-page report about the geology of an area you know well—perhaps the location of your campus or your hometown. Which of the four geological processes has played the most important role, in your opinion? Has plate tectonics played a direct role in shaping the area?

6. *Research: Human Threats to the Earth.* Write an in-depth research report, three to five pages in length, about current understanding and controversy regarding one of the following issues: global warming, ozone depletion, or the loss of species on Earth due to human activity. Be sure to address both the latest knowledge about the issue and proposals for alleviating any dangers associated with it. End your report by making your own recommendations about what, if anything, needs to be done to prevent damage to the Earth.

7. *Web Project: Life on Mars.* Follow the links from the text web site to find the latest information regarding the controversy over evidence for life in Martian meteorites. Write a one- to two-page report summarizing the current state of the controversy and your own opinion as to whether life once existed on Mars.

PART IV
STELLAR ALCHEMY

Our Star 12

TODAY, ASTRONOMY INVOLVES THE STUDY OF THE ENTIRE UNIVERSE, BUT the root of the word *astronomy* originally came from the Greek word for "star." Although we have learned a lot about the universe up to this point in the book, only now do we turn our attention to the study of the stars, the namesakes of our subject.

When we think of stars, we usually think of the beautiful points of light visible on a clear night. But the nearest and most easily studied star is visible only in the daytime—our Sun. Of course, the Sun is important to us in many more ways than as an object for astronomical study. The Sun is the source of virtually all light, heat, and energy reaching the Earth, and life on Earth's surface could not survive without it.

In this chapter, we will study the Sun in some depth. We will learn how the Sun makes life possible on Earth. Equally important, we will study our Sun as a star, so that in subsequent chapters we can more easily understand stars throughout the universe.

12.1 Why Does the Sun Shine?

Ancient peoples recognized the vital role of the Sun in their lives. Some worshiped the Sun as a god, and others created elaborate mythologies to explain its daily rise and set. Only recently, however, have we learned how the Sun provides us with light and heat.

Most ancient thinkers viewed the Sun as some type of fire, perhaps a lump of burning coal or wood. The Greek philosopher Anaxagoras (c. 500–428 B.C.) imagined the Sun to be a very hot, glowing rock about the size of the Greek peninsula of Peloponnesus (comparable in size to Massachusetts), making him one of the first people in history to believe that the heavens and the Earth are made from the same types of materials.

By the mid-1800s, the size and distance of the Sun were well known, and scientists began to address the question of how the Sun shines. Two early ideas held either that the Sun was a cooling ember that had once been much hotter or that the Sun generated energy from some type of chemical burning similar to the burning of coal. Although simple calculations showed that a cooling or chemically burning Sun could shine for a few thousand years—an age that squared well with biblically based estimates of the age of the Earth that were popular at the time—these ideas suffered from fatal flaws. If the Sun were a cooling ember, it would have been much hotter just a few hundred years earlier, making it too hot for civilization to have existed. Chemical burning was ruled out because it cannot generate enough energy to account for the rate of radiation observed from the Sun's surface.

A more plausible hypothesis of the late 1800s suggested that the Sun generates energy by contracting in size, a process called **gravitational contraction.** If the Sun were shrinking, it would constantly be converting gravitational potential energy into thermal energy, thereby keeping the Sun hot. Because of its large mass, the Sun would need to contract only very slightly each year to maintain its temperature. Unfortunately for this model, calculations showed that the Sun could last no more than about 25 million years generating energy by gravitational contraction. Geologists of the late 1800s had already established the age of the Earth to be far more than 25 million years, leaving astronomers in an embarrassing position.

Only after Einstein published his special theory of relativity, which included his discovery of $E = mc^2$, did the true energy-generation mechanism of the Sun become clear. We now know that the Sun generates energy by *nuclear fusion,* a source so efficient that the Sun can shine for 10 billion years. Because the Sun is only 4.6 billion years old today, we expect it to keep shining for some 5 billion years to come.

Our current model of solar-energy generation by nuclear fusion means that the Sun's size is generally stable, maintained by a balance between the competing forces of gravity pulling inward and pressure pushing outward. This balance is called **gravitational equilibrium** (also referred to as *hydrostatic equilibrium*). It means that, at any point within the Sun, the weight of overlying material is supported by the underlying pressure. A stack of acrobats provides a simple example of this balance (Figure 12.1). The bottom person supports the weight of everybody above him, so the pressure on his body is very great. At each higher level, the overlying weight is less, so the pressure decreases. Gravitational equilibrium in the Sun means that the pressure increases with depth, making the Sun extremely hot and dense in its central core (Figure 12.2).

TIME OUT TO THINK *The Earth's atmosphere is also in gravitational equilibrium, with the weight of upper layers supported by the pressure in lower layers. Use this idea to explain why the air gets thinner at higher altitudes.*

Interestingly, while gravitational contraction is not an important energy-generation mechanism in the Sun today, it was important in the distant past and will be important again in the distant future. Our Sun was born from a collapsing cloud of interstellar

FIGURE 12.1 An acrobat stack is in gravitational equilibrium: The lowest person supports the most weight and feels the greatest pressure, and the overlying weight and underlying pressure decrease for those higher up.

FIGURE 12.2 Gravitational equilibrium in the Sun: At each point inside, the pressure pushing up balances the weight of the overlying layers.

Table 12.1 Basic Properties of the Sun

Radius (R_{Sun})	696,000 km (about 109 times the radius of the Earth)
Mass (M_{Sun})	2×10^{30} kg (about 300,000 times the mass of the Earth)
Luminosity (L_{Sun})	3.8×10^{26} watts
Composition (by percentage of mass)	70% hydrogen, 28% helium, 2% heavier elements
Rotation rate	27 days (equator) to 31 days (poles)
Surface temperature	5,800 K (average); 4,000 K (sunspots)
Core temperature	15 million K

gas. The contraction of the cloud released gravitational potential energy, raising the interior temperature higher and higher. When the central temperature and density reached the values necessary to sustain nuclear fusion, the newly released energy provided enough pressure to halt the collapse. That is, the onset of fusion brought the Sun into gravitational equilibrium. About 5 billion years from now, when the Sun finally exhausts its nuclear fuel, the internal pressure will drop, and gravitational contraction will begin once again. As we will see later, some of the most important and spectacular processes in astronomy hinge on this ongoing "battle" between the crush of gravity and a star's internal sources of pressure.

Thus, the answer to the question "Why does the Sun shine?" is that about 4.6 billion years ago *gravitational contraction* made the Sun hot enough to sustain nuclear fusion in its core. Ever since, energy liberated by fusion has maintained the Sun's *gravitational equilibrium* and kept the Sun shining steadily, supplying the light and heat that sustain life on Earth.

12.2 Solar Properties: A Brief Overview

The rest of this chapter will discuss in detail how the Sun produces energy and how that energy travels to Earth. But first, to give a "big picture" view of the Sun, we will take a look at the Sun's basic properties (summarized in Table 12.1):

- From Earth, the Sun appears as a whitish ball of glowing gas, with a radius of nearly 700,000 kilometers.

- The average surface temperature is 5,800 K, but cooler **sunspots** often blotch its surface (Figure 12.3).

- The Sun's mass is 2×10^{30} kilograms. About 70% of this mass is hydrogen, and 28% is helium. Heavier elements make up the remaining 2%.

- The total power output of the Sun, called its **luminosity,** is 3.8×10^{26} watts. That is, every second the Sun radiates a total of 3.8×10^{26} joules of energy into space (recall that 1 watt = 1 joule/s).

Figure 12.4 summarizes the Sun's basic structure. At the Sun's center lies its **core,** the place where nuclear fusion releases energy by transforming hydrogen into helium at a temperature of 15 million K. In the **radiation zone,** just above the solar core, X-ray photons gradually carry the energy released by fusion away from the Sun's center. The radiation zone extends to about two-thirds of the Sun's radius; the next layer is the **convection zone,** in which rising bubbles of hot gas carry the Sun's energy the rest of

FIGURE 12.3 This photo of the visible surface of the Sun shows several dark sunspots.

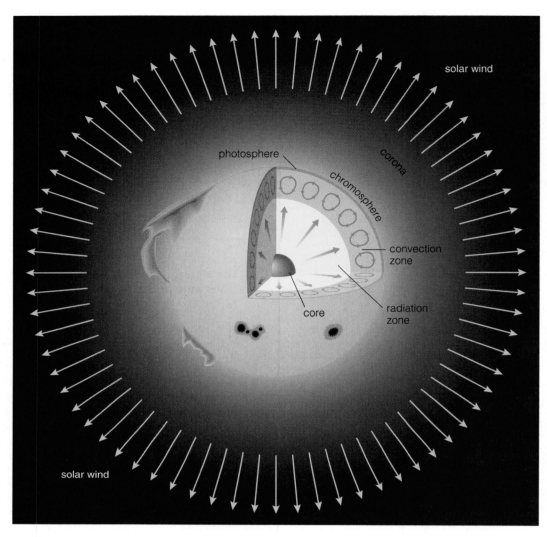

FIGURE 12.4 The basic structure of the Sun.

the way to the surface. Atop the convection zone lies the **photosphere,** the visible surface of the Sun, where the temperature is 5,800 K. Photons created in the photosphere carry the Sun's energy out into space. Two hotter, more tenuous layers sit above the photosphere: the **chromosphere,** where temperatures rise above 10,000 K, and the **corona,** where temperatures can exceed 1 million K. Charged particles moving outward from the corona form the **solar wind,** which flows past the planets to the outer reaches of the solar system. Now let's examine what goes on in each of these layers, proceeding from the inside out.

12.3 The Solar Core

The prospect of turning common metals like lead into gold enthralled medieval alchemists, but their get-rich-quick schemes never worked. There is no easy way to turn lead into gold, but it *is* possible to transmute one element or isotope into another. If a nucleus gains or loses protons, its atomic number changes and it becomes a different element. If it gains or loses neutrons, its atomic weight changes and it becomes a different isotope [Section 4.3]. The process of splitting a nucleus so that it loses protons or neutrons is called **nuclear fission.** The process of combining nuclei to make a nucleus with a greater number of protons or neutrons is called **nuclear fusion.** Human-built nuclear power plants rely on nuclear fission of uranium or plutonium. The nuclear power plant at the center of the Sun relies on nuclear fusion, turning hydrogen into helium.

Nuclear Fusion

The 15-million-K plasma in the solar core is like a "soup" of hot gas, with bare, positively charged atomic nuclei (and individual negatively charged electrons)

At low speeds, electromagnetic repulsion prevents the collision of nuclei.

At high speeds, nuclei come close enough for the strong nuclear force to bind them together.

FIGURE 12.5 Positively charged nuclei can fuse only if a high-speed collision brings them close enough for the strong nuclear force to come into play.

whizzing about at extremely high speeds. At any one time, some of these nuclei are on high-speed collision courses with each other. In most cases, electromagnetic forces deflect the nuclei, preventing actual collisions, because positive charges repel one another. However, if nuclei collide with sufficient energy, they can stick together to form a heavier nucleus (Figure 12.5).

Sticking positively charged nuclei together is not easy. The **strong force,** which binds protons and neutrons together in atomic nuclei, is the only force in nature that can overcome the electromagnetic repulsion between two positively charged nuclei. In contrast to gravitational and electromagnetic forces, which drop off gradually as the distances between particles increase (by an inverse square law), the strong force is more like glue or Velcro: It overpowers the electromagnetic force over very small distances but is insignificant when the distances between particles exceed the typical sizes of atomic nuclei. The trick to nuclear fusion therefore is to push the positively charged nuclei close enough together for the strong force to outmuscle electromagnetic repulsion.

The high pressures and temperatures in the solar core are just right for fusion of hydrogen nuclei into helium nuclei. The high temperature is important because the nuclei must collide at very high speeds if they are to come close enough together to fuse. The higher the temperature, the harder the collisions, making fusion reactions more likely at higher temperatures. The high pressure of the overlying layers is necessary because, without it, the hot plasma of the solar core would simply explode into space, shutting off the nuclear reactions.

Hydrogen Fusion in the Sun: The Proton–Proton Chain

Recall that hydrogen nuclei are just individual protons, while the most common form of helium consists of two protons and two neutrons. Thus, the overall hydrogen fusion reaction in the Sun is:

4 ^{1}H   1 ^{4}He

The sequence of steps that occurs in the Sun is called the **proton–proton chain** because it begins with collisions between individual protons (hydrogen nuclei). Each resulting helium nucleus has a mass that is slightly less (by about 0.7%) than the combined mass of the four protons that created it. Overall, fusion in the Sun converts about 600 million tons of hydrogen into 596 million tons of helium every second; the "missing" 4 million tons of matter becomes energy in accord with Einstein's formula $E = mc^2$. About 2% of this energy is carried off by *neutrinos*—minuscule particles with very little mass. The rest of the energy emerges as kinetic energy of the nuclei and radiative energy of the gamma rays. As we will see shortly, this energy slowly percolates to the solar surface, eventually emerging as the sunlight that bathes the Earth.

The Solar Thermostat

The rate of nuclear fusion in the solar core, which determines the energy output of the Sun, is very sensitive to temperature. A slight increase in temperature would cause particles to fuse faster, and a slight decrease in temperature would cause them to fuse more slowly. If the Sun's rate of fusion varied erratically, the effects on the Earth might be devastating. Fortunately, the Sun's central temperature is steady, thanks to gravitational equilibrium—the balance between the pull of gravity and the push of internal pressure.

Suppose that the core temperature rose very slightly. The rate of nuclear fusion would soar, generating excess energy that would increase the pressure. The push of this pressure would temporarily exceed the pull of gravity, causing the core to expand and cool. With this cooling, the fusion rate would drop back down. The expansion and cooling would continue until gravitational equilibrium was restored, at which point the fusion rate would be back at its original value. An opposite process restores the normal fusion rate if the core temperature drops. A decrease in core temperature would lead to decreased nuclear burning, a drop in the central pressure, and contraction of the core. As the core shrank, its temperature would rise until the burning rate returned to normal.

The response of the core pressure to changes in the nuclear fusion rate is essentially a *thermostat* that keeps the Sun's central temperature steady. Any random change in the core temperature is automatically corrected by the change in the fusion rate and the accompanying change in pressure.

While the processes involved in gravitational equilibrium prevent erratic changes in the fusion rate, they also ensure that the fusion rate gradually rises over billions of years. Theoretical models indicate that the Sun's core temperature should have increased enough to raise its fusion rate and the solar luminosity by about 30% since the Sun was born 4.6 billion years ago. The gradual increase in the solar luminosity poses a bit of a puzzle, because geological evidence shows that the Earth's temperature has remained fairly steady since the Earth finished forming over 4 billion years ago [Section 11.3]. Studying how the Sun changes with time and how the Earth responds to these changes may teach us much about how the Earth's physical and biological systems regulate the greenhouse effect.

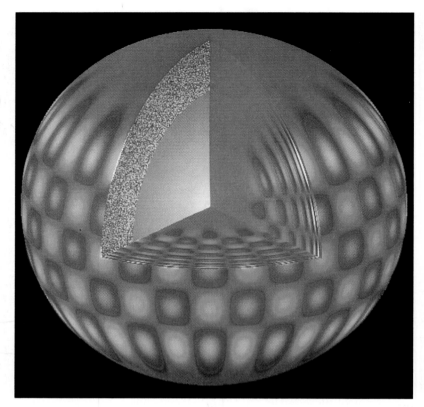

FIGURE 12.6 Vibrations on the surface of the Sun can be detected by Doppler shifts. In this schematic representation, red indicates falling gas, and blue indicates rising gas.

"Observing" the Solar Interior

The primary way we learn about the interior of the Sun and other stars is by creating *mathematical models* that use the laws of physics to predict the internal conditions. A basic model uses the Sun's observed composition and mass as inputs to equations that describe gravitational equilibrium and the solar thermostat. With the aid of a computer, we can use the model to calculate the Sun's temperature, pressure, and density at any depth. We can then predict the rate of nuclear fusion in the solar core by combining these calculations with knowledge about nuclear fusion gathered in laboratories here on Earth. Remarkably, such models correctly "predict" the radius, surface temperature, luminosity, age, and many other properties of the Sun. However, current models do not predict *everything* about the Sun correctly, so scientists are constantly working to discover what is missing from them. The fact that the models successfully predict so many observed characteristics of the Sun gives us confidence that they are on the right track and that we really do understand what is going on inside the Sun.

A second way to learn about the inside of the Sun is to observe "sun quakes"—vibrations of the Sun that are similar to the vibrations of the Earth caused by earthquakes, although they are generated very differently. Earthquakes occur when the Earth's crust suddenly shifts, generating *seismic waves* that propagate through the Earth's interior. We can learn about the Earth's interior by recording seismic waves on the Earth's surface with seismographs. Sun quakes result from waves of pressure (sound waves) that propagate deep within the Sun at all times. These waves cause the solar surface to vibrate when they reach it. Although we cannot set up seismographs on the Sun, we can detect the vibrations of the surface by measuring Doppler shifts [Section 6.4]: Light from portions of the surface that are rising toward us is slightly blueshifted, while light from portions that are falling away from us is slightly redshifted. The vibrations are relatively small but measurable (Figure 12.6). In principle, we can deduce a great deal about the solar interior by carefully analyzing these vibrations. (By analogy to seismology on Earth, this type of study of the Sun is called *helioseismology* [recall that *helios* means "sun"].) Results to date confirm that mathematical models of the solar interior are on the right track and at the same time provide data that can be used to test and improve the models further.

The Solar Neutrino Problem

One other way to study the solar interior is to observe the neutrinos coming from fusion reactions in the core. As you read this sentence, about a thousand trillion solar neutrinos will zip through your body. Fortunately, they won't do any damage, because neutrinos rarely interact with anything. Neutrinos created by fusion in the solar core fly quickly through the Sun as if passing through empty space. In fact, while an inch of lead will stop an X ray, stopping an average neutrino would require a slab of lead more than a light-year thick! Clearly, counting neutrinos is dauntingly difficult, because they stream right through any detector built to capture them.

TIME OUT TO THINK *Is the number of solar neutrinos zipping through our bodies significantly lower at night? (Hint: How does the thickness of the Earth compare with the thickness of a slab of lead needed to stop an average neutrino?)*

Nevertheless, neutrinos *do* occasionally interact with matter, and it is possible to capture a few solar neutrinos with a large enough detector, such as the "Homestake experiment," located 1,500 meters underground in a South Dakota gold mine (Figure 12.7). The designers of this experiment expected to capture an average of just one neutrino per day. However, over a period of more than two decades, they captured only about one neutrino every 3 days. That is, the Homestake experiment detected only about one-third of the predicted number of neutrinos. This disagreement between model predictions and actual observations is called the **solar neutrino problem.**

The shortfall of neutrinos found with the Homestake experiment has led to many more recent attempts to detect solar neutrinos using more sophisticated detectors. To date, all these experiments have still found fewer neutrinos than current models of the Sun predict. This discrepancy between model and experiment probably means one of two things: Either something is wrong with our models of the Sun, or something is missing in our understanding of how neutrinos behave. For the moment, many physicists and astronomers are betting that we understand the Sun just fine and that the discrepancy has to do with the neutrinos themselves. However, our understanding of nuclear fusion in the Sun will remain incomplete until we solve the puzzling mystery of the solar neutrinos.

FIGURE 12.7 This tank of dry-cleaning fluid (visible underneath the catwalk), located deep within South Dakota's Homestake mine, was a solar neutrino detector. The chlorine nuclei in the cleaning fluid turned into argon nuclei when they captured neutrinos from the Sun.

12.4 From Core to Corona

Energy liberated by nuclear fusion in the Sun's core must eventually reach the solar surface, where it can be radiated into space. The path that the energy takes to the surface is long and complex. In this section, we follow the long path that energy travels after its production in the core.

The Path Through the Solar Interior

In Chapter 6, we discussed how atoms can absorb or emit photons. In fact, photons can also interact with a charged particle, and a photon that "collides" with an electron can be deflected into a completely new direction. Deep in the solar interior, the plasma is so dense that the gamma-ray photons resulting from fusion travel only a fraction of a millimeter before colliding with an electron. Because each collision sends the photon in a random new direction, the photon bounces around the core in a haphazard way,

FIGURE 12.8 A photon in the solar interior bounces randomly among electrons, slowly working its way outward over about 1 million years.

Hot gas rises.

Cool gas descends.

FIGURE 12.9 Convection transports energy outward in the Sun's convection zone. This schematic diagram shows how a column of hot gas rises while cooler gas descends around it.

sometimes called a *random walk* (Figure 12.8). With each random bounce, the photon drifts farther and farther, on average, from its point of origin. As a result, photons from the solar core work their way outward, gradually losing energy to their surroundings. By the time the energy of fusion reaches the surface, the photons are primarily visible light.

This random bouncing of photons is the primary way by which energy moves outward through the *radiation zone,* which stretches from the core to about two-thirds of the Sun's radius (see Figure 12.4). Above this point, where the temperature has dropped to about 2 million K, the solar plasma absorbs photons more readily (rather than just bouncing them around). This is the beginning of the solar *convection zone,* where the build-up of heat resulting from photon absorption causes bubbles of hot plasma to rise upward in the process known as **convection** [Section 8.2]. Convection occurs because hot gas is less dense than cool gas. Like a hot-air balloon, a hot bubble of solar plasma rises upward through the cooler plasma above it. Meanwhile, cooler plasma from above slides around the rising bubble and sinks to lower layers, where it is heated. The rising of hot plasma and sinking of cool plasma form a cycle that transports energy outward from the top of the radiation zone to the solar surface (Figure 12.9).

The Solar Surface

The Earth has a solid crust, so its surface is well defined. In contrast, the Sun is made entirely of gaseous plasma. Defining where the surface of the Sun begins is therefore something like defining the surface of a cloud: From a distance it looks quite distinct, but up close the surface is fuzzy, not sharp. This surface layer, called the *photosphere,* is the layer from which photons finally escape into space after the million-year journey of solar energy outward from the core. Most of the energy produced by fusion in the solar core ultimately leaves the photosphere as thermal radiation [Section 6.3]. The average temperature of the photosphere is about 5,800 K, corresponding to a thermal radiation spectrum that peaks in the green portion of the visible spectrum.

Although the average temperature of the photosphere is 5,800 K, local temperatures vary significantly from place to place. The photosphere is marked throughout by the bubbling pattern of **granulation** produced by the underlying convection (Figure 12.10). Each *granule* appears bright in the center, where hot gas bubbles upward, and dark around the edges, where cool gas descends. If we made a movie of the granulation, we'd see it bubbling rather like a pot of boiling water. Just as bubbles in a pot of boiling water burst on the surface and are replaced by new bubbles, each granule lasts only a few minutes before being replaced by other granules bubbling upward.

FIGURE 12.10 Granulation of the photosphere is evident in this photo of two sunspots. Each white granule is the top of a rising column of hot gas. At the darker lines between the granules, cooler gas is descending below the photosphere.

FIGURE 12.11 (**a**) The vertical line on this photo of a sunspot and its surrounding region represents the line along which light was dispersed into a spectrum. (**b**) A portion of this spectrum is shown with a spectral line that is a single line outside the sunspot region but splits into three parts under the influence of the sunspot's strong magnetic field.

a

b

Sunspots and Magnetic Fields

Sunspots are the most striking features on the solar surface. The temperature of the plasma in sunspots is about 4,000 K, which is significantly cooler than the 5,800-K plasma of the surrounding photosphere. You may wonder how sunspots can be so much cooler than their surroundings. What keeps the surrounding hot plasma from heating the sunspots?

Something must be preventing hot plasma from entering the sunspots, and that "something" turns out to be magnetic fields. Strong magnetic fields can alter the energy levels in atoms and ions and therefore can alter the spectral lines they produce. More specifically, magnetic fields cause some spectral lines to split into two or more closely spaced lines. Thus, scientists can map magnetic fields on the Sun by studying the spectral lines in light from different parts of the solar surface (Figure 12.11a). Such maps (Figure 12.11b) reveal sunspots to be regions of intense magnetic fields.

Magnetic fields are invisible, but in principle we could visualize a magnetic field by laying out many compasses; each compass needle would point to local magnetic north. We can represent the magnetic field by drawing a series of lines, called **magnetic field lines,** connecting the needles of these imaginary compasses (Figure 12.12a). The strength of the magnetic field is indicated by the spacing of the lines: Closer lines mean a stronger field (Figure 12.12b). Because these imaginary field lines are so much easier to visualize than the magnetic field itself, we usually discuss magnetic fields by talking about how the field lines would appear. Charged particles, such as the ions and electrons in the solar plasma, follow paths that spiral along the magnetic field lines (Figure 12.12c). Thus, the solar plasma can move freely *along* magnetic field lines but cannot easily move perpendicular to them.

FIGURE 12.12 (**a**) Magnetic field lines follow the directions that compass needles would point. (**b**) Lines closer together indicate a stronger field. (**c**) Charged particles follow paths that spiral along magnetic field lines.

weaker magnetic field

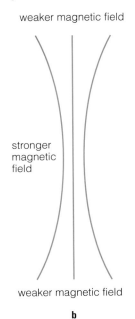

stronger magnetic field

weaker magnetic field

a

b

c

Magnetic fields
trap gas.

sunspots
T ≈ 4,500 K

T ≈ 5,800 K

T ≈ 5,800 K

convection cells

Magnetic fields of sunspots suppress convection
and prevent surrounding plasma from sliding
sideways into sunspot.

FIGURE 12.13 Pairs of sunspots are connected by tightly wound magnetic field lines.

The magnetic field lines act somewhat like elastic bands, twisted into contortions and knots by turbulent motions in the solar atmosphere. Sunspots occur where the most taut and tightly wound magnetic fields poke nearly straight out from the solar interior (Figure 12.13). Note that sunspots tend to occur in pairs connected by a loop of magnetic field lines. These tight magnetic field lines suppress convection within each sunspot and prevent surrounding plasma from sliding sideways into the sunspot. With hot plasma unable to enter the region, the sunspot plasma becomes cooler than that of the rest of the photosphere.

The magnetic field lines connecting two sunspots sometimes soar high above the photosphere, through the chromosphere, and into the corona. These vaulted loops of magnetic field sometimes appear as **solar prominences,** in which the field traps gas that may glow for days or weeks as it cools or until the field collapses. Some prominences rise to heights of more than 100,000 kilometers above the Sun's surface (Figure 12.14).

The most dramatic events on the solar surface are **flares,** which emit bursts of X rays and fast-moving charged particles into space (Figure 12.15). Flares generally occur in the vicinity of sunspots, leading us to suspect that they occur when the magnetic field lines become so twisted and knotted that they can bear the tension no longer. The magnetic field lines

FIGURE 12.14 This photo shows a large solar prominence, many times the size of Earth, which consists of glowing gas trapped by magnetic field lines arching high above the surface of the Sun.

suddenly snap like tangled elastic bands twisted beyond their limits, releasing a huge amount of energy. This energy heats the nearby plasma to 100 million K over the next few minutes to few hours, generating X rays and accelerating some of the charged particles to nearly the speed of light.

a

b

FIGURE 12.15 (**a**) Hot gas erupting from the Sun's surface during a solar flare. (Photo taken with a filter that shows only light from a particular transition in hydrogen atoms.) (**b**) This close-up photo of the Sun's surface shows a wave (dark ring) moving outward from the location of a solar flare.

The Chromosphere and Corona

The high temperatures of the chromosphere and corona perplexed scientists for decades. Temperatures gradually decline from the core to the top of the photosphere, so why does this declining trend suddenly reverse? Some aspects of this solar atmospheric heating remain a mystery, but we have at least a general explanation: The Sun's strong magnetic fields carry energy upward from the churning solar surface to the chromosphere and corona. More specifically, the rising and falling of gas in the convection zone probably shakes magnetic field lines beneath the solar surface. This shaking generates waves along the magnetic field lines that carry energy upward to the solar atmosphere. Precisely how the waves deposit their energy in the chromosphere and corona is not known, but the waves agitate the low-density plasma of these layers, somehow heating them to high temperatures.

Note that, according to this model of solar heating, the same magnetic fields that keep sunspots cool make the overlying plasma of the chromosphere and corona hot. We can test this idea observationally. The gas of the chromosphere and corona is so tenuous that we cannot see it with our eyes except during a total eclipse, when we can see the faint visible light scattered by electrons in the corona [Section 2.3]. However, the roughly 10,000-K plasma of the chromosphere emits strongly in the ultraviolet, and the million-K plasma of the corona is the source of virtually all X rays coming from the Sun. Figure 12.16 shows an X-ray image of the Sun: As the solar heating model predicts, the brightest regions of the corona tend to be directly above sunspot groups.

FIGURE 12.16 An X-ray image of the Sun reveals the corona: Brighter regions of this image correspond to regions of stronger X-ray emission. (Because X rays do not penetrate Earth's atmosphere, photos like this one must be taken from space.)

Some regions of the corona, called **coronal holes,** barely show up in X-ray images. More detailed analyses show that the magnetic field lines in coronal holes project out into space like broken rubber bands, allowing particles spiraling along them to escape the Sun altogether. These particles streaming outward from the corona constitute the *solar wind,* which blows through the solar system at an average speed of about 500 kilometers per second and has important effects on planetary surfaces, atmospheres, and magnetospheres [Section 8.4]. Somewhere beyond the orbit of Neptune, the pressure of interstellar gas must eventually halt the solar wind.

a

12.5 Solar Weather and Climate

Short-lived phenomena like sunspots, prominences, and flares constitute what we call *solar weather* or **solar activity.** You know from personal experience that the Earth's weather is notoriously unpredictable, and the same is true for the Sun: We cannot predict precisely when or where a particular sunspot or flare will appear. The Earth's *climate,* on the other hand, is quite regular from season to season. So it is with the Sun, where despite day-to-day variations the general nature and intensity of solar activity follow a predictable cycle.

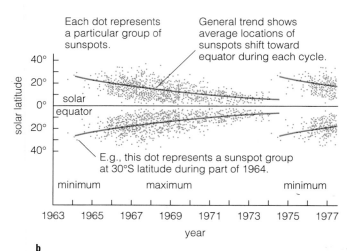

b

FIGURE 12.17 (**a**) This graph shows how the number of sunspots on the Sun changes with time. Note the approximately 11-year cycle. (**b**) This graph shows how the latitudes at which sunspot groups appear tend to shift during a single sunspot cycle.

The Sunspot Cycle

Long before we realized that sunspots were magnetic disturbances, astronomers had recognized patterns in sunspot activity. The most notable pattern is the number of sunspots visible on the Sun at any particular time. Thanks to telescopic observations of the Sun recorded by astronomers since the 1600s, we know that the number of sunspots gradually rises and falls in a **sunspot cycle** with a period of about 11 years (Figure 12.17a). At the time of **solar maximum,** when sunspots are most numerous, we may see dozens of sunspots on the Sun at one time. In contrast, we see few if any sunspots at the time of **solar minimum.** Note that the sunspot cycle is not perfectly predictable: The interval between solar maxima is sometimes as long as 15 years or as short as 7 years. More dramatically, between the years 1645 and 1715, sunspot activity virtually ceased. (This period is called the *Maunder minimum,* after E. W. Maunder, who discovered it while sifting through historical records of sunspot activity.) The

frequency of prominences and flares also follows the sunspot cycle, with these events being most common at solar maximum and least common at solar minimum.

Another feature of the sunspot cycle is a gradual change in the latitudes at which individual sunspots form and dissolve (Figure 12.17b). As a cycle begins at solar minimum, sunspots form primarily at midlatitudes (30° to 40°) on the Sun. The sunspots tend to form at lower latitudes as the cycle progresses, appearing very close to the solar equator as the next solar minimum approaches.

A less obvious feature of the sunspot cycle is that, at each solar minimum, something peculiar happens to the Sun's magnetic field: The orientation of the Sun's magnetic field appears to reverse itself. These magnetic reversals hint that the sunspot cycle is related to the generation of magnetic fields on the Sun. They also tell us that the *complete* cycle really averages 22 years, since it takes two 11-year cycles before the magnetic field is back the way it started.

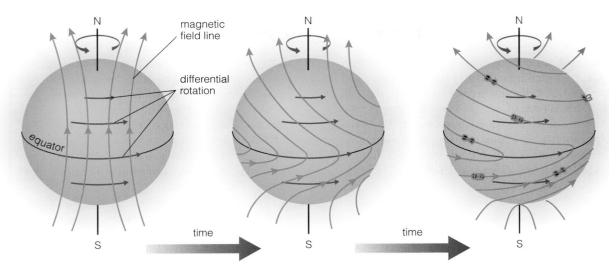

FIGURE 12.18 The Sun rotates more quickly at its equator than it does near its poles, a behavior known as differential rotation. Because gas circles the Sun faster at the solar equator, it drags the Sun's north-south magnetic field lines into a more twisted configuration. The magnetic field lines linking pairs of sunspots, depicted here as green and black blobs, trace out the directions of these stretched and distorted field lines.

What Causes the Sunspot Cycle?

The causes of the Sun's magnetic fields and the sunspot cycle are not well understood, but we believe we know the general nature of the processes involved. Convection is thought to dredge up weak magnetic fields generated in the solar interior, amplifying them as they rise. The Sun's **differential rotation**—faster at its equator than near its poles—then stretches and shapes these fields.

Imagine what happens to a magnetic field line that originally runs along the Sun's surface directly from the north pole to the south pole. At the equator, the field line circles the Sun in 27 days, but at higher latitudes the field line lags behind. Gradually, differential rotation winds the field line more and more tightly around the Sun (Figure 12.18). This process, operating at all times over the entire Sun, produces the contorted field lines that generate sunspots and other solar activity.

Making more specific statements about how the Sun's magnetic field develops and changes in time requires the aid of sophisticated computer models, which have successfully replicated such features of the sunspot cycle as changes in the number and latitude of sunspots and the magnetic field reversals that occur about every 11 years. However, much still remains mysterious, including why the period of the sunspot cycle varies and why solar activity is different from one cycle to the next.

Solar Activity and the Earth

During solar maximum, solar flares and other forms of solar activity feed large numbers of highly energetic charged particles (protons and electrons) into the solar wind. Do these forms of solar weather affect the Earth? In at least some ways, the answer is a definite yes.

The solar wind continually carries charged particles from the solar corona to the Earth, creating the shimmering light of the *aurora*. Particles streaming from the Sun after the occurrence of solar flares or other major solar storms can also have practical impacts on society. For example, these particles can hamper radio communications, disrupt electrical power delivery, and damage the electronic components in orbiting satellites. During a particularly powerful magnetic storm on the Sun in March 1989, the U.S. Air Force temporarily lost track of over 2,000 satellites, and powerful currents induced in the ground circuits of the Quebec hydroelectric system caused it to collapse for more than 8 hours.

Satellites in low-Earth orbit are particularly vulnerable during solar maximum, because the increase in solar X rays and energetic particles heats the Earth's upper atmosphere, causing it to expand. The density of the gas surrounding low-flying satellites therefore rises, exerting drag that saps their energy and angular momentum. If this drag proceeds unchecked, the satellites ultimately plummet back to Earth. Satellites in low orbits, including the Hubble Space Telescope, therefore require occasional boosts to prevent them from falling out of the sky. (Hubble is scheduled to get boosts from the Space Shuttle in 2001 and 2004.)

Connections between solar activity and the Earth's climate are much less clear. The period from 1645 to 1715, when solar activity seems to have virtually ceased, was a time of exceptionally low temperatures in Europe and North America known as the *Little Ice Age*. Did the low solar activity cause these low temperatures, or was their occurrence a coincidence? No one knows for sure. Some researchers have claimed that certain weather phenomena, such as drought cycles or frequencies of storms, are correlated with the 11- or 22-year cycles of solar activity. However, the data supporting these correlations are weak in many cases, and some other apparent correlations may be coincidental.

Part of the difficulty in linking solar activity with climate is that no one understands how the linkage might work. Although emissions of ultraviolet light, X rays, and high-energy particles increase substantially from solar minimum to solar maximum, the total luminosity of the Sun barely changes at all. (The Sun becomes slightly brighter during solar maximum, but only by about 0.01%.) Thus, if solar activity really is affecting Earth's climate, it must be through some very subtle mechanism. For example, perhaps the expansion of the Earth's upper atmosphere that occurs with solar maximum somehow causes changes in weather.

The question of how solar activity is linked to the Earth's climate is very important, because we need to know whether global warming is affected by solar activity in addition to human activity. Unfortunately, for the time being at least, we can say little about this question.

> *I say Live, Live, because of the Sun,*
> *The dream, the excitable gift.*
>
> ANNE SEXTON (1928–74)

THE BIG PICTURE

In this chapter, we have examined our Sun, the nearest star. When you look back at this chapter, make sure you understand these "big picture" ideas:

- The ancient riddle of why the Sun shines is now solved. The Sun shines with energy generated by fusion of hydrogen into helium in the Sun's core. After a million-year journey through the solar interior and an 8-minute journey through space, a small fraction of this energy reaches the Earth and supplies sunlight and heat.

- Gravitational equilibrium, the balance between pressure and gravity, determines the Sun's interior structure and maintains its steady nuclear burning rate. If the Sun were not so steady, life on Earth might not have been possible.

- The Sun's atmosphere displays its own version of weather and climate, governed by solar magnetic fields. Solar weather has important influences on the Earth.

- The Sun is important not only as our source of light and heat, but also because it is the only star near enough for us to study in great detail. In the coming chapters, we will use what we've learned about the Sun to help us understand other stars.

Review Questions

1. Briefly describe how the mechanism of *gravitational contraction* generates energy. Although this mechanism does not generate energy for the Sun today, explain how it was important in the Sun's early history.

2. What two forces are balanced in what we call *gravitational equilibrium*? Briefly describe how the idea of gravitational equilibrium implies that pressure must increase with depth in any planet or star.

3. What is the Sun made of?

4. Briefly describe the Sun's luminosity, mass, radius, and average surface temperature. Put each property in perspective by comparing it to some familiar quantity related to Earth.

5. Briefly describe the distinguishing features of each of the Sun's major layers shown in Figure 12.4.

6. What are sunspots? Why do they appear dark in pictures of the Sun?

7. Briefly distinguish between *nuclear fission* and *nuclear fusion*. Which one is used in nuclear power plants? Which one is used by the Sun?

8. Why does fusion require high temperatures and pressures? Explain why nuclei normally want to repel each other, and describe the conditions under which the *strong force* can bind nuclei together.

9. How does the idea of gravitational equilibrium explain why the Sun's core temperature and fusion rate are self-regulating, like a thermostat? Explain why the same idea ensures that the Sun's rate of fusion must gradually rise over billions of years.

10. Explain how *mathematical models* allow us to predict conditions inside the Sun. How can we be confident that the models are on the right track?

11. How are "sun quakes" similar to earthquakes? Describe how we can observe them and how they help us learn about the solar interior.

12. What is the *solar neutrino problem?*

13. Why does the energy produced by fusion in the solar core take so long to reach the solar surface?

14. What is *convection?* Describe how convection transports energy through the Sun's convection zone.

15. Describe the appearance and temperature of the Sun's photosphere. What is *granulation?* How would granulation appear in a movie?

16. Briefly describe how *magnetic field lines* are related to what we would see if we placed compasses in a magnetic field.

17. How are magnetic fields related to sunspots? Describe how magnetic fields explain each of the following: the fact that sunspot plasma is cool, the existence and appearance of *prominences,* and the dramatic explosions of *solar flares.*

18. Describe the current theory of how the chromosphere and corona are heated. What evidence supports it?

19. What are *coronal holes?* How are they related to the solar wind?

20. What do we mean by *solar activity?* Explain how it is similar to weather on short time scales and to climate on longer time scales.

21. Describe the *sunspot cycle.* What do we mean by *solar maximum* and *solar minimum?* Explain what happens to the Sun's magnetic field with each new sunspot cycle and why the complete cycle is really 22 years rather than 11 years.

22. Describe a few known ways solar activity affects the Earth. Why is it so difficult to establish links between solar activity and the Earth's climate?

Discussion Question

1. *The Sun and Global Warming.* One of the most pressing environmental issues on Earth concerns the extent to which human emissions of greenhouse gases are warming our planet. Some people claim that part or all of the observed warming over the past century may be due to changes on the Sun, rather than to anything that humans have done. Discuss how a better understanding of the Sun might help us understand the threat posed by greenhouse gas emissions. Why is it so difficult to develop a clear understanding of how the Sun affects the Earth's climate?

Problems

1. *Differential Rotation.* The Sun and Jupiter exhibit differential rotation, but Earth does not.
   a. Explain clearly what we mean by *differential rotation.*
   b. Briefly explain why it is possible for jovian planets to rotate differentially, but it is not possible for terrestrial planets to do so.
   c. Would you expect differential rotation to be common among other stars? Why or why not?

2. *The Lifetime of the Sun.* The total mass of the Sun is about 2×10^{30} kg, of which about 75% was hydrogen when the Sun formed. However, only about 13% of this hydrogen ever becomes available for fusion in the core; the rest remains in layers of the Sun where the temperature is too low for fusion.
   a. Based on the given information, calculate the total mass of hydrogen that is available for fusion over the lifetime of the Sun.
   b. Combine your results from part (a) and the fact that the Sun fuses about 600 billion kg of hydrogen each second to calculate how long the Sun's initial supply of hydrogen can last. Give your answer in both seconds and years.
   c. Given that our solar system is now about 4.6 billion years old, when will we need to worry about the Sun running out of hydrogen for fusion?

3. *Web Project: Current Solar Activity.* Find web sites describing current activity on the Sun. Where are we in the sunspot cycle right now? When is the next solar maximum or minimum expected? Have there been any major solar storms in the past few months? If so, did they have any significant effects on the Earth? Summarize your findings in a one- to two-page report.

$\int$tars

13

ON A CLEAR, DARK NIGHT, A FEW THOUSAND STARS ARE visible to the naked eye. Many more become visible through binoculars, and with a powerful telescope we can see so many stars that we could never hope to count them. Like individual people, each individual star is unique. Like the human family, all stars share much in common.

Today, we know that stars are born from clouds of interstellar gas, shine brilliantly by nuclear fusion for millions or billions of years, and then die, sometimes in dramatic ways. This chapter outlines how we study and categorize stars and how we have come to realize that stars, like people, change over their lifetime.

13.1 Snapshot of the Heavens

Imagine that an alien spaceship flies by Earth on a simple but short mission: The visitors have just 1 minute to learn everything they can about the human race. In 60 seconds, they will see next to nothing of each individual person's life. Instead, they will obtain a collective "snapshot" of humanity that shows people from all stages of life engaged in their daily activities. From this snapshot alone, they must piece together their entire understanding of human beings and their lives, from birth to death.

We face a similar problem when we look at the stars. Compared with stellar lifetimes of millions or billions of years, the few hundred years humans have spent studying stars with telescopes is rather like the aliens' 1-minute glimpse of humanity. We see only a brief moment in any star's life, and our collective snapshot of the heavens consists of such frozen moments for billions of stars. From this snapshot, we try to reconstruct the life cycles of stars while also analyzing what makes one star different from another.

Thanks to the efforts of astronomers studying this snapshot of the heavens, stars are no longer mysterious points of light in the sky. We now know that all stars form in great clouds of gas and dust. All stars begin their life with roughly the same chemical composition: About three-quarters of the star's mass at birth is hydrogen, and about one-quarter is helium, with no more than about 2% consisting of elements heavier than helium. During most of any star's life, the rate at which it generates energy depends on the same type of balance between the inward pull of gravity and the outward push of internal pressure that governs the rate of fusion in our Sun. Despite these similarities, stars appear different from one another for two primary reasons: They differ in mass, and we see different stars at different stages of their lives.

The key that finally unlocked these secrets of stars was an appropriate classification system. Before the twentieth century, humans classified stars primarily by their brightness and location in our sky. Today, astronomers classify a star primarily according to its *luminosity* and *surface temperature*. Our task in this chapter is to learn how this extraordinarily effective classification system reveals the true nature of stars and their life cycles. We begin by investigating how to determine a star's luminosity, surface temperature, and mass.

13.2 Stellar Luminosity

A star's **luminosity** is the total amount of power it radiates into space, which can be stated in *watts* [Section 12.2]. For example, the Sun's luminosity is 3.8×10^{26} watts. We cannot measure a star's lumi-nosity directly, because its brightness in our sky depends on its distance as well as its true luminosity. For example, our Sun and Alpha Centauri A (the brightest of the three stars in the Alpha Centauri system [Section 1.2]) are similar in luminosity. But Alpha Centauri A is a feeble point of light in the night sky, while our Sun provides enough light and heat to sustain life on Earth. The difference in brightness arises because Alpha Centauri A is about 270,000 times farther from Earth than is the Sun.

More precisely, we define the **apparent brightness** of any star in our sky as the amount of light reaching us *per unit area.* (A more technical term for apparent brightness is *flux*.) The apparent brightness of any light source obeys an *inverse square law* with distance, similar to the inverse square law that describes the force of gravity [Section 5.4]. If we viewed the Sun from twice the Earth's distance, it would appear dimmer by a factor of $2^2 = 4$. If we viewed it from 10 times the Earth's distance, it would appear $10^2 = 100$ times dimmer. From 270,000 times the Earth's distance, it would look like Alpha Centauri A—dimmer by a factor of $270,000^2$, or about 70 billion.

Figure 13.1 shows why apparent brightness follows an inverse square law. The same total amount of light must pass through each sphere surrounding the star. If we focus our attention on the light passing through a small square on the sphere located at 1 AU, we see that the same amount of light must pass through *four* squares of the same size on the sphere located at 2 AU. Thus, each square on the

FIGURE 13.1 The inverse square law for light. At greater distances from a star, the same amount of light passes through an area that gets larger with the square of the distance. The amount of light per unit area therefore declines with the square of the distance.

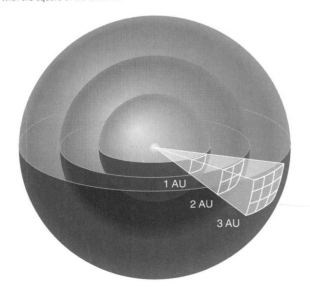

sphere at 2 AU receives only $\frac{1}{2^2} = \frac{1}{4}$ as much light as the square on the sphere at 1 AU. Similarly, the same amount of light passes through *nine* squares of the same size on the sphere located at 3 AU. Thus, each of these squares receives only $\frac{1}{3^2} = \frac{1}{9}$ as much light as the square on the sphere at 1 AU. Generalizing, we see that the amount of light received per unit area decreases with increasing distance by the square of the distance—an inverse square law.

This inverse square law leads to a very simple and important formula relating the apparent brightness, luminosity, and distance of any light source. We will call it the **luminosity–distance formula:**

$$\text{apparent brightness} = \frac{\text{luminosity}}{4\pi \times (\text{distance})^2}$$

Note that, because the standard units of luminosity are watts, the units of apparent brightness are *watts per square meter*. Since we can always measure the apparent brightness of a star, this formula provides a way to calculate a star's luminosity if we can first measure its distance, or to calculate a star's distance if we somehow know its luminosity. (The luminosity–distance formula is strictly correct only if interstellar dust does not absorb or scatter the starlight along its path to Earth.)

Although watts are the standard units for luminosity, it's often more meaningful to describe stellar luminosities in comparison to the Sun by using units of **solar luminosity:** $L_{\text{Sun}} = 3.8 \times 10^{26}$ watts. For example, Proxima Centauri, the nearest of the three stars in the Alpha Centauri system and hence the nearest star besides our Sun, is only about 0.00006 times as luminous as the Sun, or $0.00006L_{\text{Sun}}$. Betelgeuse, the bright left-shoulder star of Orion, has a luminosity of $52,000L_{\text{Sun}}$, meaning that it is 52,000 times more luminous than the Sun.

Measuring the Apparent Brightness

We can measure a star's apparent brightness by using a detector that records how much energy strikes its light-sensitive surface each second. No detector can record light of all wavelengths, so we necessarily measure apparent brightness in only some small range of the complete spectrum. Clearly, when we measure the apparent brightness in visible light, we can calculate only the star's *visible-light luminosity*. Similarly, when we observe a star with a spaceborne X-ray telescope, we measure only the apparent brightness in X rays and can calculate only the star's *X-ray luminosity*. We will use the terms **total luminosity** and **total apparent brightness** to describe the luminosity and apparent brightness we would measure *if* we could detect photons across the entire electromagnetic spectrum.

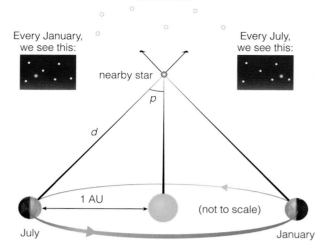

FIGURE 13.2 Parallax makes the apparent position of a nearby star shift back and forth with respect to distant stars over the course of each year. If we measure the parallax angle p in arcseconds, the distance d to the star in parsecs is $\frac{1}{p}$. The angle in this figure is greatly exaggerated: All stars have parallax angles of less than 1 arcsecond.

Measuring Distance Through Stellar Parallax

Once we have measured a star's apparent brightness, the next step in determining its luminosity is to measure its distance. The most direct way to measure the distances to stars is with *stellar parallax*, the small annual shifts in a star's apparent position caused by the Earth's motion around the Sun. You can observe parallax of your finger by holding it at arm's length and looking at it alternately with one eye closed and then the other. Astronomers measure stellar parallax by comparing observations of a nearby star made 6 months apart (Figure 13.2). The nearby star appears to shift against the background of more distant stars because we are observing it from two opposite points of the Earth's orbit. The star's **parallax angle** is defined as *half* the star's annual back-and-forth shift.

Measuring stellar parallax is difficult because stars are so far away, making their parallax angles very small. Even the nearest star, Proxima Centauri, has a parallax angle of only 0.77 arcsecond. For increasingly distant stars, the parallax angles quickly become too small to measure even with our highest-resolution telescopes. Current technology allows us to measure parallax only for stars within a few hundred light-years—not much farther than what we call our *local solar neighborhood* in the vast, 100,000-light-year-diameter Milky Way Galaxy [Section 1.2].

By definition, the distance to an object with a parallax angle of 1 arcsecond is one **parsec,** abbreviated **pc.** (The word *parsec* comes from the words *parallax* and *arcsecond.*) If we use units of arcseconds for the parallax angle, a simple formula allows us to calculate distances in parsecs:

$$d \text{ (in parsecs)} = \frac{1}{p \text{ (in arcseconds)}}$$

For example, the distance to a star with a parallax angle of $\frac{1}{2}$ arcsecond is 2 parsecs, the distance to a star with a parallax angle of $\frac{1}{10}$ arcsecond is 10 parsecs, and the distance to a star with a parallax angle of $\frac{1}{100}$ arcsecond is 100 parsecs. Astronomers often express distances in parsecs or light-years interchangeably, so you should learn to convert quickly between them. Since 1 pc = 3.26 light-years, 10 parsecs is about 30 light-years; 1,000 parsecs, or 1 kiloparsec (1 kpc), is about 3,000 light-years; and 1 million parsecs, or 1 megaparsec (1 Mpc), is about 3 million light-years.

Enough stars have measurable parallax to give us a fairly good sample of the many different types of stars. For example, we know of over 300 stars within about 33 light-years (10 parsecs) of the Sun. About half are members of binary or multiple star systems. Most are tiny, dim red stars such as Proxima Centauri—so dim that we cannot see them with the naked eye, despite the fact that they are relatively close. A few nearby stars, such as Sirius (2.7 parsecs), Vega (8 parsecs), Altair (5 parsecs), and Fomalhaut (7 parsecs), are white in color and bright in our sky, but many more of the brightest stars in the sky lie farther away. The intrinsic luminosities of stars therefore span a wide range, since so many nearby stars appear dim while many more-distant stars appear bright.

The Magnitude System

Many astronomers describe stellar brightness using the ancient **magnitude system.** The magnitude system originally classified stars according to how bright they look to our eyes—the only instruments available in ancient times. The brightest stars received the designation "first magnitude," the next brightest "second magnitude," and so on. The faintest visible stars were magnitude 6. We call these descriptions **apparent magnitudes** because they compare how bright different stars *appear* in the sky.

In modern times, the magnitude system has been extended and more precisely defined. As a result, several bright stars have apparent magnitudes of 0, which is *brighter* than magnitude 1, and the brightest star in the night sky, Sirius, has an apparent magnitude of −1. The modern magnitude system also defines **absolute magnitudes** as a way of describing stellar luminosities. A star's absolute magnitude is the apparent magnitude it would have *if* it were at a distance of 10 parsecs from Earth. For example, the Sun's absolute magnitude is about 4.8, meaning that the Sun would have an apparent magnitude between 4 and 5 if it were located at a distance of 10 parsecs—

bright enough to be visible, but not conspicuous, on a dark night.

You might want to acquaint yourself with the magnitude system, because it is still used by many astronomers, both amateurs and professionals. However, this ancient system can be difficult to use: The brighter the star, the smaller its magnitude, and the mathematical relationship between magnitudes is complicated. We generally avoid using the magnitude system in this book.

13.3 Stellar Surface Temperature

The other basic property of stars (besides luminosity) needed for modern stellar classification is surface temperature. Measuring a star's surface temperature is somewhat easier than measuring its luminosity because the star's distance does not affect the measurement. Instead, we determine surface temperature directly from the star's color or spectrum. One note of caution before we continue: We can measure only a star's *surface temperature,* not its interior temperature. When astronomers speak of the "temperature" of a star, they usually mean the surface temperature.

A star's surface temperature determines the color of light it emits. A red star is cooler than a yellow star, which in turn is cooler than a blue star. The naked eye can distinguish colors only for the brightest stars, but colors become more evident when we view stars through binoculars or a telescope (Figure 13.3). Astronomers determine the "color" of a star more precisely by comparing its apparent brightness as viewed through two different filters.

FIGURE 13.3 This specialized photograph shows the true colors of stars in Orion. Spreading each star's light into an extended cone reduces the effects of overexposure.

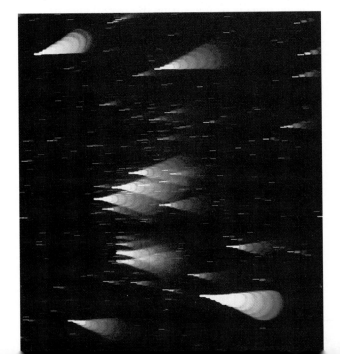

Spectral Type

The emission and absorption lines in a star's spectrum provide an independent and more accurate way to measure its surface temperature. Stars displaying spectral lines of highly ionized elements must be fairly hot, while stars displaying spectral lines of molecules must be relatively cool [Section 6.3]. Astronomers thus classify stars according to surface temperature by assigning a **spectral type** determined from the types of spectral lines present in a star's spectrum. The hottest stars, with the bluest colors, are spectral type O, followed in order of declining surface temperature by spectral types B, A, F, G, K, and M. The time-honored mnemonic for remembering this sequence, OBAFGKM, is "Oh Be A Fine Girl/Guy, Kiss Me!" Table 13.1 summarizes the characteristics of each spectral type.

Each spectral letter classification is subdivided into numbered subcategories (e.g., B0, B1, . . . , B9, A0, A1, . . .). The larger the number, the cooler the star. For example, the Sun is designated spectral type G2, which means it is slightly hotter than a G3 star but cooler than a G1 star.

TIME OUT TO THINK *Invent your own mnemonic for the OBAFGKM sequence. To help get you thinking, here are two examples: (1) Only Bungling Astronomers Forget Generally Known Mnemonics; and (2) Only Business Acts For Good, Karl Marx.*

13.4 Stellar Masses

The most important property of a star is its mass, but stellar masses are much harder to measure than luminosities or surface temperatures. The most dependable method for "weighing" a star relies on Newton's version of Kepler's third law [Section 5.4].

This law allows us to calculate the masses of orbiting objects by measuring the period and average distance (semimajor axis) of their orbit. Note that this law can be applied only when we can measure both the orbital period and semimajor axis. Thus, unless we have identified a planet orbiting a star, we can measure stellar masses only in binary star systems in which we can determine the orbital properties of the two stars.

About half of all stars orbit a companion star of some kind, and these binary star systems fall into three classes:

- A **visual binary** is a pair of stars that we can see orbiting each other. Mizar, the second star in the handle of the Big Dipper, is one example of a visual binary (Figure 13.4). Sometimes we observe a star slowly shifting position in the sky as if it were a member of a visual binary, but its com-

panion is too dim to be seen. Slow shifts in the position of Sirius, the brightest star in the sky, revealed it to be a binary star long before its companion was discovered (Figure 13.5).

- An **eclipsing binary** is a pair of stars that orbit in the plane of our line of sight (Figure 13.6). Each star occasionally passes in front of the other, causing an eclipse. When neither star is eclipsed, we see the combined light of both stars. But when one star eclipses the other, the apparent brightness of the system drops because some of the light is blocked from our view.

- If a binary system is neither visual nor eclipsing, we may be able to detect its binary nature by observing Doppler shifts in its spectral lines [Section 6.4]. If one star is orbiting another, it periodically moves toward us and away from us in its orbit, and its spectral lines show blueshifts and redshifts as a result of this motion (Figure 13.7). Because we detect the binary nature of such star systems by studying their spectra, they are called **spectroscopic binary** systems. Interestingly, each of the two stars we see when we view Mizar through a telescope is a spectroscopic binary.

Table 13.1 The Spectral Sequence

Spectral Type	Example(s)	Temperature Range	Key Absorption Line Features
O	Stars of Orion's Belt	>30,000 K	Lines of ionized helium, weak hydrogen lines
B	Rigel	30,000 K–10,000 K	Lines of neutral helium, moderate hydrogen lines
A	Sirius	10,000 K–7,500 K	Very strong hydrogen lines
F	Polaris	7,500 K–6,000 K	Moderate hydrogen lines, moderate lines of ionized calcium
G	Sun, Alpha Centauri A	6,000 K–5,000 K	Weak hydrogen lines, strong lines of ionized calcium
K	Arcturus	5,000 K–3,500 K	Lines of neutral and singly ionized metals, some molecules
M	Betelgeuse, Proxima Centauri	<3,500 K	Molecular lines strong

Brightest Wavelength (color)	Typical Spectrum
<97 nm (ultraviolet)*	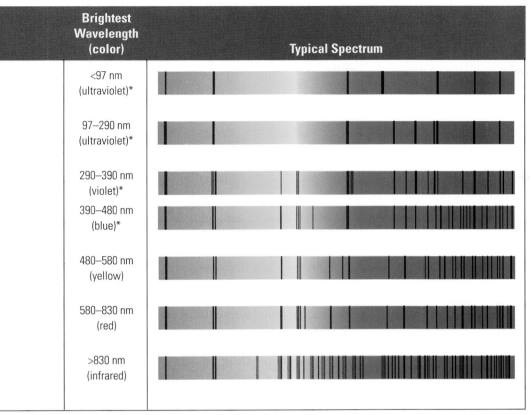
97–290 nm (ultraviolet)*	
290–390 nm (violet)*	
390–480 nm (blue)*	
480–580 nm (yellow)	
580–830 nm (red)	
>830 nm (infrared)	

*All stars above 6,000 K look more or less white to the human eye because they emit plenty of radiation at all visible wavelengths.

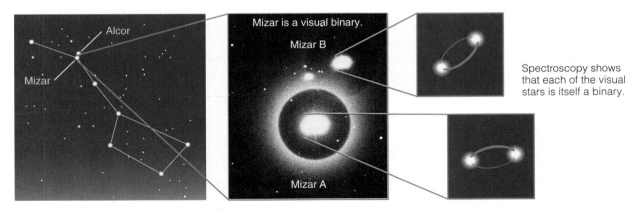

Mizar is a visual binary.

Mizar B

Alcor

Mizar

Mizar A

Spectroscopy shows that each of the visual stars is itself a binary.

FIGURE 13.4 Mizar looks like one star to the naked eye but is actually a system of four stars. Through a telescope Mizar appears to be a visual binary made up of two stars, Mizar A and Mizar B, that gradually change positions, indicating that they orbit every few thousand years. However, each of these two stars is actually a spectroscopic binary, making a total of four stars. (The star Alcor appears very close to Mizar to the naked eye but does *not* orbit it.)

| 1900 | 1910 | 1920 | 1930 | 1940 | 1950 | 1960 | 1970 |

A
B

FIGURE 13.5 Each frame represents the relative positions of Sirius A and Sirius B at 10-year intervals from 1900 to 1970. The back-and-forth "wobble" of Sirius A allowed astronomers to infer the existence of Sirius B even before the two stars could be resolved in telescopic photos.

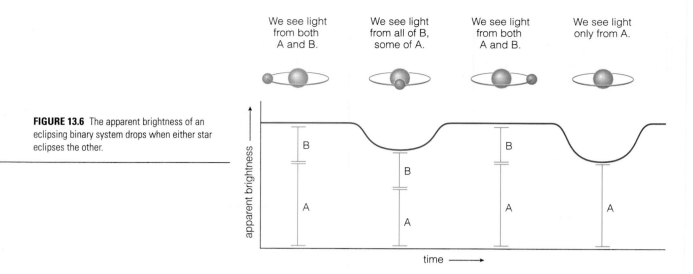

We see light from both A and B.

We see light from all of B, some of A.

We see light from both A and B.

We see light only from A.

apparent brightness

B

A

B

B

A

A

B

A

A

time ⟶

FIGURE 13.6 The apparent brightness of an eclipsing binary system drops when either star eclipses the other.

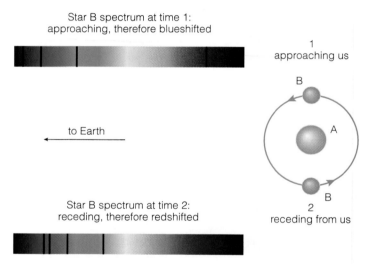

Star B spectrum at time 1: approaching, therefore blueshifted

to Earth

1 approaching us

B

A

B

2 receding from us

Star B spectrum at time 2: receding, therefore redshifted

FIGURE 13.7 The light from a star in a binary system is alternately blueshifted as it comes toward us in its orbit and redshifted as it moves away from us.

Once we have identified a binary star system, we can apply Newton's version of Kepler's third law only if we can measure both the orbital period and the separation of the two stars. Measuring orbital period is fairly easy: In a visual binary, we simply observe how long each orbit takes (or extrapolate from part of an orbit); in an eclipsing binary, we measure the time between eclipses; and in a spectroscopic binary, we measure the time it takes the spectral lines to shift back and forth. In contrast, determining the average separation of the stars in a binary system is usually much more difficult: Except in rare cases, as with visual binaries, in which we can measure the separation directly, we can calculate the separation only if

we know the actual orbital speeds of the stars. The primary technique for measuring stellar speeds relies on the Doppler effect, but Doppler shifts tell us only the portion of a star's velocity that is directly toward us or away from us. Orbiting stars generally do not move directly along our line of sight, so their actual velocities can be significantly greater than those we measure through the Doppler effect. However, the orbits of eclipsing binary stars take them directly toward us and directly away from us once each orbit, enabling us to measure their true orbital velocities. Eclipsing binaries are therefore particularly important to the study of stellar masses. As an added bonus, eclipsing binaries allow us to measure stellar radii directly: Because we know how fast the stars are moving across our line of sight as one eclipses the other, we can determine their radii by timing how long each eclipse lasts.

TIME OUT TO THINK *Suppose two orbiting stars are moving in a plane perpendicular to our line of sight. Would the spectral features of these stars appear shifted in any way? Explain.*

13.5 The Hertzsprung–Russell Diagram

During the first decade of the twentieth century, a similar thought occurred independently to astronomers Ejnar Hertzsprung, working in Denmark, and Henry Norris Russell, working in the United States at Princeton University. Each decided to make a graph plotting stellar luminosities on one axis and spectral types on the other. Such graphs, from which we can immediately read a star's luminosity and surface temperature, are now known as **Hertzsprung–Russell (H–R) diagrams.** (Mass is not shown on an H–R diagram, but later we will see how the H–R diagram helps us understand the importance of mass.) Soon after they began making their graphs, Hertzsprung and Russell uncovered some previously unsuspected patterns in the properties of stars. As we will see shortly, understanding these patterns and the H–R diagram is central to the study of stars.

A Basic H–R Diagram

Figure 13.8 displays an example of an H–R diagram. Note the following key features of the diagram's construction:

- The horizontal axis represents stellar surface temperature or, equivalently, spectral type.

- Temperature increases *from right to left* because Hertzsprung and Russell based their diagrams on the spectral sequence OBAFGKM.

- The vertical axis represents stellar luminosity, in units of the Sun's luminosity (L_{Sun}).

- Stellar luminosities span a wide range, so we keep the graph compact by making each tick mark represent a luminosity 10 times larger than represented by the prior tick mark.

- Each dot on the diagram represents the spectral type and luminosity of a single star. For example, the dot representing the Sun in Figure 13.8 corresponds to the Sun's spectral type of G2 and its luminosity of $1 L_{Sun}$.

Because luminosity increases upward on the diagram and surface temperature increases leftward, stars near the upper left are hot and luminous. Similarly, stars near the upper right are cool and luminous, stars near the lower right are cool and dim, and stars near the lower left are hot and dim.

The H–R diagram also provides direct information about stellar radii, because a star's luminosity depends on both its surface temperature and its surface area or radius. Recall that surface temperature determines the amount of power emitted by the star *per unit area*: Higher temperature means greater power output per unit area [Section 6.3]. Thus, if two stars have the same surface temperature, one can be more luminous than the other only if it is larger in size. Stellar radii therefore must increase as we go from the high-temperature, low-luminosity corner on the lower left of the H–R diagram to the low-temperature, high-luminosity corner on the upper right, as shown in Figure 13.9.

FIGURE 13.8 An H–R diagram, one of astronomy's most important tools, shows how the surface temperatures of stars, plotted along the horizontal axis, relate to their luminosities, plotted along the vertical axis.

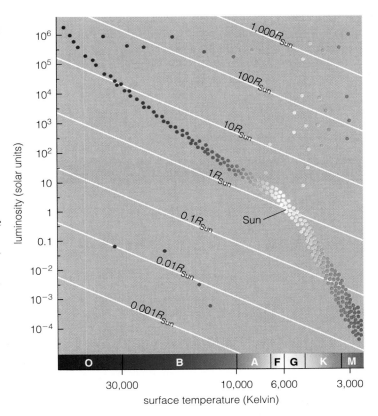

FIGURE 13.9 Stellar radii increase as we go diagonally up and right on the H–R diagram.

Patterns in the H–R Diagram

Figure 13.8 also shows that stars do not fall randomly throughout the H–R diagram but instead fall into several distinct groups:

- Most stars fall somewhere along the **main sequence,** the prominent streak running from the upper left to the lower right on the H–R diagram. Note that our Sun is a main-sequence star.

- The stars in the extreme upper right are called **supergiants** because they are very large in addition to being very bright.

- Just below the supergiants are the **giants,** which are somewhat smaller in radius and lower in luminosity.

- The stars in the lower left are small in radius and appear white in color because of their high temperature; we therefore call these stars the **white dwarfs.**

When classifying a star, astronomers generally report both the star's spectral type and a **luminosity class** that describes the region of the H–R diagram in which the star falls. Table 13.2 summarizes the luminosity classes. For example, the complete spectral classification of our Sun is G2 V: The G2 spectral type means it is yellow in color, and the luminosity class V means it is a main-sequence star. Betelgeuse is M2 I, making it a *red supergiant.* Proxima Centauri is M5 V—similar in color and surface temperature to Betelgeuse, but far dimmer because of its much smaller size. White dwarfs are usually designated with the letters *wd* rather than with a Roman numeral.

Table 13.2 Stellar Luminosity Classes

Class	Description
I	Supergiants
II	Bright giants
III	Giants
IV	Subgiants
V	Main sequence

The Main Sequence

The common trait of main-sequence stars is that, like our Sun, they are fusing hydrogen into helium in their cores. Because stars spend the majority of their lives fusing hydrogen, most stars fall somewhere along the main sequence of the H–R diagram.

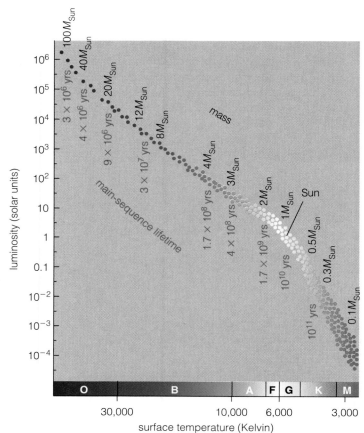

FIGURE 13.10 Along the main sequence, more massive stars are brighter and hotter but have shorter lifetimes. (Stellar masses are given in units of solar masses: $1\ M_{Sun} = 2 \times 10^{30}$ kg.)

But why do main-sequence stars span such a wide range of luminosities and surface temperatures? By measuring the masses of stars in binary systems, astronomers discovered that stellar masses increase upward along the main sequence (Figure 13.10). At the upper end of the main sequence, the hot, luminous O stars can have masses as high as 100 times that of the Sun ($100M_{Sun}$). On the lower end, cool, dim M stars may have as little as 0.1 times the mass of the Sun ($0.1M_{Sun}$). Far more stars fall on the lower end of the main sequence than on the upper end. Thus, low-mass stars are far more common than high-mass stars.

The orderly arrangement of stellar masses along the main sequence tells us that *mass* is the most important attribute of a hydrogen-burning star. Luminosity depends directly on mass because the weight of a star's outer layers determines the nuclear burning rate in its core [Section 12.3]. The nuclear burning rate, and hence the luminosity, is very sensitive to mass. For example, a $10M_{Sun}$ star on the main sequence is more than 1,000 times more luminous than the Sun.

The relationship between mass and surface temperature is a little more subtle. In general, a very luminous star must either be very large or have a very high surface temperature, or some combination of both. Stars on the upper end of the main sequence are thousands of times more luminous than the Sun but only about 10 times larger than the Sun in radius. Thus, their surfaces must be significantly hotter than the Sun's surface to account for their high luminosities. Main-sequence stars more massive than the Sun therefore have higher surface temperatures, and those less massive than the Sun have lower surface temperatures. That is why the main sequence slices diagonally from the lower right to the upper left on the H–R diagram.

Main-Sequence Lifetimes

A star has a limited supply of core hydrogen and therefore can remain as a hydrogen-fusing main-sequence star for only a limited time—the star's **main-sequence lifetime** (or *hydrogen-burning lifetime*). Like masses, stellar lifetimes vary in an orderly way as we move up the main sequence: Massive stars near the upper end of the main sequence have *shorter* lives than less massive stars near the lower end (see Figure 13.10).

Why do more massive stars live shorter lives? A star's lifetime depends on both its mass and its luminosity. Its mass determines how much hydrogen fuel the star initially contains in its core. Its luminosity determines how rapidly the star uses up its fuel. Massive stars live shorter lives because, even though they start their lives with a larger supply of hydrogen, they consume their hydrogen at a prodigious rate.

The main-sequence lifetime of our Sun is about 10 billion years [Section 12.1]. A 30-solar-mass star has roughly 30 times more hydrogen fuel than the Sun but burns it with a luminosity some 300,000 times greater. Consequently, its lifetime is roughly 30/300,000 = 1/10,000 as long as the Sun's—corresponding to a lifetime of only a few million years. Cosmically speaking, a few million years is a remarkably short time, which is one reason why massive stars are so rare: Most of the massive stars that have ever been born are long since dead. (A second reason is that lower-mass stars form in larger numbers than higher-mass stars [Section 14.2].) The fact that any massive stars exist tells us that stars form continuously in our galaxy. The massive, bright O stars we see today formed approximately at the time when our distant ancestors began to walk on two feet, long after the dinosaurs became extinct, and these stars will die long before they have a chance to complete even one orbit around the center of the galaxy.

TIME OUT TO THINK *Would you expect to find life on planets orbiting massive O stars? Why or why not? (Hint: Compare the lifetime of an O star to the amount of time that passed from the formation of our solar system to the origin of life on Earth.)*

On the other end of the scale, a 0.3-solar-mass star emits a luminosity just 0.01 times that of the Sun and consequently lives roughly 0.3/0.01 = 30 times longer than the Sun. In a universe that is now between 10 and 16 billion years old, even the most ancient of these small, dim M stars still survive and will continue to shine faintly for hundreds of billions of years to come.

Giants, Supergiants, and White Dwarfs

Giants and supergiants are stars nearing the ends of their lives because they have already exhausted their core hydrogen. Surprisingly, stars grow more luminous when they begin to run out of fuel. As we will discuss in the next chapter, a star generates energy furiously during the last stages of its life as it tries to stave off the inevitable crushing force of gravity. As ever-greater amounts of power well up from the core, the outer layers of the star expand, making it into a giant or supergiant. The largest of these celestial behemoths have radii more than 1,000 times that of the Sun. Because they are so bright, giants and supergiants can be seen even when they are not especially close to us. Many of the brightest stars visible to the naked eye are giants or supergiants; they are often identifiable by their reddish color. Nevertheless, giants and supergiants are rarer than main-sequence stars because, in our snapshot of the heavens, we catch most stars in the act of hydrogen burning and relatively few in a later stage of life.

Giants and supergiants eventually run out of fuel entirely. A giant with a mass similar to that of our Sun ultimately ejects its outer layers, leaving behind a "dead" core in which all nuclear fusion has ceased. White dwarfs are these remaining embers of former giants. They are hot because they are essentially exposed stellar cores, but they are dim because they lack an energy source and radiate only their leftover heat into space. A typical white dwarf is no larger in size than the Earth, although it may have a mass as great as that of our Sun. (Giants and supergiants with masses much larger than that of the Sun ultimately explode, leaving behind neutron stars or black holes as corpses [Section 14.4].)

Pulsating Variable Stars

Not all stars shine steadily like our Sun. Any star that significantly varies in brightness with time is called a *variable star*. A particularly important class of variable stars includes stars that have a peculiar problem with achieving the proper balance between the power

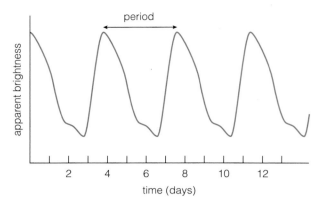

FIGURE 13.11 A typical light curve for a Cepheid variable star. Cepheids are giant, whitish stars whose luminosities regularly pulsate over periods of a few days to about a hundred days. The pulsation period of this Cepheid is about four days.

FIGURE 13.12 An H–R diagram with the instability strip highlighted.

welling up from the stellar core and the power being radiated from the stellar surface. Sometimes the photosphere of such a star is too opaque; energy and pressure build up beneath it, and the star expands in size. However, this expansion puffs the photosphere outward, making it too transparent. So much energy then escapes that the underlying pressure drops, and the star contracts again. In a futile quest for a steady equilibrium, the atmospheres of these **pulsating variable stars** alternately expand and contract, causing the star to rise and fall in luminosity. Figure 13.11 shows a typical light curve for a pulsating variable star, with the star's brightness graphed against time. Any pulsating variable star has its own particular period between peaks in luminosity, which we can discover easily from its light curve. These periods can range from as short as several hours to as long as several years.

Most pulsating variable stars inhabit a strip (called the *instability strip*) on the H–R diagram that lies between the main sequence and the red giants (Figure 13.12). A special category of very luminous pulsating variables lies at the top of this strip: the **Cepheid variables,** or **Cepheids** (so named because the first identified star of this type was the star Delta Cephei).

Cepheids fluctuate in luminosity with periods of a few days to a few months. In 1912, Henrietta Leavitt discovered that the periods of these stars are very closely related to their luminosities: The longer the period, the more luminous the star (Figure 13.13). This **period–luminosity relation** holds because larger (and hence more luminous) Cepheids take longer to pulsate in and out in size.

Once we measure the period of a Cepheid variable, we can use the period–luminosity relation to determine its luminosity. We can then calculate its distance with the luminosity–distance formula. In fact, as we'll discuss in Chapter 17, Cepheids provide our

FIGURE 13.13 The period–luminosity relation for Cepheids. The pulsation period of a Cepheid is related directly to its luminosity—the more luminous a Cepheid is, the slower it pulsates.

primary means of measuring distances to other galaxies and thus teach us the true scale of the cosmos. The next time you look at the North Star, Polaris, gaze upon it with renewed appreciation. Not only has it guided generations of navigators in the Northern Hemisphere, but it is also one of these special Cepheid variable stars.

SUBARU
7 SISTERS

FIGURE 13.14 A photo of the Pleiades, a nearby open cluster of stars. The most prominent stars in this open cluster are of spectral type B, indicating that the Pleiades are less than 100 million years old, relatively young for a star cluster. The bluish wisps of gas are the remains of the interstellar gas cloud from which these stars formed.

13.6 Star Clusters

All stars are born from giant clouds of gas. Because a single interstellar cloud can contain enough material to form many stars, stars almost inevitably form in groups. In our snapshots of the heavens, many stars still congregate in the groups in which they formed. These groups are of two basic types: modest-size **open clusters** and densely packed **globular clusters.**

Open clusters of stars are always found in the disk of the galaxy. They can contain up to several thousand stars and typically span about 30 light-years (10 parsecs). The most famous open cluster is the *Pleiades,* a prominent clump of stars in the constellation Taurus also known as the Seven Sisters (Figure 13.14).

Globular clusters are found in both the halo and the disk of our galaxy. A globular cluster can contain more than a million stars concentrated in a ball typically 60 to 150 light-years across (20 to 50 parsecs). Its innermost region can have 10,000 stars packed within just a few light-years (Figure 13.15).

Star clusters are extremely useful to astronomers for two key reasons:

1. All the stars in a cluster lie at about the same distance from Earth.

2. Cosmically speaking, all the stars in a cluster formed at about the same time (i.e., within a few million years of one another).

Astronomers can therefore use star clusters as laboratories for comparing the properties of stars, as yardsticks for measuring distances in the universe, and as timepieces for measuring the age of our galaxy.

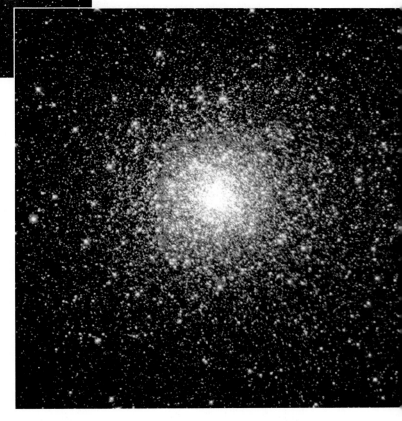

FIGURE 13.15 This globular cluster, known as M 80, is over 12 billion years old. The prominent reddish stars in this Hubble Space Telescope photo are red giant stars nearing the ends of their lives.

We can use the fact that all cluster stars form at about the same time to determine the age of a star cluster. Again, we begin by making an H–R diagram of the cluster stars. For example, Figure 13.16 shows an H–R diagram for the Pleiades. Note that most of the stars in the Pleiades fall along the standard main sequence, with one important exception: The Pleiades' stars trail away to the right of the main sequence at the upper end. That is, the Pleiades are missing the hot, short-lived O main-sequence stars. Apparently, the Pleiades are old enough for its O stars to have

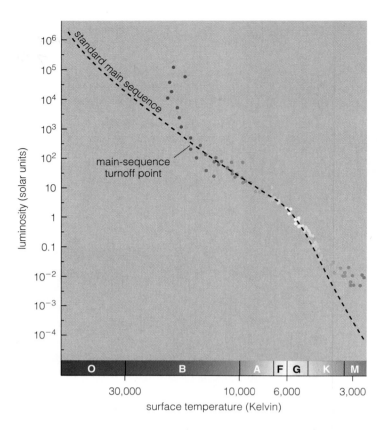

FIGURE 13.16 An H–R diagram for the stars of the Pleiades. Note that the Pleiades cluster is missing its upper main-sequence stars.

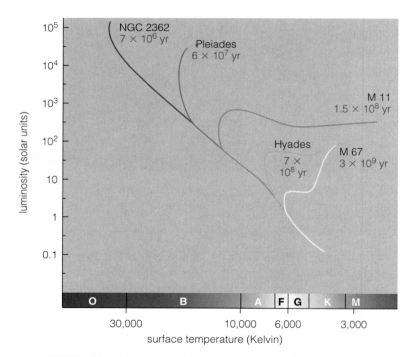

FIGURE 13.17 Each curve on this diagram represents the H–R diagram for a different cluster of stars. Each cluster has a unique main-sequence turnoff point. The older the cluster, the more of its upper-main-sequence stars are missing from its H–R diagram and the farther down the main sequence its turnoff point appears.

already ended their hydrogen-burning lives. At the same time, they are young enough for its B stars to still survive on the main sequence.

The precise point on the H–R diagram at which the Pleiades' main sequence diverges from the standard main sequence is called the **main-sequence turnoff** point. In this cluster, it occurs around spectral type B6. The main-sequence lifetime of a B6 star is roughly 60 million years, so this must be the age of the Pleiades. Any star in the Pleiades that was born with a main-sequence spectral type hotter than B6 had a lifetime shorter than 60 million years and hence is no longer found on the main sequence. Stars with lifetimes longer than 60 million years are still fusing hydrogen and hence remain as main-sequence stars.

Over the next few billion years, the B stars in the Pleiades will die out, followed by the A stars and the F stars. Thus, if we could make H–R diagrams for the Pleiades every few million years, we would find that the main sequence gradually grows shorter. Comparing the H–R diagrams of other open clusters makes this effect more apparent (Figure 13.17). In each case, we can determine the cluster's age from the lifetimes of the stars at its main-sequence turnoff point:

$$\text{age of the cluster} = \text{lifetime of stars at main-sequence turnoff point}$$

Thus, stars in a particular cluster that once resided above the turnoff point on the main sequence have already exhausted their core supply of hydrogen, while stars below the turnoff point remain on the main sequence.

TIME OUT TO THINK *Suppose a star cluster is precisely 10 billion years old. On an H–R diagram, where would you expect to find its main-sequence turnoff point? Would you expect this cluster to have any main-sequence stars of spectral type F? Would you expect it to have main-sequence stars of spectral type K? Explain. (Hint: Remember that the lifetime of our Sun is 10 billion years.)*

The technique of identifying main-sequence turnoff points is our most powerful tool for evaluating the ages of star clusters. We've learned, for example, that most open clusters are relatively young and that none are older than about 5 billion years. In contrast, the stars at the main-sequence turnoff points in globular clusters are usually less massive than our Sun (Figure 13.18). Because stars like our Sun have a lifetime of about 10 billion years and these stars have already died in globular clusters, we can conclude that globular-cluster stars are older than 10 billion years. More precise studies of the turnoff points in globular clusters, coupled with theoretical calculations of stellar lifetimes, place the ages of these clusters at between 12 and 16 billion years, making them the oldest known objects in the galaxy. In fact, globular clusters place a limit on the possible age of the universe: If stars in globular clusters are 12 billion years old, then the universe must be at least this old.

FIGURE 13.18 The H–R diagram of globular cluster Omega Centauri shows a turnoff point below the luminosity of our Sun, indicating an age of more than 10 billion years.

"All men have the stars," he answered, "but they are not the same things for different people. For some, who are travelers, the stars are guides. For others they are no more than little lights in the sky. For others, who are scholars, they are problems. For my businessman they were wealth. But all these stars are silent. You—you alone—will have the stars as no one else has them."

ANTOINE DE SAINT-EXUPERY, from *The Little Prince*

THE BIG PICTURE

We have classified the diverse families of stars visible in the night sky. Much of what we know about stars, galaxies, and the universe itself is based on the fundamental properties of stars introduced in this chapter, so make sure you understand the following "big picture" ideas:

- All stars are made primarily of hydrogen and helium, at least at the time they form. The differences between stars come about primarily because of differences in mass and age.

- Much of what we know about stars comes from studying the patterns that appear when we plot stellar surface temperatures and luminosities in

an H–R diagram. Thus, the H–R diagram is one of the most important tools of astronomers.

- Stars spend most of their lives as main-sequence stars, fusing hydrogen into helium in their cores. The most massive stars live only a few million years, while the least massive stars will survive until the universe is many times its present age.

- Much of what we know about the universe comes from studies of star clusters. Here again, H–R diagrams play a vital role. For example, H–R diagrams of star clusters allow us to determine their ages.

Review Questions

1. Briefly explain how we can learn about the lives of stars, despite the fact that stellar lives are far longer than human lives.

2. What basic composition are all stars born with? Given that all stars begin their lives with this same basic composition, why do stars differ from one another?

3. What do we mean by a star's *luminosity?* What are the standard units for luminosity? Explain what we mean when we measure luminosity in units of L_{Sun}.

4. What do we mean by a star's *apparent brightness?* What are the standard units for apparent brightness? Explain how a star's apparent brightness depends on its distance from us.

5. State the *luminosity–distance formula,* and explain why it is so important.

6. What do we mean by a star's *visible-light luminosity* or *visible-light apparent brightness?* By its *X-ray luminosity* or *X-ray apparent brightness?* Explain why we sometimes talk about such wavelength-specific measures, rather than *total luminosity* and *total apparent brightness.*

7. Briefly explain how we calculate a star's distance in parsecs by measuring its *parallax angle* in arcseconds.

8. What is the *magnitude system?* What are its origins? Briefly explain what we mean by the *apparent magnitude* and *absolute magnitude* of a star. Which one is equivalent to luminosity? Which one is equivalent to apparent brightness?

9. Briefly describe how a star's *spectral type* is related to its surface temperature. List the seven basic spectral types in order of decreasing temperature. Describe how the basic spectral types are subdivided by having numbers appended to them (e.g., A3, G5).

10. Describe and distinguish between *visual binary* systems, *eclipsing binary* systems, and *spectroscopic binary* systems.

11. Explain why we can directly measure stellar masses only in binary star systems and why eclipsing binaries are particularly important to the measurement of stellar masses.

12. Draw a sketch of a basic *Hertzsprung–Russell diagram* (*H–R diagram*). Label the *main sequence, giants, supergiants,* and *white dwarfs.* Where on this diagram do we find stars that are cool and dim? Cool and luminous? Hot and dim? Hot and bright?

13. What do we mean by a star's *luminosity class?* On your sketch of the H–R diagram, identify the regions for luminosity classes I, III, and V.

14. Explain how mass determines the luminosity and surface temperature of a main-sequence star. How do masses differ as we look along the main sequence on an H–R diagram?

15. What do we mean by a star's *main-sequence lifetime?* How do lifetimes differ as we look along the main sequence on an H–R diagram? Explain why more massive stars live shorter lives.

16. True or false? All giants, supergiants, and white dwarfs were once main-sequence stars. Explain.

17. What are *pulsating variable stars?* Why do they vary periodically in brightness?

18. What are *Cepheid variables?* Describe the *period–luminosity relation* for Cepheids, and explain how it enables us to determine distances to Cepheids that are much too far away for making parallax measurements.

19. The two basic types of star clusters are *open clusters* and *globular clusters.* Describe in general terms how these types differ in their number of stars, the age of their stars, and their location in the galaxy.

20. State two key reasons why star clusters are so important to astronomers.

21. Explain why H–R diagrams look different for star clusters of different ages. What is a *main-sequence turnoff point?* How does the location of the main-sequence turnoff point tell us the age of the star cluster?

Discussion Question

1. *Classification.* Annie Jump Cannon and other women astronomers at Harvard University made many important contributions to the area of systematic stellar classification. Why do you think rapid advances in our understanding of stars followed quickly on the heels of their efforts? Can you think of other areas in science where huge advances in understanding followed directly from improved systems of classification?

Problems

1. *Parallax and Distance.* Use the parallax formula to calculate the distance to each of the following stars. Give your answers in both parsecs and light-years.

   a. Alpha Centauri: parallax angle of 0.742″.

   b. Procyon: parallax angle of 0.286″.

2. *Stellar Data.* Consider the following data table for several bright stars. Note that M_v is absolute magnitude and m_v is apparent magnitude.

Star	M_v	m_v	Spectral Type	Luminosity Class
Aldebaran	−0.2	+0.9	K5	III
Alpha Centauri A	+4.4	0.0	G2	V
Antares	−4.5	+0.9	M1	I
Canopus	−3.1	−0.7	F0	II
Fomalhaut	+2.0	+1.2	A3	V
Regulus	−0.6	+1.4	B7	V
Sirius	+1.4	−1.4	A1	V
Spica	−3.6	+0.9	B1	V

Answer each of the following questions, including a brief explanation with each answer.

a. Which star appears brightest in our sky?

b. Which star appears faintest in our sky?

c. Which star has the greatest luminosity?

d. Which star has the smallest luminosity?

e. Which star has the highest surface temperature?

f. Which star has the lowest surface temperature?

g. Which star is most similar to the Sun?

h. Which star is a red supergiant?

i. Which star has the largest radius?

j. Which stars have finished burning hydrogen in their cores?

k. Among the main-sequence stars listed, which one is the most massive?

l. Among the main-sequence stars listed, which one has the longest lifetime?

3. *Data Tables:* Study the spectral types listed in Appendix D for the 20 brightest stars and for the stars within 12 light-years. Why do you think the two lists are so different?

4. *Research: Women in Astronomy.* Until fairly recently, men greatly outnumbered women in professional astronomy. Nevertheless, many women made crucial discoveries in astronomy throughout history. Do some research about the life and discoveries of a woman astronomer from any time period, and write a two- to three-page scientific biography.

Star Stuff

14

WE INHALE OXYGEN WITH EVERY BREATH. IRON-BEARING HEMOGLOBIN carries this oxygen through the bloodstream. Chains of carbon and nitrogen form the backbone of the proteins, fats, and carbohydrates in our cells. Calcium strengthens our bones, while sodium and potassium ions moderate communications of the nervous system. What does all this have to do with astronomy? The profound answer, recognized only in the latter half of the twentieth century, is that these elements were created by stars.

We've already discussed in general terms how the elements in our bodies came to exist: Hydrogen and helium were produced in the Big Bang, and heavier elements were created later by stars and scattered into space by stellar explosions. There, in the spaces between the stars, these elements mixed with interstellar gas and became incorporated into subsequent generations of stars.

In this chapter, we will discuss the origins of the elements in greater detail by delving into the lives of stars. As you read, keep in mind that no matter how far the stars may seem to be removed from our everyday lives, they actually are connected to us in the most intimate way possible: Without the births, lives, and deaths of stars, none of us would be here.

14.1 Lives in the Balance

The story of a star's life boils down to the story of an extended battle between two opposing forces: gravity and pressure. The most common type of pressure in stars is **thermal pressure**—the familiar type of pressure that keeps a balloon inflated and that increases when the temperature or thermal energy increases.

A star can maintain its internal thermal pressure only if it continually generates new thermal energy to replace the energy it radiates into space. This energy can come from two sources: *nuclear fusion* of light elements into heavier ones and the process of *gravitational contraction,* which converts gravitational potential energy into thermal energy [Section 12.1]. These energy-production processes operate only temporarily, although in this case "temporarily" means millions or billions of years. In contrast, gravity acts eternally, and any time gravity succeeds in shrinking a star's core the strength of gravity grows. (Recall that the force of gravity inside an object grows stronger if it either gains mass or shrinks in radius [Section 5.4].) Because a star cannot rely on thermal energy to resist gravity forever, its ultimate fate depends on whether something other than thermal pressure manages to halt the unceasing crush of gravity.

The outcome of a star's struggle between gravity and pressure depends almost entirely on its birth mass. All stars are born from spinning clumps of gas, but newborn stars can have masses ranging from less than 10% of the mass of our Sun to about 100 times that of our Sun. The most massive stars live fast and die young, proceeding from birth to explosive death in just a few million years. The lowest-mass stars consume hydrogen so slowly that they will continue to shine until the universe is many times older than it is today.

Because stars come in such a great variety of masses, we can simplify our discussion of stellar lives by dividing stars into three basic groups: **low-mass stars** are born with less than about two times the mass of our Sun, or less than 2 *solar masses* $(2M_{Sun})$ of material; **intermediate-mass stars** have birth weights between about 2 and 8 solar masses; and **high-mass stars** include all stars born with masses greater than about 8 solar masses. Both low-mass and intermediate-mass stars swell into red giants near the ends of their lives and ultimately become white dwarfs. High-mass stars also become red and large in their latter days, but their lives end much more violently. We will focus most of our discussion in this chapter on the dramatic differences between the lives of low- and high-mass stars. Because the life stages of intermediate-mass stars are quite similar to the corresponding stages of high-mass stars until the very ends of their lives, we include them in our discussion of high-mass stars.

Given the brevity of human history compared to the life of any star, you might wonder how we can know so much about stellar life cycles. As with any scientific inquiry, we study stellar lives by comparing theory and observation. On the theoretical side, we use mathematical models based on the known laws of physics to predict the life cycles of stars. On the observational side, we study stars of different mass but the same age by looking in star clusters whose ages we have determined by main-sequence turnoff [Section 13.6]. Occasionally, we even catch a star in its death throes. Theoretical predictions of the life cycles of stars now agree quite well with these observations. In the remainder of this chapter, we will examine in detail our modern understanding of the life stories of stars and how they manufacture the variety of elements—the *star stuff*—that make our lives possible.

14.2 Star Birth

A star's life begins in an interstellar cloud (Figure 14.1). Star-forming clouds tend to be quite cold—typically only 10–30 K. (Recall that 0 K is absolute zero, and temperatures on Earth are around 300 K.) They also tend to be quite dense compared to the rest of the gas between the stars, although they would qualify as a superb vacuum by earthly standards. Like the galaxy as a whole, star-forming clouds are made almost entirely of hydrogen and helium.

FIGURE 14.1 Photo of a star-forming cloud in the constellation Scorpio.

FIGURE 14.2 This false-color picture shows infrared radiation from a region within the constellation Orion. The colors correspond to the temperature of the emitting gas: Red is cooler, and white is hotter. Many of the white dots in the picture are protostars forming within molecular clouds. (Infrared Astronomical Satellite.)

The clouds that form stars are sometimes called **molecular clouds,** because their low temperatures allow hydrogen atoms to pair up to form hydrogen molecules (H_2). The relatively rare atoms of elements heavier than helium can also form molecules, such as carbon monoxide or water, or tiny, solid grains of dust. More important, the cold temperatures and relatively high densities allow gravity to overcome thermal pressure more readily in molecular clouds than elsewhere in interstellar space, initiating the gravitational collapse that gives birth to new stars.

From Cloud to Protostar

As a piece of a molecular cloud starts to collapse, gravitational contraction increases the cloud's thermal energy. However, the cloud quickly radiates this thermal energy away, preventing the pressure from building enough to resist gravity. During this early phase of collapse, the temperature remains below 100 K, and the cloud glows in long-wavelength infrared light (Figure 14.2).

The ongoing collapse increases the cloud's density, making it increasingly difficult for radiation to escape. Eventually, its central regions trap much of the thermal energy produced by gravitational contraction. Because thermal energy can no longer escape easily, the cloud's internal temperature and pressure rise dramatically. This rising pressure begins

to fight back against the crush of gravity, and the dense cloud fragment becomes a **protostar**—the seed from which a star will grow. A protostar looks starlike, but its core is not yet hot enough for fusion.

The law of conservation of angular momentum implies that a **protostellar disk** should encircle a forming protostar, similar to the *protoplanetary disk* from which the planets of our solar system formed [Section 7.3]. A collapsing cloud fragment necessarily starts with some overall angular momentum (the sum total of the angular momenta of every constituent gas particle), although it might be unmeasurable when the fragment is large. But conservation of angular momentum ensures that the fragment spins faster and faster as it collapses (Figure 14.3). The rotating cloud also flattens to form the protostellar disk as it shrinks. Protostellar disks sometimes coalesce into true planetary systems, but we do not yet know how commonly this occurs.

TIME OUT TO THINK *The term* protostellar disk *refers to any disk of material surrounding a protostar. The term* protoplanetary disk *refers to a disk of material that later produces full-fledged planets. Do you think that all protostellar disks are also protoplanetary disks? If so, why? If not, what do you think might prevent planets from forming in some protostellar disks?*

The protostellar disk probably plays a large role in eventually slowing the rotation of the protostar. The protostar's rapid rotation generates a strong magnetic field. As the magnetic field lines sweep through the protostellar disk, they transfer some of the protostar's angular momentum outward, slowing its rotation. The strong magnetic field also helps generate a strong **protostellar wind**—an outward flow of particles similar to the *solar wind* [Section 12.2]—that may carry additional angular momentum from the protostar to interstellar space.

Rotation is likely to be responsible for the formation of some binary star systems. Protostars that are unable to rid themselves of enough angular momentum spin too fast to become a stable single star and tend to split in two. These two fragments can each go on to form a separate star. If the stars are particularly close together, the resulting pair is called a **close binary** system, in which two stars coexist in close proximity, rapidly orbiting each other.

Observations show that the latter stages of a star's formation can be surprisingly violent. Besides the strong wind, many young stars also fire high-speed streams, or **jets,** of gas into interstellar space (Figure 14.4). No one knows exactly how protostars generate these jets, but two high-speed streams generally flow out along the rotation axis of the protostar, shooting

FIGURE 14.3 Artist's conception of star birth. A cloud fragment (top left) necessarily has some angular momentum, which causes it to spin faster and flatten into a protostellar disk as it collapses (center). In the late stages of collapse (bottom right), the central protostar has a strong protostellar wind and may fire jets of high-speed gas outward along its rotation axis.

FIGURE 14.4 Some protostars can be seen shooting jets of matter into interstellar space. (**a**) Schematic illustration of protostellar disk–jet structure. (**b**) Close-up view of an actual protostellar disk and jet. We are seeing the disk edge-on, as in (**a**). The disk's top and bottom surfaces, illuminated by the protostar, are shown in green, and the jets emerging along the disk's axis are shown in red. The dark central layers of the disk block our view of the protostar itself. (**c**) A wider-angle observation of a jet emanating from a protostar (left) and ramming into surrounding interstellar gas (right).

273

in opposite directions. We also sometimes see blobs of material along the jets (called *Herbig–Haro objects* after their discoverers). These blobs appear to be collections of gas swept up as the jet plows into the surrounding interstellar material. Together, winds and jets play an important role in clearing away the cocoon of gas that surrounds a forming star, eventually allowing the star to emerge into view.

A Star Is Born

A full-fledged star is born when the core temperature of the protostar exceeds 10 million K—hot enough for hydrogen fusion to operate efficiently by the *proton–proton chain* [Section 12.3]. The ignition of fusion halts the protostar's gravitational collapse. The new star's interior structure stabilizes, with the thermal energy generated by fusion maintaining the balance between gravity and pressure that we call *gravitational equilibrium* [Section 12.1]. The star is now a hydrogen-burning, *main-sequence* star [Section 13.5].

The length of time from the formation of a protostar to the birth of a main-sequence star depends on the star's mass. Remember that massive stars do everything faster. The collapse of a high-mass proto-

star may take only a million years or less. Collapse into a star like our Sun takes about 50 million years, and collapse into a small star of spectral type M may take more than a hundred million years. Thus, the most massive stars in a young star cluster may live and die before the smallest stars finish their prenatal years.

We can summarize the transitions that occur during star birth with a special type of H–R diagram. Recall that a standard H–R diagram shows luminosities and surface temperatures for many different stars [Section 13.5]. In contrast, this special H–R diagram shows part of a **life track** (also called an *evolutionary track*) for a single star; for reference, the diagram also shows the standard main sequence. Each point along a star's life track represents its surface temperature and luminosity at some moment during its life. Figure 14.5 shows a life track for the prenatal period in the life of a $1M_{Sun}$ star like our Sun.

Stars of different masses behave similarly during their prenatal periods. Figure 14.6 illustrates life tracks on an H–R diagram for the prenatal stages of several stars of different masses.

FIGURE 14.5 The life track of a $1M_{Sun}$ star from protostar to main-sequence star.

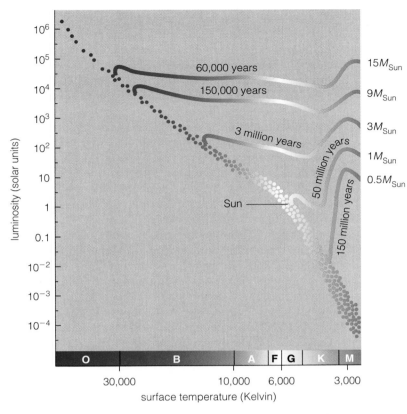

FIGURE 14.6 Life tracks from protostar to the main sequence for stars of different masses.

less massive than the Sun. With the passing of time, the balance tilts even more in favor of the low-mass stars as the high-mass stars die away.

The masses of stars have limits. Theoretical models indicate that stars above about $100M_{Sun}$ generate power so furiously that gravity cannot contain their internal pressure. Such stars effectively tear themselves apart and drive their outer layers into space. Observations confirm the absence of stars much larger than $100M_{Sun}$—if any stars this massive existed nearby, they would be so luminous that we would easily detect them. On the other end of the scale, calculations show that the central temperature of a protostar with less than $0.08M_{Sun}$ never climbs above the 10-million-K threshold needed for efficient hydrogen fusion. Instead, a strange effect called **degeneracy pressure** halts the gravitational contraction of the core before hydrogen burning can begin.

Degeneracy pressure arises from a quantum mechanical law called the *exclusion principle,* which prevents two identical electrons (or protons or neutrons) from occupying the same place at the same time. The following analogy might help you visualize how degeneracy pressure works. Imagine that a group of people are in an auditorium filled with hundreds of folding chairs. As long as the chairs greatly outnumber the people, it's rare that two people ever fight over the same chair. However, if the number of people is nearly equal to the number of chairs, it's hard to pack them any tighter because people aren't allowed to share seats.

In the context of this analogy, protostars with masses above $0.08M_{Sun}$ are like auditoriums with many more available chairs than people, so the people (particles) can easily squeeze into a smaller section of the auditorium. But the cores of protostars with masses below $0.08M_{Sun}$ are like auditoriums with so few chairs that the people (particles) fill nearly all of them. Because there are no extra chairs available, the people (particles) cannot squeeze into a smaller section of the auditorium. This resistance to squeezing is why degeneracy pressure halts gravitational contraction. Note that the degeneracy pressure arises *only* because particles have no place else to go. Thus, unlike thermal pressure, *degeneracy pressure has nothing to do with temperature.*

Because degeneracy pressure halts the collapse of a protostellar core with less than $0.08M_{Sun}$ before fusion becomes self-sustaining, the result is a "failed

TIME OUT TO THINK *Explain in your own words what we mean by a* life track *for a star. Why do we say that Figures 14.5 and 14.6 show only pre–main-sequence life tracks? In general terms, predict the appearance on Figure 14.6 of pre–main-sequence life tracks for a $25M_{Sun}$ star and for a $0.1M_{Sun}$ star.*

Stellar Birth Weights

A single group of molecular clouds can contain thousands of solar masses of gas, which is why stars generally are born in clusters. We do not yet fully understand the processes that govern the clumping and fragmentation of these clouds into protostars with a wide array of masses. However, we can observe the results. In a newly formed star cluster, stars with low masses greatly outnumber stars with high masses. For every star with a mass between 10 and 100 solar masses, there are typically 10 stars with masses between 2 and 10 solar masses, 50 stars with masses between 0.5 and 2 solar masses, and a few hundred stars with masses below $0.5M_{Sun}$. Thus, although the Sun lies nearly in the middle of the overall *range* of stellar masses, most stars in a new star cluster are star" that slowly radiates away its internal thermal energy, gradually cooling with time. Such objects, called **brown dwarfs,** occupy a fuzzy gap between what we call a planet and what we call a star. (Note that $0.08M_{Sun}$ is about 80 times the mass of Jupiter.) Because degeneracy pressure does *not* rise and fall with temperature, the gradual cooling of a brown dwarf's interior does not weaken its degeneracy pressure. In the constant battle of any "star" to resist the crush of gravity, brown dwarfs are winners, albeit dim ones: Their degeneracy pressure will not diminish with time, so gravity will never gain the upper hand.

14.3 Life as a Low-Mass Star

In the grand hierarchy of stars, our Sun ranks as rather mediocre. We should be thankful for this mediocrity. If the Sun had been a high-mass star, it would have lasted only a few million years, dying before life could have arisen on Earth. Instead, the Sun has shone steadily for nearly 5 billion years and will continue to do so for about 5 billion more. In this section, we investigate the lives of low-mass stars like our Sun.

Slow and Steady

Low-mass stars spend their main-sequence lives fusing hydrogen into helium in their cores slowly and steadily via the *proton–proton chain* [Section 12.3]. As in the Sun, the energy released by nuclear fusion in the core may take a million years to reach the surface, where it finally escapes into space as the star's luminosity. The energy moves outward from the core through the random bounces of photons from one electron to another and convection [Section 12.4].

Convection is the dominant process in the outer one-third of the Sun's interior. More massive stars have hotter interiors and shallower convection zones. Lower-mass stars have cooler interiors and deeper convection zones. In very low mass stars, the convection zone extends all the way down to the core. The highest-mass stars have no convection zone at all near their surface, but they have a convective core because they produce energy so furiously (Figure 14.7).

A low-mass star gradually consumes its core hydrogen, converting it into helium over a period of billions of years. In the process, the declining number of independent particles in the core (four independent protons fuse into just one independent helium nucleus) causes the core to shrink and heat very gradually, pushing the luminosity of the main-sequence star slowly upward as it ages. But the most dramatic changes occur when nuclear fusion exhausts the hydrogen in the star's core.

Red Giant Stage

The energy released by hydrogen fusion during a star's main-sequence life maintains the thermal pressure that holds gravity at bay. But when the core hydrogen is finally depleted, nuclear fusion ceases. With no fusion to supply thermal energy, the core

FIGURE 14.7 Among main-sequence stars, convection zones extend deeper in lower-mass stars. High-mass stars have convective cores but have no convection zones near their surfaces.

high-mass star

$1M_{Sun}$ star

very low mass star

pressure can no longer resist the crush of gravity, and the core begins to shrink more rapidly.

Surprisingly, the star's outer layers expand outward while the core is shrinking. On an H–R diagram, the star's life track moves almost horizontally to the right as the star grows in size to become a **subgiant** (Figure 14.8). Then, as the expansion of the outer layers continues, the star's luminosity begins to increase substantially, and the star slowly becomes a **red giant.** For a $1M_{Sun}$ star, this process takes about a billion years, during which the star's radius increases about 100-fold and its luminosity grows by an even greater factor. (Like all phases of stellar lives, the process occurs faster for more massive stars and slower for less massive stars.) This process may at first seem paradoxical: Why does the star grow bigger and more luminous at the same time that its core is shrinking?

We can find the answer by considering the interior structure of the star. The core is now made of helium—the "ash" left behind by hydrogen fusion—but the surrounding layers still contain plenty of fresh hydrogen. Gravity shrinks both the *inert* (nonburning) helium core and the surrounding *shell* of hydrogen, and the shell soon becomes hot enough to sustain hydrogen fusion (Figure 14.9). In fact, the shell becomes so hot that this **hydrogen shell burning** proceeds at a higher rate than core hydrogen fusion did during the star's main-sequence life. The result is that the star becomes more luminous than ever before. The thermal pressure generated by the hydrogen shell burning overcomes gravity in the star's upper layers, pushing them to expand outward. What was once a fairly dim main-sequence star balloons into a luminous red giant. Note that the red giant is large on the outside, but most of its mass is buried deep in a shrunken stellar core.

The situation grows more extreme as long as the helium core remains inert. Recall that, in a main-sequence star like the Sun, a rise in the fusion rate causes the core to inflate and cool until the fusion rate drops back down; in Chapter 12, we referred to this self-correcting process

FIGURE 14.8 The life track of a $1M_{Sun}$ star on an H–R diagram from the end of its main-sequence life until it becomes a red giant.

FIGURE 14.9 After a star ends its main-sequence life, its inert helium core contracts while hydrogen shell burning begins. The high rate of fusion in the hydrogen shell forces the star's upper layers to expand outward.

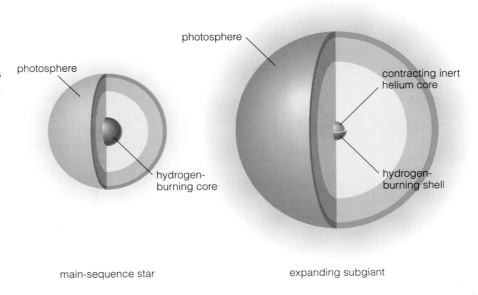

as the solar *thermostat* [Section 12.3]. But thermal energy generated in the hydrogen-burning shell of a red giant cannot do anything to inflate the inert core that lies underneath. Instead, newly produced helium keeps adding to the mass of the helium core, amplifying its gravitational pull and shrinking it further. The hydrogen-burning shell shrinks along with the core, growing hotter and denser. The fusion rate in the shell consequently rises, feeding even more helium ash to the core. The star is caught in a vicious circle. The core and shell continue to shrink in size and grow in luminosity, while thermal pressure continues to push the star's upper layers outward. This cycle breaks down only when the inert helium core reaches a temperature of about 100 million K, at which point helium nuclei can fuse together. Throughout the expansion phase, **red-giant winds** carry away much more matter than the winds from main-sequence stars, but at much slower speeds.

TIME OUT TO THINK *Before you read on, briefly summarize why a star grows larger and brighter after it exhausts its core hydrogen. When does the growth of a red giant finally halt, and why? Would a star's red giant stage be different if the temperature required for helium fusion were around 200 million K, rather than 100 million K? Why?*

Helium Burning

Recall that fusion occurs only when two nuclei come close enough together for the attractive *strong force* to overcome electromagnetic repulsion [Section 12.3]. Helium nuclei have two protons (and two neutrons) and hence a greater positive charge than the single proton of a hydrogen nucleus. The greater charge means that helium nuclei repel one another more strongly than hydrogen nuclei. **Helium fusion** therefore occurs only when nuclei slam into one another at much higher speeds than needed for hydrogen fusion. Therefore, helium fusion requires much higher temperatures. The helium fusion process converts three helium nuclei into one carbon nucleus:

Energy is released because the carbon-12 nucleus has a slightly lower mass than the three helium-4 nuclei, and the lost mass becomes energy in accord with $E = mc^2$.

In a low-mass star, helium ignites very suddenly in a process known as a **helium flash,** which rapidly inflates the core. This core expansion pushes the

hydrogen-burning shell outward, lowering its temperature and its burning rate. The result is that, even though the star now has core helium fusion *and* hydrogen shell burning taking place simultaneously, the total energy production falls from its peak during the red giant phase. The reduced total energy output of the star reduces its luminosity, thereby allowing its outer layers to contract from their peak size during the red giant phase (Figure 14.10). As the outer layers contract, the star's surface temperature also increases somewhat.

Because the helium-burning star is now smaller and hotter than it was as a red giant, its life track on the H–R diagram drops downward and to the left (Figure 14.11a). Note that the helium cores of all low-mass stars fuse helium into carbon at about the same rate, so these stars all have about the same luminosity. However, the outer layers of these stars can have different masses, depending on how much mass they lost through their red-giant winds. Stars that lost more mass end up with smaller radii and higher surface temperatures and hence are farther to the left on the H–R diagram. In a cluster of stars, those stars that are currently in their helium-burning phase therefore all have about the same luminosity but differ in surface temperature. Thus, on an H–R diagram for a cluster of stars, the helium-burning stars are arranged along a **horizontal branch** (Figure 14.11b).

Last Gasps

It is only a matter of time until a horizontal-branch star fuses all its core helium into carbon. The core helium in a low-mass star will run out in about a hundred million years. When the core helium is exhausted, fusion turns off, and the core begins to shrink once again under the crush of gravity.

The basic processes that changed the star from a main-sequence star into a red giant now resume, but this time it is helium fusion that ignites in a shell around an inert carbon core. Meanwhile, the hydrogen shell still burns atop the helium layer. Both shells contract along with the inert core, driving their temperatures and fusion rates much higher. The luminosity of this double-shelled star grows greater than ever, and its outer layers swell to an even huger size. On the H–R diagram, the star's life track once again turns upward (Figure 14.12).

The furious burning in the helium and hydrogen shells cannot last long—maybe a few million years or less for a $1 M_{Sun}$ star. The star's only hope of extending its life lies with the carbon core, but this is a false hope in the case of low-mass stars. Carbon fusion is possible only at temperatures above about 600 million K. Before the core of a low-mass star ever reaches

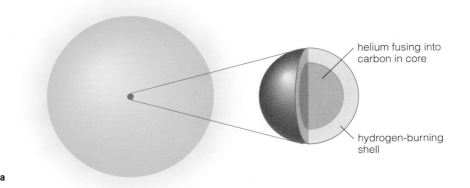

helium fusing into
carbon in core

hydrogen-burning
shell

a

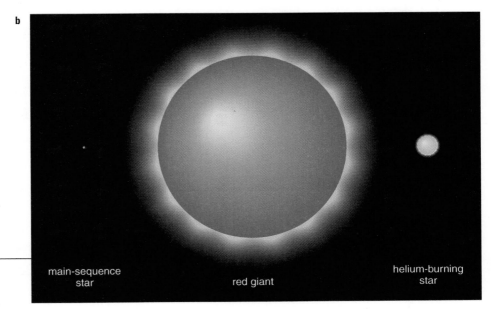

b

FIGURE 14.10 (**a**) Core structure of a helium-burning star. (**b**) Approximate relative sizes of a low-mass star as a main-sequence star, a red giant, and a helium-burning star. The scale is not precise; in particular, the size of the main-sequence star is even smaller compared to the others.

main-sequence
star

red giant

helium-burning
star

a

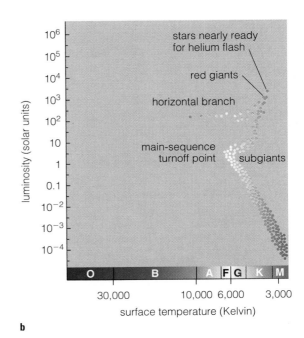

b

FIGURE 14.11 (**a**) After helium flash, a low-mass star's surface shrinks and heats, so the star's life track moves downward and to the left on the H–R diagram. (**b**) This H–R diagram plots the luminosity and surface temperature of individual stars in a cluster (i.e., it does *not* show life tracks).

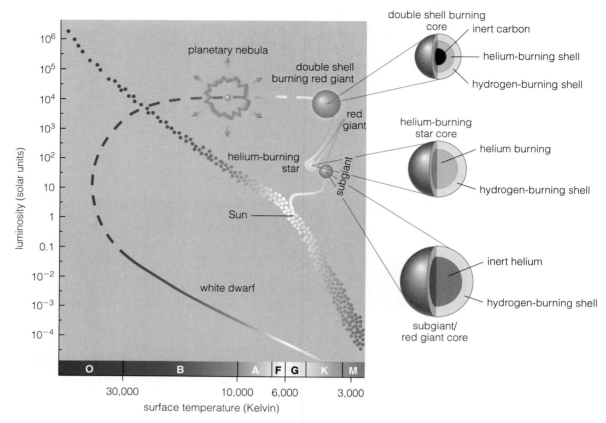

FIGURE 14.12 The life track of a $1 M_{Sun}$ star from main-sequence star to white dwarf. The dashed line represents the rapid transition from planetary nebulae to white dwarf as the ejected outer layers dissipate, revealing the hot core.

such a lofty temperature, degeneracy pressure halts its gravitational collapse.

For a low-mass star with a carbon core, the end is near. The huge size of the dying star means that it has a very weak grip on its outer layers. As the star's luminosity and radius keep rising, matter flows from its surface at increasingly high rates. Meanwhile, strong convection dredges up carbon from the core, enriching the surface of the star with carbon. Red giants whose atmospheres become especially carbon-rich in this way are called **carbon stars.** Carbon stars have cool, low-speed stellar winds, and some of the gas atoms in these winds stick together in micro-scopic clusters, forming small, solid particles of dust. These dust particles continue their slow drift with the stellar wind into interstellar space, where they become **interstellar dust grains.** The process of particulate formation is very similar to the formation of smoke particles in a fire. Thus, in a sense, carbon stars are the most voluminous polluters in the universe. However, this "carbon smog" is essential to life: Most of the carbon in your body (and in all life on Earth) was manufactured in carbon stars and blown into space by their stellar winds.

TIME OUT TO THINK *Suppose the universe contained only low-mass stars. Would elements heavier than carbon exist? Why or why not?*

Before a low-mass star dies, it treats us to one last spectacle. Through winds and other processes, the star ejects its outer layers into space. The result is a huge shell of gas expanding away from the inert, degenerate carbon core. The exposed core is still very hot and therefore emits intense ultraviolet radiation that ionizes the gas in the expanding shell, causing it to glow brightly as what we call a **planetary nebula.** Despite their name, planetary nebulae have nothing to do with planets. The name comes from the fact that nearby nebulae look disk-shaped (like planets) through a small telescope. Through a larger telescope, more detail is visible. The famous Ring Nebula in the constellation Lyra is a planetary nebula, as is the Helix Nebula in the southern sky (Figure 14.13).

The glow of the planetary nebula fades as the exposed core cools and the ejected gas disperses into space. The nebula will disappear within a million years, leaving behind the cooling carbon core. On the H–R diagram, the life track now represents this "dead" core (see Figure 14.12). At first the life track heads to the left, because the core is initially quite hot and luminous. But the life track soon veers downward and to the right as the remaining ember cools and fades. You already know these naked, inert cores by another name: *white dwarfs* [Section 13.5].

a Ring Nebula

b Helix Nebula

FIGURE 14.13 Planetary nebulae occur when low-mass stars in their final death throes cast off their outer layers of gas, as seen in these photos. The hot core that remains ionizes and energizes the richly complex envelope of gas surrounding it.

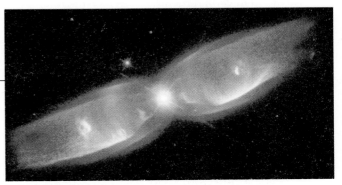

c Twin Jet Nebula

In the ongoing battle between gravity and a star's internal pressure, white dwarfs are a sort of stalemate. As long as no mass is added to the white dwarf from some other source (such as a companion star in a binary system), neither the strength of gravity nor the strength of the degeneracy pressure that holds gravity at bay will ever change. Thus, a white dwarf is little more than a decaying corpse that will cool for the indefinite future.

The Fate of Life on Earth

The life stages of low-mass stars are immensely important to the Earth, because we orbit a low-mass star—the Sun. Fortunately, the Sun should cause few problems for Earth over the next 5 billion years. It will gradually brighten during its remaining time as a main-sequence star, but it has already been doing so for the past 4 billion years with no serious consequences for life on Earth. However, somewhere around the year A.D. 5,000,000,000, the hydrogen in the solar core will run out. The Sun will then expand to a subgiant about three times as large and twice as bright as it is today. This increased luminosity will probably begin to evaporate the oceans, perhaps leading to a runaway greenhouse effect that could raise Earth's average temperature to something simi-

lar to that of Venus and probably hotter [Section 11.3]. If any life survives, it will have to be very tolerant of heat.

Things will only get worse as the Sun grows into a red giant over the next several hundred million years. Just before helium flash, the Sun will be about 100 times larger in radius and over 1,000 times more luminous than it is today. Earth's surface temperature will exceed 1,000 K, and the oceans will long since have boiled away. By this time, any surviving humans will need to have found a new home. Saturn's moon Titan [Section 9.5] might not be a bad choice: Its surface temperature will have risen from well below freezing today to about the present temperature of Earth. The Sun will shrink somewhat after helium burning begins, providing a temporary lull in incineration while the Sun spends 100 million years as a helium-burning star.

Anyone who survives the Sun's helium-burning phase will need to prepare for one final disaster. After exhausting its core helium, the Sun will expand again during its last million years. Its luminosity will soar to thousands of times what it is today, and it will grow so large that solar flames might lap at the Earth's surface. Then it will eject its outer layers as a planetary nebula that will engulf Jupiter and Saturn and drift on past Pluto into interstellar space. If the Earth is not destroyed, its charred surface will be cold and dark in the faint, fading light of the white dwarf that the Sun has become. From then on, the Earth will be little more than a chunk of rock circling the corpse of our once-brilliant star.

14.4 Life as a High-Mass Star

Human life would be impossible without both low-mass stars and high-mass stars. The long lives of low-mass stars allow evolution to proceed for billions of years, but only high-mass stars produce the full array of elements on which life depends. Fusion of elements heavier than helium to produce elements heavier than carbon requires extremely high temperatures in order to overcome the larger electromagnetic repulsion of more highly charged nuclei. Reaching such temperatures requires an extremely strong crush of gravity on a star's core—a crush that occurs only under the immense weight of the overlying gas in high-mass stars.

In the final stages of their lives, the highest-mass stars proceed to fuse increasingly heavy elements until they have exhausted all possible fusion sources. When fusion ceases, gravity drives the core to implode suddenly, which, as we will soon see, causes the star to self-destruct in the titanic explosion we call a *supernova*. The fast-paced life and cataclysmic death of a high-mass star—surely one of the great dramas of the universe—are the topics of this section.

Brief but Brilliant

During its main-sequence life, the strong gravity of a high-mass star compresses its hydrogen core to higher temperatures than we find in lower-mass stars. You already know that the rate of fusion via the proton–proton chain increases substantially at higher temperatures, but the even hotter core temperatures of high-mass stars enable protons to slam into carbon, oxygen, or nitrogen nuclei as well as into other protons. Although carbon, nitrogen, and oxygen make up less than 2% of the material from which stars form in interstellar space, this 2% is more than enough to be useful in a stellar core. The carbon, nitrogen, and oxygen act as *catalysts* for hydrogen fusion, making it proceed at a far higher rate than would be possible by the proton–proton chain alone. (A *catalyst* is something that aids the progress of a reaction without being consumed in the reaction.) The lives of high-mass stars are truly brief but brilliant.

The chain of reactions that leads to hydrogen fusion in high-mass stars is called the **CNO cycle;** the letters *CNO* stand for carbon, nitrogen, and oxygen, respectively. Just as in the proton–proton chain [Section 12.3], four hydrogen nuclei go in while one helium-4 nucleus comes out. The amount of energy generated in each reaction cycle therefore is the same as in the proton–proton chain: It is equal to the difference in mass between the four hydrogen nuclei and the one helium nucleus, multiplied by c^2. The CNO cycle is simply another, faster way to accomplish hydrogen fusion.

The escalated fusion rates in high-mass stars generate remarkable amounts of power. Many more photons stream from the photospheres of high-mass stars than from the Sun, and many more photons are bouncing around inside. Although photons have no mass, they act like particles and carry *momentum* [Section 5.1], which they can transfer to anything they hit, imparting a very slight jolt. The combined jolts from the huge number of photons streaming outward through a high-mass star apply a type of pressure called **radiation pressure.** This type of pressure can have dramatic effects on high-mass stars. In the most massive stars, radiation pressure is more important than thermal pressure in keeping gravity at bay and can drive strong, fast-moving stellar winds.

TIME OUT TO THINK *We have now discussed three types of pressure: thermal pressure, degeneracy pressure, and radiation pressure. Before continuing, briefly review each type of pressure and describe the conditions under which each is important.*

Advanced Nuclear Burning

The exhaustion of core hydrogen in a high-mass star sets in motion the same processes that turn a low-mass star into a red giant, but the transformation proceeds much more quickly. The star develops a hydrogen-burning shell, and its outer layers begin to expand outward. At the same time, the core contracts, and this gravitational contraction releases energy that raises the core temperature until it becomes hot enough to fuse helium into carbon. In stars of more than 2 solar masses, the onset of fusion is gradual—there is no helium flash as in low-mass stars.

A high-mass star fuses helium into carbon relatively quickly, leaving it with an inert carbon core after just a few hundred thousand years or less. Once again, the absence of fusion leaves the core without a thermal energy source to fight off the crush of gravity. The inert carbon core shrinks, the crush of gravity intensifies, and the core pressure, temperature, and density all rise. Meanwhile, a helium-burning shell forms between the inert core and the hydrogen-burning shell. The star's outer layers swell again.

Up to this point, the life stories of intermediate-mass stars ($2–8M_{Sun}$) and high-mass stars ($>8M_{Sun}$) are very similar. However, degeneracy pressure prevents the cores of intermediate-mass stars from reaching the temperatures required to burn carbon or oxygen into anything heavier. These stars eventually blow away their upper layers and finish their lives as white dwarfs. The rest of a high-mass star's life, on the other hand, is unlike anything that a low-mass star or an intermediate-mass star ever experiences.

The crush of gravity in a high-mass star is so overwhelming that degeneracy pressure never comes into play in the collapsing carbon core. The gravitational contraction of the core continues, and the core temperature soon reaches the 600 million K required to fuse carbon into heavier elements. Carbon fusion provides the core with a new source of energy that restores the balance versus gravity, but only temporarily. In the highest-mass stars, carbon burning may last only a few hundred years. When the core carbon is depleted, the core again begins to collapse, shrinking and heating until it can fuse a still-heavier element. The star is engaged in the final phases of a desperate battle against the ever-strengthening crush of gravity. Each successive stage of core nuclear burning proceeds more rapidly than prior stages. In a field in which we usually think in terms of millions or billions of years, we are suddenly dealing with time scales that are short even for humans.

The nuclear reactions in the star's final stages of life become quite complex, and many different reactions may take place simultaneously. The simplest sequence of fusion stages involves **helium capture**—the fusing of helium nuclei into progressively heavier elements. This process, which adds two protons with each reaction, tends to create elements with even numbers of protons.

Each time the core depletes the elements it is fusing, it shrinks and heats until it becomes hot enough for other fusion reactions. Meanwhile, a new type of shell burning ignites between the core and the overlying shells of fusion. Near the end, the star's central region resembles the inside of an onion, with layer upon layer of shells burning different elements (Figure 14.14). During the star's final few days, iron begins to pile up in the star's core.

Despite the dramatic events taking place in its interior, the high-mass star's outer appearance changes only slowly. As each stage of core fusion ceases, the surrounding shell burning intensifies and further inflates the star's outer layers. Each time the core flares up again, the outer layers may contract a bit.

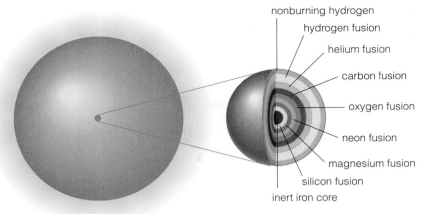

nonburning hydrogen
hydrogen fusion
helium fusion
carbon fusion
oxygen fusion
neon fusion
magnesium fusion
silicon fusion
inert iron core

FIGURE 14.14 The multiple layers of nuclear burning in the core of a high-mass star during the final days of its life.

FIGURE 14.15 Life tracks on the H–R diagram from main-sequence star to red supergiant for selected high-mass stars. (Based on models from A. Maeder and G. Meynet.)

The result is that the star's life track zigzags across the top of the H–R diagram (Figure 14.15). In very massive stars, the core changes happen so quickly that the outer layers don't have time to respond, and the star progresses steadily toward becoming a red supergiant. Betelgeuse, the upper-left shoulder star of Orion, is the best-known red supergiant star. Its radius is over 800 solar radii, or about four times the distance from the Sun to Earth. We have no way of knowing what stage of nuclear burning is now taking place in Betelgeuse's core. Betelgeuse may have a few thousand years of nuclear burning still ahead, or it may be piling up iron in its core as you read. If the latter is the case, then sometime in the next few days we will witness the most dramatic event that ever occurs in the universe. (This statement assumes we've taken into account the light-travel time.)

Iron: Bad News for the Stellar Core

As a high-mass star develops an inert core of iron, the core continues shrinking and heating while iron continues to pile up from nuclear burning in the surrounding shells. If iron were like the other elements in prior stages of nuclear burning, this core contraction would stop when iron fusion ignited. However, iron is unique among the elements in a very important way: It is the one element from which it is *not* possible to generate any kind of nuclear energy.

To understand why iron is unique, remember that only two basic processes can release nuclear energy: *fusion* of light elements into heavier ones, and *fission* of very heavy elements into not-so-heavy ones. Recall that hydrogen fusion converts four protons (hydrogen nuclei) into a helium nucleus that consists of two protons and two neutrons. Thus, the total number of *nuclear particles* (protons and neutrons combined) does not change. However, this fusion reaction generates energy (in accord with $E = mc^2$) because the *mass* of the helium nucleus is less than the combined mass of the four hydrogen nuclei that fused to create it—despite the fact that the *number* of nuclear particles is unchanged.

In other words, fusing hydrogen into helium generates energy because helium has a lower *mass per nuclear particle* than hydrogen. Similarly, fusing three helium-4 nuclei into one carbon-12 nucleus generates energy because carbon has a lower mass per nuclear particle than helium—which means that some mass disappears and becomes energy in this fusion reaction. In fact, the decrease in mass per nuclear particle from hydrogen to helium to carbon is part of a general trend shown in Figure 14.16. The mass per nuclear particle tends to decrease as we go from light elements to iron, which means that fusion of light nuclei into heavier nuclei generates energy. Note that this trend reverses beyond iron: The mass per nuclear particle tends to increase as we look to still heavier elements. As a result, the only way to generate nuclear energy from elements heavier than iron is through fission into lighter elements. For example, uranium has a greater mass per nuclear parti-

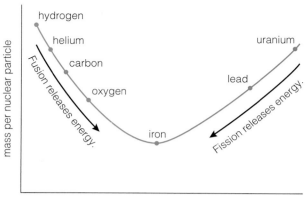

FIGURE 14.16 Overall, the average mass per nuclear particle declines from hydrogen to iron and then increases. Selected nuclei are labeled to provide reference points. (This graph shows the most general trends only; a more detailed graph would show numerous up-and-down bumps superimposed on the general trend.)

cle than lead, so some mass is converted into energy when uranium nuclei split apart and ultimately leave behind lead as a by-product.

Iron itself has the lowest mass per nuclear particle of all nuclei and therefore cannot release energy by either fusion or fission. Thus, once the matter in a stellar core turns to iron, it can burn no further. Unable to generate thermal energy or pressure by nuclear burning, the iron core's only hope of resisting the crush of gravity lies with degeneracy pressure. But iron keeps piling up in the core until degeneracy pressure can no longer support it. What ensues is the ultimate nuclear-waste catastrophe.

TIME OUT TO THINK *How would the universe be different if hydrogen, rather than iron, had the lowest mass per nuclear particle? Why?*

Supernova

The degeneracy pressure that briefly supports the inert iron core arises because the laws of quantum mechanics prohibit electrons from getting too close together. But once gravity pushes the electrons past the quantum mechanical limit, they can no longer exist freely. The electrons disappear by combining with protons to form neutrons, releasing neutrinos in the process (Figure 14.17). The electron degeneracy pressure suddenly vanishes, and gravity has free rein.

In a fraction of a second, an iron core with a mass comparable to that of our Sun and a size larger than that of Earth collapses into a ball of neutrons just a few kilometers across. The collapse halts only because the neutrons have a degeneracy pressure of their own. The entire core then resembles a giant atomic nucleus. If you recall that ordinary atoms are made almost entirely of empty space [Section 4.3] and that

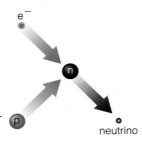

FIGURE 14.17 During the final, catastrophic collapse of a high-mass stellar core, electrons and protons combine to form neutrons, accompanied by the release of neutrinos.

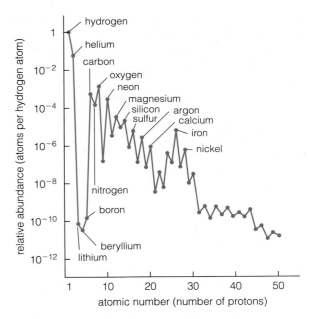

FIGURE 14.18 This graph shows the observed relative abundances of elements in the galaxy in comparison to the abundance of hydrogen. For example, the abundance of nitrogen is about 10^{-4}, which means that there are about $10^{-4} = 0.0001$ times as many nitrogen atoms in the galaxy as hydrogen atoms.

almost all their mass is in their nuclei, you'll realize that a giant atomic nucleus must have an astoundingly high density.

The gravitational collapse of the core releases an enormous amount of energy—more than a hundred times what the Sun will radiate over its entire 10-billion-year lifetime! This energy drives the outer layers off into space in a titanic explosion—a **supernova.** The ball of neutrons left behind is called a **neutron star.** In some cases, the remaining mass may be so large that gravity also overcomes neutron degeneracy pressure, and the core continues to collapse until it becomes a *black hole.*

The precise mechanism of the explosion is not yet clear. Current models of supernovae suggest that the most important process involves the neutrinos that are formed when electrons and protons combine to make neutrons. Although these ghostly particles rarely interact with anything [Section 12.3], so many are produced when the core implodes that they propel the star's upper layers into space at speeds of up to 10,000 km/s, heating it so that it shines with dazzling brilliance. For about a week, a supernova blazes as powerfully as 10 billion Suns, rivaling the luminosity of a moderate-size galaxy. The ejected gases slowly cool and fade in brightness over several months. The debris from the supernova carries with it elements produced in the star's nuclear furnace, as well as additional elements created when some of the neutrons produced during the core collapse slam into other nuclei. This debris may eventually be incorporated into a new generation of stars.

The Origin of Elements

Before we leave the subject of massive star life cycles, it's useful to consider the evidence that indicates we actually understand the origin of the elements. We cannot see inside stars, so we cannot directly observe elements being created in the ways we've discussed. However, the signature of nuclear reactions in massive stars is written in the patterns of elemental abundances across the universe.

For example, if massive stars really produce heavy elements (that is, elements heavier than hydrogen

and helium) and scatter these elements into space when they die, the total amount of these heavy elements in interstellar gas should gradually increase with time (because additional massive stars have died). Thus, we should expect stars born recently to contain a greater proportion of heavy elements than stars born in the distant past, because they formed from interstellar gas that contained more heavy elements. Stellar spectra confirm this prediction: Older stars do indeed contain smaller amounts of heavy elements than younger stars. For very old stars in globular clusters, elements besides hydrogen and helium typically make up as little as 0.1% of the total mass. In contrast, about 2–3% of the mass of young stars that formed in the recent past is in the form of heavy elements.

We gain even more confidence in our model of elemental creation when we compare the abundances of different elements. For example, because helium-capture reactions add two protons at a time, we expect nuclei with even numbers of protons to outnumber those with odd numbers of protons that fall between them. Sure enough, even-numbered nuclei such as carbon, oxygen, and neon are relatively abundant (Figure 14.18). Similarly, because certain elements heavier than iron are made only by rare fusion reactions shortly before and during a supernova, we expect these elements to be extremely rare. Again, this prediction made by our model of nuclear creation is verified by observations.

FIGURE 14.19 The Crab Nebula is the remnant of a supernova observed in A.D. 1054. The speed at which the Crab Nebula is growing indicates that the original explosion occured around A.D. 1100. Chinese observers recorded a "guest star" near this location on July 4, 1054, undoubtedly the supernova that created this nebula.

Supernova Observations

The study of supernovae owes a great debt to astronomers of many different epochs and cultures. Careful scrutiny of the night skies allowed the ancients to identify several supernovae whose remains still adorn the heavens. The most famous example concerns the Crab Nebula in the constellation Taurus. The Crab Nebula is a **supernova remnant**—an expanding cloud of debris from a supernova explosion (Figure 14.19). A spinning neutron star lies at the center of the Crab Nebula, providing evidence that supernovae really do create neutron stars. Photographs taken years apart show that the nebula is growing larger at a rate of several thousand kilometers per second. Calculating backward from its present size, we can trace the nebula's birth to somewhere near A.D. 1100.

No supernova has been seen in our own galaxy since 1604, but today astronomers routinely discover supernovae in other galaxies. The nearest of these extragalactic supernovae, and the only one near enough to be visible to the naked eye, burst into view in 1987. Because it was the first supernova detected that year, it was given the name **Supernova 1987A.** Supernova 1987A was the explosion of a star in the *Large Magellanic Cloud,* a small galaxy that orbits the Milky Way and is visible only from southern latitudes. The Large Magellanic Cloud is about 150,000 light-years away, so the star really exploded some 150,000 years ago.

As the nearest supernova witnessed in four centuries, Supernova 1987A provided a unique opportunity to study a supernova and its debris in detail. Older photographs of the Large Magellanic Cloud allowed astronomers to determine precisely which star exploded (Figure 14.20). It turned out to be a blue star, not the red supergiant expected when core fusion has ceased. The most likely explanation is that the star's outer layers were unusually thin and warm near the end of its life (perhaps because some of its matter spilled onto a close binary companion), changing its appearance from that of a red supergiant to a blue one. The surprising color of the pre-explosion

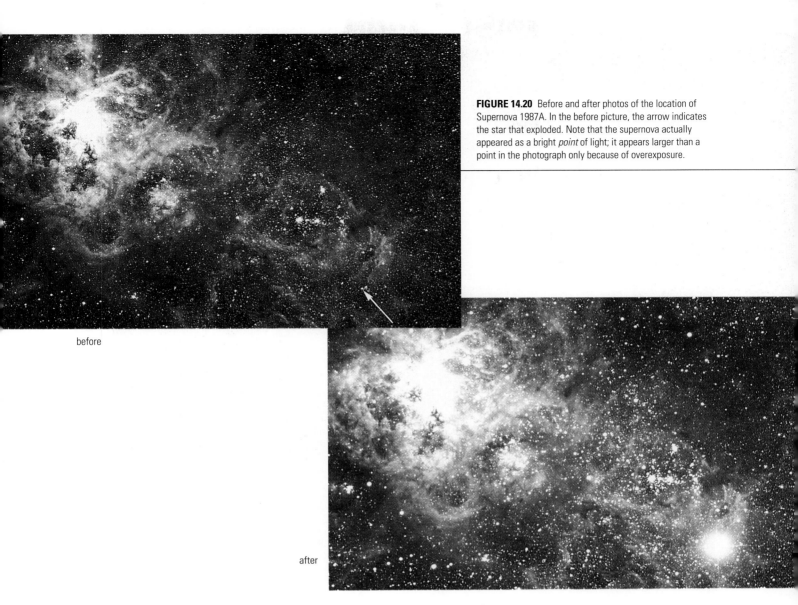

before

after

FIGURE 14.20 Before and after photos of the location of Supernova 1987A. In the before picture, the arrow indicates the star that exploded. Note that the supernova actually appeared as a bright *point* of light; it appears larger than a point in the photograph only because of overexposure.

star demonstrates that we still have much to learn about supernovae. Reassuringly, most other theoretical predictions of stellar life cycles were well matched by observations of Supernova 1987A.

One of the most remarkable findings from Supernova 1987A was that bursts of neutrinos coming from its direction were recorded by two neutrino detectors, one in Japan and one in Ohio. These data confirmed that the explosion released most of its energy in the form of neutrinos, suggesting that we are correct in believing that the stellar core undergoes sudden collapse to a ball of neutrons.

TIME OUT TO THINK *When Betelgeuse explodes as a supernova, it will be more than 10 times brighter than the full moon in our sky. If Betelgeuse had exploded a few hundred or a few thousand years ago, do you think it could have had any effect on human history? How do you think our modern society would react if we saw Betelgeuse explode tomorrow?*

Summary of Stellar Lives

We have seen that the primary factor determining how a star lives its life is its mass. Low-mass stars live long lives and die in planetary nebulae, leaving behind white dwarfs. High-mass stars live short lives and die in supernovae, leaving behind neutron stars and black holes. Both types of stars are crucial to life. Near the ends of their lives, low-mass stars can become carbon stars, which are the source of most of the carbon in our bodies. High-mass stars produce the vast array of other chemical elements on which life depends. Figure 14.21 summarizes the life cycles of stars of different masses.

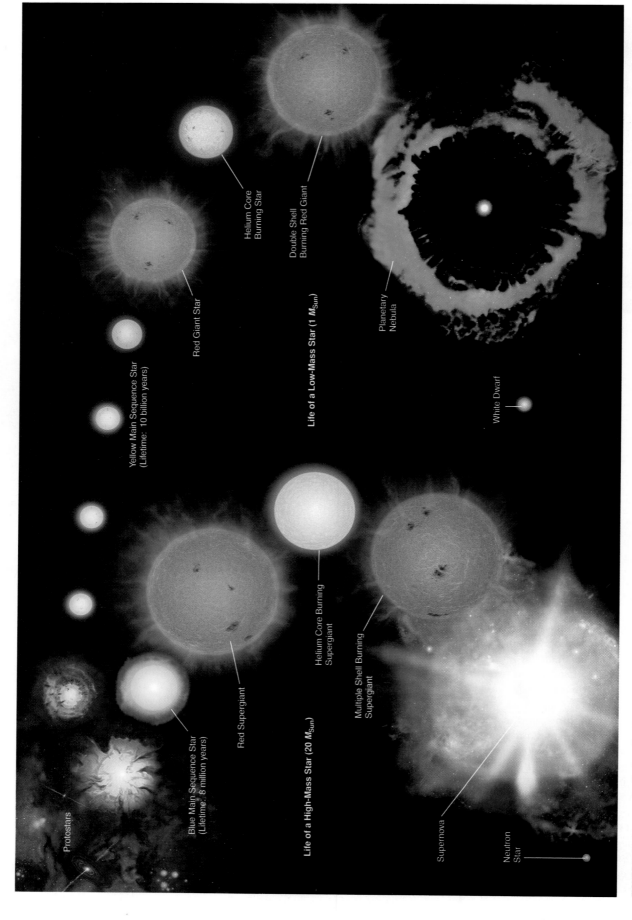

FIGURE 14.21 Summary of stellar lives. The life stages of a high-mass star (on the left) and a low-mass star (on the right) are depicted in clockwise sequences beginning with the protostellar stage in the upper left corner.

14.5 The Lives of Close Binary Stars

For the most part, stars in binary systems proceed from birth to death as if they were isolated and alone. The exceptions are close binary stars. Algol, the "demon star" in the constellation Perseus, consists of two stars that orbit each other closely: a $3.7M_{Sun}$ main-sequence star and a $0.8M_{Sun}$ subgiant. A moment's thought reveals that something quite strange is going on. The stars of a binary system are born at the same time and therefore must both be the same age. We know that more massive stars live shorter lives, and therefore the more massive star must exhaust its core hydrogen and become a subgiant before the less massive star does. How, then, can Algol's less massive star be a subgiant while the more massive star is still burning hydrogen as a main-sequence star?

This so-called *Algol paradox* reveals some of the complications in ordinary stellar life cycles that can arise in close binary systems. The two stars in close binaries are near enough to exert significant tidal forces on each other [Section 5.5]. The gravity of each star attracts the near side of the other star more strongly than it attracts the far side. The stars therefore stretch into football-like shapes rather than remaining spherical. In addition, the stars become *tidally locked* so that they always show the same face to each other, just as the Moon always shows the same face to Earth.

During the time that both stars are main-sequence stars, the tidal forces have little effect on their lives. But when the more massive star (which exhausts its core hydrogen sooner) begins to expand into a red giant, gas from its outer layers can spill over onto its companion. This **mass exchange** occurs when the giant grows so large that its tidally distorted outer layers succumb to the gravitational attraction of the smaller companion star. The companion then begins to gain mass at the expense of the giant.

The solution to the Algol paradox should now be clear (Figure 14.22). The $0.8M_{Sun}$ subgiant *used to be* much more massive. As the more massive star, it was the first to begin expanding into a red giant. But, as it expanded, so much of its matter spilled over onto its companion that it is now the less massive star.

FIGURE 14.22 Artist's conception of the development of the Algol close binary system.

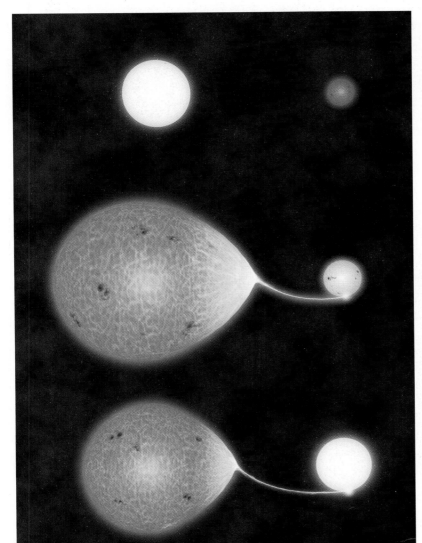

Algol, shortly after its birth. The higher-mass main-sequence star (left) evolved more quickly than its lower-mass companion (right).

Algol, onset of mass transfer. When the more massive star expanded into a red giant, it began losing some of its mass to its main-sequence companion.

Algol today. As a result of the mass transfer, the red giant has shrunk to a subgiant, and the main-sequence star on the right is now the more massive of the two stars.

The future may hold even more interesting events for Algol. The $3.7M_{Sun}$ star is still gaining mass from its subgiant companion. Thus, its life cycle is actually accelerating as its increasing gravity raises its core hydrogen fusion rate. Millions of years from now, it will exhaust its hydrogen and begin to expand into a red giant itself. At that point, it can begin to transfer mass *back* to its companion. Even stranger things can happen in other mass-exchange systems, particularly when one of the stars is a white dwarf or a neutron star. But that is a topic for the next chapter.

I can hear the sizzle of newborn stars, and know anything of meaning, of the fierce magic emerging here. I am witness to flexible eternity, the evolving past, and know I will live forever, as dust or breath in the face of stars, in the shifting pattern of winds.

JOY HARJO, *Secrets from the Center of the World*

THE BIG PICTURE

In this chapter, we answered the question of the origin of elements that we first discussed in Chapter 1. As you look back over this chapter, remember these "big picture" ideas:

- Virtually all elements besides hydrogen and helium were forged in the nuclear furnaces of stars. Carbon can be released from low-mass stars near the ends of their lives (carbon stars), and many other elements are released into space by massive stars in supernova explosions.

- The tug-of-war between gravity and pressure determines how stars behave from the time of their birth in a cloud of molecular gas to their sometimes violent death.

- Low-mass stars like our Sun live long lives and die with the ejection of a planetary nebula, leaving behind a white dwarf.

- High-mass stars live fast and die young, exploding dramatically as supernovae.

- Close binary stars can exchange mass, altering the usual course of stellar evolution.

Review Questions

1. What is *thermal pressure?* What are the two energy sources that can help a star maintain its internal thermal pressure?

2. Distinguish between the mass ranges for low-mass stars, intermediate-mass stars, and high-mass stars. Why is it useful to group stars into these mass ranges when we discuss the lives of stars?

3. What is a *molecular cloud?* Briefly describe the process by which a *protostar* forms from gas in a molecular cloud. Why are protostars surrounded by *protostellar disks?*

4. What is a *close binary?* Under what conditions does a close binary form?

5. Describe some of the activity seen in protostars, such as strong *protostellar winds* and *jets*.

6. What do we mean by a star's *life track* on an H–R diagram? How does an H–R diagram that shows life tracks differ from a standard H–R diagram?

7. Describe the observed trends in the number of stars that are born with different masses. What types of stars are most common? What types are least common?

8. What is *degeneracy pressure?* How does it differ from thermal pressure? Explain why degeneracy pressure can support a stellar core against gravity even when the core becomes very cold.

9. What is a *brown dwarf?*

10. What are the two basic ways by which energy can move outward from a star's core to its surface? Explain why stars of different masses have convection zones of different depths, and briefly describe the interior structure of stars of different masses.

11. What happens to the core of a star when it exhausts its hydrogen supply? Why does *hydrogen shell burning* begin around the inert core?

12. Explain why a star expands into a *subgiant* and then a *red giant,* even while its core is contracting. What is a *red-giant wind?*

13. Why does *helium fusion* require much higher temperatures than hydrogen fusion? Briefly describe the overall reaction by which helium fuses into carbon.

14. Explain why a star's overall radius shrinks from its peak size as a red giant after helium fusion begins. Why do helium-burning stars in a star cluster all fall on a *horizontal branch* on an H–R diagram?

15. Describe what happens to a low-mass star after it exhausts its core helium supply. What makes a star become a *carbon star?* How is the existence of carbon stars important to life on Earth?

16. What is a *planetary nebula*? What happens to the core of a star after a planetary nebula occurs?

17. Briefly describe how the Sun will change, and how Earth will be affected by these changes, over the next several billion years.

18. What is the *CNO cycle*? Why are high-mass stars able to use the CNO cycle, while low-mass stars do not?

19. What is *radiation pressure*? Why is it more important in high-mass stars than in low-mass stars?

20. Why does continued nuclear burning occur in high-mass stars but not in low-mass stars?

21. What special feature of iron nuclei determines how massive stars end their lives?

22. Describe the mechanism of a *supernova*. What is left behind after a supernova?

23. Summarize some of the observational evidence supporting our ideas about how the elements formed and showing that supernovae really occur.

24. What is the *Algol paradox*? What is its resolution? Use your answer to summarize how the lives of stars in close binaries can differ from the lives of single stars.

Discussion Question

1. *Humanity in* A.D. *5,000,000,000.* Do you think it is likely that humanity will survive until the Sun begins to expand into a red giant 5 billion years from now? Why or why not? If the human race does survive, how do you think people in A.D. 5,000,000,000 will differ from people today? What do you think they will do when faced with the impending death of the Sun? Debate these questions, and see if you and your friends can come to any agreement on possible answers.

Problems

1. *Homes to Civilizations.* We do not yet know how many stars have Earth-like planets, nor do we know the likelihood that such planets might harbor advanced civilizations like our own. However, some stars can probably be ruled out as candidates for advanced civilizations. For example, given that it took a few billion years for humans to evolve on Earth, it seems unlikely that advanced life would have had time to evolve around a star that is only a few million years old. For each of the following types of star, decide whether you think it is possible that it could harbor an advanced civilization, and explain your reasoning in one or two paragraphs.

   a. A $10M_{Sun}$ main-sequence star.

   b. A carbon star.

   c. A $1.5M_{Sun}$ red giant.

   d. A $1M_{Sun}$ horizontal-branch star.

   e. A red supergiant.

2. *Rare Elements.* Lithium, beryllium, and boron are elements with atomic number 3, 4, and 5, respectively. But despite their being three of the five simplest elements, Figure 14.18 shows that they are rare compared to many heavier elements. Suggest a reason for their rarity. (*Hint:* Consider the process by which helium fuses into carbon.)

3. *Future Skies.* As a red giant, the Sun's angular size in the Earth's sky will be about 30°. What will sunset and sunrise be like? About how long will they take? Do you think the color of the sky will be different from what it is today? Explain.

4. *Massive Star Instability.* If you study the life track of a high-mass star on the H–R diagram, you will see that it passes through the *instability strip* discussed in Chapter 13. What happens to the star during the time it resides in the instability strip? What would we call this type of star during this time? Explain. Based on the fact that stars of this type are massive, explain why they are so relatively rare.

5. *Web Project: Coming Fireworks in Supernova 1987A.* Astronomers believe that the show from Supernova 1987A is not yet over. In particular, sometime between now and about 2010, the expanding cloud of gas from the supernova is expected to ram into surrounding material, and the heat generated by the impact is expected to create a new light show. Using the Web, learn more about how Supernova 1987A is changing and what we might expect to see from it in the future. Summarize your findings in a one- to two-page report.

6. *Web Project: Picturing Star Birth and Death.* Photographs of stellar birthplaces (i.e., molecular clouds) and death places (e.g., planetary nebulae and supernova remnants) can be strikingly beautiful, but only a few such photographs are included in this chapter. Search the Web for additional photographs of these types; look not only for photos taken in visible light, but also for false-color photographs made from observations in other wavelengths of light. Put each photograph you find into a personal on-line journal, along with a one-paragraph description of what the photograph shows. Try to compile a journal of at least 20 such photographs.

The Bizarre Stellar Graveyard

15

WELCOME TO THE AFTERWORLD OF STARS, THE FASCINATING DOMAIN OF white dwarfs, neutron stars, and black holes. To scientists, these dead stars are ideal laboratories for testing the most extreme predictions of general relativity and quantum theory. To most other people, the eccentric behavior of stellar corpses demonstrates that the universe is stranger than they had ever imagined. Dead stars behave in unusual and unexpected ways that challenge our minds and stretch the boundaries of what we believe is possible. Stars that have finished nuclear burning have only one hope of staving off the crushing power of gravity: the strange quantum mechanical effect of degeneracy pressure. But even this strange pressure cannot save the most massive stellar cores, which collapse into oblivion as black holes. Prepare to be amazed by the eerie inhabitants of the stellar graveyard!

15.1 A Star's Final Battle

In the previous chapter, we saw that an ongoing "battle" between the inward crush of gravity and the outward push of pressure governs a star's life from the time when it first begins to form in an interstellar cloud to the time when it finally exhausts its nuclear fuel. Throughout most of this time, the pressure that holds gravity at bay is *thermal pressure*, which results from the heat produced as the star fuses light elements into heavier ones in its core. But the nuclear fuel eventually runs out. In the end, after the star dies in a planetary nebula or supernova, the fate of the stellar corpse lies in the outcome of a final battle between gravity and pressure—but this time the source of the pressure is the quantum mechanical effect called *degeneracy pressure*.

We have already discussed how degeneracy pressure successfully resists the crush of gravity in the stellar corpses known as white dwarfs and neutron stars, as well as in several other cases (e.g., brown dwarfs and inert stellar cores). Degeneracy pressure arises when subatomic particles are packed as closely as the laws of quantum mechanics allow [Section 14.2]. More specifically, white dwarfs are supported against the crush of gravity by **electron degeneracy pressure,** in which the pressure arises from densely packed electrons. In neutron stars, it is *neutrons* that are packed tightly together, thereby generating **neutron degeneracy pressure.**

White dwarfs and neutron stars would be strange enough if the story ended here, but it does not. Sometimes, gravity wins the battle with degeneracy pressure, and the stellar corpse collapses without end, crushing itself out of existence and forming a *black hole.* A black hole is truly a hole in the universe. If you enter a black hole, you leave our observable universe and can never return. In the rest of this chapter, we will study the bizarre properties and occasional catastrophes of the stellar corpses known as white dwarfs, neutron stars, and black holes.

15.2 White Dwarfs

A **white dwarf** is the inert core left over after a star has ceased nuclear burning, so its composition reflects the products of the star's final nuclear-burning stage. The white dwarf left behind by a $1M_{Sun}$ star like our Sun will be made mostly of carbon, the product of the star's final helium-burning stage. The cores of very low mass stars never become hot enough to fuse helium and thus end up as helium white dwarfs. Some intermediate-mass stars progress to carbon burning but do not create any iron (and hence do not explode as supernovae). These stars leave behind white dwarfs containing large amounts of oxygen or even heavier elements.

Despite the ordinary-sounding compositions of white dwarfs, their degenerate matter is unlike anything ever seen on Earth. Gravity and electron degeneracy pressure have battled to a draw in white dwarfs. Electron degeneracy pressure finally halts the collapse of a $1M_{Sun}$ stellar corpse when it shrinks to about the size of the Earth. If you recall that our Earth is smaller than a typical sunspot, it should be clear that it's no small feat to pack the entire mass of the Sun into the volume of the Earth. The density of such a white dwarf is so high that a pair of standard dice made from its material would weigh about 5 tons.

The White Dwarf Limit

Theoretical calculations show that the mass of a white dwarf cannot exceed a **white dwarf limit** of about $1.4M_{Sun}$, commonly called the *Chandrasekhar limit* after its discoverer. The limit comes about because the electron speeds are higher in more massive white dwarfs, and these speeds approach the speed of light in white dwarfs with masses approaching $1.4M_{Sun}$. Neither electrons nor anything else can travel faster than the speed of light, so electrons can do nothing to halt the crush of gravity in a stellar corpse with a mass above the white dwarf limit. Such an object must inevitably collapse into a compact ball of neutrons, at which point *neutron* degeneracy pressure can stop the crush of gravity.

Strong observational evidence supports this theoretical limit on the mass of a white dwarf. Many known white dwarfs are members of binary systems, and hence their masses can be measured [Section 13.4]. In every observed case, the white dwarfs have masses below $1.4M_{Sun}$, just as we would expect from theory.

White Dwarfs in Close Binary Systems

Left to itself, a white dwarf will never succumb to the crush of gravity because its electron degeneracy pressure does not lessen even as the white dwarf cools. However, white dwarfs in close binary systems do not necessarily rest in peace.

A white dwarf in a close binary system can gain substantial quantities of mass if its companion is a main-sequence or giant star. When a clump of mass first spills over from the companion to the white dwarf, it has some small orbital velocity. The law of conservation of angular momentum dictates that the clump must orbit faster and faster as it falls toward the white dwarf's surface. The infalling matter therefore forms a whirlpool-like disk around the

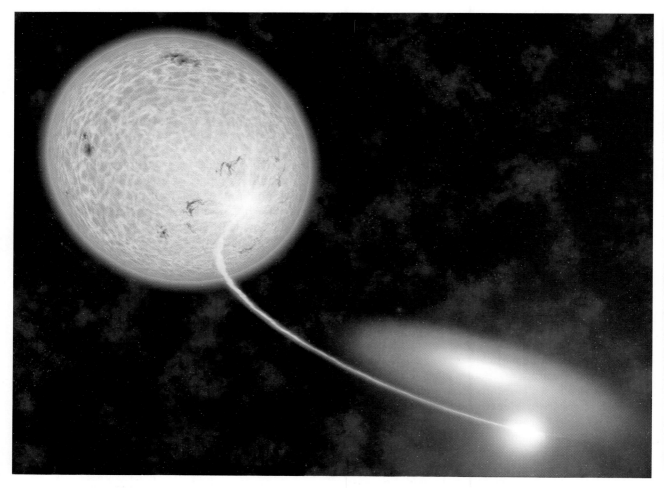

FIGURE 15.1 This artist's conception shows how mass spilling from a companion star (left) toward a white dwarf (right) forms an accretion disk around the white dwarf. The white dwarf itself is in the center of the accretion disk—too small to see on this scale. Matter streaming onto the disk creates a hot spot at the point of impact.

white dwarf (Figure 15.1). The process in which material falls onto another body is called *accretion* [Section 7.4], so this rapidly rotating disk is called an **accretion disk.**

TIME OUT TO THINK *Explain why the infalling matter forms a disk around the white dwarf (as opposed to making, say, a spherical distribution of matter). (Hint: See Sections 7.3 and 14.2.) How is an accretion disk similar to a protoplanetary disk? How is it different?*

Accreting gas gradually spirals inward through the accretion disk and eventually falls onto the surface of the white dwarf. This happens because the individual gas particles in the accretion disk, like anything else that orbits a massive body, obey Kepler's laws [Section 5.3]. Gas in the inner parts of the accretion disk moves faster than gas in the outer parts; because of these differences in orbital speed, gas in any particular part of the accretion disk "rubs" against slower-moving gas just outside of it. This "rubbing" generates friction and heat in the same way that rubbing your palms together makes them warm. The friction slowly causes the orbits of individual gas particles to decay, making the orbits smaller and smaller until they fall onto the white dwarf surface. Meanwhile, additional gas spilling over from the companion constantly replenishes the accretion disk.

Accretion can provide the "dead" white dwarf with a new energy source. Theory predicts that the heat generated by friction should make the accretion disk hot enough to radiate optical and ultraviolet light, and sometimes even X rays. Thus, although accretion disks are far too small to be seen directly, we should be able to detect their intense ultraviolet or X-ray radiation. Searches for this radiation have turned up strong evidence for accretion disks around

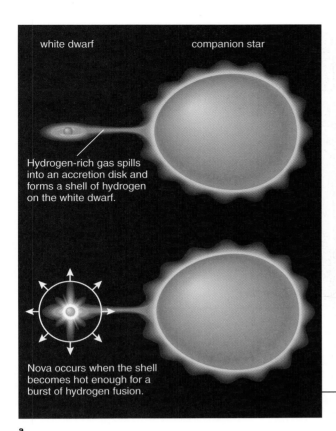

white dwarf   companion star

Hydrogen-rich gas spills into an accretion disk and forms a shell of hydrogen on the white dwarf.

Nova occurs when the shell becomes hot enough for a burst of hydrogen fusion.

a

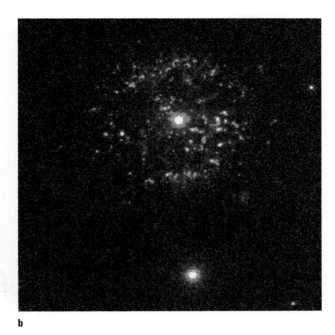

b

FIGURE 15.2 (**a**) Diagram of the nova process. (**b**) Hubble Space Telescope image showing blobs of gas ejected from the nova T Pyxidis. The bright spot at the center of the blobs is the binary star system that generated the nova.

many white dwarfs. In some cases, the brightness of these systems is highly variable—sudden increases in brightness by a factor of 10 or more may persist for a few days and then fade away, only to repeat a few weeks or months later. Such brightening probably arises when instabilities in the accretion disk cause some of the matter to fall suddenly onto the white dwarf surface, with an accompanying release of gravitational potential energy. (This type of brightening is sometimes called a *dwarf nova*.)

Accreting white dwarfs occasionally flare up even more dramatically. Remember that a white dwarf is "dead" because it has no hydrogen left to fuse. This situation changes in an accreting white dwarf. The gas spilling onto an accreting white dwarf comes from the upper layers of its companion star and thus is composed mostly of hydrogen. A thin shell of fresh hydrogen builds up on the surface of the white dwarf as more and more material rains down from the accretion disk. The pressure and temperature at the bottom of this shell increase as the shell grows, and hydrogen fusion ignites when the temperature exceeds 10 million K. The white dwarf suddenly blazes back to life as its hydrogen shell burns. This thermonuclear flash causes the binary system to shine for a few glorious weeks as a **nova,** which can radiate as

brightly as 100,000 Suns (Figure 15.2a). The nova generates heat and pressure, ejecting most of the material that has accreted onto the white dwarf. This material expands outward, creating a *nova remnant* that sometimes remains visible years after the nova explosion (Figure 15.2b).

Accretion resumes after a nova explosion subsides, so the entire process can repeat itself. The time between successive novae in a particular system depends on the rate at which hydrogen accretes on the white dwarf surface and on how highly compressed this hydrogen becomes. The compression of hydrogen is greatest for the most massive white dwarfs, which have the strongest surface gravities. In some cases, novae have been observed to repeat after just a few decades. More commonly, accreting white dwarfs may have to wait 10,000 years between nova outbursts.

Note that, according to our modern definitions, a nova and a supernova are quite different events: A nova is a relatively minor detonation of hydrogen fusion on the surface of a white dwarf in a close binary, while a supernova is the total explosion of a star. However, the word *nova* simply means "new," and historically a nova referred to any star that appeared to the naked eye where none was visible before.

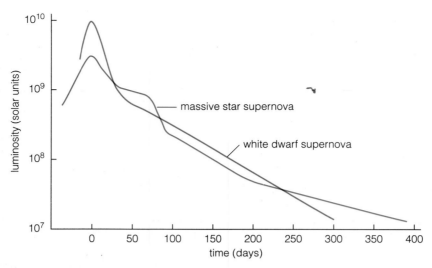

FIGURE 15.3 Typical light curves for white dwarf and massive star supernovae.

Because supernovae generate far more power than novae—the light of 10 billion Suns in a supernova versus 100,000 Suns in a nova—a very distant supernova can appear as bright in our sky as a nova that is relatively close. Thus, we could not distinguish between novae and supernovae prior to modern times.

White Dwarf Supernovae

Each time a nova occurs, the white dwarf ejects some of its mass. Each time a nova subsides, the white dwarf begins to accrete matter again. Theoretical models cannot yet tell us whether the net result is a gradual increase or decrease in the white dwarf's mass. Nevertheless, in at least some cases, accreting white dwarfs in binary systems continue to gain mass as time passes. If such a white dwarf gains enough mass, it can one day reach the $1.4 M_{Sun}$ white dwarf limit. This day is the white dwarf's last.

The electron degeneracy pressure that has been supporting the white dwarf gives out as soon as it reaches the $1.4 M_{Sun}$ limit. Gravity is suddenly unopposed, and the white dwarf begins to collapse. The gravitational potential energy released in the collapse quickly heats the white dwarf, which soon reaches the temperature needed for carbon fusion. Because the white dwarf material is degenerate, carbon fusion ignites almost instantly throughout the star. This "carbon bomb" detonation is similar to the helium flash that occurs in low-mass red giants but releases far more energy [Section 14.3]. The white dwarf explodes completely in what we will call a **white dwarf supernova** (also known as a Type Ia supernova).

TIME OUT TO THINK *Why do we expect that most collapsing white dwarfs undergo carbon fusion, as opposed to fusion of hydrogen, helium, or some other element?*

A white dwarf supernova shines as brilliantly as a supernova that occurs at the end of a high-mass star's life [Section 14.4]. To distinguish the two types, we'll refer to the latter as a **massive star supernova** (also known as a Type Ib or Type II supernova). Both white dwarf supernovae and massive star supernovae result in the destruction of a star. A massive star supernova is thought inevitably to leave behind either a neutron star or a black hole, but nothing remains after the "carbon bomb" detonation of a white dwarf supernova.

Astronomers can distinguish between white dwarf and massive star supernovae by studying their light. Because white dwarfs contain almost no hydrogen, the spectra of white dwarf supernovae show no spectral lines of hydrogen. In contrast, massive stars usually have plenty of hydrogen in their outer layers at the time they explode, so hydrogen lines are prominent in the spectra of most massive star supernovae. A second way to distinguish the two types of supernovae is to plot *light curves* that show how their luminosity fades with time. Figure 15.3 contrasts typical light curves for white dwarf and massive star supernovae. Note that both types of supernovae reach a peak luminosity of about 10 billion Suns ($10^{10} L_{Sun}$), but white dwarf supernovae fade steadily, while massive star supernovae fade in two distinct stages.

Not only are white dwarf supernovae dramatic, but they also provide one of the primary means by which we measure large distances in the universe. Massive star supernovae differ in intrinsic brightness, but white dwarf supernovae always occur in white dwarfs that have just reached the $1.4M_{Sun}$ limit. The light curves of all white dwarf supernovae therefore look amazingly similar, and their maximum luminosities are nearly identical. This fact is extremely useful: Once we know the true luminosity of one white dwarf supernova, we essentially know the luminosities of them all. Thus, whenever we discover a white dwarf supernova in a distant galaxy, we can determine the galaxy's distance by using the luminosity–distance formula [Section 13.2].

15.3 Neutron Stars

White dwarfs with densities of 5 tons per teaspoon may seem incredible, but neutron stars are stranger still. A **neutron star** is the ball of neutrons created by the collapse of the iron core in a massive star supernova. Typically just 10 kilometers across yet more massive than the Sun, neutron stars are essentially giant atomic nuclei, with two important differences: (1) They are made almost entirely of neutrons, and (2) gravity, not the strong force, is what binds them together.

The force of gravity at the surface of a neutron star is truly awe-inspiring. Escape velocity is one-half the speed of light. The strong gravity causes photons to emerge with a *gravitational redshift* that increases their wavelengths by about 15%, just as predicted by Einstein's general theory of relativity. If the Earth were visited by a neutron star, it would be squashed into a shell no thicker than your thumb on the surface of the neutron star.

Pulsars

Theorists first speculated about neutron stars in the 1930s. But until observational proof of their existence arrived in 1967, most astronomers assumed that neutron stars were too strange to exist. The proof came largely through the efforts of Jocelyn Bell. In October 1967, she detected *pulses* of radio waves coming from the direction of the constellation Cygnus at precise 1.337301-second intervals (Figure 15.4). No known astronomical object pulsated so regularly. In fact, the pulsations came at such precise intervals that they were nearly as reliable for measuring time as the most precise human-made clocks. For a while, the mysterious source of the radio waves

FIGURE 15.4 About 20 seconds of data from the first pulsar, discovered by Jocelyn Bell in 1967. Arrows mark the pulses, which come precisely 1.337301 seconds apart.

was dubbed "LGM" for Little Green Men, only half-jokingly. Today we refer to such rapidly pulsing radio sources as **pulsars.**

The mystery of pulsars was soon resolved. By the end of 1968, astronomers had found two smoking guns: Pulsars sat at the centers of both the Crab Nebula and the Vela Nebula, the gaseous remnants of supernovae. The pulsars are neutron stars left behind by the supernova explosions.

The pulsations arise because the neutron star is spinning rapidly as a result of the conservation of angular momentum: As an iron core collapses into a neutron star, its rotation rate must increase as it shrinks in size. The collapse also bunches the magnetic field lines running through the core far more tightly, greatly amplifying the strength of the magnetic field. Shortly after the supernova event, the magnetic field of the remaining neutron star is a trillion times stronger than the Earth's. These intense magnetic fields somehow direct beams of radiation out along the magnetic poles, although we do not yet know exactly how the radiation is generated. If a neutron star's magnetic poles do not align with its rotation axis (just as the Earth's magnetic poles do not coincide with its geographic poles), the beams of radiation sweep round and round (Figure 15.5). Like lighthouses, these neutron stars actually emit a fairly steady beam of light, but we see pulses of light with each sweep of the beam past our location.

Pulsars gradually slow their rotation, and their magnetic field weakens, so eventually they no longer "pulse." Also, some spinning neutron stars may be oriented so that their beams do not sweep past our location. Thus, all pulsars are neutron stars, but not all neutron stars are pulsars.

TIME OUT TO THINK *Suppose we do not see pulses from a particular neutron star and hence do not call it a pulsar. Is it possible that a civilization living in some other star system would see this neutron star as a pulsar? Explain.*

FIGURE 15.5 (**a**) A pulsar is a rapidly rotating neutron star that beams radiation along its magnetic axis. (**b**) This artwork likens a pulsar (top) to a lighthouse (bottom). If a pulsar's radiation beams are not aligned with its rotation axis, they will sweep through space. Each time one of these beams sweeps across Earth, we see a pulse of radiation.

We know that pulsars must be neutron stars, because no other massive object could spin so fast. A white dwarf, for example, can spin no faster than about once per second; an increase in spin would tear it apart because its surface would be rotating faster than the escape velocity. Yet pulsars have been discovered that rotate as fast as 625 times per second. Only an object as extremely small and dense as a neutron star could spin so fast without breaking apart.

Neutron Stars in Close Binary Systems

Like their white dwarf brethren, neutron stars in close binary systems can brilliantly burst back to life. As is the case with white dwarfs, gas overflowing from a companion star can create a hot, swirling accretion disk around the neutron star. However, in the neutron star's mighty gravitational field, infalling matter releases an amazing amount of gravitational potential energy. Dropping a brick onto a neutron star would liberate as much energy as an atomic bomb. Because the gravitational energies of accretion disks around neutron stars are so tremendous, they are much hotter and more luminous than those around white dwarfs. The high temperatures in the inner regions of the accretion disk make it radiate powerfully in X rays. Some close binaries with neutron stars emit 100,000 times more energy in X rays than our Sun emits in all wavelengths of light combined. Due to this intense X-ray emission, close binaries that contain accreting neutron stars are often called **X-ray binaries.** Their existence confirms many of the strange properties of neutron stars. Today we know of hundreds of X-ray binaries in the Milky Way. The **all-sky map** in Figure 15.6 shows that X-ray binaries are concentrated in the disk, just like most of our galaxy's stars, gas, and dust.

Like accreting white dwarfs that occasionally erupt into novae, accreting neutron stars sporadically erupt with enormous luminosities. Hydrogen-rich material from the companion star builds up on the surface of the neutron star. Pressures at the bottom of this accreted hydrogen shell, only a meter thick, maintain steady fusion. Helium accumulates beneath the hydrogen-burning shell, and helium fusion suddenly ignites when the temperature builds to 100 million K. The helium burns rapidly to carbon and heavier elements, generating a burst of energy that flows from the neutron star in the form of X rays. These **X-ray bursters** typically flare every few hours

FIGURE 15.6 This all-sky map shows the distribution of X-ray binaries in all directions in the sky. Notice that most X-ray binaries, represented by green dots, are found in the disk of the Milky Way Galaxy, which runs horizontally across the center of this map. To visualize how the entire map corresponds to what we see in the sky, recall that the disk of the Milky Way forms a band of light that wraps a full 360° around our sky. You could reconstruct the circular band we see in the sky by cutting out this picture and attaching the left end to the right end to form a circular strip. The points directly at the top and bottom of this map correspond to the two opposite directions that are each 90° away from the band of the Milky Way in our sky—just as the points at the top and bottom of a flat map of the Earth represent the North and South poles, which are each 90° away from the Earth's equator.

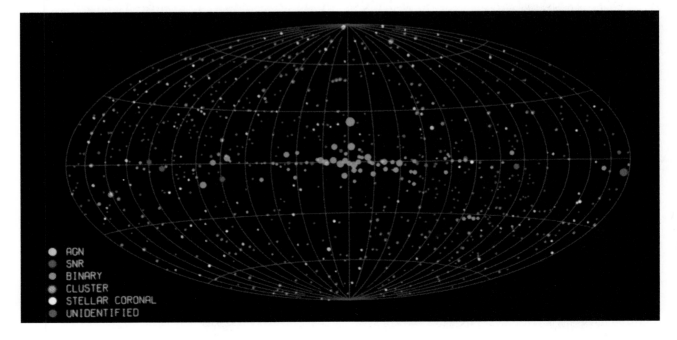

AGN
SNR
BINARY
CLUSTER
STELLAR CORONAL
UNIDENTIFIED

to every few days. Each burst lasts only a few seconds, but during that short time the system radiates power equivalent to 100,000 Suns, all in X rays. Within a minute after a burst, the X-ray burster cools back down and resumes accreting.

15.4 Black Holes: Gravity's Ultimate Victory

We know that white dwarfs cannot exceed $1.4M_{Sun}$ because gravity overcomes electron degeneracy pressure above that mass. The mass of a neutron star has a similar limit. The precise *neutron star limit* is not known, but it certainly lies below $3M_{Sun}$. A collapsing stellar core that weighs more than $3M_{Sun}$ faces the ultimate oblivion: becoming a **black hole.**

A black hole originates in the collapse of the iron core that forms just prior to the supernova of a very high mass star. Any star born with more than about $10M_{Sun}$ dies in a supernova, but most of the star's mass is blown into space by the explosion. As a result, the collapsed cores left behind by most supernovae are neutron stars. However, theoretical models show that the most massive stars might not succeed in blowing away all their upper layers. If enough matter falls back onto the core, raising its mass above the neutron star limit, then neutron degeneracy pressure will not be able to fend off gravity.

A core whose mass exceeds the neutron star limit will continue to collapse catastrophically. Then, when the core has only a fraction of a second left, gravity plays its cruelest trick. Usually, the gravitational potential energy released as a star collapses boosts its temperature and pressure enough to fight off gravity. However, in a star destined to become a black hole, the enhanced temperature and pressure just make gravity stronger. According to Einstein's theory of relativity, energy is equivalent to mass ($E = mc^2$) and thus must exert some gravitational attraction. The gravity of pure energy usually is negligible, but not in a stellar core collapsing beyond the neutron star limit. Here the energy associated with the temperature and pressure concentrated in the tiny core acts like additional mass, hastening the collapse. To the best of our current understanding, *nothing* can halt the crush of gravity. The core collapses without end, forming a black hole. Gravity has achieved its ultimate triumph.

A Hole in the Universe

The idea of a black hole was first suggested in the late 1700s by British philosopher John Mitchell and French physicist Pierre Laplace. It was already known from Newton's laws that the escape velocity from any object depends only on its mass and size:

More compact objects of a particular mass have higher escape velocities [Section 5.6]. Mitchell and Laplace speculated about objects so compact that their escape velocity exceeded the speed of light. Because they worked in a time before it was known that light always travels at the same speed, they assumed that light emitted from such an object would behave like a rock thrown upward, slowing to a stop and falling back down.

Einstein's work showed that black holes are considerably more bizarre. He found that space and time are not distinct, as we usually think of them, but instead are bound up together as four-dimensional **spacetime.** Moreover, in his general theory of relativity, Einstein showed that what we perceive as gravity arises from *curvature of spacetime*. The concept "curvature of spacetime" is quite strange, because we cannot even visualize four dimensions, let alone visualize their curvature. We can draw an analogy to spacetime curvature with a "rubber sheet" diagram in which we show how different masses would affect a stretched rubber sheet. In this analogy, a black hole is a region in which spacetime is stretched so far that it becomes a bottomless pit (Figure 15.7). Keep in mind that a black hole is not really shaped like a funnel; the illustration is only a two-dimensional analogy. Black holes are actually spherical, and they are black because not even light can escape from them.

A black hole really is a *hole* in the observable universe, and the boundary between the inside of the black hole and the universe outside is called the **event horizon.** Within the event horizon, the escape velocity exceeds the speed of light, so nothing—not even light—can ever get out. The event horizon gets its name because information can never reach us from events that occur within it. The radius of the event horizon is known as the **Schwarzschild radius,** named for Karl Schwarzschild (1873–1916), who computed it from Einstein's general theory of relativity in 1916. The Schwarzschild radius of a black hole depends only on its mass: The larger the mass, the larger the Schwarzschild radius.

There is no way to tell what has fallen into a black hole in the past. A black hole that forms from the collapse of a stellar core has the mass of that core, but no recognizable material remains. Because stellar cores normally rotate, conservation of angular momentum dictates that black holes should be rotating rapidly when they form. Aside from mass and angular momentum, the only other measurable property of a black hole is its electrical charge, and most black holes are probably electrically neutral. Any other information carried by objects that plunge into a black hole is irrevocably lost from the universe.

radial distance

radial distance ▼

event horizon

FIGURE 15.7 (**a**) A two-dimensional representation of "flat" spacetime. The circumference of each circle is 2π times its radius. (**b**) A two-dimensional representation of the "curved" spacetime around a black hole. The black hole's mass distorts spacetime, causing it to curve sharply in the vicinity of the event horizon. The funnel-shaped surface represents how radial distances (measured along the curved surface) between circles centered on the black hole become larger than they would be in flat spacetime.

Voyage to a Black Hole

Imagine that you are a pioneer of the future, making the first visit to a black hole. You've selected a black hole with a mass of $10M_{Sun}$ and a Schwarzschild radius of 30 km. As your spaceship approaches the black hole, you fire its engines to put the ship on a circular orbit a few thousand kilometers above the event horizon. Note that this orbit will be perfectly stable—there is no need to worry about getting "sucked in."

Your first task is to test Einstein's general theory of relativity. This theory predicts that time should run more slowly as the force of gravity grows stronger. It also predicts that light coming out of a strong gravitational field should show a redshift, called a *gravitational redshift,* that is due to gravity rather than to the Doppler effect. You test these predictions with the aid of two identical clocks whose numerals glow with blue light. You keep one clock aboard the ship and push the other one, with a small rocket attached, directly toward the black hole (Figure 15.8). The small rocket automatically fires its engines just enough so that the clock falls only gradually toward the event horizon. Sure enough, the clock on the rocket ticks more slowly as it heads toward the black hole, and its light becomes increasingly redshifted. When the clock reaches a distance of about 10 km above the event horizon, you see it ticking only half as fast as the clock on your spaceship, and its numerals are red instead of blue.

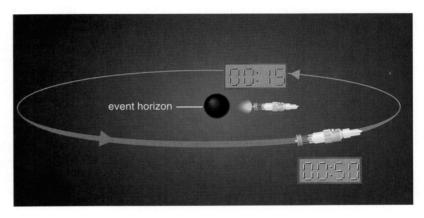

FIGURE 15.8 Time runs more slowly on the clock nearer to the black hole, and gravitational redshift makes its glowing blue numerals appear red from your spaceship.

The rocket has to expend fuel rapidly to keep the clock hovering in the strong gravitational field, and it soon runs out of fuel. The clock plunges toward the black hole. From your safe vantage point inside the spaceship, you see the clock ticking more and more slowly as it falls. However, you soon need a radio telescope to "see" it, as the light from the clock face shifts from the red part of the visible spectrum, through the infrared, and on into the radio. Finally, its light is so far redshifted that no conceivable telescope could detect it. Just as the clock vanishes from view, you see that the time on its face has frozen to a stop.

Curiosity overwhelms the better judgment of one of your colleagues. He hurriedly climbs into a spacesuit, grabs the other clock, resets it, and jumps out of the airlock on a trajectory aimed straight for the black hole. Down he falls, clock in hand. He watches the clock, but because he and the clock are traveling together, its time seems to run normally and its numerals stay blue. From his point of view, time seems to neither speed up nor slow down. In fact, he'd say that *you* were the one with the strange time, as he would see your time running increasingly fast and your light becoming increasingly blueshifted. When the clock reads, say, 00:30, he and his clock pass through the event horizon. There is no barrier, no wall, no hard surface. The event horizon is a mathematical boundary, not a physical one. From his point of view, the clock keeps ticking. He is inside the event horizon, the first human being ever to leave our universe.

Back on the spaceship, you watch in horror as your overly curious friend plunges to his death. Yet, from your point of view, he will *never* cross the event horizon. You'll see time come to a stop for him and his clock just as he vanishes from view due to the huge gravitational redshift of light. When you return home, you can play a videotape for the judges at your trial, proving that your friend is still a part of our universe. Strange as it may seem, all of this is true according to Einstein's theory. From your point of view, your friend takes *forever* to cross the event horizon; from his point of view, it is but a moment's plunge before he leaves the universe.

FIGURE 15.9 Tidal forces would be lethal near a black hole formed by the collapse of a star. The black hole would pull more strongly on the astronaut's feet than on his head, stretching him lengthwise and squeezing him from side to side.

Common Misconceptions: Black Holes Don't Suck

What would happen if our Sun suddenly became a black hole? For some reason, it has become part of our popular culture for most people to believe that Earth and the other planets would be inevitably "sucked in" by the black hole. This is not true. Although the sudden disappearance of light and heat from the Sun would be bad news for life, the Earth's orbit would not change. Newton's law of gravity tells us that the allowed orbits in a gravitational field are ellipses, hyperbolas, and parabolas [Section 5.4]; note that "sucking" is not on the list! A spaceship would get into trouble only if it came so close to a black hole—within about three times its Schwarzschild radius—that gravity would deviate significantly from Newton's law. Otherwise, a spaceship passing near a black hole would simply swing around it on an ordinary orbit (ellipse, parabola, or hyperbola). In fact, because most black holes are so small—typical Schwarzschild radii are smaller than any star or planet, and smaller even than most asteroids—a black hole is actually one of the most difficult things in the universe to fall into by accident.

The truly sad part of this story is that your friend did not live to experience the crossing of the event horizon. The force of gravity he felt as he approached the black hole grew so quickly that it actually pulled much harder on his feet than on his head, simultaneously squeezing him from side to side (Figure 15.9). In essence, your friend was stretched in the same way the oceans are stretched by the tides, except that the *tidal force* near the black hole is trillions of times stronger than the tidal force of the Moon on the Earth [Section 5.5]. No human could survive it.

If he had thought ahead, your friend might have waited to make his jump until you visited a much larger black hole, like one of the *supermassive black holes* thought to reside in the center of many galaxies [Section 17.6]. A 1-billion-M_{Sun} black hole has a Schwarzschild radius of 3 billion kilometers—about the distance from our Sun to Uranus. Although the gravitational forces at the event horizon of all black holes are equally great, the larger size of the supermassive black hole makes its tidal forces much weaker and hence nonlethal. Your friend could safely plunge through the event horizon. Again, from your point of view, the crossing would take forever, and you would see time come to a stop for him just as he vanished from sight because of the gravitational redshift. Again, he would experience time running normally, and he would see time in the outside universe running increasingly fast as he approached the event horizon. Unfortunately, anything he learned as he watched

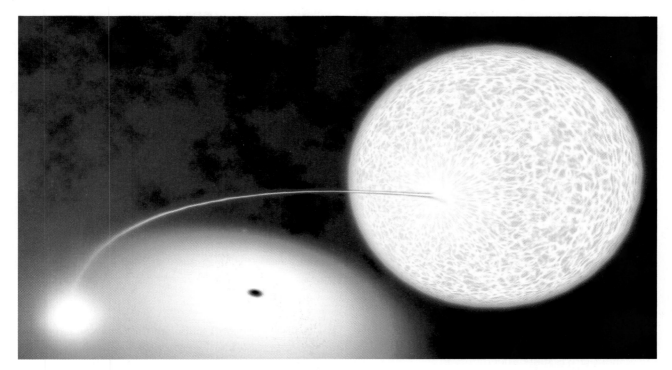

FIGURE 15.10 Artist's conception of the Cygnus X-1 system. The X rays come from the high-temperature gas in the accretion disk surrounding the black hole.

the future of the universe unfold would do him little good as he plunged to oblivion inside the black hole.

Singularity and the Limits to Knowledge

The center of a black hole is thought to be a place where gravity crushes all matter to an infinitely tiny and infinitely dense point called a **singularity.** This singularity is the point at which all the mass that created the black hole resides.

We can never know what really happens inside a black hole, because no information can ever emerge from within the event horizon. Nevertheless, we can use Einstein's theory of relativity to predict conditions inside the black hole, as long as we don't try to describe conditions too close to the singularity. The singularity itself is more puzzling, because the equations of modern physics yield conflicting predictions when we try to apply them to an infinitely compressed mass. General relativity predicts that spacetime should grow infinitely curved as it enters the pointlike singularity. Quantum physics predicts that, as a consequence of the uncertainty principle, spacetime should fluctuate chaotically in regions smaller than 10^{-35} meter across. No current theory adequately accommodates these divergent claims. We will not fully understand how a singularity behaves until we have developed a quantum theory of gravity that encompasses both general relativity and quantum mechanics. This uncertainty in our current

knowledge is a gold mine for science fiction writers, who speculate about using black holes for exotic forms of travel through spacetime.

Evidence for Black Holes

Have we discovered any black holes yet? We think so, but the evidence is indirect. Black holes emit no light. They are impossible to see, so we must look for their effects on surrounding matter. Black holes in close binaries should be among the easiest to identify. (Supermassive black holes in the center of galaxies should also be relatively easy to identify; we will discuss the evidence that these exist in Chapter 17.) Gas overflowing a black hole's stellar companion will form a hot, X ray–emitting accretion disk similar to the disks that circle accreting neutron stars. The X rays can escape because the disk emits them from well outside the event horizon.

We strongly suspect that a few X-ray binaries contain black holes rather than neutron stars. The trick is to tell the difference, since the accretion disk is likely to be just as hot and to emit about as many X rays whether it circles a neutron star or a black hole. However, we can distinguish between the two possibilities if we can measure the mass of the accreting object. One of the most promising *black hole candidates* is in an X-ray binary called Cygnus X-1 (Figure 15.10). This system contains an extremely bright star with an estimated mass of $18M_{Sun}$; based

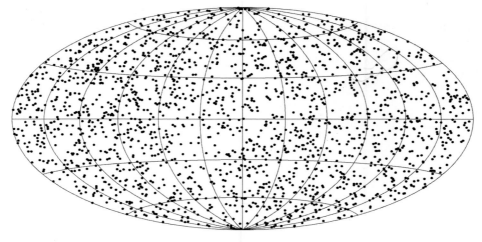

FIGURE 15.11 An all-sky diagram showing the distribution of gamma-ray bursts. Notice that the distribution appears random.

on Doppler shifts of its spectral lines, this star orbits an unseen companion with a mass of about $10M_{Sun}$. Although there is some uncertainty in these mass estimates, the mass of the invisible accreting object clearly exceeds the $3M_{Sun}$ neutron star limit. Moreover, careful studies of variations in the system's X-ray emission indicate that this massive accreting object must be very small in size—far too small to be an ordinary star. Thus, based on our current knowledge, the accreting object in Cygnus X-1 cannot be anything other than a black hole.

15.5 The Mystery of Gamma-Ray Bursts

In the early 1960s, the United States began launching a series of top-secret satellites designed to look for gamma rays emitted by nuclear bomb tests. The satellites soon began detecting occasional bursts of gamma rays, typically lasting a few seconds. It took several years for military scientists to become convinced that these **gamma-ray bursts** were coming from space, not from some sinister human activity. They publicized the discovery in 1973.

An increasingly large armada of satellites launched over the next two decades detected hundreds of other gamma-ray bursts. However, not only was their origin unknown, but it was very difficult to tell what direction the bursts came from. The problem is that gamma rays are very difficult to focus. A detector can record that it has been hit by a gamma ray but can provide little information on the direction of the gamma ray. With no specific proof, most astronomers assumed that gamma-ray bursts, like X-ray bursts, came from explosive events associated with neutron stars in X-ray binaries.

In 1991, NASA launched the *Compton Gamma Ray Observatory,* or *Compton* for short, which carried an array of eight detectors designed expressly to study gamma-ray bursts. By comparing the data recorded by all eight detectors, scientists could determine the direction of a gamma-ray burst within about 1°. The results were stunning. Compton detected gamma-ray bursts at a rate of about one per day and compiled a catalog of more than a thousand gamma-ray bursts within a few years. The results, plotted on an all-sky map, are shown in Figure 15.11. A quick comparison with the all-sky map of X-ray binaries in Figure 15.6 makes one thing very clear: The gamma-ray bursts are *not* concentrated in the disk of the Milky Way as the X-ray binaries are and thus must not be associated with neutron stars in the disk of our galaxy.

So where do gamma-ray bursts come from? Careful analysis of Figure 15.11 confirms what your eyes probably tell you: The gamma-ray bursts come from completely random directions in the sky. This fact seems to rule out the possibility that they come from anywhere in the Milky Way Galaxy. If the bursts came from objects distributed spherically about the Milky Way Galaxy, we would see a concentration of them in the direction of the galactic center. (Remember that we are located more than halfway out from the center of our galaxy.) The most reasonable conclusion is that gamma-ray bursts originate far outside our own galaxy. Additional evidence for this conclusion came in 1997, when astronomers first observed the afterglow of a gamma-ray burst in other wavelengths. The higher resolution possible in these other wavelengths allowed astronomers to pinpoint the burst's origin in a distant galaxy. Because gamma-ray bursts are coming from great distances, their apparent brightnesses imply that they are *by far* the most powerful bursts of energy that ever occur in the universe. For a few seconds, a gamma-ray burst releases as much total energy as a *hundred quadrillion* (10^{17}) Suns, or some 10,000 times the total luminosity of the brightest galaxy.

Astronomers still do not know what causes such massive outbursts of energy. One hypothesis suggests that the bursts come from the collision of *two* neutron stars in a binary system. Rough calculations suggest that such neutron star collisions should happen *somewhere* in the universe about once each day, which is consistent with the one-per-day frequency

of detected gamma-ray bursts. However, why a neutron star collision would produce the peculiar spectra and light curves of gamma-ray bursts is not clear, and many scientists have raised theoretical objections to this model. An alternative hypothesis suggests that gamma-ray bursts are somehow related to supernovae, but the evidence remains inconclusive.

Thus, we are left with one of the greatest mysteries in all of science: Gamma-ray bursts are the most powerful events in the universe, but we have virtually no idea what causes them.

Now, my suspicion is that the universe is not only queerer than we suppose, but queerer than we can suppose.

J. B. S. HALDANE, *Possible Worlds,* 1927

THE BIG PICTURE

We have now seen what happens to stars after they die. What a mind-bending experience! Nevertheless, try to keep these "big picture" ideas straight in your head:

- Despite the strange nature of stellar corpses, clear evidence exists for white dwarfs and neutron stars, and the case for black holes is very strong.

- White dwarfs, neutron stars, and black holes can all have close stellar companions from which they accrete matter. These binary systems produce some of the most spectacular events in the universe, including novae, white dwarf supernovae, and X-ray bursters.

- Black holes are truly holes in the observable universe that strongly warp space and time around them. The nature of black hole singularities remains beyond the frontier of current scientific understanding.

- Gamma-ray bursts were once thought to be related to neutron stars in our galaxy, but recent evidence indicates that this idea is wrong. Their origin is one of the greatest mysteries in the universe.

Review Questions

1. What is the difference between *electron degeneracy pressure* and *neutron degeneracy pressure?* Which type supports a white dwarf? Why are neutron stars so much smaller in size than white dwarfs?

2. With degeneracy pressure, how are the speeds of electrons (or neutrons) related to the mass of the degenerate object? Explain why this relationship between speed and mass implies upper limits on the possible masses of white dwarfs and neutron stars. What is the *white dwarf limit?* What is the *neutron star limit?*

3. Contrast the densities of white dwarf material with the densities of ordinary objects on Earth.

4. What is an *accretion disk?* Under what conditions does an accretion disk form? Explain how the accretion disk provides a white dwarf with a new source of energy that we can detect from Earth.

5. Describe the process of a *nova.* Why do novae only occur in close binary star systems? Can the same star system undergo a nova event more than once? Explain.

6. Contrast the process of a *white dwarf supernova* with that of a *massive star supernova.* Observationally, how can we distinguish between these two types of supernovae?

7. What is a *gravitational redshift,* and how does it affect the light we see from neutron stars?

8. What is a *pulsar?* Briefly describe how they were discovered and how they get their name. How do we know that pulsars must be neutron stars?

9. Explain why neutron stars in close binary systems are often called *X-ray binaries.*

10. What is an *X-ray burster?* How is the process of an X-ray burst similar to that of a nova? How do X-ray bursts differ from novae observationally?

11. Briefly explain the process by which the core of a very high mass star can collapse to form a *black hole.*

12. What is the *event horizon* of a black hole? How does it get its name? How is it related to the *Schwarzschild radius?*

13. Suppose you are orbiting a black hole. Describe how you will see the passage of time on an object approaching the black hole. What will you notice about light from the object? Explain why an object falling toward the black hole will eventually fade from view yet never cross the event horizon from your point of view.

14. Suppose you are falling into a black hole. How will you perceive the passage of your own time? How will you perceive the passage of time in the universe around you? Briefly explain why your trip will be lethal if the black hole is relatively small in mass, and why you may survive to cross the event horizon of a supermassive black hole.

15. What do we mean by the *singularity* of a black hole? How do we know that our current theories are inadequate to explain what happens at the singularity?

16. Briefly describe the observational evidence supporting the idea that Cygnus X-1 contains a black hole.

17. What are gamma-ray bursts? Why are they so mysterious?

Discussion Question

1. *Black Holes in Popular Culture.* Phrases such as "it disappeared into a black hole" are now common in popular culture. Give a few examples in which the term *black hole* is used in popular culture but is not meant to be taken literally. In what ways are these uses correct in their analogies to real black holes? In what ways are they incorrect? Why do you think such an esoteric scientific idea as that of a black hole has so captured the public imagination?

Problems

1. *Life Stories of Stars.* Write a one- to two-page short story for each of the following scenarios. Each story should be detailed and scientifically correct, but also creative. That is, it should be entertaining while at the same time proving that you understand stellar evolution. Be sure to state whether "you" are a member of a binary system.
   a. You are a white dwarf of $0.8M_{Sun}$. Tell your life story.
   b. You are a neutron star of $1.5M_{Sun}$. Tell your life story.

2. *Web Project: Gamma-Ray Bursts.* Go to the text web site and seek out the latest information about gamma-ray bursts. Write a one- to two-page essay on recent discoveries and how they may shed light on the mystery of gamma-ray bursts.

PART V

Galaxies and Beyond

16

Our Galaxy

In PREVIOUS CHAPTERS, WE SAW HOW STARS FORGE NEW ELEMENTS AND expel them into space. We also studied how interstellar gas clouds enriched with these stellar by-products form new stars and planetary systems. These processes do not occur in isolation. Instead, they are part of a dynamic system that acts throughout our Milky Way Galaxy.

You are probably familiar with the idea that all living species on Earth interact with one another and with the land, water, and air to form a large, interconnected ecosystem. In a similar way, but on a much larger scale, our galaxy is a self-contained system that cycles matter from stars into interstellar space and back into stars again. The birth of our solar system and the evolution of life on Earth would not have been possible without this "galactic ecosystem."

In this chapter, we will study our Milky Way Galaxy. We will investigate the galactic processes that maintain an ongoing cycle of stellar life and death, examine the structure and motion of the galaxy, and explore the mysteries of the galactic center. Through it all, we will see that we are not only "star stuff" but "galaxy stuff"— the product of eons of complex recycling and reprocessing of matter and energy in the Milky Way Galaxy.

16.1 The Milky Way Revealed

On a dark night, you can see a faint band of light slicing across the sky through several constellations, including Sagittarius, Cygnus, Perseus, and Orion. This band of light looked like a flowing ribbon of milk to the ancient Greeks, and we now call it the *Milky Way*. In the early 1600s, Galileo used his telescope to prove that the light of the Milky Way comes from a myriad of individual stars. Together these stars make up the kind of stellar system we call a *galaxy*, echoing the Greek word for "milk," *galactos*.

Today we know that our Milky Way Galaxy holds over 100 billion stars and is just one among tens of billions of galaxies in the universe. If we could stand outside our galaxy, we would see it as a flat **disk** of stars with a bright central **bulge**, spectacular **spiral arms,** and a dimmer, rounder **halo** surrounding everything (Figure 16.1). A few hundred **globular clusters** of stars [Section 13.6] circle our galaxy's center in orbits extending tens of thousands of light-years into the halo. Two entire galaxies, known as the Large and Small Magellanic Clouds, orbit at even greater distances of some 150,000 light-years (50 kpc). Although the Magellanic Clouds are relatively small for galaxies (a few billion stars each), they are far larger than globular clusters (which typically contain a few hundred thousand stars).

Our knowledge of the Milky Way's true size and shape was long in coming. Clouds of interstellar gas and dust known collectively as the **interstellar medium** fill the galactic disk, obscuring our view when we try to peer directly through it. The smoggy nature of the interstellar medium hides most of our galaxy from us and long fooled astronomers into believing that we lived near our galaxy's center. Astronomer Harlow Shapley finally proved otherwise in the 1920s, when he demonstrated that the Milky Way's globular clusters orbit a point tens of thousands of light-years from our Sun. He concluded that this point, not our Sun, must be the center of the galaxy. We now know that our Sun lies near the outskirts of the galactic disk, about 28,000 light-years (8.5 kpc) from its center.

Our own galaxy can be difficult to study not only because it is dusty, but also because we see it from the inside. The Milky Way's dustiness is no longer such a hindrance, because technologies developed in the past few decades allow us to observe the Milky Way's radio and infrared light. These wavelengths penetrate the enshrouding interstellar medium, enabling us to see into regions of the galaxy previously obscured from view. Still, determining our galaxy's structure from our location within the galactic disk

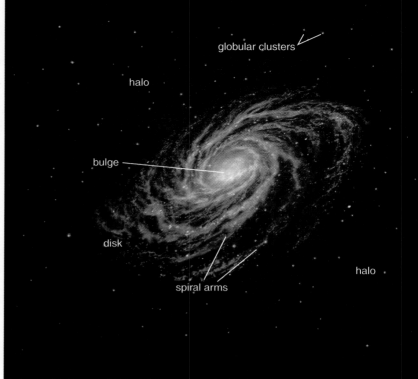

FIGURE 16.1 Artist's conception of the Milky Way viewed from its outskirts.

is somewhat like trying to draw a picture of your house without ever leaving your bedroom. Just as it is easier to draw pictures of other houses that you can see out the window, it is easier to measure the sizes and shapes of other galaxies. Nevertheless, the Milky Way is still the only galaxy whose inner workings we can examine up close.

We once thought of the Milky Way as a band of light, and later we saw it as a collection of stars. But now we see it revealed as a dynamic and complex system. In the rest of this chapter, we will see that our galaxy in many ways is like a forest of stars whose ecology is shaped by the cycle of stellar life and death and whose structure is determined by gravity and the majestic rotation of the galactic disk.

16.2 The Star–Gas–Star Cycle

Stars have formed, fused atomic nuclei, and exploded throughout the history of our galaxy. Generations of stars continually recycle the same galactic matter through their cores, gradually raising the overall abundance of elements made by fusion. Elements heavier than helium, usually called **heavy elements** by astronomers, now constitute about 2% of the galaxy's gaseous content; the overall composition of the galaxy is about 70% hydrogen, 28% helium, and 2% heavy elements. The process of adding to the abundance of heavy elements is called **chemical enrich-**

ment, and it is an inevitable by-product of continual star formation. However, the synthesis of elements is only one part of the galactic ecocycle. If the galaxy did not reincorporate these elements into new stars and their accompanying planetary systems, the mineral riches fused in the bellies of stars would be wasted.

TIME OUT TO THINK *Based on the idea of chemical enrichment, which types of stars must contain a higher proportion of heavy elements: stars in globular clusters or stars in open clusters? (Hint: Recall from Chapter 13 that stars in globular clusters are all very old, while stars in open clusters are relatively young.)*

Capturing the newly made elements released by stars is not an easy task. Were it not for the interstellar medium, the new heavy elements released by a supernova would fly straight out of the Milky Way into intergalactic space. Instead, the blobs of matter expelled from the supernova collide with the interstellar medium, slow down, and eventually stop.

Supernovae of one star after another stir and heat the interstellar medium while feeding it new heavy elements. These elements eventually blend with the older, less chemically enriched hydrogen gas in the vicinity. However, before new stars can form, this gas must cool and form clouds. The cooling of interstellar gas into clouds of molecular hydrogen takes millions of years. These clouds subsequently give birth to new stars more highly enriched in heavy elements, thus completing the star–gas–star cycle (Figure 16.2). Let's look at the stages of this cycle more closely.

FIGURE 16.2 A pictorial representation of the star–gas–star cycle. (For descriptions of individual photos, proceeding counterclockwise from lower right, see Figures 16.4, 16.5, 16.11a, 16.8, 16.10, and 16.12.)

atomic-hydrogen clouds

molecular clouds

hot bubbles

Star–gas–star cycle

star formation

supernovae and stellar winds

stellar burning/ heavy-element formation

FIGURE 16.3 A dying low-mass star returns gas to the interstellar medium in a planetary nebula, as viewed by the Hubble Space Telescope.

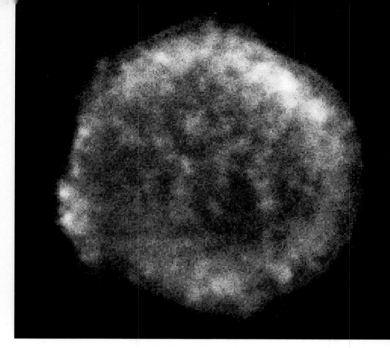

FIGURE 16.5 This false-color image shows X-ray emission from hot electrons in a young supernova remnant—the remnant from the supernova observed by Tycho Brahe in 1572.

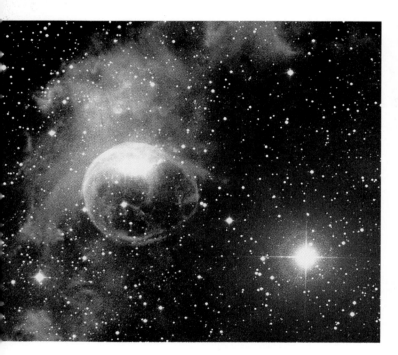

FIGURE 16.4 This photo shows a bubble of hot, ionized gas blown by a wind from the hot star at its center.

Gas from Stars

All stars return much of their original mass to interstellar space in two ways: through stellar winds that blow throughout their lives, and through "death events" of planetary nebulae (for low-mass stars) or supernovae (for high-mass stars). Low-mass stars generally have weak stellar winds while they are on the main sequence, but their winds grow stronger and carry more material into space when they be-

come red giants. By the time a low-mass star like the Sun ends its life with the ejection of a planetary nebula [Section 14.3], it has returned almost half its original mass to the interstellar medium (Figure 16.3).

High-mass stars lose mass much more dynamically and explosively. The powerful winds from supergiants and massive O and B stars recycle large amounts of matter into the galaxy. At the ends of their lives, these stars explode as supernovae. The high-speed gas ejected into space by these winds and supernovae sweeps up surrounding interstellar material, excavating a **bubble** of hot, ionized gas around the exploding star. Although the bubble in Figure 16.4 looks much like a soap bubble, it is actually an expanding shell of hot gas. The glowing surface is the edge of the expanding bubble, where gas piles up as the bubble sweeps outward through the interstellar medium. These hot, tenuous bubbles are quite common, filling roughly 20–50% of the Milky Way's disk. However, they are not always easy to detect. While some are hot enough to emit profuse amounts of X rays and others emit strongly in visible light, many others are evident only through radio emission from the shells of atomic hydrogen gas that surround them.

Shock Waves and Supernova Remnants

Supernovae generate **shock waves**—waves of pressure created by gas moving faster than the speed of sound. A shock wave sweeps up surrounding gas as it travels, creating a "wall" of hot, fast-moving gas on its leading edge. When we observe a *supernova remnant*, we are seeing the aftermath of its shock wave [Section 14.4]. Figure 16.5 shows a young supernova remnant

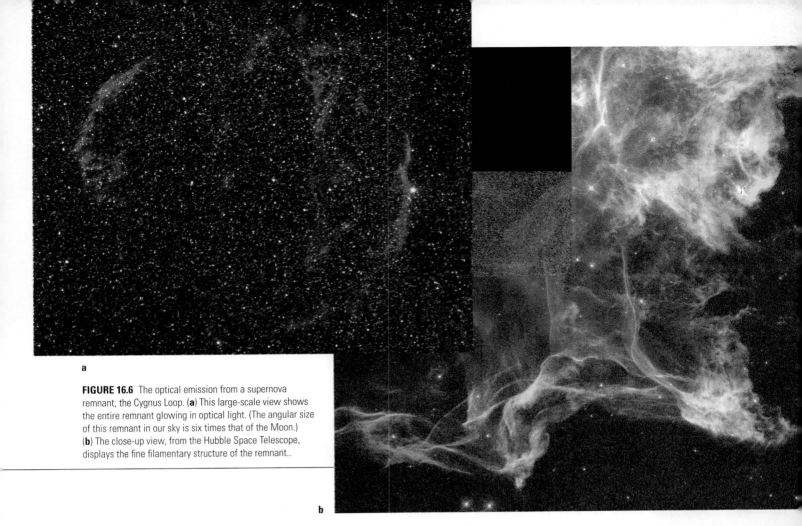

a

FIGURE 16.6 The optical emission from a supernova remnant, the Cygnus Loop. (**a**) This large-scale view shows the entire remnant glowing in optical light. (The angular size of this remnant in our sky is six times that of the Moon.) (**b**) The close-up view, from the Hubble Space Telescope, displays the fine filamentary structure of the remnant..

b

whose shocked gas is hot enough to emit X rays. In contrast, older supernova remnants are cooler because their shock waves have swept up more material and must share their energy among more particles (Figure 16.6). Eventually, the shocked gas radiates away most of its original energy, and the expanding wall of gas slows to subsonic speeds. As the energy dissipates and the gas cools, the supernova's cargo of new elements merges with the surrounding interstellar medium.

Multiple supernovae from a single star cluster can combine forces to create a huge cavity of hot gas over a thousand light-years (300 pc) wide called a **superbubble.** In many places in our galactic disk, we see what appear to be elongated bubbles extending from young clusters of stars to distances of 3,000 light-years (1 kpc) or more above the disk. These probably are places where superbubbles have grown so large that they cannot be contained within the disk of the Milky Way. Once the superbubble breaks out of the disk, where nearly all of the Milky Way's gas resides, nothing remains to slow its expansion except gravity. Such a *blowout* is in some ways similar to a volcanic eruption, but on a galactic scale: Hot plasma erupts from the disk and shoots high into the galactic halo (Figure 16.7).

FIGURE 16.7 This image shows a supercomputer simulation of a superbubble blowout (blue = coldest gas; red = hottest gas). The cool gaseous disk of the simulated galaxy runs along the bottom of the picture. Supernova explosions at the center of this disk have ejected a plume of hot gas that rises high above the disk and into the hot halo.

Atomic Hydrogen Gas

The hot, ionized gas in bubbles and superbubbles is dynamic and widespread, but it is a relatively small fraction of the gas in the Milky Way. Atomic hydrogen gas is much more common and emits a spectral line with a wavelength of 21 cm in the radio portion of the electromagnetic spectrum. We see the radio emission from this **21-cm line** coming from all directions, telling us that about 5 billion solar masses of atomic hydrogen gas is distributed throughout the galactic disk.

Matter remains in the atomic hydrogen stage of the star–gas–star cycle for millions of years. Gravity slowly draws blobs of this gas together into tighter clumps, which radiate energy more efficiently as they grow denser. The blobs therefore cool and contract, forming smaller clouds of cool atomic hydrogen. This process of shrinkage and coagulation takes a much longer time than the other steps in the journey from star death to star birth. The slowness of the transition from warm gas to cool clouds accounts for the large amount of gaseous matter in the atomic hydrogen stage of the star–gas–star cycle.

Remember that, although we speak of clouds of *hydrogen*, all interstellar material actually has a composition of about 70% hydrogen, 28% helium, and 2% heavy elements. Some of the heavy elements in regions of atomic hydrogen are in the form of tiny, solid **dust grains:** flecks of carbon and silicon minerals that resemble particles of smoke and form in the winds of red-giant stars [Section 14.3]. Dust grains remain in the interstellar medium until they are destroyed by a passing shock wave or incorporated into a protostar. Although dust grains make up only about 1% of the mass of the atomic hydrogen clouds, they are responsible for the absorption of visible light that prevents us from seeing through the disk of the galaxy.

Molecular Clouds

As the temperature drops further in the center of a cool cloud of atomic hydrogen, hydrogen atoms com-

FIGURE 16.8 Image of a molecular cloud in the constellation Orion, showing its complex structure. The picture was made by measuring Doppler shifts of emission lines from carbon monoxide molecules in different locations. The colors indicate gas motions: Relative to the cloud as a whole, bluer parts are moving toward us and redder parts are moving away from us.

bine into molecules, making a **molecular cloud.** Molecular clouds are the coldest, densest collections of gas in the interstellar medium. The total mass of molecular clouds in the Milky Way is somewhat uncertain, but it is probably about the same as the total mass of atomic hydrogen gas—about 5 billion solar masses. Throughout much of this molecular gas, temperatures hover only a few degrees above absolute zero, and gas densities are a few hundred molecules per cubic centimeter.

Molecular hydrogen (H_2) is by far the most abundant molecule in molecular clouds, but it is difficult to detect. As a result, most of what we know about molecular clouds comes from observing spectral lines of molecules that make up only a tiny fraction of a cloud's mass. The most abundant of these molecules is carbon monoxide (CO, also a common ingredient in car exhaust), which produces strong emission lines in the radio portion of the spectrum when it is at the 10–30-K temperatures of molecular clouds (Figure 16.8). Among the more familiar molecules in molecular clouds are water (H_2O), ammonia (NH_3), and alcohol (CH_3OH).

Gravitational forces in molecular clouds gather molecules into the compact *cores* that eventually become protostars, and the final stages of star formation can be quite disruptive. Recall that protostars often have violent jets of gas spurting outward (Section 14.2). The turbulence of these jets stirs up nearby regions of the molecular cloud, probably preventing other stars from forming in the vicinity of the growing protostar, at least for a while (Figure 16.9).

Completion of the Cycle

Once a few stars form in a cluster, their radiation begins to erode the surrounding gas in the molecular cloud. Ultraviolet photons from high-mass stars heat and ionize the gas, and winds and radiation pressure push the ionized gas away. This kind of feedback prevents much of the gas in a molecular cloud from turning into stars, sometimes with spectacular results (Figure 16.10).

FIGURE 16.9 Jets from protostars disrupt nearby regions of molecular clouds. This photo shows a jet emanating from the protostar on the left. All along the jet in the center and on the far right of the photo, gas shocked by the jet is glowing in optical light.

FIGURE 16.10 A portion of the Eagle Nebula as seen by the Hubble Space Telescope. The dark, lumpy columns are molecular clouds. Off to the side (outside the range of the picture), newly formed massive stars glow with ultraviolet radiation. This radiation sears the surface of the molecular clouds, destroying molecules and stripping electrons from atoms. As a result, matter "evaporates" from the molecular clouds and joins the hotter ionized gas encircling them.

We have arrived back where we started in the star–gas–star cycle. The most massive stars now forming in the Eagle Nebula will explode within a few million years, filling the region with hot gas and newly formed heavy elements. Farther in the future, this gas will once again cool and coalesce into molecular clouds, forming new stars, new planets, and maybe even new civilizations.

Despite the recycling of matter from one generation of stars to the next, the star–gas–star cycle cannot go on forever. With each new generation, some of the galaxy's gas becomes permanently locked away in brown dwarfs that never return material to space and in stellar corpses left behind when stars die (white dwarfs, neutron stars, and black holes). The interstellar medium therefore is slowly running out of gas, and the rate of star formation will gradually taper off over the next 50 billion years or so. Eventually, star formation will cease.

Putting It All Together: The Distribution of Gas in the Milky Way

As we look at different regions of the galaxy, we see various stages of the star–gas–star cycle playing themselves out. Because the cycle proceeds so slowly compared to a human lifetime, each stage appears to us as a snapshot. We therefore see the interstellar medium in a wide variety of manifestations, ranging from the tenuous million-degree gas of bubbles to the cold, dense gas of molecular clouds. Table 16.1 summarizes the different states in which we see interstellar gas in the galactic disk.

Figure 16.11 shows seven views of the disk of the Milky Way Galaxy. Each view represents a panorama made by photographing the Milky Way's disk in every direction from Earth. You can visualize how one of these views corresponds to the sky by imagining cutting it out, bringing its ends together to form a circu-

lar band, and then lining up the band with the Milky Way on a model of the celestial sphere. Note that each view shows the Milky Way as it appears in a different set of wavelengths of light and thus reveals different features of the galactic disk.

- Figure 16.11a shows variations in the intensity of radio emission from the 21-cm line of atomic hydrogen. Thus, it maps the distribution of atomic hydrogen gas, demonstrating that this gas fills much of the galactic disk.

- Figure 16.11b shows variations in the intensity of radio emission lines from carbon monoxide (CO) and therefore maps the distribution of molecular clouds. Note that these cold, dense clouds are concentrated in a narrow layer near the midplane of the galactic disk.

- Figure 16.11c shows variations in the intensity of infrared emission (at wavelengths of 60–100 micrometers) from interstellar dust grains. Note that the regions of strongest emission from dust correspond to the locations of molecular clouds in Figure 16.11b.

- Figure 16.11d shows infrared light from stars at wavelengths that penetrate clouds of gas and dust (1–4 micrometers). Thus, this image shows how our galaxy would look if there were no dust blocking our view. The galactic bulge is clearly evident at the center.

- Figure 16.11e shows the galactic disk in visible light, just as it appears in the night sky. (Of course, only part of the Milky Way is above the horizon at any one time.) Because visible light cannot penetrate interstellar dust, the dark blotches correspond closely to the bright patches of molecular radio emission and infrared dust emission in Figures 16.11b and c.

Table 16.1 Typical States of Gas in the Interstellar Medium

State of Gas	Primary Constituent	Approximate Temperature	Approximate Density (atoms per cm³)	Description
Hot bubbles	Ionized hydrogen	1,000,000 K	0.01	Pockets of gas heated by supernova shock waves
Warm atomic gas	Atomic hydrogen	10,000 K	1	Fills much of galactic disk
Cool atomic clouds	Atomic hydrogen	100 K	100	Intermediate stage of star–gas–star cycle
Molecular clouds	Molecular hydrogen	30 K	300	Regions of star formation
Molecular cloud cores	Molecular hydrogen	60 K	10,000	Star-forming clouds

a 21-cm radio emission from atomic hydrogen gas.

b Radio emission from carbon monoxide reveals molecular clouds.

c Infrared (60–100 μm) emission from interstellar dust.

d Infrared (1–4 μm) emission from stars that penetrates most interstellar material.

e Visible light emitted by stars is scattered and absorbed by dust.

f X-ray emission from hot gas bubbles (diffuse blobs) and X-ray binaries (pointlike sources).

g Gamma-ray emission from collisions of fast-moving particles with atomic nuclei in interstellar clouds.

FIGURE 16.11 Panoramic views of the Milky Way in different bands of the spectrum. The center of the galaxy, which lies in the direction of the constellation Sagittarius, is in the center of each strip. The rest of each strip shows all other directions in the Milky Way disk as seen from Earth. (Imagine attaching the left and right ends of each strip to form a circular band that corresponds to the 360° band of the Milky Way in our sky.)

■ Figure 16.11f shows the distribution of X-ray light from the galactic disk. The pointlike blotches in this view are mostly X-ray binaries [Section 15.3], but the rest of the X-ray emission comes primarily from hot gas bubbles. Note that the hot gas is less concentrated toward the midplane than the atomic and molecular gas, because it tends to rise into the halo. (The prominent yellow blob on the lower right is the Vela supernova remnant.)

■ Figure 16.11g shows gamma-ray emission from the Milky Way. Most of the gamma-ray emission is produced by collisions between fast-moving particles and atomic nuclei in interstellar clouds. Such collisions happen most frequently where gas densities are highest, so the gamma-ray emission corresponds closely to the locations of molecular and atomic gas. (Gamma rays from the pulsar at the center of the Vela supernova remnant are also prominent on the lower right.)

TIME OUT TO THINK *Carefully compare and contrast the different views of the Milky Way's disk in Figure 16.11. Why do regions that appear dark in some views appear bright in others? What kinds of general patterns do you notice?*

16.3 Galactic Environments

The star–gas–star cycle has operated continuously since the Milky Way's birth, yet new stars are not spread evenly across the galaxy. Some regions seem much more fertile than others. Galactic environments rich in molecular clouds tend to spawn new stars easily, while gas-poor environments do not. A quick tour of some characteristic galactic environments will help you spot where the action is.

Out in the Halo

A census of stars in the Milky Way's *halo* would turn up many senior citizens and very few newborns. Most of the halo stars are old, red, and dim and much smaller in mass than our Sun. Halo stars also contain far fewer heavy elements than our Sun, sometimes having heavy-element proportions as low as 0.02% (in contrast to about 2% in the Sun). The relative lack of chemical enrichment in the halo indicates that its stars formed early in the galaxy's history—before many supernovae had exploded, adding heavy elements to star-forming clouds.

The halo's gas content corroborates this view. The halo is virtually gas-free compared to the disk, with very few detectable molecular clouds. Apparently, the bulk of the Milky Way's gas settled into the disk long ago. The halo environment is a place where lack of gas caused star formation to cease long ago. Now only very old stars still survive, and new stars are rarely born.

TIME OUT TO THINK
How does the halo of our galaxy resemble the distant future fate of the galactic disk? Why?

Our Neighborhood

Our own stellar neighborhood is more active than the halo and typifies much of the galactic disk. Within about 33 light-years (10 pc), we know of over 300 stars. Most are dim, red, spectral type M stars. A few, including Sirius, Vega, Altair, and Fomalhaut, are bright, white stars younger than our Sun. While stars of many different ages and different proportions of heavy elements are in our neighborhood, no very massive, short-lived stars (i.e., spectral type O or B) are present. We therefore infer that stars form periodically in our neighborhood but that no star clusters have formed here recently.

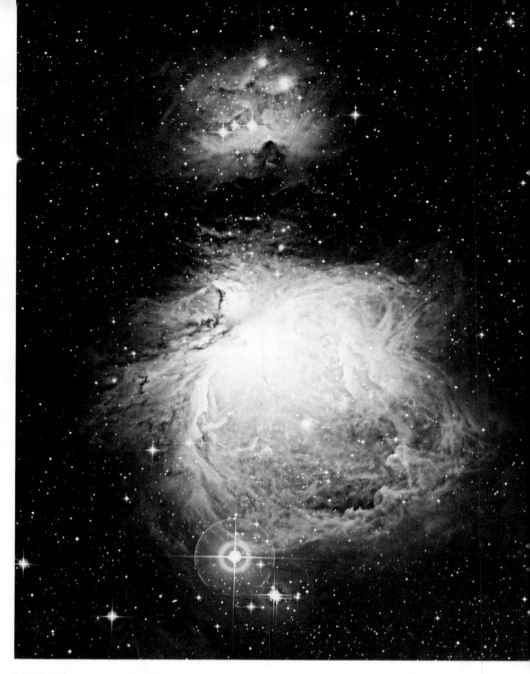

FIGURE 16.12 A photo of the Orion Nebula, an ionization nebula energized by ultraviolet photons from hot stars.

Hot-Star Hangouts

The hot spots in our galaxy are the neighborhoods of high-mass stars. Because hot, massive stars live fast and die young, they never get a chance to move very far from their birthmates. Thus, we find them in star clusters close to the molecular clouds from which they formed. These environments are highly active and extraordinarily picturesque.

We find colorful, wispy blobs of glowing gas known as **ionization nebulae** throughout the galactic disk, particularly in the spiral arms. (Ionization nebulae are sometimes called *emission nebulae* or *H II regions*.) The energy that powers ionization nebulae comes from neighboring hot stars that irradiate them with ultraviolet photons. These photons ionize and excite the atoms in the nebulae, causing them to emit light. The Orion Nebula, about 1,400 light-years away in the "sword" of the constellation Orion, is among the most famous. Few astronomical objects can match its spectacular beauty (Figure 16.12).

Most of the red light in an ionization nebula comes from excited hydrogen atoms, but the blue and black tints in some nebulae have a different origin. Starlight reflected from dust grains produces

FIGURE 16.13 The blue tints in this nebula in the constellation Orion are produced by reflected light. This kind of nebula is called a *reflection nebula*.

FIGURE 16.14 A photo of the Horsehead Nebula and its surroundings.

the blue colors, because interstellar dust grains scatter blue light much more readily than red light (Figure 16.13). (This effect is similar to the scattering of sunlight in our atmosphere that makes the sky blue [Section 8.4].) The black regions of nebulae are dark, dusty gas clouds that block our view of the stars beyond them. Figure 16.14 shows a multicolored nebula characteristic of a hot-star neighborhood.

TIME OUT TO THINK *In Figure 16.14, identify the red ionized regions, the blue reflecting regions, and the dark obscuring regions. Briefly explain the origin of the colors in each region.*

16.4 The Milky Way in Motion

The internal motions of the Milky Way range from the chaos of stars soaring on randomly oriented orbits through the halo to the stately rotation of the galactic disk. These motions, combined with the law of gravity, can tell us how mass is distributed within the Milky Way. When we decipher these movements, we find that the matter binding stars and gas to our galaxy extends far beyond its visible reaches and that the galaxy's spiral arms are propagating waves of new star formation. Let's investigate the motion of the Milky Way in more detail.

Orbits in the Disk and Halo

If you could stand outside the Milky Way and watch it for a few billion years, the disk would resemble a huge merry-go-round. All the stars in the disk, including our Sun, orbit the center in the same direction. The disk stars are spread over a thickness of about 1,000 light-years—which is quite thin in comparison to the 100,000-light-year diameter of the disk (Figure 16.15). (The disk has some thickness because disk stars bob up and down in response to the disk's gravity as they orbit, rather like horses on a merry-go-round.)

The orbits of stars in the halo and bulge are much less organized than the orbits of stars in the disk. Individual bulge and halo stars travel around the galactic center on more or less elliptical paths, but the orientations of these paths are relatively random. Neighboring halo stars can circle the galactic center in opposite directions. They swoop from high above

the disk to far below it and back again, plunging through the disk at velocities so high that the disk's gravity hardly alters their trajectories. Near the Sun, we see several fast-moving halo stars in the midst of hasty excursions through the local region of the disk. One such star is Arcturus, the fourth-brightest star in the night sky.

TIME OUT TO THINK *Is there much danger that the Sun or Earth will someday be hit by a halo star swooping through the disk of the galaxy? Why or why not? (Hint: Think about the typical distances between stars, as illustrated by use of the 1-to-10-billion scale in Chapter 1.)*

The Sun's orbital path around the galaxy is called the **solar circle,** and its radius is our 28,000-light-year distance from the galactic center. By measuring the speeds of globular clusters relative to the Sun, we've determined that the Sun and its neighbors orbit the center of the Milky Way at a speed of about 220 km/s. Even at this speed, one circuit of the Sun around the solar circle takes 230 million years. Early dinosaurs ruled the Earth when it last visited this side of the galaxy.

Orbits and Galactic Mass

Recall that Newton's law of gravity determines how quickly objects orbit one another. This fact, embodied in Newton's version of Kepler's third law, allows us to determine the mass of a relatively large object when we know the period and average distance of a much smaller object in orbit around it [Section 5.4]. A closely related law, which we will call the *orbital velocity law,* allows us to "weigh" the galaxy using the Sun's orbital velocity and its distance from the galactic center. The orbital velocity law is an equation that allows us to calculate the mass of the galaxy within the Sun's orbit by knowing only the Sun's orbital velocity and its distance from the galactic center.

Note that the orbital velocity law tells us only how much mass lies *within* the Sun's orbit. It does not tell us the mass of the entire galaxy, because matter lying outside the Sun's orbit has very little effect on the Sun's orbital velocity. Every part of the galaxy exerts gravitational forces on the Sun as it orbits, but the net force from matter outside the Sun's orbit is relatively small because the pulls from opposite sides of the galaxy virtually cancel one another. In contrast, the net gravitational forces from mass within the Sun's orbit all pull the Sun in the same direction—toward the galactic center. Thus, the Sun's orbital velocity responds almost exclusively to the gravitational pull of matter inside its orbit. Substituting the Sun's 28,000-light-year distance and 220-km/s orbital velocity into the orbital velocity law,

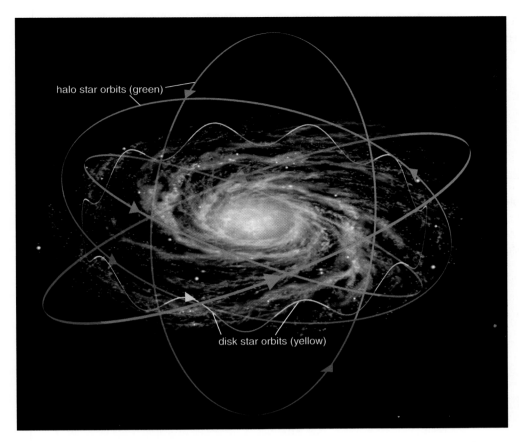

FIGURE 16.15 This painting shows typical orbits of stars in the Milky Way Galaxy. Stars in the disk all orbit the galactic center in the same plane and the same direction. (They also bob slightly up and down in response to the disk's gravity; here these bobs have been exaggerated to make them visible.) In contrast, stars in the halo may orbit in any direction around the galactic center. The overall result is that disk star orbits look very organized while halo star orbits appear random.

we find that the total amount of mass within the solar circle is about 2×10^{41} kilograms, or about 100 billion solar masses.

The Distribution of Mass in the Milky Way

Just as we can use the orbit of the Sun to determine the mass of the galaxy within the solar circle, we can use the orbital motion of any other star to measure the mass of the Milky Way within the star's own orbital circle. In principle, we could determine the complete distribution of mass in the Milky Way by applying the orbital velocity law to the orbits of stars at every different distance from the galactic center. In practice, interstellar dust obscures our view of disk stars beyond a few thousand light-years, making it very difficult to measure stellar velocities. However, radio waves penetrate this dust, so we can see the 21-cm line from atomic hydrogen gas no matter where the gas is located in the galaxy. Such studies allow us to build a map of atomic hydrogen clouds. It is such maps that reveal our galaxy's large-scale spiral structure and also its overall pattern of rotation.

We can get a sense of the distribution of mass in the Milky Way by making a diagram called a **rotation curve,** which plots *rotational velocity* against *distance from the center.* As a simple example of the concept, let's construct a rotation curve for a merry-go-round. Every object on a merry-go-round goes around the center with the same rotational period, but objects farther from the center move in larger circles. Thus, objects farther from the center move at faster speeds, and the rotation curve for a merry-go-round is a straight line that rises steadily outward (Figure 16.16a). In contrast, the rotation curve for our solar system drops off with distance from the Sun because inner planets orbit at faster speeds than outer planets (Figure 16.16b). This drop-off in speed with distance occurs because virtually all the mass of the solar system is concentrated in the Sun. The gravitational force holding a planet in its orbit decreases with distance from the Sun, and a smaller force means a lower orbital speed. The rotation curve of any astronomical system whose mass is concentrated toward the center therefore drops steeply.

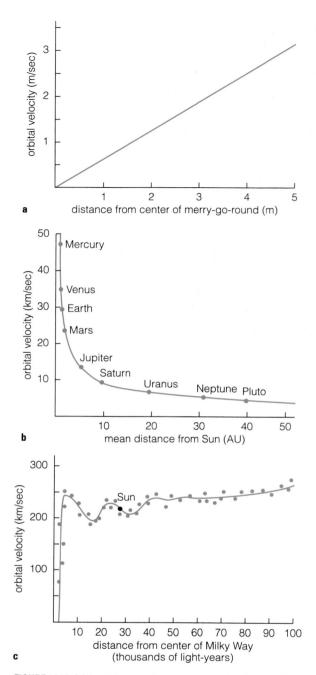

FIGURE 16.16 (**a**) A rotation curve for a merry-go-round. (**b**) The rotation curve for the planets in our solar system. (**c**) The rotation curve for the Milky Way Galaxy.

the solar system. Thus, unlike with the solar system, most of the mass of the Milky Way must *not* be concentrated at its center. Instead, the orbits of progressively more distant hydrogen clouds must encircle more and more mass. The solar circle encompasses about 100 billion solar masses, but a circle twice as large surrounds twice as much mass, and a larger circle surrounds even more mass. Because of the difficulty involved in finding clouds to measure on the outskirts of the galaxy, we have not yet found the "edge" of this mass distribution.

The flatness of the Milky Way's rotation curve came as an immense surprise to astronomers because it implies that most of our galaxy's mass must lie well beyond our Sun, tens of thousands of light-years from the galactic center. In fact, a more detailed analysis suggests that most of this mass is distributed in the galactic halo and that the halo might outweigh all the disk stars *combined* by a factor of 10. However, aside from the light of the relatively small number of halo stars, we have detected very little other radiation coming from this enormous amount of mass. Because we see so little light coming from outside the Sun's orbit, most of the halo's huge mass cannot be in the form of orbiting stars. The nature of this mass remains unknown—which means that we do not yet know the nature of the vast majority of matter in our own galaxy. We call this mysterious mass **dark matter** because it does not emit any light that we have yet detected. The nature of this dark matter is one of the greatest mysteries in astronomy today. We will investigate it in much more depth in Chapter 18.

Spiral Arms

The disks of spiral galaxies like the Milky Way display sweeping spiral arms that can stretch tens of thousands of light-years from their bulges. At first glance, spiral arms look as if they ought to rotate with the stars, like the fins of a giant pinwheel in space. However, because stars near the center of the galaxy take less time to complete an orbit than more distant stars, the arms would gradually wind themselves up if they really did rotate with the stars in the disk. By the time a spiral galaxy was a few billion years old, the spiral arms would look like a tightly wound coil. Because we do not see such tightly coiled spiral arms, the explanation for spiral structure must be more complex.

Detailed images show the spiral arms to be home to most of the young, bright, blue stars in a galaxy (Figure 16.17). Clusters of these hot, short-lived massive stars populate the spiral arms. Between the arms, we find many fewer hot stars. We also see enhanced amounts of molecular and atomic gas in the spiral arms, and streaks of interstellar dust often obscure the inner sides of the arms themselves

The rotation curve for the Milky Way Galaxy is shown in Figure 16.16c. Each individual dot represents the distance from the galactic center and the orbital speed of a particular star or cloud of atomic hydrogen. The curve running through the dots represents a "best fit" to the data. Note that, beyond the inner few thousand light-years, the orbital velocities remain approximately flat. This behavior contrasts sharply with the steeply declining rotation curve of

FIGURE 16.17 NGC 2997, a spiral galaxy with two prominent spiral arms. Note how blue the spiral arms are compared with the yellower tones of the bulge and the regions of the disk between the arms. Because blue stars live only for a few million years, the relative blueness of these spiral arms tells us that stars must be forming more actively here than elsewhere in the galaxy.

(Figure 16.18). All these clues suggest that spiral arms result from waves of star formation that propagate through the disks of spiral galaxies.

Theoretical models suggest that disturbances called **spiral density waves** are the source of our galaxy's spiral arms. These waves are generated by some sort of disturbance such as a close encounter with another galaxy, and they do not carry stars or interstellar gas with them as they move through the disk. The spiral arms themselves are actually waves of enhanced density that move through the disk. Stars and gas at the location of any wave crest are packed slightly more densely than normal, making the wave crest a location of enhanced gravitational attraction within the disk. This gravitational attraction pulls the stars ahead of the crest more closely together; after the crest passes, the stars move apart into looser groups.

Interstellar clouds react even more strongly to the gravitational attraction of the wave crest. Gas clouds collide with one another most frequently in the densest part of the wave, collecting into groups. Because gas clouds crowd so thickly into the crest of the wave, stars form more profusely in spiral arms than elsewhere in the disk. A froth of young stars and ionization nebulae thus marks the crests of spiral density waves.

TIME OUT TO THINK *Perhaps you have noticed what happens when the scene of an accident blocks one side of a highway. Curious drivers on the other side of the highway slow down to check out the accident, and traffic backs up behind them. After the drivers have passed by, they speed back up, and traffic thins out. How is this disturbance in the flow of traffic similar to a spiral density wave?*

Dust lane arises on inner edge of spiral arm where gas clouds crowd together.

Young blue stars are found on outer edge of spiral arm.

Ionization nebulae arise where newly forming blue stars are ionizing gas clouds.

FIGURE 16.18 The relationship of dust, gas, and new star clusters in a spiral arm.

The bluish hue of the spiral arms comes from their prolific star formation. Massive, short-lived, blue stars die out quickly as the wave crest passes by, long before the next spiral wave sweeps through. These luminous hot stars are therefore found close to the spiral arms in which they formed. Longer-lived yellow and red stars live through the passage of many spiral density waves and therefore are distributed relatively evenly throughout the galactic disk.

16.5 The Mysterious Galactic Center

The center of the Milky Way Galaxy lies in the direction of the constellation Sagittarius. This region of the sky does not look particularly special to our unaided eyes; however, if we could remove the interstellar dust that obscures our view, the elliptical shape of the galaxy's central bulge would fill Sagittarius, and its brilliance would be one of the night sky's most spectacular sights.

Because the Milky Way is relatively transparent to long-wavelength radiation, we can use radio and infrared telescopes to peer into its heart (Figure 16.19). Deep in the galactic center, we find swirling clouds of gas and a cluster of several million stars. Bright radio emission traces out the magnetic fields that thread this turbulent region. At the core of it all sits a source of bright radio emission named Sagittarius A* (pronounced "Sagittarius A-star"), or Sgr A* for short, that is quite unlike any other radio source in our galaxy.

The motions of the gas and stars in Sgr A* indicate that it contains a few million solar masses within a region no larger than about 3 light-years (1 pc) across. The star cluster in Sgr A* cannot account for all that mass. Many astronomers therefore suspect that Sgr A* contains a black hole weighing around a million solar masses.

Other astronomers are skeptical. The most plausible black-hole candidates in our galaxy—those in binary systems like Cygnus X-1 [Section 15.4]—accrete matter from their companions. The resulting accretion disks radiate brightly in X rays. If Sgr A* truly is a giant black hole, we might expect it to be accreting some of the surrounding gas from the galactic center. Because of its large mass, it would have a large accretion disk and would shine brightly in X rays. (Many other galaxies have tremendously luminous X-ray sources at their centers that most astronomers believe really are accretion disks circling enormous

FIGURE 16.19 The galactic center at various wavelengths. (**a**) In the optical band, dusty gas obscures our view of the galactic center, marked by the 0.65° green square. (**b**) In this infrared image, roughly the size of the green square in (**a**), we can see through the dust to the star cluster at the center. (**c**) This close-up radio image of the galactic center shows the strange environment of Sgr A*; the bright filaments trace magnetic field lines full of charged particles presumably energized by the central black hole.

black holes [Section 17.6].) These X rays would shine through the dusty gas of our galaxy, and we would be able to detect them easily. However, we see only relatively faint X-ray emission from Sgr A*, casting doubt on the giant-black-hole hypothesis.

It remains possible that Sgr A* contains a huge black hole that simply has run out of gas to accrete.

Even more intriguingly, the black hole might be consuming the high-energy radiation from the accreting matter before this radiation can escape. Nevertheless, Sgr A* does not behave like other black holes whose accretion disks are more apparent. Because the black hole at the center of our galaxy is so well hidden, the nature of Sgr A* remains mysterious.

The infinitude of creation is great enough to make a world, or a Milky Way of worlds, look in comparison with it what a flower or an insect does in comparison with the Earth.

IMMANUEL KANT

THE BIG PICTURE

In this chapter, we have explored our galaxy, focusing on the recycling of gas that continually produces new stars and on the motions of the Milky Way that reveal its mass. When you review this chapter, pay attention to these "big picture" ideas:

■ The inability of visible light to penetrate deeply through interstellar gas and dust concealed the true nature of our galaxy until recent times. Modern astronomical instruments reveal the Milky Way as a dynamic system of stars and gas that continually gives birth to new stars and planetary systems.

■ Stellar winds and explosions make interstellar space a violent place. Hot gas tears through the pervasive atomic hydrogen gas in the galactic disk, leaving expanding pockets of hot plasma and fast-moving clouds in its wake. All this violence might seem quite dangerous, but it performs the great service of mixing essential new elements throughout the Milky Way.

■ We can use the orbital speeds of stars and gas clouds around the center of the galaxy to determine the distribution of mass in the Milky Way. To our great surprise, we've discovered that most of the mass of the galaxy lies in the halo, not in the galactic disk where most of the stars are located. The nature of this *dark matter* is one of the greatest mysteries in astronomy today.

■ Although the elements from which we are made were forged in stars, we could not exist if stars were not organized into galaxies. The Milky Way Galaxy acts as a giant recycling plant, converting gas expelled from each generation of stars into the next and allowing some of the heavy elements to solidify into planets like our own.

Review Questions

1. Draw simple sketches of our galaxy as it would appear face-on and edge-on, identifying the *disk, bulge, halo,* and *spiral arms.* What kind of star clusters are found in the halo?

2. What do we mean by the *interstellar medium?* How does it obscure our view of most of the galaxy? How can we see through it?

3. What do we mean by *heavy elements?* How much of the Milky Way's gas is in the form of heavy elements? Where are heavy elements made?

4. What is the *star–gas–star cycle,* and how does it lead to *chemical enrichment?* Use the idea of chemical enrichment to explain why stars that formed early in the history of the galaxy contain a smaller proportion of heavy elements than stars that formed more recently.

5. Distinguish between the following forms of gas: ionized hydrogen, atomic hydrogen, and molecular hydrogen.

6. What are *bubbles?* How are they made? What kinds of light do they emit?

7. What is a *shock wave?* What does a shock wave do as it sweeps through the interstellar medium?

8. What is a *superbubble?* How is it made? Describe what happens in a blowout in which a superbubble breaks out of the galactic disk.

9. What is the wavelength of the radio emission line characteristic of atomic hydrogen gas?

10. What are *dust grains?* Where do they form?

11. How do *molecular clouds* form? How do we observe them?

12. Will the star–gas–star cycle continue forever? Why or why not?

13. Describe how the panoramic strips in Figure 16.11 correspond to the Milky Way that we see in our night sky. Briefly summarize what we learn from these panoramas in different wavelengths.

14. Describe the stars we find in the *halo* in terms of mass, luminosity, color, and proportion of heavy elements. Why do halo stars have these characteristics?

15. Describe the kinds of stars found in the Sun's neighborhood.

16. What is an *ionization nebula?* Why are such nebulae found near massive young stars? What produces the striking red color of these nebulae? What produces the blue and black tints in some nebulae?

17. Contrast the general patterns of the orbits of stars in the disk and the orbits of stars in the halo.

18. At what speed does the Sun travel around the *solar circle?* About how long does one orbit take? Briefly explain how we can use the characteristics of the Sun's orbit to determine the mass of the Milky Way contained within the solar circle. What is the result?

19. Briefly describe how, in principle, we can use the orbital characteristics of stars at many distances from the galactic center to determine the distribution of mass in the Milky Way. Why, in practice, do we determine the overall pattern of the galaxy's rotation primarily from clouds of atomic hydrogen gas?

20. What is a *rotation curve?* Describe the rotation curve of the Milky Way, and explain how it indicates that most of the Milky Way's mass lies outside the solar circle. What do we mean by *dark matter?*

21. How do we know that spiral arms are not simple "pinwheels" as they appear at first glance? What features do we see in images of spiral arms that suggest that they result from waves of star formation propagating through galaxies?

22. Describe a *spiral density wave.* What happens to gas clouds inside the crest of such a wave? How might such a wave account for the nature of spiral arms?

23. What is Sgr A*? What evidence suggests that it contains a massive black hole?

Discussion Question

1. *Galactic Ecosystem.* This chapter has likened the star–gas–star cycle in our Milky Way to the ecosystem that sustains life on Earth. Here on our planet, water molecules cycle from the sea to the sky to the ground and back to the sea. Our bodies convert atmospheric oxygen molecules into carbon dioxide, and plants convert the carbon dioxide back into oxygen molecules. How are the cycles of matter on Earth similar to the cycles of matter in the galaxy? How do they differ? Do you think the term *ecosystem* is appropriate to discussions of the galaxy?

Problems

1. *Unenriched Stars.* Suppose you discovered a star made purely of hydrogen and helium. How old do you think it would be? Explain your reasoning.

2. *Enrichment of Star Clusters.* The gravitational pull of an isolated globular cluster is rather weak—a single supernova explosion can blow all the interstellar gas out of a globular cluster. How might this fact be related to observations indicating that stars ceased to form in globular clusters long ago? How might it be related to the fact that globular clusters are so deficient in elements heavier than hydrogen and helium? Summarize your answers in one or two paragraphs.

3. *High-Velocity Star.* The average speed of stars *relative* to the Sun in the solar neighborhood is about 20 km/s (i.e., the speed at which we see stars moving toward or away from the Sun—*not* their orbital speed around the galaxy). Suppose you discover a star in the solar neighborhood that is moving relative to the Sun at a much higher speed, say 200 km/s. What kind of orbit does this star probably have around the Milky Way? In what part of the galaxy does it spend most of its time? Explain.

4. *Research: Discovering the Milky Way.* Humans have been looking at the Milky Way since long before recorded history, but only in the past century did we verify the true shape of the galaxy and our location within it. Learn more about how conceptions of the Milky Way developed through history. What names did different cultures give the band of light they saw? What stories did they tell about it? How have ideas about the galaxy changed in the past few centuries? Try to locate diagrams that illustrate these changes. Write a two- to three-page summary of your findings.

5. *Web Project: Images of the Star–Gas–Star Cycle.* Explore the web to find pictures of nebulae and other forms of interstellar gas in different stages of the star–gas–star cycle. Assemble the pictures into a sequence that tells the story of interstellar recycling, with a one-paragraph explanation accompanying each image.

6. *Web Project: The Galactic Center.* Search the Web for recent images of the galactic center, along with information about whether the center hides a massive black hole. Present a two- to three-page report, with pictures, giving an update on current knowledge about the center of the Milky Way Galaxy.

A Universe of Galaxies

17

FAR BEYOND THE MILKY WAY, WE SEE MANY OTHER GALAXIES—SOME similar to our own and some very different—scattered throughout space to the very limits of the observable universe. This fabulous sight inspires some fundamental questions: How far away are the galaxies? How old is the universe? How big is it? Such questions might have seemed ridiculously speculative a century ago. Today, we believe we know the answers to all three questions with respectable accuracy.

Edwin Hubble, the man for whom the Hubble Space Telescope is named, provided the key discovery when he proved conclusively that galaxies exist beyond the Milky Way. The distances he measured revealed an astonishing fact: The more distant a galaxy is, the faster it moves away from us. Hubble's discovery dealt a mortal blow to the traditional belief in a static, eternal, and unchanging universe. The motions of the galaxies away from one another imply instead that the entire universe is expanding and that its age is finite.

In this chapter, we will get acquainted with these islands in space, the galaxies of stars. We will discuss how we know just how far away they are. Because some of these galaxies are billions of light-years away, we are seeing them as they were billions of years ago. Thus, by studying their light, we are gleaning clues about the history of the universe as it is written in the stars and the gas of the distant galaxies.

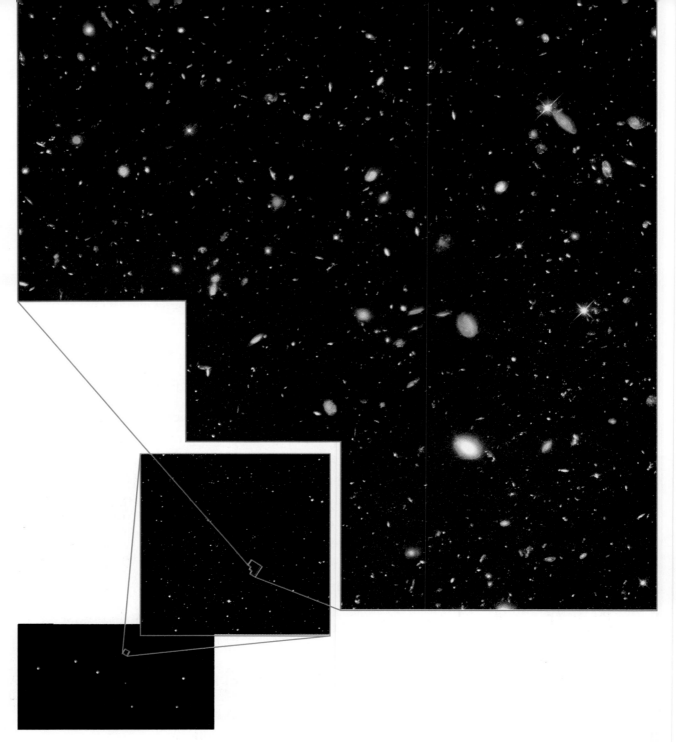

FIGURE 17.1 The Hubble Deep Field, an image composed of 10 days of exposures taken with the Hubble Space Telescope. Some of the galaxies pictured are three-quarters of the way across the observable universe. The field itself is located in the Big Dipper.

17.1 Islands of Stars

Figure 17.1 shows an amazing image of a tiny slice of the sky taken by the Hubble Space Telescope. The telescope pointed toward a single direction in the sky, collecting all the light it could for 10 days. If you held a grain of sand at arm's length, the angular size of the grain would match the angular size of every-thing in this picture. This slice of the sky is jam-packed with galaxies of many sizes, colors, and shapes. Some look large, some small. Some are red-dish, some whitish. Some appear round, and some appear flat. Small galaxies greatly outnumber large ones, yet the large ones produce most of the light in the universe. Counting the galaxies in this slice of the sky and multiplying by the number of such slices it would take to cover the entire sky, we find that the observable universe contains over 80 billion galaxies.

We would like to tell the life stories of these galaxies as completely as we told the life stories of stars (Chapter 14), but we cannot. Too many aspects of **galaxy evolution**—the formation and development of galaxies—remain mysterious. Galaxies are more challenging to study than stars both because they are far more complex and because most of them are so distant. Nevertheless, galaxy evolution is one of the most active research areas in astronomy, and each year brings new discoveries.

In broad terms, the story we seek should tell us how the hydrogen and helium gas in the early universe evolved into the galaxies we see today. Even this general story still presents puzzles, such as why some galaxies turned out flattened and gas-rich like our own while others did not. But we also need to explain some of the stranger aspects of galaxies. For example, some galaxies have extremely bright centers, known as **active galactic nuclei,** that sometimes outshine all the stars in the galaxy combined. In some cases, these active galactic nuclei appear to be firing powerful jets of material millions of light-years into space. We now believe that the "engines" for active galactic nuclei are enormous black holes and that their tremendous power comes from gigantic accretion disks formed by gas spilling into the black holes. However, we do not yet know where these black holes came from or how their accretion disks got started.

In the rest of this chapter, we will discuss what we currently know about galaxies and their origins. We will begin by outlining how astronomers classify galaxies and will go on to investigate how we measure their distances and estimate their ages. Along the way, we will learn how the first measurements of galaxy distances revolutionized the way we now look at the universe—as an expanding realm within which the story of creation is still unfolding. After laying this groundwork, we will examine what is currently known about galaxy evolution and conclude with a discussion of active galactic nuclei and the evidence for giant black holes at the hearts of galaxies.

17.2 Galaxy Types

Astronomers classify galaxies into three major categories. **Spiral galaxies** look like flat, white disks with yellowish bulges at their centers. The disks are filled with cool gas and dust, interspersed with hotter ionized gas as in the Milky Way, and usually display beautiful spiral arms. **Elliptical galaxies** are redder, more rounded, and often longer in one direction than in the other, like a football. Compared with spiral galaxies, elliptical galaxies contain very little cool

gas and dust, though they often contain very hot, ionized gas. Galaxies that appear neither disklike nor rounded are classified as **irregular galaxies.** The sizes of all three types of galaxy span a wide range, from dwarf galaxies containing as few as 100 million (10^8) stars to giant galaxies with more than 1 trillion (10^{12}) stars.

TIME OUT TO THINK *Take a moment and try to classify the larger galaxies in Figure 17.1. How many appear spiral? Elliptical? Irregular? Do the colors of galaxies seem related to their shapes?*

Spiral Galaxies

Like the Milky Way, other spiral galaxies also have a thin disk extending outward from a central bulge (Figure 17.2). The bulge of a spiral galaxy merges smoothly into a halo that can extend to a radius of over 100,000 light-years (30 kpc). Together, the bulge and halo of a spiral galaxy make up its **spheroidal component,** so named because of its rounded

FIGURE 17.2 Spiral galaxies. (**a**) M 83, a face-on spiral with a small bulge. (**b**) NGC 891, an edge-on spiral considered to be very similar to the Milky Way—note the dusty disk.

a

b

FIGURE 17.3 NGC 1365, a barred spiral galaxy.

FIGURE 17.4 NGC 5078, a lenticular galaxy. Most lenticular galaxies have less cool gas than this one, which displays a prominent black streak of interstellar dust.

FIGURE 17.5 An unusually tightly packed group of galaxies called Stephan's Quintet. (The galaxy on the lower left is counted as part of the "quintet," but it is not a true member of the group. Instead it is a foreground galaxy that happens to lie along the line of sight.)

shape. Although no clear boundary divides the two parts of the spheroidal component, astronomers usually consider stars within about 10,000 light-years (3 kpc) of the center to be members of the bulge and those outside this radius to be members of the halo.

The **disk component** of a spiral galaxy slices directly through the halo and bulge. The disk of a large spiral galaxy like the Milky Way extends 50,000 light-years (15 kpc) or more from the center. The disks of all spiral galaxies contain an interstellar medium of gas and dust, but the amounts and proportions of the interstellar medium in molecular, atomic, and ionized forms differ from one spiral galaxy to the next. Spiral galaxies with large bulges generally have less interstellar gas and dust than those with small bulges.

Not all galaxies with disks are standard spiral galaxies. Some spiral galaxies appear to have a straight bar of stars cutting across the center, with spiral arms curling away from the ends of the bar. Such galaxies are known as *barred spiral galaxies* (Figure 17.3). Other galaxies have disks but do not appear to have spiral arms (Figure 17.4). These are called *lenticular galaxies* because they look lens-shaped when seen edge-on (*lenticular* means "lens-shaped"). Because they have disks, lenticular galaxies are somewhat like spiral galaxies without arms. But they might more appropriately be considered an intermediate class between spirals and ellipticals—they tend to have less cool gas than normal spirals, but more than ellipticals.

About 75–85% of large galaxies in the universe are spiral or lenticular. Spiral galaxies are often found in loose collections of several galaxies, called **groups,** that extend over a few million light-years (Figure 17.5). Our Local Group is one example, with two large spirals: the Milky Way and the Great

FIGURE 17.6 A portion of the Virgo Cluster of galaxies, the nearest large galaxy cluster to the Milky Way.

Galaxy in Andromeda (M31) [Section 1.1]. Lenticular galaxies are particularly common in **clusters** of galaxies, which can contain hundreds and sometimes thousands of galaxies extending over more than 10 million light-years (Figure 17.6).

Elliptical Galaxies

The major difference between elliptical and spiral galaxies is that ellipticals lack a significant disk component (Figure 17.7). Thus, an elliptical galaxy has only a spheroidal component and looks much like the bulge and halo of a spiral galaxy. (In fact, elliptical galaxies are sometimes called *spheroidal galaxies*.) Most of the interstellar medium in large elliptical galaxies consists of low-density, hot, X ray–emitting gas—rather like the gas in bubbles and superbubbles in the Milky Way [Section 16.2]. Elliptical galaxies usually contain very little dust or cool gas, although they are not completely devoid of either. Some have cold, relatively small gaseous disks rotating at their centers that might be the remnants of a collision with a spiral galaxy.

Elliptical galaxies appear to be more social than spiral galaxies: They are much more common in large clusters of galaxies than outside clusters. Elliptical galaxies make up about half the large galaxies in the cores of clusters, while they represent only about 15% of the large galaxies found outside clusters. Ellipticals are also more common among small galaxies. Small elliptical galaxies with less than a billion stars, called **dwarf elliptical galaxies,** are often found near larger spiral galaxies. At least 10 dwarf elliptical galaxies belong to the Local Group.

FIGURE 17.7 Elliptical galaxies. (**a**) M 87, a giant elliptical galaxy in the Virgo Cluster. (**b**) Leo I, a dwarf elliptical galaxy in the Local Group.

a

b

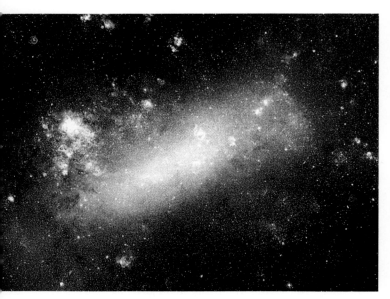

FIGURE 17.8 The Large Magellanic Cloud, a small irregular galaxy that orbits the Milky Way.

Classifying galaxies helps us identify the patterns underlying their enormous variety, but classification alone tells us little about the size and luminosity of a galaxy. No equivalent to the Hertzsprung–Russell diagram exists for galaxies—they are much more complex than stars. Thus, in order to know how large and luminous a particular galaxy is, we must first measure its distance.

17.3 Measuring Cosmic Distances

Measuring cosmic distances is one of the most fundamental and challenging tasks we face when trying to understand galaxies and the universe as a whole. Our determinations of astronomical distances depend on a chain of methods that begins with knowing distances in our own solar system. At each link in this chain, we establish standards that help us forge the next link. Because each link depends on preceding links, any inaccuracies accumulate as we move up the chain. Nevertheless, we now can measure the distances to the farthest galaxies to an accuracy within better than 20%. The distance chain begins with radar measurements of the solar system's size. The second link involves parallax measurements of the distances to nearby stars [Section 13.2]. We will now follow the rest of this chain, link by link, to the outermost reaches of the observable universe.

Irregular Galaxies

A small percentage of the large galaxies we see nearby fall into neither of the two major categories. This irregular class of galaxies is a miscellaneous class, encompassing small galaxies such as the Magellanic Clouds and "peculiar" galaxies that appear to be in disarray (Figure 17.8). These blobby star systems are usually white and dusty, like the disks of spirals. Telescopic observations probing deep into the universe show that distant galaxies are more likely to be irregular in shape than those nearby. Because the light of more distant galaxies was emitted longer ago in the past, these observations tell us that irregular galaxies were more common when the universe was younger.

Standard Candles

We can always measure an object's apparent brightness, so we can use the luminosity–distance formula to calculate luminosity if we know the distance, or to calculate distance if we know the luminosity [Section 13.2]. For example, suppose we see a distant street

---

THINKING ABOUT . . .

Hubble's Galaxy Classes

Edwin Hubble invented a system for classifying galaxies that remains widely used. It assigns the letter *E*, followed by a number, to each elliptical galaxy. The larger the number, the flatter the galaxy. An E0 galaxy is round, and an E7 galaxy is highly elongated. A spiral galaxy is assigned an uppercase *S*, or *SB* if it has a bar, and a lowercase *a*, *b*, or *c*. The lowercase letter indicates the size of the bulge and the dustiness of the disk. An Sa galaxy has a large bulge and a modest amount of dusty gas; an SBc galaxy has a bar, a small bulge, and lots of dusty gas. The lenticular galaxies are designated S0, signifying their inter-

mediate spot between spirals and ellipticals. Irregular galaxies are designated Irr.

Astronomers had once hoped that the classification of galaxies might yield deep insights, just as the classification of stars did in the early twentieth century. The Hubble classification scheme itself was suspected for a time to be an evolutionary sequence in which galaxies flattened and spread out as they aged. Unfortunately for astronomers, galaxies turn out to be far more complex than stars, and classification schemes like this one have not led to easy answers about the nature of galaxies.

lamp and know that all street lamps of its type put out 1,000 watts of light. Then, once we measure the apparent brightness of the street lamp (in units of watts/m²), we can calculate its distance with the luminosity–distance formula.

An object such as a street lamp, for which we are likely to know the true luminosity, represents what astronomers call a **standard candle.** The term *standard candle* is meant to suggest a light source of a known, standard luminosity. Unlike light bulbs, astronomical objects do not come marked with wattage, so an astronomical object can serve as a standard candle only if we have some way of knowing its true luminosity without first measuring its apparent brightness and distance. Fortunately, many astronomical objects meet this requirement. For example, any star that is a twin of our Sun—that is, a main-sequence star with spectral type G2—should have about the same luminosity as the Sun. Thus, if we measure the apparent brightness of a Sun-like star, we can use the luminosity–distance formula to calculate its distance by assuming that it has the same luminosity we have measured for the Sun (3.8×10^{26} watts).

There is always some uncertainty in distances calculated by applying the luminosity–distance formula, because no astronomical object is a perfect standard candle. The challenge of measuring great astronomical distances comes down to the challenge of finding the objects that make the best standard candles. The more confidently we know an object's true luminosity, the less uncertain the distance we calculate with the luminosity–distance formula.

Main-Sequence Fitting

We can use Sun-like stars as standard candles because we know that they are similar to the Sun and because we can measure the Sun's luminosity quite easily. However, Sun-like stars are relatively dim, and we cannot detect them at great distances. To measure distances beyond 1,000 light-years or so, we need brighter standard candles. An obvious first choice is to use brighter main-sequence stars. However, before we can use any main-sequence star as a standard candle, we must first have some way of knowing its true luminosity. The key to understanding this process is to remember that we can use the luminosity–distance formula in two ways:

1. First, we identify a star cluster that is close enough for us to determine its distance by parallax and plot its H–R diagram. Because we know the distances to the cluster stars, we can use the luminosity–distance formula to establish their true luminosities from their apparent brightnesses.

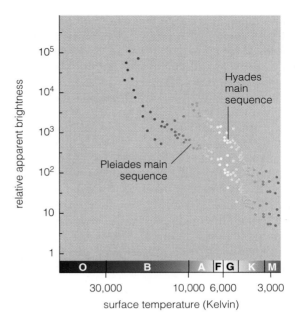

FIGURE 17.9 Comparisons of the apparent brightnesses of stars in the Hyades Cluster with the apparent brightnesses of those in the Pleiades Cluster show that the Pleiades are 2.7 times farther away because they are $2.7^2 = 7.3$ times dimmer.

2. Then we can look at stars in other clusters that are too far away for parallax measurements and measure their apparent brightnesses. If we assume that main-sequence stars in other clusters have the same true luminosities as their counterparts in the nearby cluster, we can calculate their distances with the luminosity–distance formula.

The nearest star cluster with a well-populated main sequence, the *Hyades Cluster* in the constellation Taurus, has long been crucial to this technique. We can measure the true distance of the Hyades Cluster through its parallax. Knowing this distance allows us to determine the true luminosities of all its stars with the luminosity–distance formula. We can find the distances to other star clusters by comparing the apparent brightnesses of their main-sequence stars with those in the Hyades Cluster and assuming that all main-sequence stars of the same color have the same luminosity (Figure 17.9). This technique of determining distances by comparing main-sequence stars in different star clusters is called **main-sequence fitting.**

Cepheid Variables

Main-sequence fitting works well for measuring distances to star clusters throughout the Milky Way, but not for measuring distances to other galaxies. Most main-sequence stars are too faint to be seen in other galaxies, even with our largest telescopes. Instead, we need very bright stars to serve as standard candles for distance measurements beyond the Milky Way, and the most useful bright stars are the Cepheid variables [Section 13.5]. Recall that Cepheids are pulsating variable stars that follow a simple period–luminosity relation: The longer the time period between peaks in brightness, the greater the luminosity of the Cepheid variable star (Figure 17.10). All Cepheids that vary with a particular period have very nearly the same luminosity (within about 10%), so Cepheids effectively scream out their true luminosities when we measure their periods. "Hey, everybody, my luminosity is 10,000 solar luminosities!" is how we would translate the message of a Cepheid variable whose brightness peaks every 30 days. Thus, Cepheid variables are the primary standard candles used to determine distances to nearby galaxies.

Cepheids played a key role in the discovery that the Milky Way is only one of billions of galaxies in the universe. Before the 1920s we had no way of measuring the distances to other galaxies, and many astronomers suspected that these swirling collections of stars were located within the Milky Way. In 1924, Edwin Hubble used the 100-inch telescope atop southern California's Mount Wilson (Figure 17.11)—the largest telescope in the world at the time—to discover Cepheid variables in the Andromeda Galaxy by comparing photographs of the galaxy taken days apart. He used the period–luminosity relation to determine the luminosities of these stars and then used these luminosities in the luminosity–distance formula to compute their distances. His

FIGURE 17.11 Edwin Hubble at the Mount Wilson Observatory.

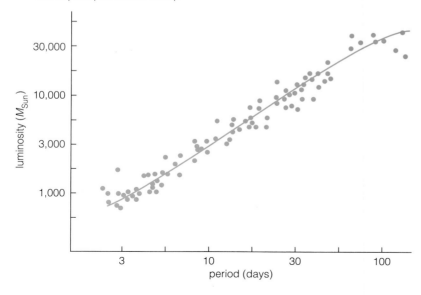

FIGURE 17.10 Cepheid period–luminosity relation. Note that all Cepheids of a particular period have very nearly the same luminosity.

distance measurements proved that the Andromeda Galaxy sat far beyond the outer reaches of stars in the Milky Way, demonstrating that it is a separate galaxy. This single stroke of scientific discovery dramatically changed our view of the universe. Rather than inhabiting a universe that ended with the Milky Way, we suddenly knew that vast numbers of galaxies lay beyond the Milky Way's borders.

Hubble's Law

Hubble's determination of the distance to the Andromeda Galaxy assured him a permanent place in the history of astronomy, but he didn't stop there. Hubble proceeded to measure the distances to many more galaxies. Within just a few years, Hubble made one of the most astonishing discoveries in the history of science: that the universe is expanding.

Astronomers had known since the 1910s that the spectra of most of the faint smudges on the sky known as the "spiral nebulae" tended to be redshifted. In other words, each of the emission and absorption lines in their spectra appeared at longer (redder) wavelengths than expected. Recall that redshifts occur when the object emitting the radiation is moving away from us [Section 6.4]. But, since Hubble had not yet proved that the "spiral nebulae" were distant galaxies, no one understood the true significance of their motions.

In 1929, after measuring distances to a number of "spiral nebulae," Hubble announced his conclusion: The more distant a galaxy, the greater its redshift and hence the faster it is moving away from us. Hubble's original assertion was based on an amazingly small sample of galaxies. Even more incredibly, he had grossly underestimated the luminosities of his standard candles. The "brightest stars" he had been using as standard candles were really entire clusters of bright stars. Fortunately, Hubble was both bold and lucky. Subsequent studies of much larger samples of galaxies showed that they are indeed receding from us, but they are even farther away than Hubble thought.

We can express the idea that more distant galaxies move away from us faster with a very simple formula, now known as **Hubble's law:**

$$v = H_0 \times d$$

where v stands for velocity (sometimes called a *recession velocity*), d stands for distance, and H_0 (pronounced "H-naught") is a number called **Hubble's constant.** Astronomers generally quote the Hubble constant in strange-sounding units of *kilometers per second per megaparsec* (km/s/Mpc). The easiest way to explain these units is by example. Suppose Hubble's constant has a value of $H_0 = 100$ km/s/Mpc. Then, according to Hubble's law, we would expect a galaxy located at distance $d = 10$ Mpc (32.5 million light-years) to be moving away from us at a speed of:

$$v = H_0 \times d = 100 \; \frac{km/s}{Mpc} \times 10 \; Mpc = 1,000 \; km/s$$

Dividing both sides of Hubble's law by H_0 puts it into a form in which we can use a galaxy's velocity to determine its distance:

$$d = v / H_0$$

In principle, applying this law is one of the best ways to determine distances to galaxies. We find the velocity of a distant galaxy from its redshift; then we divide this velocity by H_0 to find its distance. However,

we encounter two major practical difficulties in finding distances in this way:

1. Galaxies do not obey Hubble's law perfectly because they can have velocities relative to one another caused by gravitational tugs that Hubble's law does not take into account.

2. Even when galaxies do obey Hubble's law well, the distances we find with it are only as accurate as our best measurement of Hubble's constant.

The first problem makes Hubble's law difficult to use for nearby galaxies. Within the Local Group, for example, Hubble's law does not work at all: The galaxies in the Local Group are gravitationally bound together with the Milky Way and therefore are not moving away from us in accord with Hubble's law. However, Hubble's law works fairly well for more distant galaxies: The recession speeds of galaxies at large distances are so great that any motions caused by the gravitational tugs of neighboring galaxies are tiny in comparison.

The second problem means that, even for distant galaxies, we can know only relative distances until we pin down the true value of H_0. For example, Hubble's law tells us that a galaxy moving away from us at 20,000 km/s is twice as far away as one moving at 10,000 km/s, but we can determine the actual distances of the two galaxies only if we know H_0. Because of its importance in measuring cosmic distances, the quest to measure H_0 accurately has been one of the main missions of the Hubble Space Telescope.

TIME OUT TO THINK *Suppose a galaxy is moving away from us at 10,000 km/s. Use Hubble's law in the form $d = v/H_0$ to calculate its distance (in Mpc) if $H_0 = 100$ km/s/Mpc. What is its distance if $H_0 = 50$ km/s/Mpc?*

Seeking Distant Standards

The process of measuring H_0 is rather like the process of calibrating a scale. We can calibrate a scale by, for example, making sure that it reads 10 pounds when we place a 10-pound weight on it, 20 pounds when we place a 20-pound weight on it, and so on. Once it is properly calibrated, we can use it to weigh objects whose weights are not known in advance. In a similar way, astronomers calibrate Hubble's law by making sure that it gives correct distances for galaxies whose distances we already know from applying some other method. The superior resolving power of the Hubble Space Telescope enables us to measure the periods and apparent brightnesses of Cepheid variable stars in galaxies up

a

b

FIGURE 17.12 Hubble Space Telescope observations of (**a**) the center of galaxy M 100 in the Virgo Cluster and (**b**) one of its Cepheid variable stars varying in brightness over several weeks.

April 23   May 4   May 9

May 16   May 20   May 31

to 100 million light-years (30 Mpc) away, but this is not quite far enough (Figure 17.12). At this distance, random gravitational tugs can still cause noticeable deviations from Hubble's law. Nevertheless, these Cepheids can help us by enabling us to calibrate even brighter standard candles, such as supernovae.

White-dwarf supernovae are thought to be exploding white-dwarf stars that have reached the 1.4-solar-mass limit and should all have nearly the same luminosity [Section 15.2]. Thus, white-dwarf supernovae should be good standard candles. We can use Cepheids to find the distances to galaxies in which white-dwarf supernovae have occurred, and once we know the galaxy's distance, the luminosity–distance formula provides the supernova's true luminosity. This technique has allowed researchers using the Hubble Space Telescope to determine the true luminosities of several historical white-dwarf supernovae. As expected, they are all about the same, and we can now use them as reliable standard candles. Because white-dwarf supernovae are so bright—about 10 billion solar luminosities at their peak—we can use them to measure distances to galaxies billions of light-years away (Figure 17.13).

Galaxies themselves can also be used as standard candles. Astronomers studying spiral galaxies have discovered a close relationship between their total luminosities and the rotation speeds of their disks: The faster a spiral galaxy's rotation speed, the more luminous it is (Figure 17.14). This relationship, called the **Tully–Fisher relation** (after its discover-

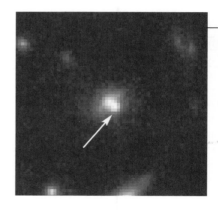

FIGURE 17.13 Photo of a white-dwarf supernova in a galaxy halfway across the observable universe, recorded in March 1996 by the Hubble Space Telescope.

ers), holds because both luminosity and rotation speed depend on the galaxy's mass. A galaxy's luminosity depends on the number of stars it contains, which is related to the total amount of matter within it, and the total amount of matter determines a galaxy's rotation speed [Section 16.4]. Once we have measured the rotation speed of a spiral galaxy, the Tully–Fisher relation tells us its true luminosity, which makes the galaxy itself a standard candle that we can use to determine its distance.

Despite the new calibrations of Hubble's law made possible by the Hubble Space Telescope, some uncertainties remain. These uncertainties should decrease as we measure accurate distances to more and more galaxies. As of 1999, the true value of H_0 appears to be somewhere between 55 and 75 km/s/Mpc.

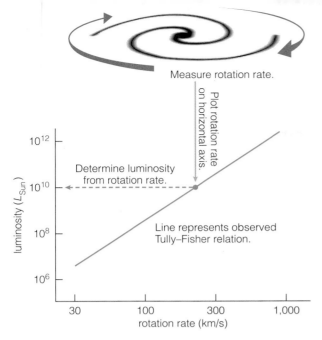

FIGURE 17.14 A schematic diagram of the Tully–Fisher relation. (The precise relation differs among different subclasses of spiral galaxies.)

Summary: The Distance Chain

Distance measurements are fundamental to our understanding of galaxies and the universe. Here we briefly summarize the chain of measurements that allows us to determine ever-greater distances (Figure 17.15). Note that with each link in the distance chain uncertainties become somewhat greater. Thus, although we know the Earth–Sun distance at the base of the chain extremely accurately, distances to the farthest reaches of the universe remain uncertain by about 20%.

- **Radar ranging:** We measure distances within the solar system by bouncing radio waves off planets. Then, with the aid of some geometry, we use these radar measurements to determine the Earth–Sun distance.

- **Parallax:** We measure the distances to nearby stars by observing how their positions change relative to the background stars as the Earth moves around the Sun. These distances thus rely on our knowledge of the Earth–Sun distance.

- **Main-sequence fitting:** We know the distance to the Hyades Cluster in our Milky Way Galaxy through parallax. Comparing the apparent brightness of its main-sequence stars to the brightness of those in other clusters gives us the distances to these other star clusters in our galaxy.

- **Cepheid variables:** By studying Cepheids in star clusters with distances measured by main-sequence fitting, we learn the precise period–luminosity relation for Cepheids. When we find a Cepheid in a more distant star cluster or galaxy, we can determine its true luminosity by measuring the period between its peaks in brightness and then use this true luminosity to determine the distance.

- **Distant standards:** By measuring distances to relatively nearby galaxies with Cepheids, we learn the true luminosities of white-dwarf supernovae and the Tully–Fisher relation between the luminosities and rotation speeds of spiral galaxies. Thus, supernovae and entire galaxies become useful standard candles for measuring great distances in the universe.

- **Hubble's law:** Distances measured to galaxies with white-dwarf supernovae and the Tully–Fisher relation allow us to calibrate Hubble's law and measure Hubble's constant, H_0. Once we know H_0, we can determine a galaxy's distance directly from its redshift.

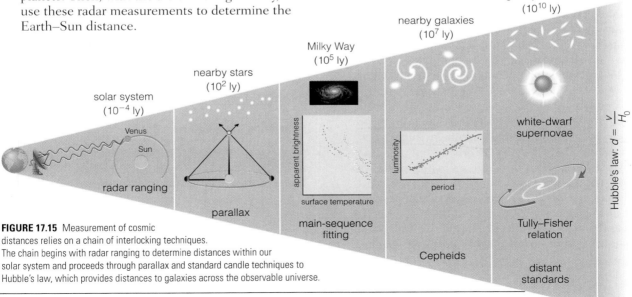

FIGURE 17.15 Measurement of cosmic distances relies on a chain of interlocking techniques. The chain begins with radar ranging to determine distances within our solar system and proceeds through parallax and standard candle techniques to Hubble's law, which provides distances to galaxies across the observable universe.

17.4 Measuring Cosmic Ages

When we look deep into space, as in Figure 17.1, we are also looking back through time. Remember that light travels at a finite speed. Light from the most distant galaxies we can see began its journey to Earth when the universe was a small fraction of its current age. We are therefore viewing these most distant galaxies as they were when they were young, but the light we are detecting is ancient. Because a deep image like this records galaxies at many different distances, it contains information about galaxies of many different ages. For example, we see galaxies 1 billion light-years away as they were 1 billion years ago and galaxies 2 billion light-years away as they were 2 billion years ago. The history of galaxies across the universe is therefore encoded in such images, but to decipher the clues they hold we need to know how the universe's history is related to its expansion.

Universal Expansion

The fact that galaxies all across the universe are moving away from one another in accordance with Hubble's law implies that galaxies must have been closer together in the past. Tracing this convergence back in time, we reason that all the matter in the observable universe started very close together and that the entire universe came into being at a single moment. Ever since that moment, the universe has been expanding, and all the while the gravitational pull of each galaxy on every other galaxy has been working to slow the expansion rate. In the densest regions of the universe, gravity has won out over expansion. The Local Group no longer expands, nor do other clusters of galaxies, but the expansion continues on all larger scales.

When people first hear about the expansion of the universe, they are frequently tempted to think of the universe as a ball of galaxies expanding into a void. This impression is mistaken. The universe expands, but it is not expanding into anything. Each point moves away from every other point, and the universe looks about the same no matter where you stand. (In this book, whenever we speak about the universe as a whole, we are assuming that matter in the universe is evenly distributed on scales much larger than superclusters of galaxies. This assumption is known as the *Cosmological Principle*. It is impossi-

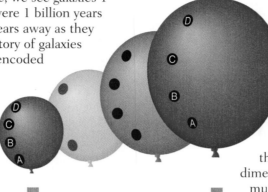

FIGURE 17.16 As the balloon expands, dots uniformly move apart.

ble to prove that the Cosmological Principle holds outside the observable regions of the universe, but within the observable universe matter does appear to be distributed more evenly as we look at increasingly larger scales [Section 18.5].)

Back in Chapter 1, we likened the expanding universe to a raisin cake rising as it bakes [Section 1.3], but a cake has edges and a center. A better analogy would be something that can expand but that has no center and no edges—like the surface of a balloon. Just as you'll never encounter an edge if you sail around the world, there is no edge to a balloon's surface. And, while a balloon has a center inside it, the center is not part of the surface. Thus, like the universe, the surface of a balloon has no center and no edges. The only significant problem with this analogy is that, because we want to represent the entire universe with the two-dimensional surface of a balloon, we must imagine flattening the universe into only two dimensions. We attach plastic polka dots to the surface of the balloon to represent flattened clusters of galaxies and blow into the balloon to represent the expansion of the universe (Figure 17.16).

As the real universe expands, the distances between clusters of galaxies grow. However, the clusters themselves do not grow because gravity binds each cluster's galaxies together. In a similar way, as the surface of the balloon expands, the polka dots move apart, but the dots themselves don't grow. Suppose that, 1 second after you begin blowing into the balloon, dots A, B, C, and D are spaced 1 cm apart along a line. After 2 seconds, the distances between neighboring dots grow to 2 cm, and after 3 seconds to 3 cm. If you recorded the distances of the other dots from Dot B once each second, your observations would look like this:

Measuring Distances from Dot B

Time	Dot A Distance	Dot B Distance	Dot C Distance	Dot D Distance
1 s	1 cm	0 cm	1 cm	2 cm
2 s	2 cm	0 cm	2 cm	4 cm
3 s	3 cm	0 cm	3 cm	6 cm

Miniature scientists on Dot B might be tempted to think they are at the center of some explosion. After all, they see every other dot moving away from them, and the farthest dots move fastest. However, miniature scientists on Dot C would record a very similar set of observations:

Measuring Distances from Dot C

Time	Dot A Distance	Dot B Distance	Dot C Distance	Dot D Distance
1 s	2 cm	1 cm	0 cm	1 cm
2 s	4 cm	2 cm	0 cm	2 cm
3 s	6 cm	3 cm	0 cm	3 cm

The miniature scientists on Dot C would also be tempted to think they are at the center of some explosion. Eventually, the miniature scientists would realize that the balloon itself is expanding. Just as we cannot point to a single spot on a balloon's surface and say "the balloon is expanding from here," we cannot identify a single point in space from which the universe is expanding. The universe has no center. Likewise, neither the balloon nor the universe has any edge.

The balloon analogy, while helpful, has its limitations. For example, the balloon has a finite total surface area, but the universe might be infinite in extent. In addition, you might be tempted to say that the balloon is expanding away from a point inside the balloon or that the surface of the balloon has an inside edge and an outside edge. We notice these features of the balloon because the two-dimensional surface of the balloon exists within a three-dimensional space. But the third dimension that we perceive so easily would make no sense to a two-dimensional miniature scientist restricted to the surface of the balloon.

Hubble Constant and Age

Let's return for a moment to the miniature scientists living on Dot B of the expanding balloon. If you look back at the table showing distances from Dot B, you'll see that the distance to Dot C grows by 1 cm every second. Thus, the scientists on Dot B will measure Dot C to be moving away from them at a speed of 1 cm/s. Similarly, they will see the distance to Dot D growing by 2 cm every second, so Dot D is moving away from them at a speed of 2 cm/s. If they made their measurements 3 seconds after the balloon began expanding, when Dot C was 3 cm away and Dot D was 6 cm away, they would conclude that the

following formula relates the distances and velocities of other dots:

$$v \, (\text{cm/s}) = \tfrac{1}{3} \, \text{cm/s/cm} \times d \, (\text{cm})$$

where v and d are the velocity and distance, respectively, of any dot.

TIME OUT TO THINK *Confirm that this formula gives the correct values for the speeds of dots C and D, as seen from Dot B, 3 seconds after the balloon began expanding. How fast would a dot located 9 cm from Dot B move, according to the scientists on Dot B?*

If the miniature scientists think of their balloon as a bubble, they might call the number relating distance to velocity—the term $\tfrac{1}{3}$ cm/s/cm in the above formula—the "bubble constant." An especially insightful miniature scientist might flip over the "bubble constant" and find that it was exactly equal to the time since the balloon started expanding. That is, the "bubble constant" of $\tfrac{1}{3}$ cm/s/cm tells them that the balloon has been expanding for 3 seconds. Perhaps you see where we are heading.

Just as the inverse of the "bubble constant" tells the miniature scientists that their balloon has been expanding for 3 seconds, the inverse of the Hubble constant tells us how long the universe has been expanding. Current estimates that place the value of Hubble's constant between 55 and 75 km/s/Mpc would put the age of the universe between about 12 and 18 billion years. Intriguingly, the oldest stars in globular clusters appear to be somewhere between 12 and 16 billion years old [Section 13.6]. These stars could not possibly have existed before the birth of the universe! The best available evidence as of 1999 suggests that the universe is somewhere between 11 and 16 billion years old. If you are reading this book at a much later date, it is likely that the age of the universe is known within more tightly constrained limits.

Lookback Time

All the galaxies we observe must be younger than the universe itself. Pictures of the farthest galaxies show us how galaxies looked when they were only a few billion years old; pictures of the closest show galaxies nearly as old as the universe. However, when we try to specify exactly how the distances of galaxies relate to their ages, we run into complications.

Imagine photons of light from a supernova in a distant galaxy, traveling toward us at the speed of light. If the supernova occurred in a galaxy 400 million light-years away, we are seeing a supernova that occurred 400 million years ago. But what do we mean by the distance of 400 million light-years? Because the universe is expanding, the distance between Earth and the supernova is greater today than it was at the time of the supernova event. Are we talking about the distance from the supernova to Earth when the supernova exploded, or are we referring to the distance between the supernova and Earth when the light arrived here? Do we mean that photons traveled through 400 million light-years of space?

Because the distances between galaxies are always changing, it is easier to speak about faraway galaxies in terms of the time it has taken for their light to reach us—400 million years in the case of the galaxy with the supernova. We call this time the **lookback time** to the supernova. In other words, a distant object's lookback time is the difference between the current age of the universe and the age of the universe when the light left the object. This quantity is much more meaningful than an actual distance when we discuss very distant objects that are moving away from us with the expansion of the universe.

FIGURE 17.17 As the universe expands, photon wavelengths stretch like the wavy lines on this expanding balloon.

TIME OUT TO THINK *Explain why there is no question about meaning when we talk about distances to objects within the Local Group, but distances become difficult to define when we talk about objects much farther away.*

An object's lookback time is directly related to its redshift. We have seen that the redshifts of galaxies tell us how quickly they are moving away from us. In the context of an expanding universe, redshifts have an additional, more fundamental interpretation. Let's return one final time to the balloon analogy. Suppose you drew wavy lines on the balloon's surface to represent light waves. As the balloon inflates, these wavy lines stretch out, and their wavelengths increase (Figure 17.17). This stretching closely resembles what happens to photons in an expanding universe. The expansion of the universe stretches out all the photons within it, shifting them to longer, redder wavelengths. We call this effect a **cosmological redshift**.

In a sense, we have a choice when we interpret the redshift of a distant galaxy: We can think of the redshift as being caused either by the Doppler effect as the galaxy moves away from us or by a photon-stretching, cosmological redshift. However, as we look to very distant galaxies, the ambiguity in the meaning of distance also makes it difficult to specify precisely what we mean by a galaxy's speed. Thus, it becomes preferable to interpret the redshift as being due to photon stretching in an expanding universe. From this perspective, it is better to think of space itself as expanding, carrying the galaxies along for the ride, than to think of the galaxies as projectiles flying through a static universe. The cosmological redshift of a galaxy thus tells us how much space has expanded during the lookback time to that galaxy.

Capitalizing on this relationship between redshift and lookback time, astronomers can use the redshifts of galaxies to assemble "family albums" showing galaxies of many different ages (Figure 17.18). The photos in such an album are pictures of galaxies with lookback times ranging from a few million light-years to billions of light-years. Each picture shows one galaxy at a single stage in its life, and we can arrange these pictures to make a family album for galaxies of each type. Subtracting a galaxy's lookback time from the age of the universe tells us about how old the universe was at the time the light left the galaxy. We cannot state the precise ages of these galaxies because we do not know exactly when they formed. However, galaxies whose light reaches us from when the universe was less than 3 billion years old are certainly in their childhood, and those whose light reaches us from when the universe was more than 6 billion years old are considerably more mature.

TIME OUT TO THINK *You may notice that we use the term* today *in a very broad sense. For example, if we look at a relatively nearby galaxy—one located, say, 20 million light-years away—we see it as it was 20 million years ago. In what sense is this "today"? (Hint: How does 20 million years compare to the age of the universe?)*

The Horizon of the Universe

What lies beyond the youngest, most distant galaxies? When we began our discussion of the expanding universe, we stressed that the universe does not have an edge. Yet the universe does have a horizon, a place beyond which we cannot see. This **cosmological horizon** is a boundary in time, not in space. It exists because we cannot see back to a time before the universe began. The lookback time to the cosmological horizon is equal to the age of the universe.

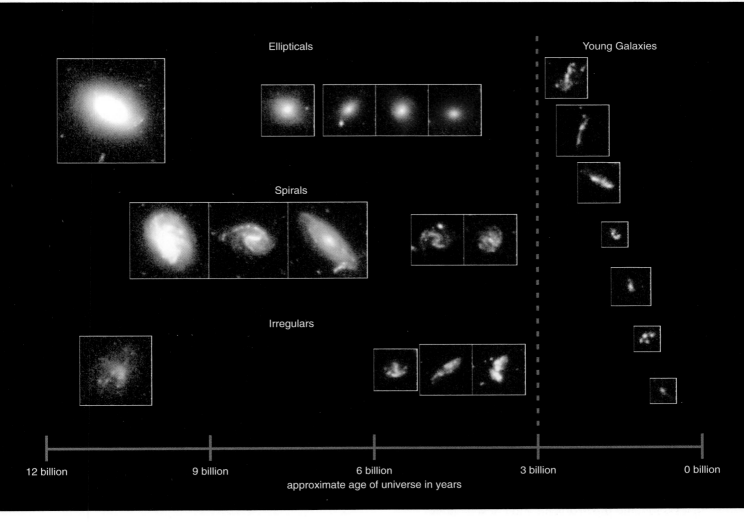

Ellipticals

Spirals

Irregulars

Young Galaxies

| 12 billion | 9 billion | 6 billion | 3 billion | 0 billion |

approximate age of universe in years

FIGURE 17.18 Family albums for spiral, elliptical, and irregular galaxies at different ages along with some very young, distant galaxies. (Younger galaxies appear smaller because they are more distant.)

For reasons discussed in Chapter 19, we cannot see matter at the cosmological horizon. But if we could, we would see it as it looked at the very beginning of time. One year from now, we would see this same matter as it looked 1 year after the beginning—which means that the horizon will have moved somewhat farther from us. We would therefore be able to see some new matter next year that we cannot see today. In fact, our horizon is continually expanding outward at the speed of light. Our observable universe therefore grows larger in radius by 1 light-year with each passing year, continually encompassing more and more matter.

Our quest to measure the distances and ages of galaxies has brought us to the very limits of the observable universe. Now it is time to examine what these galaxies have been doing for the past 10 billion or more years.

17.5 Galaxy Evolution

We pieced together the life stories of stars by observing relatively nearby stars of many different ages, but we cannot do the same for galaxies. Because the youngest galaxies are so distant, our current instruments are not sensitive enough to reveal the whole history of galaxy evolution. The galaxies pictured in Figure 17.18 extend back to fairly young ages, but not all the way back to the infancy of galaxies. Even the superior resolving power and sensitivity of the Hubble Space Telescope cannot reveal what young galaxies look like in detail. Thus, we are left having to weave several different lines of less direct evidence into a single coherent story of galaxy evolution. In this section, we will review the story of galaxy evolution as we currently understand it, keeping in mind that this story remains incomplete.

Modeling Galaxy Birth

Because our telescopes cannot yet see back to the time when galaxies formed their first stars, we must use theoretical modeling to study the earliest stages in galaxy evolution. The most successful models for galaxy formation presume the following:

- Hydrogen and helium gas filled all of space more or less uniformly when the universe was very young—say, in the first million years after its birth.

- This uniformity was not quite perfect, and certain regions of the universe were ever so slightly denser than others.

Beginning from these assumptions, which are supported by a mounting body of evidence (discussed in Chapter 19), we can model galaxy formation using well-established laws of physics. The models show that the regions of enhanced density originally expanded along with the rest of the universe. However, the slightly greater pull of gravity in these regions gradually slowed their expansion. Within about a billion years, the expansion of these denser regions halted and reversed, and the material within them began to contract into **protogalactic clouds,** the clouds of matter that eventually went on to form galaxies.

Stars must have begun to form almost immediately, but current models suggest that explosions of the first stars slowed the rate of star formation shortly thereafter. Protogalactic clouds initially cooled as they contracted, radiating away their thermal energy, and the first generation of stars grew from the densest, coldest clumps of gas. The most massive of these stars lived and died within just a few million years, a short time compared to the time required for the collapse of a protogalactic cloud into a mature galaxy. The supernovae of these massive stars generated shock waves that heated the surrounding interstellar gas, slowing the collapse of young galaxies and the rate at which new stars formed within them.

Old Spheroids, Young Disks

The colors of galaxies provide a crucial clue to what happened next. Recall that the disks of spiral galaxies appear whitish with flecks of blue. This indicates that the stars in the disks of spiral galaxies, sometimes referred to as the *disk population* (or *Population I*), include hot, blue stars as well as cool, red ones. Because only short-lived, massive stars look blue, we conclude that star formation must be an ongoing process in galactic disks—just as it is in the disk of the Milky Way. In contrast, the *spheroidal population*

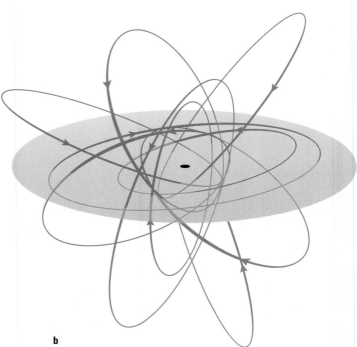

FIGURE 17.19 Characteristic stellar orbits in (**a**) disk component and (**b**) spheroidal component. Stars in the disk tend to orbit in the same direction around the center of the galaxy. (The bit of up and down motion is caused by the gravity of the disk itself.) Stars in the bulge and halo also orbit the center of the galaxy, but the orientations of their orbits are more random.

of stars (or *Population II*) found in elliptical galaxies and in the bulges and halos of spiral galaxies looks reddish in color. The absence of blue stars among the spheroidal population indicates that most of the stars in these spheroids are old and that their star formation must have ceased long ago—just as is the case in the halo of the Milky Way [Section 16.3].

The characteristic motions of stars in disks and spheroids provide another important clue to their nature (Figure 17.19). Disk stars generally orbit in orderly circles in the same direction, like the rotation of the Milky Way's disk [Section 16.4]. Stars of the spheroidal population in elliptical galaxies and in the bulges and halos of spiral galaxies tend to orbit in

A protogalactic cloud contains only hydrogen and helium gas.

Halo stars begin to form as the protogalactic cloud collapses.

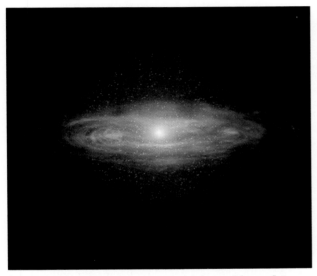

Conservation of angular momentum ensures that the remaining gas flattens into a spinning disk.

Billions of years later, the star–gas–star cycle supports ongoing star formation within the disk. The lack of gas in the halo precludes further star formation outside the disk.

FIGURE 17.20 Schematic model of galaxy formation.

random directions. Astronomers reason that, because the motions of their stars are so different, disks and spheroids must have formed in very different ways.

According to the most basic model for galaxy formation from a protogalactic cloud, the stars of the spheroidal population formed first, while the cloud remained blobby in shape with little or no measurable rotation. The orbits of these stars around the center of the galaxy could have had any orientation, accounting for the randomly oriented orbits of spheroidal population stars that we see today. Later, at least in a spiral galaxy, conservation of angular mo-

mentum caused the remaining gas to flatten into a spinning disk as it contracted under the force of gravity (Figure 17.20). Stars that form within this spinning gas disk are born on orbits moving at the same speed and in the same direction as their neighbors and thus become the disk population stars.

TIME OUT TO THINK *How does the disk of a spiral galaxy resemble a protoplanetary disk? How are the processes that form galactic disks and protoplanetary disks similar?*

Why Do Galaxies Differ?

On the whole, the idea that galaxies grow from contracting protogalactic clouds and form their stars over billions of years is consistent with what we know from studying the Milky Way and from our limited observations of very young galaxies. However, this simple model does not explain why galaxies come in such a great variety of shapes and sizes. We now believe that both the initial conditions of the protogalactic cloud and later interactions with other galaxies affected the development of all types of galaxies, leading to the variety we see today. To help differentiate the roles of initial conditions and subsequent interactions, we will consider a question about galaxy evolution that is still unanswered: Why do spiral galaxies have gas-rich disks while elliptical galaxies do not?

Two potential explanations for the differences between spirals and ellipticals trace a galaxy's current appearance back to the properties of the protogalactic cloud from which it formed. The first explanation concerns angular momentum differences among protogalactic clouds. If a protogalactic cloud has enough angular momentum, it will rotate quickly as it collapses, and the galaxy it produces will tend to form a disk. The resulting galaxy would therefore be a spiral galaxy. If a protogalactic cloud has little or no net angular momentum, its gas might not form a disk at all, and the resulting galaxy would be elliptical.

The second scenario hinges on the ability of the protogalactic gas to cool and form stars. Remember that elliptical galaxies are most common in clusters of galaxies. The density of matter in these clusters was probably higher than elsewhere in the universe when galaxies first formed. Thus, the preponderance of elliptical galaxies in clusters suggests that elliptical galaxies formed from protogalactic clouds with higher densities than those of the clouds from which spirals formed. The higher gas densities in these proto-elliptical clouds would have enabled them to radiate energy more efficiently and to cool more quickly. This cooling would have allowed gravity to collapse clumps of gas into stars more quickly, using up all the gas before the cloud had time to collapse into a disk. The resulting galaxy would look elliptical. In contrast, lower-density, proto-spiral clouds would have had a slower rate of star formation, leaving plenty of gas to form a disk as the cloud collapsed. The resulting spiral galaxy would have a gas-filled disk in which the star–gas–star cycle could maintain ongoing star formation.

Some evidence for this latter scenario comes from a few giant elliptical galaxies at very large distances (Figure 17.21). These galaxies look very red even after we have accounted for their large redshifts. They apparently have no blue or white stars at all, indicating that new stars no longer form

FIGURE 17.21 The redness of this distant elliptical galaxy, LBDS 53W091, indicates that its stars all formed during the first billion years or so after the birth of the universe. Its two red companions to the left are probably also elliptical galaxies with similar star-formation histories.

within these galaxies—even though we are seeing them as they were when the universe was only a few billion years old. This finding supports the idea that all the stars in these galaxies formed almost simultaneously, leaving no time for a disk to develop.

These two scenarios, in which the formation of a gas-rich disk depends on the angular momentum or density of the protogalactic cloud, probably describe important parts of the overall story. However, they ignore one key fact: Galaxies rarely evolve in perfect isolation. Recall that, in our scale-model solar system in Chapter 1, we made the Sun the size of a grapefruit. On this scale, the nearest star was like another grapefruit a few thousand kilometers away. Because the average distances between stars are so huge compared to the sizes of stars, direct star–star collisions are extremely rare. However, if we rescale the universe so that our entire galaxy is the size of a grapefruit, the Andromeda Galaxy is like another grapefruit only about a meter away, and a few smaller galaxies lie considerably closer. Thus, the average distances between galaxies are not tremendously larger than the sizes of galaxies, and collisions between galaxies are inevitable.

Such collisions are spectacular events that unfold over hundreds of millions of years (Figure 17.22). In our short lifetime, we can at best see an instantaneous snapshot of a collision in progress, as evidenced by the distorted shape of the colliding galaxies. Galactic collisions must have been even more frequent in the past, when the universe was smaller and galaxies were even closer together. Photographs of galaxies at a variety of distances confirm that distorted-looking galaxies—probably galaxy collisions in progress—were more common in the early universe than they are today (Figure 17.23).

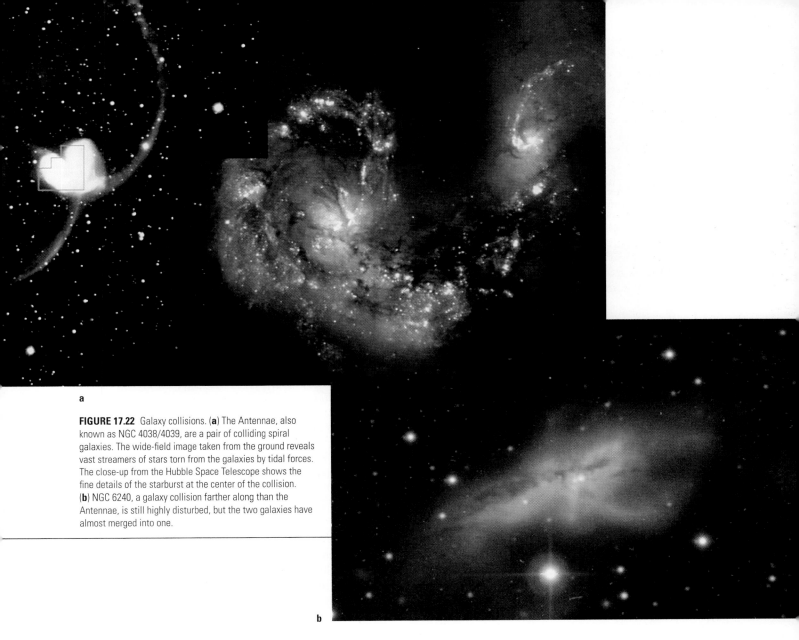

a

FIGURE 17.22 Galaxy collisions. (**a**) The Antennae, also known as NGC 4038/4039, are a pair of colliding spiral galaxies. The wide-field image taken from the ground reveals vast streamers of stars torn from the galaxies by tidal forces. The close-up from the Hubble Space Telescope shows the fine details of the starburst at the center of the collision. (**b**) NGC 6240, a galaxy collision farther along than the Antennae, is still highly disturbed, but the two galaxies have almost merged into one.

b

Age of Universe: 2–4 billion years

Age of Universe: 5–7 billion years

FIGURE 17.23 Hubble Space Telescope photos of distorted young galaxies. The larger number of distorted galaxies in the past suggests that collisions between galaxies were more common during the universe's first few billion years.

T=12

Two simulated spiral galaxies approach each other on a collision course.

T=24

The first encounter disrupts the two galaxies and sends them into orbit around each other.

T=36

Gravitational forces between the galaxies tear out long streamers of stars.

T=48

Because the first disruptive encounter saps kinetic energy from the galaxies, they cannot escape each other.

T=60

The second encounter is more direct than the first; the galaxies collide head-on and begin to merge.

T=72

The single galaxy resulting from the collision and merger is an elliptical galaxy surrounded by debris.

FIGURE 17.24 Several stages in a supercomputer simulation of a collision between two spiral galaxies. The whole sequence spans about 1.5 billion years. (Each unit on the time counter "T=" corresponds to 25 million simulated years.)

We can learn much more about collisions between galaxies with the aid of computer models that simulate interactions between galaxies. Such models allow us to "watch" a collision that takes hundreds of millions of years in nature unfold in a few minutes on a computer screen. These computer models show that a collision between two spiral galaxies can create an elliptical galaxy (Figure 17.24). Tremendous tidal forces between the colliding galaxies tear apart the two disks, randomizing the orbits of their stars. Meanwhile, a large fraction of their gas sinks to the center of the collision and rapidly forms new stars. When the cataclysm finally settles down, the merger of the two spirals has produced a single elliptical galaxy; little gas is left for a disk, and the orbits of the stars have random orientations.

Observational evidence confirms the idea that at least some elliptical galaxies are the result of collisions and subsequent mergers. The fact that elliptical galaxies dominate the galaxy populations at the cores of dense clusters of galaxies, where collisions should be most frequent, may mean that some of the spirals once present became ellipticals through colli-

sions. Stronger evidence comes from structural details of elliptical galaxies, which often attest to a violent past. Some elliptical galaxies have stars and gas clouds in their core that orbit differently than the other stars in the galaxy, suggesting that they are leftover pieces of galaxies that merged in a past collision. Other elliptical galaxies are surrounded by shells of stars that probably formed from stars stripped out of smaller galaxies that once strayed too close and were destroyed (Figure 17.25).

The most decisive evidence that collisions affect the evolution of some elliptical galaxies comes from observations of the **central dominant galaxies** found at the centers of many dense clusters. These giant elliptical galaxies apparently grew to huge sizes by consuming other galaxies through collisions. Central dominant galaxies frequently contain several tightly bound clumps of stars that probably once were the centers of individual galaxies later swallowed by the giant central galaxy (Figure 17.26). This process of galactic cannibalism can create galaxies over 10 times more massive than the Milky Way, making them the largest galaxies in the universe.

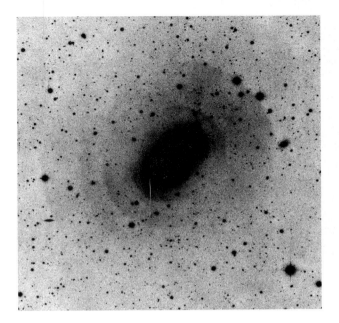

FIGURE 17.25 Shells of stars around the elliptical galaxy NGC 3923.

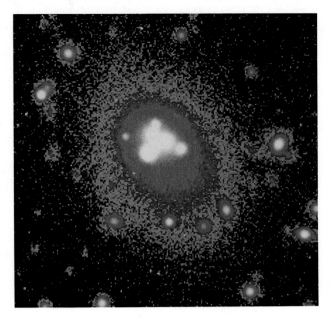

FIGURE 17.26 The central galaxy of the cluster Abell 3827, shown here in false color, has multiple clumps of stars that probably once were the centers of individual galaxies.

In summary, we have identified several scenarios that may determine whether a galaxy becomes an elliptical or a spiral galaxy. Two of these scenarios depend on initial conditions: One suggests that elliptical galaxies form when protogalactic clouds have low angular momentum, and the other suggests that elliptical galaxies form when high density leads to rapid star formation. But we have also found that interactions such as collisions between galaxies can drastically influence the lives of galaxies and very likely cause some spiral galaxies to become elliptical. Each of these mechanisms probably plays a role in galaxy evolution, and we do not yet know whether any one is more important than the others. Irregular galaxies, which make up only a small percentage of large galaxies, probably arise from a variety of different and unusual circumstances, but they probably are affected by the same processes that affect spirals and ellipticals.

Starburst Galaxies

Most of the galaxies in the present-day universe now appear comfortably settled, steadily forming new stars at modest rates. The Milky Way itself produces an average of about one new star per year. At this rate, the Milky Way won't exhaust the interstellar gas in its disk until long after the Sun has died. Certain galaxies are not so thrifty. Some form new stars at rates exceeding 100 stars per year, more than 100 times the star-formation rate in the Milky Way.

These **starburst galaxies** cannot possibly sustain such a torrid pace of star formation for long: At their current rates of star formation, they would consume all their interstellar gas in just a few hundred million years—a relatively short time compared to the 10 billion or more years since galaxies first formed. Starburst galaxies are thus passing through an important but temporary phase in the evolution of at least some galaxies.

A star-formation rate 100 times that of the Milky Way also means an occurrence of supernovae at 100 times the Milky Way's rate. Just as in the Milky Way, each supernova in a starburst galaxy generates a shock wave that creates a bubble of hot gas [Section 16.2]. The shock waves from several nearby supernovae quickly overlap and blend into a much larger superbubble. This is where the story usually ends here in the Milky Way, but in a starburst galaxy the drama is just beginning. Supernovae continue to explode inside the superbubble, adding to its thermal and kinetic energy. When the superbubble starts to break through the disrupted gaseous disk, it expands even faster. Hot gas erupts into intergalactic space, creating a *galactic wind*.

Galactic winds consist of low-density but extremely hot gas, typically with temperatures of 10–100 million Kelvin. They do not emit much visible light, but they do generate X rays. X-ray telescopes in orbit have detected pockets of X-ray emission surrounding the disks of some starburst galaxies, presumably coming from the outflowing

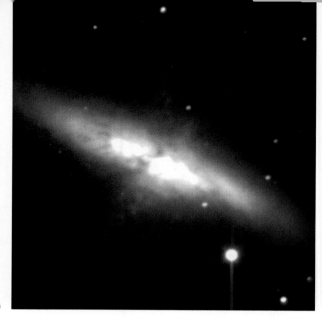

FIGURE 17.27 **(a)** This photographic image shows the visible light from the starburst galaxy NGC 253; the contour lines represent regions of strong X-ray emission. Note that the X rays appear on either side of the disk, suggesting that they come from hot gas blown out of the disk. **(b)** This photograph of the starburst galaxy M 82 shows violently disturbed gas (red) poking out from above and below the disk where hot gas is escaping.

galactic wind (Figure 17.27a). Sometimes we also see the glowing remnants of a punctured superbubble extending out into space (Figure 17.27b).

Many of the most luminous starburst galaxies look violently disturbed. They are neither flat disks with symmetrical spiral arms nor smoothly rounded balls of stars like elliptical galaxies. Streamers of stars are strewn everywhere. Such galaxies are filled with dusty molecular clouds, and deep inside them we see two distinct clumps of stars that look like two different galactic centers. Their appearance suggests that such starbursts result from a collision between two gas-rich spiral galaxies. The collision apparently compresses the gas inside the colliding galaxies, leading to the burst of star formation.

At least some large starbursts represent a short-lived phase in the life of galaxies that occurs when two spiral galaxies collide. The causes of smaller-scale starbursts are less clear: Some may result from direct collisions, while others may be caused by close encounters with other galaxies. The Large Magellanic Cloud is currently undergoing a period of rapid star formation, perhaps because of the tidal influence of the Milky Way. But no matter what their exact causes turn out to be, starburst galaxies clearly represent an important piece in the overall puzzle of galaxy evolution.

Loose Ends

All in all, we seem to be well on our way to understanding how and why protogalactic clouds evolve into spiral and elliptical galaxies and why we should expect a few galaxies, such as colliding galaxies and starburst galaxies, to have irregular appearances. Nevertheless, a few puzzles remain. For example, while the oldest halo stars in our Milky Way have very low fractions of heavy elements, these fractions are not zero. Where did the heavy elements in these stars come from? They probably came from an even earlier generation of stars, but where are the remnants of this first generation today? An even more fundamental question relates to the formation of the protogalactic clouds themselves. We said that these clouds formed in regions of slight density enhancement in the early universe, but what caused these density enhancements? This last question is one of the major puzzles in astronomy, and we'll revisit it in Chapters 18 and 19.

17.6 Quasars and Other Active Galactic Nuclei

Collisions between galaxies may be spectacular, but some galaxies display even more incredible phenomena: extreme amounts of radiation, and sometimes powerful jets of material, emanating from deep in their centers. We generally refer to these unusually bright galactic centers as **active galactic nuclei** and reserve the term **quasar** for the very brightest of them. The brightest quasars shine more powerfully than 1,000 galaxies the size of the Milky Way.

The glory days of quasars are long past. The fact that we find quasars primarily at great distances tells us that these blazingly luminous objects arose during an early stage in the evolution of at least some galaxies. Because we find no nearby quasars and relatively few nearby galaxies with less luminous active galactic nuclei, we conclude that the objects that shine as quasars in young galaxies must become dormant as the galaxies age. Thus, many nearby galaxies—we don't yet know which ones—that now look quite normal must have centers that once shone brilliantly as quasars.

What could possibly drive the incredible luminosities of quasars, and why did quasars fade away? A growing body of evidence points to a single answer: The energy output of quasars comes from gigantic accretion disks surrounding **supermassive black holes**—black holes with masses millions to billions of times that of our Sun. But before we study how these incredible powerhouses work, let's investigate the evidence that points to their existence.

The Discovery of Quasars

In the early 1960s, a young professor at the California Institute of Technology named Maarten Schmidt was busy identifying cosmic sources of radio-wave emission. Usually, the radio sources turned out to be normal-looking galaxies, but one day he discovered a major mystery. A radio source called 3C 273 looked like a blue star through a telescope but had strong emission lines at wavelengths that did not appear to correspond to any known chemical element. After months of puzzlement, Schmidt suddenly realized that the emission lines were not coming from an unfamiliar element at all. Instead, they were emission lines of hydrogen hugely redshifted from their normal wavelengths.

Schmidt computed the distance to 3C 273 using Hubble's law; then he plugged this distance into the luminosity–distance formula [Section 13.2]. What he found was astonishing: 3C 273 has a luminosity of about 10^{39} watts, well over a trillion (10^{12}) times that of our Sun—making it hundreds of times more powerful than the entire Milky Way Galaxy. Discoveries of similar but even more distant objects soon followed. Because the first few of these objects were strong sources of radio emission that looked like stars through visible-light telescopes, they were named "quasi-stellar radio sources," or quasars for short. Later, astronomers learned that most quasars are not such powerful radio emitters, but the name has stuck.

For many years, a debate raged among astronomers over whether Hubble's law could really be used to determine quasar distances. Some argued that quasars might have high redshifts for other reasons and therefore might be much nearer to us than Hubble's law would suggest. The vast majority of astronomers now consider this debate settled. Improved images show that quasars are indeed the centers of extremely distant galaxies and often are members of very distant galaxy clusters.

Most quasars lie more than halfway to the cosmological horizon. The lines in typical quasar spectra are shifted to more than three times their rest wavelengths, which tells us that the light from these quasars emerged when the universe was less than a third of its present age. Light from the farthest known quasars began its journey when the universe was only about a tenth of its present age.

The extraordinary energy output of quasars emerges across an unusually wide swath of the electromagnetic spectrum. Quasars radiate approximately equal amounts of power from infrared wavelengths all the way through to gamma rays; they also produce strong emission lines. By comparison, most stars and galaxies emit primarily visible light. The wide spread of photon energies coming from quasars implies that they contain matter with a wide range of temperatures. How does this matter manage to radiate such a large luminosity?

Evidence from Nearby Active Galactic Nuclei

Quasars are difficult to study in detail because they are so far away. Luckily, some quasarlike objects are much closer to home. About 1% of present-day galaxies—that is, galaxies we see nearby—have less powerful active galactic nuclei that look very much like weak quasars (such galaxies are called *Seyfert galaxies*). The spectra of these active nuclei range from infrared to gamma rays, just like those of quasars. The only real distinction between nearby active galactic nuclei and quasars is their luminosity. Generally, the power outputs of quasars swamp those of the galaxies that contain them, making the surrounding galaxy hard to detect (Figure 17.28). That

FIGURE 17.28 The bright quasar in the nucleus of the galaxy just left of center outshines the rest of the galaxy, which appears disturbed, as if it has recently undergone a collision.

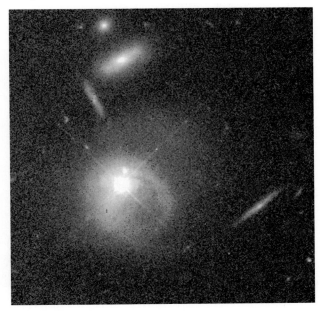

is why quasars look "quasi-stellar." The galaxies surrounding dimmer active galactic nuclei are much easier to see because their centers are not so overwhelmingly bright.

The light-emitting regions of active galactic nuclei are so small that even the sharpest images do not resolve them. Our best visible-light images show only that active galactic nuclei must be smaller than 100 light-years (30 parsecs) across. Radio-wave images made with the aid of interferometry show that these nuclei are even smaller: less than 3 light-years (1 parsec) across. But rapid changes in the luminosities of some active galactic nuclei point to an even smaller size.

To understand how variations in luminosity give us clues about an object's size, imagine that you are a master of the universe and you want to signal one of your fellow masters a billion light-years away. An active galactic nucleus would make an excellent signal beacon, because it is so bright. However, suppose the smallest nucleus you can find is 1 light-year across. Each time you flash it on, the photons from the front end of the source reach your fellow master a full year before the photons from the back end. Thus, if you flash it on and off more than once a year, your signal will be smeared out. Similarly, if you find a source that is 1 light-day across, you can transmit signals that flash on and off no more than once a day. If you want to send signals just a few hours apart, you need a source no more than a few light-hours across.

Occasionally, the luminosity of an active galactic nucleus doubles in a matter of hours. The fact that we see a clear signal indicates that the source must be less than a few light-hours across. In other words, the incredible luminosities of active galactic nuclei and quasars are apparently being generated in a volume of space not much bigger than our solar system.

Radio Galaxies and Jets

In the early 1950s, a decade before the discovery of quasars, radio astronomers noticed that certain galaxies emit unusually large quantities of radio waves. Today we believe that these **radio galaxies** are another class of celestial powerhouse that is closely related to quasars. Upon close inspection, we find that much of the radio emission comes not from the galaxies themselves, but rather from pairs of huge **radio lobes,** one on either side of the galaxy (Figure 17.29).

The radio waves from the lobes are produced by electrons and protons spiraling around magnetic field lines at nearly the speed of light. Astronomers of the 1950s guessed that this might be the source of the radio emission, but the implied amounts of energy in the lobes seemed implausibly large until the discovery of quasars. British theoretician Martin Rees led the way in showing that radio galaxies contain active galactic nuclei. He argued in 1971 that radio galaxies must have powerful **jets** spurting from their nuclei that transport energy to their lobes. That is, although the lobes lie far outside the visible galaxy, he asserted that the ultimate source of their energy would be found in the galaxy's tiny nucleus, which must therefore be as powerful as a quasar. A few years later, detailed radio images began to support his claims.

Today, radio telescopes resolve the structure of radio galaxies in vivid detail. At the center of a typical radio galaxy sits a tiny radio core—the active galactic nucleus of the radio galaxy—less than a few light-years across. Two jets of plasma shoot out of the core in opposite directions. Frequently, only the jet tilted in our direction is visible. Using time-lapse radio images taken several years apart, we can track the

FIGURE 17.29 Radio image of the radio galaxy Cygnus A taken with the Very Large Array in New Mexico. The distance between the lobes is about 400,000 light-years, several times larger than the extent of the galaxy in visible light.

lobe

hot spot

hot spot

core

jet

lobe

motions of various plasma blobs in the jets. Some of these blobs move at close to the speed of light. The lobes lie at the ends of the jets, sometimes as much as a million light-years from the core.

Quasars and radio galaxies may be much more similar than they appear. Indeed, many quasars, including 3C 273, have core–jet–lobe radio structures reminiscent of radio galaxies. For example, Figure 17.30 shows a series of pictures of a jet in which a blob of plasma is moving outward from a quasar at close to the speed of light. Moreover, the active galactic nuclei of many radio galaxies seem to be concealed beneath donut-shaped rings of dark molecular clouds (Figure 17.31). It may be that such structures look like quasars when they are oriented so that we can see the active galactic nucleus at the center, and like the cores of radio galaxies when the ring of dusty gas dims our view of the central object.

Supermassive Black Holes

Astronomers have worked hard to envision physical processes that might explain how radio galaxies, quasars, and other active galactic nuclei release so

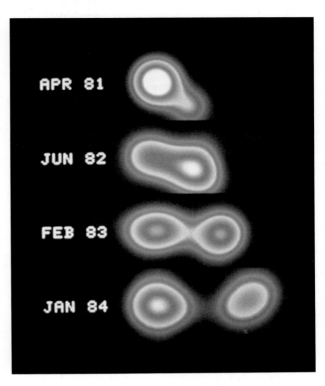

FIGURE 17.30 These images of the radio jet in the quasar 3C 345, taken over a period of several years, show a blob of plasma (right) moving away from the core (left) at close to the speed of light.

FIGURE 17.31 Artist's conception of the central region of a radio galaxy. The active galactic nucleus, obscured by a ring of dusty clouds, lies at the point from which the jets emerge. If viewed along a direction closer to the jet axis, the active nucleus would not be obscured and would look more like a quasar.

351

FIGURE 17.32 Artist's conception of an accretion disk surrounding a supermassive black hole. Note that this picture represents only the very center of an object like that shown in Figure 17.31.

much energy within such small central volumes. Only one explanation seems to fit: The energy comes from matter falling into a supermassive black hole (Figure 17.32). Gravity converts the potential energy of the infalling matter into kinetic energy. Collisions between infalling particles convert the kinetic energy into thermal energy, and photons carry this thermal energy away. As in X-ray binaries, we expect that the infalling matter swirls through an accretion disk before it disappears beneath the event horizon of the black hole [Section 15.4].

This method of energy generation can be awesomely efficient. The gravitational potential energy lost by a chunk of matter falling into a black hole is equivalent to its mass-energy, $E = mc^2$. As much as 10–40% of this energy can emerge as radiation before the matter crosses the event horizon. Thus, accretion by black holes is far more efficient at producing light than is nuclear fusion, which converts less than 1% of mass-energy into photons. Note that, as with accretion into the black holes formed in supernovae, the light is coming not from the black hole itself but rather from the hot gas surrounding it.

A variety of mechanisms explain why quasars and other active galactic nuclei radiate energy across the electromagnetic spectrum. The hot gas in and above the accretion disk produces copious amounts of ultra-violet and X-ray photons. This radiation ionizes surrounding interstellar gas, energizing intense ionization nebulae that emit visible light. The emission lines produced by these nebulae are the same ones Maarten Schmidt used to measure the first quasar redshifts. The infrared light may come from rings of molecular clouds that encircle the active galactic nucleus (see Figure 17.31). Dust grains in these molecular clouds can absorb the high-energy light and reemit it as infrared light. This ring of dense clouds could well be the reservoir for the accretion disk, supplying the matter that the giant black hole eventually consumes. Finally, the radio emission from active galactic nuclei comes from the fast-moving electrons that we sometimes see jetting from these nuclei at nearly the speed of light.

Although it's easy to understand why matter swirls inward through the accretion disk toward the black hole, explaining the powerful jets emerging from active galactic nuclei is more challenging. One plausible model for jet production relies on the twisted magnetic fields thought to accompany accretion disks. As an accretion disk spins, it pulls the magnetic field lines that thread it around in circles. Charged particles start to fly outward along the field lines like beads on a twirling string. The particles careening along the field lines accelerate to speeds

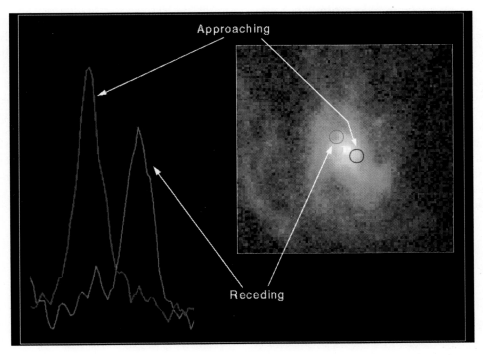

FIGURE 17.33 Doppler shifts of the emission lines in the elliptical galaxy M 87 indicate a 2–3-billion-solar-mass black hole.

near the speed of light, forming a jet that shoots out into space. Some jets blast all the way through the galaxy's interstellar medium and penetrate into the much less dense intergalactic gas. The hot spots at the ends of the lobes in radio galaxies (see Figure 17.29) are the places where the jets are currently ramming into the intergalactic gas. When particles traveling down a jet hit the hot spot, they are deflected into the surrounding radio lobe like water from a firehose hitting a wall. The particles then fill the lobe with energy, generating powerful radio emission.

The supermassive black hole theory explains many of the observed features of quasars and other active galactic nuclei, but it is incomplete. We do not yet know what would create such giant black holes, nor do we know why quasars eventually run out of gas and stop shining. The preponderance of quasars during the first several billion years of the universe suggests that the formation of supermassive black holes is somehow linked to galaxy formation, perhaps through collisions between galaxies. Some scientists have suggested that clusters of neutron stars resulting from extremely dense starbursts at the centers of galaxies might somehow coalesce to form an enormous black hole, but these speculations are still unverified. The origins of supermassive black holes remain mysterious.

Do supermassive black holes really drive the tremendous activity of radio galaxies, quasars, and other active galactic nuclei? The idea that such monsters even exist has been hard for some astronomers to swallow. Proving that we have found a black hole is tricky. Black holes themselves do not emit any light, so we need to infer their existence from the ways in which they alter their surroundings. In the vicinity of a black hole, matter should be orbiting at high speed around something invisible.

The case for supermassive black holes grew much stronger during the 1990s. The relatively nearby galaxy M 87 features a bright nucleus and a jet that emits both radio and visible light. Thus, it was already a prime black-hole suspect when astronomers pointed the Hubble Space Telescope at its core in 1994 (Figure 17.33). The spectra they gathered showed blueshifted emission lines on one side of the nucleus and redshifted emission lines on the other. This pattern of Doppler shifts is the characteristic signature of orbiting gas: On one side of the orbit the gas is coming toward us and hence is blueshifted, while on the other side it is moving away from us and is redshifted. The magnitude of these Doppler shifts shows that the gas, located up to 60 light-years (18 parsecs) from the center, is orbiting something invisible at a speed of hundreds of kilometers per second. This high-speed orbital motion indicates that the central object has a mass some 2–3 billion times that of our Sun. A supermassive black hole is the only thing we know of that could be so massive while remaining unseen.

Observations of NGC 4258, another galaxy with a visible jet, delivered even more persuasive evidence just 1 year later. A ring of molecular clouds orbits the nucleus of this galaxy in a circle less than 1 light-year in radius. We can pinpoint these clouds because they amplify the microwave emission lines of water molecules, generating beams of microwaves very similar to laser beams. The Doppler shifts of these emission lines allow us to determine the orbits of the clouds very precisely. Their orbital motion tells us that the clouds are circling a single, invisible object—presumably a black hole—with a mass of 36 million solar masses.

We may never be 100% certain that we have discovered black holes in other galaxies. The best we can do is rule out all other possibilities. However, the hypothesis that gigantic black holes lie at the cores of quasars, nearby active galactic nuclei, and radio galaxies is withstanding the tests of time and thousands of observations.

Reality provides us with facts so romantic that imagination itself could add nothing to them.

JULES VERNE

THE BIG PICTURE

The picture could hardly get any bigger than it has in this chapter. Looking through both space and time, we have seen nearly to the limits of the observable universe and have found that it is filled with a wide variety of galaxies. As you look back, keep sight of these "big picture" ideas:

- Our measurements of the distances to faraway galaxies rely on a chain of distance measurements that begins with the Earth–Sun distance measured with the aid of radar ranging and continues with parallax and various standard-candle techniques. The distance chain allowed Hubble to measure distances to galaxies and thus to discover the relationship known as Hubble's law and the expansion of the universe.

- As the universe expands, it carries the matter within it along for the ride. From the current expansion rate and from estimates of the overall strength of gravity in the universe and the ages of the oldest stars, we infer that the universe is somewhere between about 11 and 16 billion years old.

- Although we do not yet know the complete story of galaxy evolution, we are rapidly learning more. We know that galaxies grow from protogalactic clouds of gas, but collisions with neighboring galaxies have probably affected many galaxies.

- The tremendous energy outputs of quasars and other active galactic nuclei, including those of radio galaxies, are probably powered by gas accreting onto supermassive black holes. The centers of many present-day galaxies must still contain the supermassive black holes that once enabled them to shine as quasars.

Review Questions

1. Distinguish between *spiral galaxies* and *elliptical galaxies* in terms of their shapes and colors. What are *irregular galaxies*?

2. Distinguish between the *disk component* and the *spheroidal component* of a spiral galaxy. Which component includes the galaxy's spiral arms? Which includes its bulge? Which includes its halo?

3. What is a *group* of galaxies? What is a *cluster* of galaxies?

4. What is the major difference between an elliptical galaxy and a spiral galaxy? How is an elliptical galaxy similar to the halo of the Milky Way?

5. What do we mean by a *standard candle?* Briefly explain how, once we identify an object as a standard candle, we can use the luminosity–distance formula to find its distance.

6. Describe the technique of *main-sequence fitting.* Why isn't this technique useful for measuring distances to other galaxies?

7. Why are Cepheid variables good standard candles? Briefly explain how Edwin Hubble used his discovery of a Cepheid in Andromeda to prove that the "spiral nebulae" were actually entire galaxies.

8. What is *Hubble's law?* How did Hubble discover it? Explain the meaning of Hubble's law when it is expressed mathematically as $v = H_0 \times d$. What are the units of *Hubble's constant, H_0*?

9. Describe how we can use Hubble's law to determine the distance to a distant galaxy. What practical difficulties limit the use of Hubble's law for measuring distances?

10. What makes white-dwarf supernovae good standard candles? Briefly describe how we have learned the true luminosities of white-dwarf supernovae.

11. What is the *Tully–Fisher relation?* How does it enable us to use entire spiral galaxies as standard candles?

12. What is our current best estimate of Hubble's constant? Summarize all the links in the distance chain that allow us to estimate distances to the farthest reaches of the universe.

13. In what ways is the surface of an expanding balloon a good analogy to the universe? In what ways is this analogy limited? Explain why a miniature scientist living in a polka dot on the balloon would observe all other dots to be moving away, with more distant dots moving away faster.

14. What is a *lookback time?* Why does it make more sense to talk about a distant object in terms of its lookback time than in terms of its distance?

15. What do we mean by a *cosmological redshift?* How does our interpretation of a distant galaxy's redshift differ if we think of it as a cosmological redshift rather than a Doppler shift?

16. What two assumptions underlie theoretical models of galaxy formation? According to these models, what processes eventually slow the collapse of *protogalactic clouds?*

17. Distinguish between disk-population stars and spheroidal-population stars, in terms of their ages, masses, colors, and orbits. Briefly describe how the gravitational contraction of a protogalactic cloud explains the different characteristics of these populations.

18. Briefly explain why we expect that collisions between galaxies should be relatively common, while collisions between stars are extremely rare. Why should galaxy collisions have been more common in the past than they are today?

19. Briefly describe how a collision between two spiral galaxies might lead to the creation of a single elliptical galaxy. What evidence supports this scenario for the formation of elliptical galaxies?

20. Briefly describe the discovery of *quasars.* What evidence convinced astronomers that the high redshifts of quasars really do imply great distances? Why can we learn more about quasars by studying nearby *active galactic nuclei?*

21. Briefly explain how we can use variations in luminosity to set limits on the size of an object's emitting region. For example, if an object doubles its luminosity in 1 hour, how big can it be?

22. What is a *radio galaxy?* Describe *jets* and *radio lobes*. Why do we think that their ultimate energy sources lie in quasarlike galactic nuclei?

23. Briefly explain the general picture of how *supermassive black holes* enable quasars to produce their huge luminosities. Summarize the evidence supporting the idea that supermassive black holes lie at the center of radio galaxies, quasars, and other active galactic nuclei.

Discussion Questions

1. *Cosmology and Philosophy.* One hundred years ago, many scientists believed that the universe was infinite and eternal, with no beginning and no end. When Einstein first developed his general theory of relativity, he found it predicted that the universe should be either expanding or contracting. He believed so strongly in an eternal and unchanging universe that he modified the theory, a modification he would later call his "greatest blunder." Why do you think Einstein and others assumed that the universe had no beginning? Do you think that a universe with a definite beginning in time, some 15 billion or so years ago, has any important philosophical implications? Explain.

2. *The Case for Supermassive Black Holes.* The evidence for supermassive black holes at the center of galaxies is strong. However, it is very difficult to prove absolutely that they exist because the black holes themselves emit no light. We can only infer their existence from their powerful gravitational influences on surrounding matter. How compelling do you find the evidence? Do you think astronomers have proved the case for black holes beyond a reasonable doubt? Defend your opinion.

Problems

1. *Classifying Galaxies.* Figure 17.1 shows a wide variety of galaxies. Devise your own scheme for classifying them by color, and describe it. Estimate the fraction of galaxies that fall into each color class. Devise and describe your own scheme for classifying galaxies by shape. Estimate the fraction that fall into each shape class. Describe any relationships you see between the shapes and the colors of these galaxies.

2. *Distance Measurements.* The techniques astronomers use to measure distances are not so different from the ones you use every day. Describe how parallax measurements are similar to your visual depth perception. Describe how standard candle measurements are similar to the way you estimate the distance to an oncoming car at night.

3. *Standard Candles.* List at least three qualities that would tend to make a type of astronomical object useful as a standard candle. Explain why each quality is important if we hope to use objects of this type to measure distances in the universe.

4. *Life Story of a Spiral.* Imagine that you are a spiral galaxy. Describe your life history from birth to the present day. Your story should be detailed and scientifically consistent, but also creative. That is, it should be entertaining while at the same time incorporating current scientific ideas about the formation of spiral galaxies.

5. *Life Story of an Elliptical.* Imagine that you are an elliptical galaxy. Describe your life history from birth to the present. There are several possible scenarios for the formation of elliptical galaxies, so choose one and stick to it. Be creative while also incorporating scientific ideas that demonstrate your understanding.

6. *Research Project: Hubble's Constant.* Measurements of Hubble's constant are rapidly growing more reliable. Find at least two measurements of H_0 determined in the past 2 years. Describe how these values were measured. What is the quoted uncertainty of each? By how much do the values differ?

7. *Web Project: Future Missions.* The subject of galaxy evolution is a very active area of research. Using the Web, look for information on current and future NASA missions involved in investigating galaxy evolution. How big are the planned telescopes? What wavelengths will they look at? When will they be launched? Assemble a journal from your Web research, with pictures and a paragraph on each mission you find. If possible, make your journal a Web page and include links to the relevant web sites.

8. *Research: The Quasar Controversy.* For many years, some astronomers argued that quasars were not really as distant as Hubble's law implies. Research the history of the discovery of quasars and the debates that followed. What evidence led some astronomers to think that quasars might be nearer than Hubble's law suggested? Why did most astronomers eventually conclude that quasars really are far away? Write a one- to two-page report summarizing your findings.

18 Dark Matter

THE MAJORITY OF THE MATTER IN THE UNIVERSE, THE STUFF THAT BINDS galaxies together with the force of its own gravity, is too dark to see. We know that this matter exists because we can detect its gravitational influences on other things, but what is this so-called *dark matter*? Do armies of Jupiter-size bodies, too dim for us to see at a distance, populate the voids between the stars? Could black holes be responsible? Or is dark matter an entirely new form of matter, still undiscovered here on Earth? We don't yet know the answer. Incredibly, we still haven't identified the most common form of matter in the universe, making dark matter one of the greatest cosmic mysteries.

In this chapter, we will investigate why most astronomers believe that dark matter exists. We'll see how it affects structures the size of galaxies and larger. We'll investigate current hypotheses about the nature of dark matter. And we'll see how the fate of the universe rests on the question of just how much dark matter there is.

18.1 The Mystery of Dark Matter

We first encountered dark matter when we studied the Milky Way's rotation in Chapter 16. We saw that atomic hydrogen clouds lying farther from the galactic center than our Sun orbit the galaxy at unexpectedly high speeds. We concluded that much of our galaxy's mass must lie beyond the distance of the Sun's orbit around the galactic center, distributed throughout the galaxy's halo [Section 16.4]. Yet most of the Milky Way's *light* comes from stars lying closer to the galaxy's center than does our Sun. Together, these facts imply that the halo contains large amounts of matter, but this matter emits so little light that we cannot see it. Thus, we call it *dark matter*.

Does dark matter really exist? To claim that the galaxy is filled with a form of matter that we cannot yet identify may seem strange, but only two possible explanations can account for the high orbital velocities of the atomic hydrogen clouds: Either their velocities are caused by the gravitational attraction of unseen matter, or we are doing something wrong when we apply the law of gravity to understand their orbits. The latter possibility implies that we do not understand how gravity operates on galaxy-size scales, but our theories of gravity successfully account for many other cosmic phenomena. Thus, the majority of astronomers believe that we correctly understand how gravity works and that dark matter must really exist.

The evidence that the Milky Way contains dark matter qualifies as an interesting surprise on its own, but it is also a glimpse into a more profound mystery. If our suspicions about dark matter are correct, then the luminous part of the Milky Way's disk must be rather like the tip of an iceberg, marking only the center of a much larger clump of mass (Figure 18.1). The total mass of dark matter in the Milky Way might be 10 times greater than the mass of visible stars. Each galaxy in the universe is similar to our own in this respect, meaning that we probably cannot see most of the matter in the universe. This matter's darkness makes it challenging to study, but its overwhelming dominance of the universe makes it extremely important. Dark matter, by virtue of its mighty gravitational pull, seems to be the glue that holds galaxies and clusters of galaxies together.

Dark matter has even more profound implications for the fate of the universe. The universe might continue to expand forever, or it might someday stop expanding and begin to collapse. We do not yet know which fate awaits our universe because the answer depends on the overall density of matter—and we cannot determine the overall density until we first determine how much dark matter is out there. Thus,

FIGURE 18.1 Dark matter distribution relative to luminous matter in a spiral galaxy.

dark matter

luminous matter

the very fate of the universe hinges on the total amount of dark matter.

In this chapter, we will presume that we understand gravity correctly and consequently that dark matter pervades the universe. If our understanding of gravity is correct, then we already know quite a lot about dark matter, despite the fact that its composition remains a mystery. We do not know the total amount of dark matter in the universe, but we can measure the amounts in individual galaxies and clusters of galaxies from its gravitational effects. We can even make some reasonable guesses as to what it will turn out to be. We'll begin by discussing how we "discover" dark matter and conclude by discussing what our current knowledge predicts about the ultimate fate of the universe.

18.2 Dark Matter in Galaxies

The claim that dark matter far outweighs the visible matter in the universe might seem farfetched, but it rests on fundamental physical laws. Newton's laws of motion and gravity are among the most trustworthy tools in science. We have used them time and again to measure masses of celestial objects. We found the masses of the Earth and the Sun by applying Newton's version of Kepler's third law [Section 5.4]. We used this same law to calculate the masses of stars in binary star systems, revealing the general relationships between the masses of stars and their outward appearances. Newton's laws have also told us the masses of things we can't see directly, such as the masses of neutron stars in X-ray binaries and of black holes in active galactic nuclei. These laws have proved extremely reliable in many different applications. Thus, when Newton's laws tell us that dark matter exists, we are not inclined to ignore them.

We determine the amount of dark matter in a galaxy from the galaxy's mass and luminosity: We can infer the total mass in stars from the galaxy's luminosity, so whatever additional mass remains unaccounted for must be dark. Measuring a galaxy's luminosity is relatively easy. We simply point a telescope at the galaxy in question, measure its apparent brightness, and calculate its luminosity from the luminosity–distance formula [Section 13.2]. Measuring the mass is more complicated. We would like to apply Newton's laws in the form of the orbital velocity law from Chapter 16 to matter orbiting as far from the galaxy's center as possible. However, the matter farthest from the center of a galaxy is extremely dim. For all we know, the matter at the edge of a galaxy might be completely dark.

Weighing Galaxies

We can weigh a spiral galaxy by measuring the gravitational effects of the galaxy's mass on the orbits of objects in its disk. Even beyond the point in the disk at which starlight fades into the blackness of intergalactic space, we can still see radio waves from atomic hydrogen gas. We can therefore use Doppler shifts of the 21-cm emission line of atomic hydrogen [Section 16.2] to determine how quickly this gas moves toward us or away from us (Figure 18.2). Galaxies beyond the Local Group have cosmological redshifts affecting all their spectral lines. However, on one side of a spiral galaxy the gas is rotating away from us, so its 21-cm line is redshifted a little more than the redshift of the galaxy as a whole. On the other side, the 21-cm line is blueshifted relative to the redshift of the galaxy as a whole because the gas is rotating toward us. From the Doppler shifts of these clouds, we can construct a *rotation curve*—a plot showing orbital velocities of gas clouds and stars—just as we did for the Milky Way [Section 16.4]. A rotation curve contains all the information we need to measure the mass contained within the orbits of the outermost gas clouds.

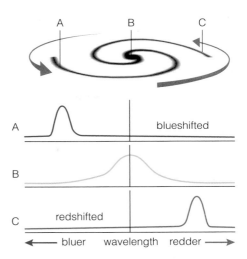

FIGURE 18.2 Measuring the rotation of a spiral galaxy with the 21-cm line of atomic hydrogen. Blueshifted lines on the left side of the disk show how fast that side is rotating toward us. Redshifted lines on the right side show how fast that side is rotating away from us.

TIME OUT TO THINK *As a brief review, draw a rotation curve for our solar system. How does the solar system rotation curve differ from the rotation curve of the Milky Way? What does that tell us about the distribution of matter in the Milky Way? Why? (Hint: See Chapter 16 for review.)*

The rotation curves of most spiral galaxies turn out to be remarkably flat as far out as we can see, which means that the orbital speeds of gas clouds remain roughly constant with increasing distance from the galactic center (Figure 18.3). We saw in Chapter 16 that our Milky Way's rotation curve remains more or less flat from well within the radius of the Sun's orbit (28,000 light-years, or 8.5 kpc) to beyond twice that radius. The rotation curves of some other spiral galaxies are flat to well beyond 160,000 light-years (50 kpc).

Because the orbital speeds of gas clouds tell us the amount of mass contained within their orbital paths, the flat rotation curves imply that a great deal of matter lies far from the galactic center [Section 16.4]. In particular, if we imagine drawing bigger

FIGURE 18.3 Actual rotation curves of four spiral galaxies.

and bigger circles around a galaxy, the flat rotation curves imply that we must keep encircling more and more matter. The mass of a spiral galaxy with gas clouds orbiting at 200 km/s at a distance of 160,000 light-years from its center must be at least 500 billion (5×10^{11}) solar masses.

In elliptical galaxies, we can study only the motions of stars, but the basic result remains the same—a great deal of dark matter exists well outside the galaxies' cores.

Mass-to-Light Ratio

We can determine how much dark matter a galaxy contains by comparing the galaxy's measured mass to its luminosity. For example, the Milky Way contains about 90 billion (9×10^{10}) solar masses of material within the Sun's orbit. However, the total luminosity of stars within this region is only about 15 billion (1.5×10^{10}) solar luminosities. Thus, on average, it takes 6 solar masses of matter to produce 1 solar luminosity of light in this region of the galaxy (15 billion × 6 = 90 billion). We therefore say that the Milky Way's **mass-to-light ratio** within the Sun's orbit is about 6 solar masses per solar luminosity. This fact tells us that most matter is dimmer than our Sun, which is not surprising since we know that most stars are smaller and dimmer than our Sun. We find similar mass-to-light ratios for the inner regions of most other spiral galaxies.

TIME OUT TO THINK *Note that, because the Sun has a mass of $1 M_{Sun}$ and a luminosity of $1 L_{Sun}$, its mass-to-light ratio is $1 M_{Sun}/1 L_{Sun} = 1$ solar mass per solar luminosity. What is the mass-to-light ratio of a $1 M_{Sun}$ red giant with a luminosity of $100 L_{Sun}$? What is the mass-to-light ratio of a $1 M_{Sun}$ white dwarf with a luminosity of $0.001 L_{Sun}$?*

Elliptical galaxies contain virtually no luminous high-mass stars, so their stars are less bright on average than the stars in a spiral galaxy. In the inner regions of elliptical galaxies, the orbits of stars indicate a mass-to-light ratio of about 10 solar masses per solar luminosity—nearly double the mass-to-light ratio in the central region of the Milky Way. This is not surprising given the fact that elliptical galaxies have dimmer stars.

The surprise in mass-to-light ratios—and one of the key pieces of evidence for the existence of dark matter—comes when we look to the outer reaches of galaxies. As we look farther from a galaxy's center, we find a lot more mass but not much more light. The mass-to-light ratios for entire spiral galaxies can be as high as 50 solar masses per solar luminosity, and the overall ratios in some dwarf galaxies can be

even higher. We are forced to conclude that stars alone cannot account for the amount of mass present in galaxies. For example, if a galaxy's overall mass-to-light ratio is 60 solar masses per solar luminosity and its stars account for only 6 solar masses per solar luminosity, then the remaining 90% of the galaxy's mass must be dark.

18.3 Dark Matter in Clusters

The problem of dark matter in astronomy is not particularly new. Back in the 1930s, astronomer Fritz Zwicky was already arguing that clusters of galaxies held enormous amounts of this mysterious stuff (Figure 18.4). Few of his colleagues paid attention. Later observations supported Zwicky's claims. We can now weigh clusters of galaxies in three different ways: by measuring the speeds of galaxies orbiting the center of the cluster, by studying the X-ray emission from hot gas between the cluster galaxies, and by observing how the clusters bend light as *gravitational lenses*. All three techniques indicate that clusters contain huge amounts of dark matter. Let's investigate each of these techniques more closely.

Orbiting Galaxies

Zwicky was one of the first astronomers to think of galaxy clusters as huge swarms of galaxies bound together by gravity. It seemed natural to him that galaxies clumped closely in space should all be orbiting one another, just like the stars in a star cluster. He therefore assumed that he could measure cluster masses by observing galaxy motions and applying the orbital velocity law.

Armed with a spectrograph, Zwicky measured the redshifts of the galaxies in a particular cluster and used these redshifts to calculate the speeds at which the individual galaxies are moving away from us. He determined the velocity of the cluster as a whole by averaging the velocities of its individual galaxies. He could then estimate the speed of a galaxy around the cluster—that is, its orbital speed—by subtracting this average velocity from the individual galaxy's velocity. Finally, he plugged these orbital speeds into an appropriate form of the orbital velocity law to estimate the cluster's mass and compared this mass to the luminosity of the cluster. To his surprise, Zwicky found that clusters of galaxies have huge mass-to-light ratios: They hold hundreds of solar masses for each solar luminosity of radiation they emit. He concluded that most of the matter within these clusters must be almost entirely dark. Many astronomers disregarded Zwicky's result, believing that he must have done something wrong to arrive at such a strange answer.

FIGURE 18.4 Fritz Zwicky.

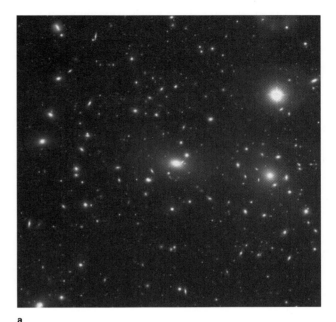

Coma Cluster
0.5-2.0 keV

0.5 Degree

b

FIGURE 18.5 (**a**) The Coma Cluster of galaxies seen in visible light. Virtually every pictured object here is a galaxy. (**b**) A map of X-ray emission from the Coma Cluster. Hot, X ray–emitting gas bound to the cluster by gravity fills the spaces between the galaxies.

Today, far more sophisticated measurements of galaxy orbits in clusters confirm Zwicky's original finding. If we measure the radial velocities of many galaxies in a cluster, we can estimate the average orbital velocity of all the cluster's galaxies. Cluster masses found in this way imply lots of dark matter. The visible portions of a cluster's galaxies represent less than 10% of the cluster's mass.

Intracluster Medium

A second method for weighing a cluster of galaxies relies on X-ray observations of the hot gas that fills the space between the galaxies in the cluster (Figure 18.5). This gas, also known as the **intracluster medium** (*intra* means "within"), is so hot that it emits primarily X rays and therefore went undetected until the 1960s, when X-ray telescopes were finally launched above the Earth's atmosphere. The temperature of this gas is tens of millions of degrees in many clusters and can exceed 100 million degrees in the largest clusters. Even though the intracluster medium is invisible in optical light, its mass often exceeds the mass of all the visible stars in all the cluster's galaxies combined. The largest clusters of galaxies contain up to five times more matter in the form of hot gas than in the form of stars.

The temperature of the hot intracluster gas depends on the mass of the cluster itself, enabling us to use X-ray telescopes to measure the masses of galaxy clusters. The intracluster medium in most clusters is nearly in a state of *gravitational equilibrium*—that is, the outward gas pressure balances gravity's inward pull [Section 12.1]. In this state of balance, the average kinetic energies of the gas particles are determined primarily by the strength of gravity and hence by the amount of mass within the cluster. Because the temperature of a gas reflects the average kinetic energies of its particles, the gas temperatures we measure with X-ray telescopes tell us the average speeds of the X ray–emitting particles. We can then use these speeds and the orbital velocity law to weigh the cluster.

The results obtained with this method agree well with the results found by studying the orbital motions of the cluster's galaxies. Again, we find that mass-to-light ratios in clusters of galaxies generally exceed 100 solar masses per solar luminosity. Even after accounting for the hot intracluster gas, it is clear that clusters must contain huge quantities of dark matter binding all the galaxies together.

Gravitational Lensing

We have so far relied exclusively on methods derived from Newton's laws, such as the orbital velocity law, to measure galaxy and cluster masses. These laws keep telling us that the universe holds far more matter than we can see, a difficult result for some people to accept. Can we trust these laws to tell the truth? Orbital motions were our only tool for measuring masses until quite recently. Now we have another tool: *gravitational lensing.*

Gravitational lensing occurs because masses distort spacetime—the "fabric" of the universe [Section 14.4]. Massive objects can therefore act as **gravitational lenses** that bend light beams passing nearby. This prediction of Einstein's theory of general relativity was first verified in 1919 during an eclipse of the Sun. Because the light-bending angle of a gravitational lens depends on the mass of the object doing the bending, we can measure the masses of objects by observing how strongly they distort light paths.

Figure 18.6 shows a particularly striking example of how a cluster of galaxies can act as a gravitational lens. Many of the yellow elliptical galaxies concentrated toward the center of the picture belong to the cluster, but at least one of the galaxies pictured does not. At several positions on various sides of the central clump of yellow galaxies you will notice multiple images of the same blue galaxy. Each one of these images, whose sizes differ, looks like a distorted oval with an off-center smudge.

The blue galaxy seen in these multiple images lies almost directly behind the center of the cluster, at a much farther distance. Multiple images arise because photons traveling from this blue galaxy to Earth do not follow straight paths. Instead, the cluster's gravity bends their paths as they pass through the cluster, enabling light from this galaxy to arrive at Earth from a few slightly different directions. Each alternative path produces a separate, distorted image of the blue

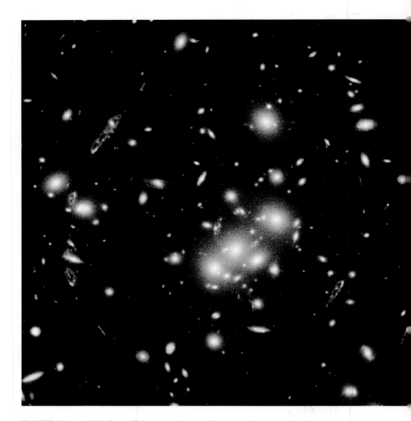

FIGURE 18.6 Hubble Space Telescope picture of a galaxy cluster acting as a gravitational lens. The yellow elliptical galaxies are cluster members, and the odd, blue ovals are multiple images of a single galaxy that lies almost directly behind the cluster's center.

galaxy, making this one galaxy look like several galaxies (Figure 18.7).

Multiple images of a gravitationally lensed galaxy are rare, occurring only when a distant galaxy lies directly behind the lensing cluster, but single images of gravitationally lensed galaxies behind clusters are quite common. Figure 18.8 shows a more typical example. This picture shows numerous normal-looking galaxies and several arc-shaped galaxies. The oddly curved galaxies are not members of the cluster, nor are they really curved. They are normal galaxies lying far beyond the cluster whose images have been distorted by the cluster's gravity.

Careful analyses of the distorted images in pictures of clusters like these enable us to weigh the clusters without resorting to the orbital velocity law. Instead, Einstein's theory of general relativity tells us how massive these clusters must be to generate the observed distortions. It is reassuring that cluster masses derived in this way generally agree with those derived from galaxy velocities and X-ray temperatures. These completely different methods all indicate that clusters of galaxies hold very substantial amounts of dark matter.

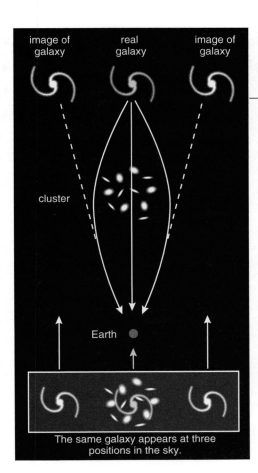

image of galaxy   real galaxy   image of galaxy

cluster

Earth

The same galaxy appears at three positions in the sky.

FIGURE 18.7 A cluster's powerful gravity bends light paths from background galaxies to Earth. If light can arrive from several different directions, we see multiple images of the same galaxy.

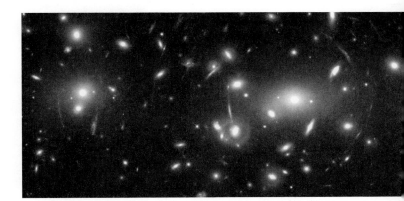

FIGURE 18.8 Hubble Space Telescope picture of the cluster Abell 2218. The thin, elongated galaxies are the images of background galaxies distorted by the cluster's gravity.

TIME OUT TO THINK *Should the fact that we have three different ways of measuring cluster masses give us greater confidence that we really do understand gravity and that dark matter really does exist? Why or why not?*

18.4 Dark Matter: Ordinary or Extraordinary?

We do not yet know what the dark matter in galaxies and clusters of galaxies is composed of, but our educated guesses fit into two basic categories. First, some or all of the dark matter could be *ordinary,* made of protons, neutrons, and electrons. In that case, the only unusual thing about dark matter is that it is dim; otherwise, it is made of the same stuff as all the "bright matter" that we can see. The second possibility is that some or all of the dark matter is *extraordinary,* made of particles that we have yet to discover. Protons and neutrons belong to a category of particles called *baryons,* so ordinary matter is sometimes called *baryonic matter.* By extension, extraordinary matter is called *nonbaryonic matter.*

Ordinary Dark Matter: MACHOs

Matter need not be extraordinary to be dark. In astronomy, "dark" merely means not as bright as a normal star and therefore not visible across vast distances of space. Your body is dark matter. Everything you own is dark matter. Earth and the planets are also dark matter. The process of star formation itself leaves behind dark matter in the form of *brown dwarfs,* those dim "failed stars" that are not quite massive enough to sustain nuclear fusion [Section 14.2]. Some scientists believe that trillions of faint red stars, brown dwarfs, and Jupiter-size objects left over from the Milky Way's formation still roam our galaxy's halo, providing much of its mass. They fancifully term these objects **MACHOs,** for *massive compact halo objects.*

Because MACHOs are too faint for us to see directly, astronomers who wish to detect them must resort to clever techniques. One innovative way to detect faint starlike objects in the Milky Way's halo takes advantage of gravitational lensing. If trillions of these MACHOs really exist, every once in a while a MACHO should drift across our line of sight to a

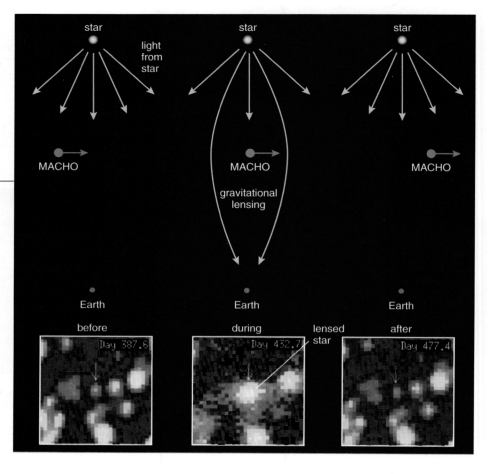

FIGURE 18.9 A lensing event in which a MACHO passing in front of a more distant star temporarily makes the star appear brighter, revealing the presence of the dark MACHO. Multiple images may be present, but the separation between them is too small to see with a telescope.

more distant star. When a MACHO lies almost directly between us and the farther star, the MACHO's gravity will focus the star's light directly toward the Earth. The distant star thus will appear much brighter than usual for several days or weeks as the MACHO passes in front of it (Figure 18.9). We cannot see the MACHO itself, but the duration of the lensing event reveals its mass.

Gravitational lensing events such as these are rare; they happen to about one star in a million each year. To detect MACHO lensing events, we therefore must monitor huge numbers of stars. Current large-scale monitoring projects record numerous lensing events annually. These events demonstrate that MACHOs do indeed populate our galaxy's halo, but probably not enough of them to account for all the Milky Way's dark matter. Something else lurks unseen in the outer reaches of our galaxy. Maybe it is also normal matter made from baryons, but maybe not.

Extraordinary Dark Matter: WIMPs

A more exotic possibility is that most of the dark matter in galaxies and clusters of galaxies is not made of ordinary, baryonic matter at all. Let's begin to ex-

plore this possibility by taking another look at those nonbaryonic particles we discussed in Section 12.3: neutrinos. These unusual particles are dark by their very nature, because they have no electrical charge and hence cannot emit electromagnetic radiation of any kind. Moreover, they are never bound together with charged particles in the way that neutrons are bound in atomic nuclei, so their presence cannot be revealed by associated light-emitting particles. In fact, particles like neutrinos interact with other forms of matter through only gravity and a force called the *weak force*. For this reason, they are said to be *weakly interacting particles*. If you recall that trillions of neutrinos from the Sun are passing through your body at this very moment without doing any damage, you can see that the name *weakly interacting* fits well.

The dark matter in galaxies cannot be made of neutrinos, because these tiny particles travel through the universe at enormous speeds and can easily escape a galaxy's gravitational pull. (However, as we'll discuss shortly, neutrinos might contribute to the amount of dark matter outside galaxies.) But what if other weakly interacting particles exist that are similar to neutrinos but considerably heavier? They too would evade direct detection, but they would move more slowly and could collect in galaxies, adding mass without adding light. Such hypothetical particles are called **WIMPs,** for *weakly interacting massive particles.* WIMPs could make up most of our galaxy's mass, but they would be completely invisible in all wavelengths of light.

It might surprise you that scientists would suspect the universe to be filled with particles they haven't yet discovered. However, WIMPs could also explain why dark matter doesn't behave like the visible matter in galaxies. During the early stages of the Milky Way's formation, the ordinary (baryonic) matter settled toward our galaxy's center and then flattened into a disk. Meanwhile, the dark matter stayed where it was, out in the galaxy's halo. This resistance of the dark matter to settling is exactly what we would expect from weakly interacting particles. Because WIMPs do not emit electromagnetic radiation, they cannot radiate away their energy. They are therefore stuck orbiting out at large distances and cannot collapse with the rest of the protogalactic cloud.

TIME OUT TO THINK *What do you think of the idea that much of the universe is made of as-yet-undiscovered particles? Can you think of other instances in the history of science in which the existence of something was predicted before it was discovered?*

18.5 Structure Formation

Dark matter remains enigmatic, but we are learning more about its role in the universe every year. Because galaxies and clusters of galaxies seem to contain much more dark matter than luminous matter, we believe that dark matter's gravitational pull must be the primary force holding these structures together. Thus, we strongly suspect that the gravitational attraction of dark matter is what pulled galaxies and clusters together in the first place.

Growth of Structure

Stars, galaxies, and clusters of galaxies are all **gravitationally bound systems,** meaning that their gravity is strong enough to hold them together. In most of the gravitationally bound systems we have discussed so far, gravity has completely overwhelmed the expansion of the universe. That is, while the universe as a whole is expanding, space is *not* expanding within our solar system, our galaxy, or our Local Group of galaxies.

Our best guess at how galaxies formed, briefly outlined in Section 17.5, envisions them as growing from slight density enhancements in the early universe.

The expansion of space hindered galaxy formation for millions of years after the birth of the universe. No galaxies arose until gravity pulled the expanding hydrogen and helium gas that filled the universe into protogalactic clouds in which stars could begin to form. If dark matter is indeed the most common form of mass in galaxies, it must have provided most of this gravitational attraction. Regions with the highest densities of dark matter, and therefore the strongest gravity, would have gradually drawn in the surrounding gas. Some of this gas formed stars, but most of the galaxy remained dark. According to this model, the luminous matter in each galaxy must now be nestled inside the larger cocoon of dark matter that initiated the galaxy's growth.

The formation of a cluster of galaxies probably echoes the formation of a galaxy. Early on, all the galaxies that will eventually constitute the cluster are flying apart with the expansion of the universe. But the gravity of the dark matter associated with the cluster eventually reverses the trajectories of these galaxies. The galaxies ultimately fall back inward and start orbiting each other randomly, like the stars in the halo of our galaxy.

Some clusters of galaxies apparently have not yet finished forming: Their immense gravity continues to draw in new members. For example, the nearby Virgo Cluster of galaxies appears to be tugging on our own Local Group. Right now the Local Group is still moving away from the Virgo Cluster, but not as quickly as a simple application of Hubble's law would predict. Recall that Hubble's law lets us calculate the speed at which universal expansion carries us away from the Virgo Cluster [Section 17.3], but it does not account for any gravitational effects. The discrepancy between the predictions of Hubble's law and our actual velocity away from the Virgo Cluster is about 400 km/s. That is, we are moving away from the Virgo Cluster 400 km/s more slowly than we would be as a result of the expansion of the universe alone. Astronomers call this 400-km/s deviation from Hubble's law a **peculiar velocity,** but it's really not so peculiar: It is simply the effect of the gravitational attraction pulling us back toward Virgo against the flow of universal expansion. The overall effect is rather like that of swimming upstream against a strong current: You move "upstream" relative to other objects floating in the current, but you're still headed "downstream" relative to the shore because of the strong current. The speed of the current is like the speed of expansion of the universe, and your swimming speed "upstream" is your peculiar velocity.

Just as the Earth's gravity slows a rising baseball and eventually turns it around and pulls it back toward the ground, the gravitational tug of the Virgo Cluster may eventually turn the galaxies of our Local Group around and pull them into the cluster. Similar

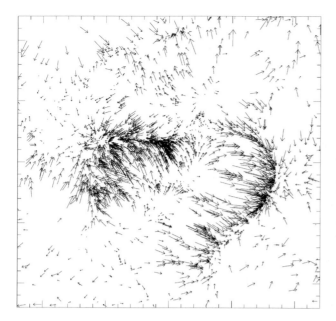

FIGURE 18.10 Peculiar velocities of galaxies flowing into superclusters. Each arrow shows the peculiar velocity of a galaxy inferred from a combination of observations and modeling. The Milky Way is at the center of the picture, and not all galaxies are shown. The area pictured is about 600 million light-years from side to side. Note how the galaxies tend to flow into regions where the density of galaxies is already high. These vast, high-density regions are probably superclusters in the process of formation.

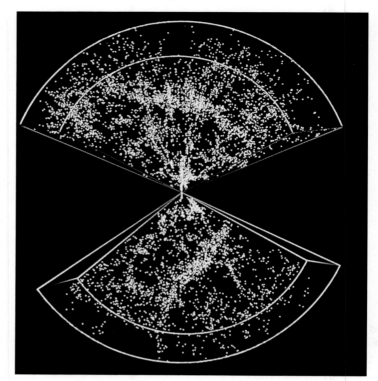

FIGURE 18.11 This "slice" shows galaxies in a thin, fanlike swath extending outward from the Earth. Note that galaxies are not scattered randomly but instead seem to lie along sheets and strings interspersed with voids that contain very few galaxies. The lower swath is similar to the upper swath, but it points in a different direction.

processes are taking place on the outskirts of other large clusters of galaxies, where we see many galaxies with large peculiar velocities pointing in the direction of the cluster. Their peculiar velocities indicate that the cluster's gravity is pulling on them. Eventually, some or maybe all of these galaxies will fall into the cluster. Thus, many clusters are still attracting galaxies, adding to the hundreds they already contain. On even larger scales, clusters themselves have surprisingly large peculiar velocities, hinting that they are parts of even bigger gravitationally bound systems, called **superclusters,** that are just beginning to form (Figure 18.10).

Large-Scale Structures

Beyond about 300 million light-years (100 Mpc), we can no longer measure distances accurately enough to determine peculiar velocities. We can estimate the distances to faraway galaxies only by measuring their redshifts and applying Hubble's law. Today we have redshift measurements for thousands upon thousands of such galaxies. When we convert the redshifts to distance estimates, we can create three-dimensional

maps of the universe. Such maps have been made for only a few "slices" of the sky so far, but they are already revealing **large-scale structures** much vaster than clusters of galaxies.

One of the most famous depictions of large-scale structure is the "slice of the universe" pictured in Figure 18.11. Many astronomers collaborated to measure the redshifts of all the galaxies brighter than an agreed-upon limit in this particular slice of the sky. This project dramatically revealed the complex structure of our corner of the universe. Each dot in the picture represents an entire galaxy of stars. The arrangement of the dots reveals huge sheets of galaxies spanning many millions of light-years. Clusters of galaxies are located at the intersections of these sheets. Between the sheets of galaxies lie giant empty regions called **voids.**

Some of the structures we see in the universe are amazingly large. The so-called *Great Wall* of galaxies stretches across an expanse measuring some 180 million light-years (50 Mpc) side to side. Immense structures such as these apparently have not yet collapsed into randomly orbiting, gravitationally bound systems. They haven't had enough time. The universe

$z = 6.62$ | $z = 4.41$ | $z = 3.32$

$z = 2.41$ | $z = 1.83$ | $z = 1.49$

FIGURE 18.12 Frames from a supercomputer simulation of structure formation. As time progresses, the higher-density portions of the universe, shown in black, continue to attract more and more matter. Note how similar the filamentary structures in this simulation are to the slice of the universe in Figure 18.11.

may still be growing structures on ever larger scales, even though it is billions of years old.

Most astronomers believe that all these large-scale structures grew from slight density enhancements in the early universe, just as galaxies did. Galaxies, clusters, superclusters, and the Great Wall probably all started as mildly high-density regions of different sizes. The voids in the distribution of galaxies probably started as mildly low-density regions.

If this picture of structure formation is correct, then all the structures we see in today's universe mirror the original distribution of dark matter very early in time. Supercomputer models of structure formation in the universe can now simulate the growth of galaxies, clusters, and larger structures from very tiny density enhancements as the universe evolves. The results of these models look remarkably similar to the slice of the universe in Figure 18.11, bolstering our confidence in this scenario (Figure 18.12). However, the models do not tell us *why* the universe started with these slight density enhancements—that is a topic for the next chapter. Nevertheless, it seems increasingly clear that these "lumps" in the early universe were the seeds of all the marvelous structures we see today.

18.6 Fire or Ice?

Some say the world will end in fire,
Some say in ice.
From what I've tasted of desire
I hold with those who favor fire.
But if it had to perish twice,
I think I know enough of hate
To say that for destruction ice
Is also great
And would suffice.

ROBERT FROST

We now arrive at one of the ultimate questions in astronomy: How will the universe end? Edwin Hubble's work established that the galaxies in the universe are rapidly flying away from one another [Section 17.3], but the gravitational pull of each galaxy on every other galaxy acts to slow the expansion. Will the expansion of the universe ever slow to a halt? Will gravity ever reverse the expansion and pull everything back together in a cataclysmic crunch? The ultimate fate of the universe boils down to a simple question: Does the universe have enough kinetic energy to escape its own gravitational pull?

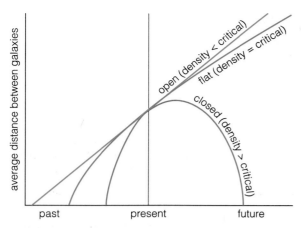

FIGURE 18.13 The expansion rate and fate of the universe depend on whether its density is greater than or less than the critical density. A rising curve means that the universe is expanding, and a falling curve means that it is contracting. The universe is expanding at present. If the universe is open or flat, it will continue to expand forever. If it is closed, it will someday stop expanding and begin contracting.

Hubble's constant essentially tells us the kinetic energy of the universe, but we do not yet know the overall strength of the universe's gravitational pull. The strength of this pull depends on the *density* of matter in the universe. The greater the density, the greater the overall strength of gravity and the higher the likelihood that gravity will someday halt the expansion. Precise calculations show that gravity will win out over expansion if the current density of the universe exceeds a seemingly minuscule 10^{-29} gram per cubic centimeter. This density is roughly equivalent to a few hydrogen atoms in a volume the size of a closet. But the universe might not be even this dense—we simply do not yet know.

The precise density marking the dividing line between eternal expansion and eventual contraction is called the **critical density.** Figure 18.13 illustrates the three basic possibilities:

- If the mass density of the universe is larger than the critical density, the collective gravity of all its matter will eventually halt the expansion and reverse it. The galaxies will come crashing back together, and the entire universe will end in a fiery "big crunch." We call this type of universe *closed*.

- If the mass density of the universe is smaller than the critical density, the collective gravity of all its matter cannot halt the expansion, and the universe will keep growing forever into the future. In this kind of universe, called an *open* universe, the galaxies will never come back together. The universe will cool into icy darkness.

- If the density of the universe equals the critical density, the collective gravity of all its matter is exactly the amount needed to balance the expansion. The universe will never collapse, but it will expand more and more slowly as time progresses. Such a precisely balanced universe is sometimes called a *flat* universe.

TIME OUT TO THINK *Do you think that one of the potential fates of the universe is preferable to the others? If so, why? If not, why not?*

Counting all the luminous matter in the universe shows that the mass contained in stars falls far short of the critical density. The visible parts of galaxies contribute less than 1% of the mass density needed to halt the universe's expansion. The fate of the universe thus rests with the dark matter. Is there enough dark matter to halt the expansion of the universe?

Recall that we can estimate the amount of dark matter by looking at the mass-to-light ratio. It turns out that for dark matter to contribute enough mass for the universe to have the critical density, the average mass-to-light ratio must be approximately 1,000 solar masses per solar luminosity throughout the universe. Clusters of galaxies have mass-to-light ratios of a few hundred solar masses per solar luminosity, still a few times less than the ratio needed for a closed universe. If the proportion of dark matter in the universe at large is similar to that in clusters, then the universe is open, with a density currently equal to about 25% of the critical density. For the universe to be closed, even more dark matter must lie beyond the boundaries of galaxy clusters.

Amazingly, tiny ghostlike neutrinos might supply enough mass to raise the universe's density to the critical density. The masses of individual neutrinos are still too small to measure, but what neutrinos lack in mass they might make up in sheer numbers—neutrinos are quite likely the most common particles in the universe other than photons. Because neutrinos zip through the universe so quickly, they do not necessarily collect into clusters of galaxies. However, they might gravitate into large-scale structures, where their influence would be measurable.

The peculiar velocities of galaxies allow us to probe the distribution of dark matter in large-scale structures, but studies of these departures from Hubble's law remain inconclusive. Some researchers have found that large-scale structures have even higher mass-to-light ratios than clusters of galaxies, indicating that the universe's mass density could be close to the critical value. As of 1999, however, most other studies hold the line near the 25% value we infer from clusters. The question of the universe's fate thus remains unanswered, with an eternally expanding universe looking like the odds-on favorite.

This is the way the world ends
This is the way the world ends
This is the way the world ends
Not with a bang but a whimper.

FROM *The Hollow Men* BY T. S. ELIOT

THE BIG PICTURE

In this chapter, we have found that there may be much more to the universe than meets the eye. Dark matter too dim for us to see seems to far outweigh the stars. Here are some key "big picture" points to remember about this mysterious matter:

■ Either dark matter exists, or we do not understand how gravity operates across galaxy-size distances. We have many reasons to have confidence in our understanding of gravity, so the majority of astronomers believe that dark matter is real.

■ Measurements of the mass and luminosity of galaxies and galaxy clusters indicate that they contain far more mass in dark matter than in stars.

■ Despite the fact that dark matter is by far the most abundant form of mass in the universe, we still have little idea what it is.

■ Superclusters, walls, and voids much larger than clusters of galaxies extend many millions of light-years across the universe. Each of these structures probably began as a very slight enhancement in the density of dark matter early in time, and these enormous structures are still in the process of forming.

■ Dark matter holds the key to the fate of the universe. If there is enough of it, its gravity will cause the expansion of the universe to halt and reverse someday. Current indications are that there is not enough dark matter to halt the expansion, in which case the universe will continue to expand forever.

Review Questions

1. In what sense is dark matter "dark"?

2. What evidence suggests that the Milky Way contains dark matter? What alternative explanation might be possible? Why do most astronomers prefer the idea of dark matter to the alternative explanation?

3. Briefly explain why it is more difficult to measure a galaxy's mass than to measure its luminosity.

4. Briefly describe how we construct rotation curves for spiral galaxies and how these curves lead us to conclude that spiral galaxies contain dark matter.

5. What is a *mass-to-light ratio?* Explain why higher mass-to-light ratios imply more dark matter.

6. How do the mass-to-light ratios of inner regions of galaxies compare to the mass-to-light ratios we find when we look farther from a galaxy's center? What does that tell us about dark matter in galaxies?

7. Briefly describe the three different ways of measuring the mass of a cluster of galaxies. What do the results tell us about dark matter in galaxy clusters?

8. What do we mean by the *intracluster medium* in galaxy clusters? Why wasn't it discovered prior to the 1960s? How does its temperature tell us about the mass of a cluster?

9. What is gravitational lensing? Why does it occur? How can we use it to estimate the masses of lensing objects?

10. Briefly explain what we mean by baryonic and nonbaryonic matter. Which type are you made of?

11. What do we mean by *MACHOs?* Describe several possibilities for the nature of MACHOs.

12. How can gravitational lensing events tell us about MACHOs? Based on current evidence, can MACHOs explain all the dark matter in the Milky Way? Why or why not?

13. What do we mean by a gravitationally bound system? Why isn't space expanding within systems such as our solar system or the Milky Way?

14. What are *peculiar velocities?* What causes them?

15. Briefly describe the appearance of the universe on the largest scales, including the distribution of galaxies in great sheets and the voids between the sheets.

16. How is the fate of the universe related to its density?

17. Explain why the total amount of dark matter in the universe holds the key to its final fate. According to current evidence about the amount of dark matter, what is the fate of the universe?

Discussion Questions

1. *Dark Matter or Revised Gravity.* One possible explanation for the large mass-to-light ratios we measure in galaxies and clusters is that we are currently using the wrong law of gravity to measure the masses of very

large objects. If we really do misunderstand gravity, then many fundamental theories of physics, including Einstein's theory of general relativity, will need to be revised. Which explanation for these mass-to-light ratios do you find more appealing, dark matter or revised gravity? Explain why. Why do you suppose most astronomers find dark matter more appealing?

2. *Our Fate.* Scientists, philosophers, and poets alike have speculated on the fate of the universe. How would you prefer the universe as we know it to end, in a big crunch or through eternal expansion? Explain the reasons behind your choice.

Problems

1. *Mass-to-Light Ratio.* Suppose you discovered a galaxy with a mass-to-light ratio of 0.1 solar mass per solar luminosity. Would you be surprised? Explain why or why not. What would this measurement say about the nature of the stars in this galaxy?

2. *Web Project: Gravitational Lenses.* Gravitational lensing occurs in numerous astronomical situations. Com-

pile a catalog of examples from the Web. Try to find pictures of lensed stars, quasars, and galaxies. Supply a one-paragraph explanation of what is going on in each picture.

3. *Research Project: Fate of the Universe.* We are continually learning more about the amount of dark matter in the universe, and the question of whether the universe is open or closed is therefore constantly revisited. Search for the most recent information about dark matter and the fate of the universe, preferably by finding news articles on this topic published within the past year. Write a one- to three-page report on your findings. Be sure to state whether, according to the most recent evidence, the universe appears to be open or closed.

4. *Research Project: The Nature of Dark Matter.* Find and study recent reports on new ideas about the possible nature of dark matter. Write a one- to three-page report that summarizes the latest ideas about what dark matter is made of.

The Beginning of Time

THE UNIVERSE HAS BEEN EXPANDING FOR OVER 10 BILLION YEARS. DURING that time, matter collected into galaxies. Stars formed in those galaxies, producing heavy elements that were recycled into later generations of stars. One of these late-coming stars formed about 4.6 billion years ago, in a remote corner of a galaxy called the Milky Way. This star was born with a host of planets that formed in a flattened disk surrounding it. One of these planets soon became covered with life that gradually evolved into more and more complex forms. Today, the most advanced life-forms on this planet, human beings, can look back on this series of events and marvel at how the universe created conditions suitable for life.

To this point in the book, we have discussed how the matter in the early universe collected into galaxies, stars, planets, and ultimately people. We have seen that we are star stuff, and galaxy stuff as well. More to the point, we now recognize that we are the product of more than 10 billion years of cosmic evolution. We have one question left to address: Where did the *matter itself* come from?

To answer this ultimate question, we must go beyond the most distant galaxies and even beyond what we can see near the horizon of the universe. We must go back not only to the origins of matter and energy but to the beginning of time itself.

19.1 Running the Expansion Backward

Hubble's discovery that the universe is expanding led to the insight that all things very likely sprang into being at a single moment in time, in an event that we have come to call the *Big Bang*. Today, powerful telescopes allow us to view how galaxies have changed over the last 10 billion years, and at great distances we see young galaxies still in the process of forming [Section 17.4]. These observations confirm that the universe is gradually aging, just as we should expect if the entire universe really was born some 10–16 billion years ago.

Unfortunately, we cannot see back to the very beginning of time. Light from the most distant galaxies shows us what the universe looked like when it was a few billion years old. Beyond these galaxies, we have not yet found any objects shining brightly enough for us to see them. But we face an even more fundamental problem. The universe is filled with a faint glow of radiation that appears to be the remnant heat of the Big Bang. This faint glow comes from a time when the universe was about 300,000 years old, and it prevents us from seeing directly to earlier times. Thus, just as we must rely on theoretical modeling to determine what the Earth or the Sun is like on the inside, we must use modeling to investigate what happened in the universe in its first 300,000 years.

Fortunately, calculating conditions in the early universe is straightforward if we consider some fundamental principles. We know that the expanding universe is cooling and becoming less dense as it grows. Therefore, the universe must have been hotter and denser in the past. Calculating exactly how hot and dense the universe must have been when it was more compressed is similar to calculating the temperatures and densities in a car engine as the pistons compress the fuel mix—except that the conditions become much more extreme. Figure 19.1 shows just how hot the universe was during its earliest moments, according to such calculations. To understand these early moments, we thus need to know how matter and energy behave under such extraordinarily hot conditions.

Advances made in modern physics during the latter half of the twentieth century enable us to predict the behavior of matter and energy under conditions far more extreme than in the centers of the hottest stars. When we apply our understanding of modern physics to the early universe, we find something so astonishing that it almost sounds preposterous: Today, we may understand the behavior of matter and energy well enough to describe what was happening in the universe just *one ten-billionth* (10^{-10}) of a second after the Big Bang. And that's not all. While our understanding of physics is less

FIGURE 19.1 Temperature of the universe from the Big Bang to the present (10^{10} years $\approx 3 \times 10^{17}$ seconds).

certain under the more extreme conditions that prevailed even earlier, we have some ideas about what the universe was like when it was a mere 10^{-35} second old, and perhaps a glimmer of what it was like at the age of just 10^{-43} second. These tiny fractions of a second are so small that, for all practical purposes, we are studying the very moment of creation.

19.2 A Scientific History of the Universe

The **Big Bang theory**—the scientific theory of the universe's earliest moments—presumes that all we see today, from Earth to the cosmic horizon, began as an incredibly tiny, hot, and dense collection of matter and radiation. It describes how expansion and cooling of this unimaginably intense mixture of particles and photons could have led to the present universe of stars and galaxies, and it explains several aspects of today's universe with impressive accuracy. Our ideas about the earliest moments of the universe are quite speculative because they depend on aspects of physics that we do not yet fully comprehend. However, the evidence supporting this theory grows stronger and stronger as we focus on progressively later moments in time. We will discuss the evidence supporting the Big Bang theory later in this chapter. First, in order to help you understand the significance of the evidence, let's examine the story of creation according to this theory.

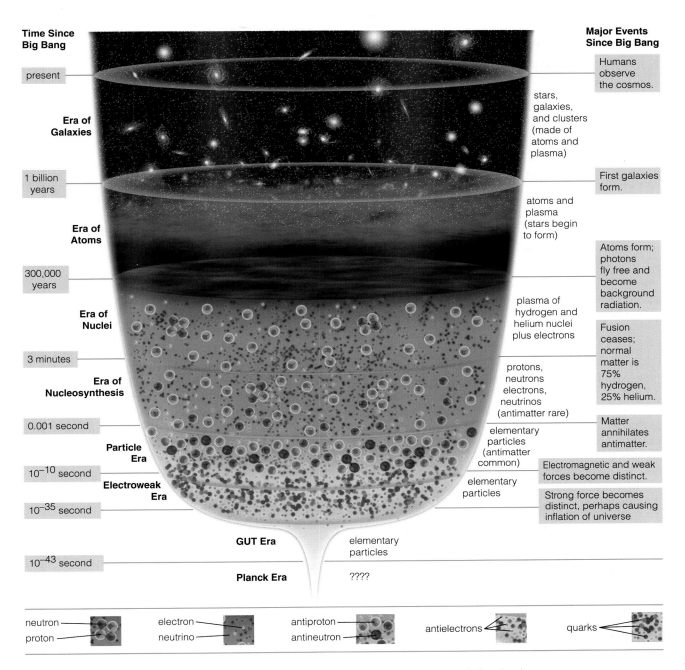

Time Since Big Bang

present

Era of Galaxies

1 billion years

Era of Atoms

300,000 years

Era of Nuclei

3 minutes

Era of Nucleosynthesis

0.001 second

Particle Era

10^{-10} second

Electroweak Era

10^{-35} second

GUT Era

10^{-43} second

Planck Era

Major Events Since Big Bang

Humans observe the cosmos.

stars, galaxies, and clusters (made of atoms and plasma)

First galaxies form.

atoms and plasma (stars begin to form)

Atoms form; photons fly free and become background radiation.

plasma of hydrogen and helium nuclei plus electrons

Fusion ceases; normal matter is 75% hydrogen, 25% helium.

protons, neutrons electrons, neutrinos (antimatter rare)

Matter annihilates antimatter.

elementary particles (antimatter common)

Electromagnetic and weak forces become distinct.

elementary particles

Strong force becomes distinct, perhaps causing inflation of universe

elementary particles

????

neutron — proton

electron — neutrino

antiproton — antineutron

antielectrons

quarks

FIGURE 19.2 This diagram summarizes the eras of the universe. The names of the eras and their ending times are indicated on the left, and the state of matter during each era and the events marking the end of each era are indicated on the right. The times given in this chapter are probably accurate to the nearest power of 10.

Figure 19.2 summarizes the story by dividing the overall history of the universe into a series of *eras*, or time periods. Each era is distinguished from the next by some major change in the conditions of the universe. You'll find it easiest to keep track of the various eras if you refer back to this figure as we discuss each era in detail.

Before we delve into the story of creation, we must discuss one important difference between the early universe and the present. The early universe was so hot that photons could transform themselves into matter, and vice versa, in accordance with Einstein's formula $E = mc^2$ [Section 4.2]. Physicists can produce reactions that create and destroy matter in their laboratories, but these reactions are now relatively rare in the universe at large.

One example of such a reaction is the creation or destruction of an *electron–antielectron pair.* When two photons collide with a total energy greater than twice the mass-energy of an electron (i.e., the

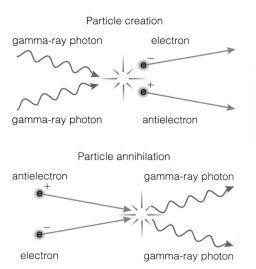

Particle creation

gamma-ray photon   electron

e−

e+

gamma-ray photon   antielectron

Particle annihilation

antielectron   gamma-ray photon

e+

e−

electron   gamma-ray photon

FIGURE 19.3 Electron–antielectron creation and annihilation.

electron's mass times c^2), they can create two brand-new particles: a negatively charged electron and its positively charged twin, the *antielectron,* also known as a *positron* (Figure 19.3). The electron is a particle of **matter,** and the antielectron is a particle of **antimatter.** The reaction that creates an electron–antielectron pair also runs in reverse: When an electron and an antielectron meet, they *annihilate* each other totally, transforming all their mass-energy back into photon energy.

Similar reactions can produce or destroy any particle–antiparticle pair, such as a proton and antiproton or neutron and antineutron. The early universe was therefore filled with an extremely dynamic blend of photons, matter, and antimatter, converting furiously back and forth. However, despite all these vigorous reactions, describing conditions in the early universe is straightforward because we can calculate the proportions of the various forms of radiation and matter from the universe's temperature and density at any time.

The First Instant

The first 10^{-43} second of the universe's history is called the **Planck era** after physicist Max Planck, one of the founders of the science of quantum mechanics. According to the laws of quantum mechanics and general relativity, random mass and energy fluctuations were so large during the Planck era that our current theories are powerless to describe what might have been happening. The problem is that we do not yet have a theory that links quantum mechanics and general relativity. In a sense, quantum mechanics is our successful theory of the very small (subatomic particles), and general relativity is our successful theory of the very big (the large-scale

structure of spacetime). Perhaps someday we will be able to merge these theories of the very small and the very big into a single "theory of everything." Until that happens, science cannot describe the universe's first instant.

The GUT Era

Understanding the transition that marked the beginning of the next era requires thinking in terms of the *forces* that operate in the universe. Everything that happens in the universe today is governed by four distinct forces: *gravity, electromagnetism,* the *strong force,* and the *weak force.* Gravity dominates large-scale action, and electromagnetism dominates chemical and biological reactions. The strong and weak forces determine what happens within atomic nuclei: The strong force binds nuclei together, and the weak force mediates nuclear reactions such as fission and fusion.

These forces were not so distinct in the early universe. Physicists have shown that the electromagnetic and weak forces lose their separate identities under conditions of very high temperature or energy and merge together into a single *electroweak force.* Physicists also believe that the electroweak force and the strong force ultimately lose their separate identities at even higher energies (Figure 19.4). The theories that predict this merger of the electroweak and strong forces are called **grand unified theories,** or **GUTs** for short; their merger is usually referred to as the *GUT force.* According to these theories, when the universe entered the **GUT era** at an age of 10^{-43} second, only two forces operated in the universe: gravity and the GUT force.

The GUT era lasted but a tiny fraction of a second, coming to an end when the universe had cooled to 10^{27} K at an age of 10^{-35} second. Grand unified theories predict that the strong force separated from the GUT force at this point, leaving three forces operating in the universe: gravity, the strong force, and the electroweak force. As we will discuss later in the chapter, it is possible that this separation of the strong force released an enormous amount of energy. This energy release would have caused the universe to undergo a sudden and dramatic expansion that we call **inflation.** In the space of a mere 10^{-33} second, pieces of the universe the size of an atomic nucleus may have grown to the size of our solar system!

The Electroweak Era

The end of the GUT era marks the beginning of the **electroweak era,** so named because the electromagnetic and weak forces were still unified into the electroweak force. Intense radiation filled all of space, as it had since the Planck era, spontaneously producing matter and antimatter particles of many different

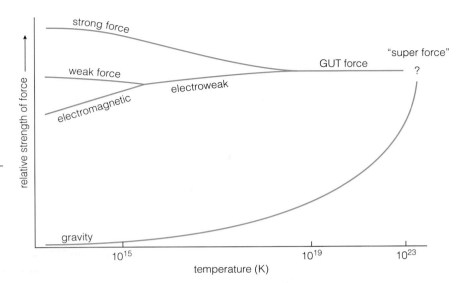

FIGURE 19.4 The four forces are distinct at low temperatures but merge at very high temperatures.

sorts that immediately reverted back to energy. The universe continued to expand and cool throughout the electroweak era, dropping to a temperature of 10^{15} K when it reached an age of 10^{-10} second. This temperature is still 100 million times hotter than the temperature in the core of the Sun, but it was low enough for the electromagnetic and weak forces to separate from the electroweak force. After this instant (10^{-10} s), all four forces were forever distinct in the universe.

The end of the electroweak era marks an important transition not only in the universe, but also in human understanding of the universe. The theory that unified the weak and electromagnetic forces, developed in the 1970s, predicted the emergence of new types of particles (called the *W* and *Z*, or *weak, bosons*) when temperatures rose above the 10^{15} K that pervaded the universe when it was 10^{-10} second old. In 1983, experiments performed in a huge particle accelerator called CERN reached energies equivalent to such high temperatures for the first time. The new particles showed up just as predicted, produced from the extremely high energy in accord with $E = mc^2$. Thus, we have direct experimental evidence concerning the conditions in the universe at the end of the electroweak era. We do not have any direct experimental evidence for prior to that time. Thus, our theories concerning the earlier parts of the electroweak era and the GUT era are much more speculative than our theories describing the universe from the end of the electroweak era to the present.

The Particle Era

Spontaneous creation and annihilation of particles continued throughout the next era, which we call the **particle era** because tiny particles of matter were as numerous as photons. These particles included electrons, neutrinos, and *quarks*—the building blocks

of heavier particles such as protons, antiprotons, neutrons, and antineutrons. Near the end of the particle era, when the universe was about 0.0001 second old, the temperature grew too cool for quarks to exist on their own, and quarks combined in groups of three to form protons and neutrons. The particle era came to an end when the universe reached an age of 1 millisecond (0.001 s) and the temperature had fallen to 10^{12} K. At this point, it was no longer hot enough to spontaneously produce protons and antiprotons (or neutrons and antineutrons) from pure energy.

If the universe had contained equal numbers of protons and antiprotons (or neutrons and antineutrons) at the end of the particle era, all of the pairs would have annihilated each other, creating photons and leaving essentially no matter in the universe. In today's universe, photons outnumber protons by over a billion to one and appear to outnumber antiprotons by a much larger factor. Thus, we conclude that protons must have slightly outnumbered antiprotons at the end of the particle era: For every billion antiprotons, there were about a billion and one protons. Thus, for each 1 billion protons and antiprotons that annihilated each other, creating photons, a single proton was left over. This seemingly slight excess of ordinary matter over antimatter makes up all the matter in the present-day universe. That is, the protons (and neutrons) left over from when the universe was 0.001 second old are the very ones that make up our bodies.

The Era of Nucleosynthesis

So far, everything we have discussed occurred within the first 0.001 second of the universe's existence—a time span shorter than the time it takes you to blink an eye. At this point, the protons and neutrons left over after the annihilation of antimatter attempted

Einstein's Greatest Blunder

Shortly after Einstein completed his general theory of relativity in 1915, he found that it predicted that the universe could not be standing still: The mutual gravitational attraction of all the matter would make the universe collapse. However, Einstein thought at the time that the universe should be eternal and static, so he decided to alter his equations. In essence, he inserted a "fudge factor," which he called the *cosmological constant,* that acted as a repulsive force to counteract the attractive force of gravity. Had he not been so convinced that the universe should be standing still, he might instead have come up with the correct explanation for why the universe is not collapsing: because it is still expanding from the event of its birth. After Hubble discovered that the universe is expanding, Einstein called his invention of the cosmological constant the greatest blunder of his career.

Despite Einstein's disavowal, variations both on the idea of an eternal universe and on the cosmological constant continued to crop up over the years.

One of the cleverest ideas, developed in the late 1940s, was called the *steady state theory* of the universe. This theory accepted the fact that the universe is expanding but rejected the idea of a Big Bang and instead postulated that the universe is infinitely old. This idea may seem paradoxical at first: If the universe has been expanding forever, shouldn't every galaxy be infinitely far away from every other galaxy? The steady state theory suggested that new galaxies continually form in the gaps that open up as the universe expands, thereby keeping the average distance between galaxies the same at all times. In a sense, the steady state theory said that the creation of the universe is an ongoing and eternal process rather than having happened all at once with a Big Bang.

Two key discoveries ruled out the steady state theory as an alternative to the Big Bang. First, the 1965 discovery of the cosmic background radiation matched a prediction of the Big Bang theory but could not be adequately explained by the steady state

---

to fuse into heavier nuclei. However, the heat of the universe remained so high that most nuclei broke apart as fast as they formed. This dance of fusion and break-up marked the **era of nucleosynthesis.**

The era of nucleosynthesis ended when the universe was about 3 minutes old. By this time, the density in the expanding universe had dropped so much that fusion no longer occurred, even though the temperature was still about a billion Kelvin (10^9 K)—much hotter than the temperature at the center of the Sun today. When fusion ceased, about 75% of the mass of the ordinary (baryonic) matter in the universe remained as individual protons, or hydrogen nuclei. The other 25% of this mass had fused into helium nuclei, with trace amounts of deuterium (hydrogen with a neutron) and lithium. Except for the small amount of matter that stars later forged into heavier elements, the chemical composition of the universe remains unchanged today.

The Era of Nuclei

At the end of the era of nucleosynthesis, the universe consisted of a very hot plasma of hydrogen nuclei, helium nuclei, and free electrons. This basic picture held for about the next 300,000 years as the universe continued to expand and cool. The fully ionized nuclei moved independently of electrons during this

period (rather than being bound with electrons in neutral atoms), which we call the **era of nuclei.** Throughout this era, photons bounced rapidly from one electron to the next, just as they do deep inside the Sun today, never managing to travel far between collisions. Any time a nucleus managed to capture an electron to form a complete atom, one of the photons quickly ionized it.

The era of nuclei came to an end when the expanding universe was about 300,000 years old, at which point the temperature had fallen to about 3,000 K—roughly half the temperature of the Sun's surface today. Hydrogen and helium nuclei finally captured electrons for good, forming stable, neutral atoms for the first time. With electrons now bound into atoms, the universe suddenly became transparent. Photons, formerly trapped among the electrons, began to stream freely across the universe. We still see these photons today as the *cosmic background radiation,* which we will discuss shortly.

The Era of Atoms and the Era of Galaxies

We've already discussed the rest of the story in earlier chapters. The end of the era of nuclei marked the beginning of the **era of atoms,** when the universe consisted of a mixture of neutral atoms and theory. Second, the steady state theory predicts that the universe looks about the same at all times, which is inconsistent with observations showing that galaxies at great distances look younger than nearby galaxies. As a result of these predictive failures, the steady state theory is no longer taken seriously by most astronomers.

More recently, astronomers have begun to take the idea of a cosmological constant more seriously. In the mid-1990s, a few observations suggested that the ages of the oldest stars are slightly older than the age of the universe derived from Hubble's constant [Section 17.4]. Clearly, stars cannot be older than the universe. If these observations were being interpreted correctly, the universe had to be older than Hubble's constant implies. Recall that we estimate the age of the universe by running the expansion backward and by assuming that the rate of expansion has either remained steady or slowed due to gravity. If instead the expansion rate has accelerated, so that the universe is expanding faster today than it was in the past, then the age of the universe is greater than that ordinarily found from Hubble's constant. What could cause the expansion of the universe to accelerate over time? The repulsive force represented by a cosmological constant, of course.

Although further study of the troubling observations suggests that the stars probably are *not* older than the age of the universe derived from Hubble's constant, other observations also suggest a possible need for a cosmological constant. As of 1999, measurements of distances to high-redshift galaxies (using white-dwarf supernovae as standard candles) suggest that the expansion *is* accelerating. A cosmological constant could account for this startling finding, but we'll need many more observations before we'll know whether it is correct. In either case, scientists are attempting to use variations of the cosmological constant to explain some of the remaining puzzles associated with the Big Bang theory, such as the precise nature of inflation. Einstein's greatest blunder, it seems, just won't go away.

plasma. Thanks to the slight density enhancements present in the universe at this time and the gravitational attraction of dark matter, the atoms and plasma slowly assembled into protogalactic clouds [Section 18.5]. Stars formed in these clouds, transforming the gas clouds into galaxies. The first full-fledged galaxies had formed by the time the universe was about 1 billion years old, beginning what we call the **era of galaxies.**

The era of galaxies continues to this day. Generation after generation of star formation in galaxies steadily builds elements heavier than helium and incorporates them into new star systems. Some of these star systems develop planets, and on at least one of these planets life burst into being a few billion years ago. And here we are, thinking about it all. Describing both the follies and the achievements of the human race, Carl Sagan once said, "These are the things that hydrogen atoms do—given 15 billion years of cosmic evolution."

19.3 Evidence for the Big Bang

Like any scientific theory, the Big Bang theory is a model designed to explain a set of facts. If it is close to the truth, it should be able to make predictions about the real universe that we can verify through observations or experiments. The Big Bang model has gained wide scientific acceptance for two key reasons:

- The Big Bang model predicts that the radiation that began to stream across the universe at the end of the era of nuclei should still be present today. Sure enough, we find that the universe is filled with what we call **cosmic background radiation.** Its characteristics precisely match what we expect according to the Big Bang model.

- The Big Bang model predicts that some of the original hydrogen in the universe should have fused into helium during the era of nucleosynthesis. Observations of the actual helium content of the universe closely match the amount of helium predicted by the Big Bang model. Fusion of hydrogen to helium in stars could have produced only about 10% of the observed helium.

The Cosmic Background Radiation

The first major piece of evidence supporting the Big Bang theory arrived in 1965. Arno Penzias and Robert Wilson, two physicists working at Bell Laboratories in New Jersey, were calibrating a sensitive microwave antenna designed for satellite communications (Figure 19.5). (*Microwaves* fall within the radio portion

FIGURE 19.5 Arno Penzias and Robert Wilson with the Bell Labs microwave antenna.

Meanwhile, physicists at nearby Princeton University were busy calculating the expected characteristics of the radiation left over from the heat of the Big Bang. They concluded that, if the Big Bang had really occurred, this radiation should be permeating the entire universe and should be detectable with a microwave antenna. On a fateful airline trip home from an astronomical meeting, Penzias sat next to an astronomer who told him of the Princeton calculations. The Princeton group soon met with Penzias and Wilson to compare notes. The "noise" in the Bell Labs antenna was not an embarrassment after all. Instead, it was the cosmic background radiation—and the first strong evidence that the Big Bang had really happened. Penzias and Wilson received the 1978 Nobel Prize in physics for their discovery of the cosmic background radiation.

The cosmic background radiation consists of photons arriving at Earth directly from the end of the era of nuclei, when the universe was about 300,000 years old. Because neutral atoms could remain stable for the first time, they captured most of the electrons in the universe. With no more free electrons to block them, the photons from that epoch have flown unobstructed through the universe ever since (Figure 19.6). When we observe the cosmic background radiation, we essentially are seeing back to a time when the universe was only 300,000 years old. Surprisingly, it does not take a particularly powerful telescope to "see" this radiation. In fact, you can pick it up with an ordinary television antenna. If you set an antenna-fed television (i.e., *not* cable or satellite TV) to a channel for which there is no local station, you will see a screen full of static "snow." About 1%

of the electromagnetic spectrum.) Much to their chagrin, they kept finding unexpected "noise" in every measurement they made with the antenna. The noise was the same no matter where they pointed their antenna, indicating that it came from all directions in the sky and ruling out the possibility that it came from any particular astronomical object or from any place on the Earth. Embarrassed by their inability to explain the noise, Penzias and Wilson prepared to "bury" their discovery at the end of a long scientific paper about their antenna.

FIGURE 19.6 Photons frequently collided with free electrons during the era of nuclei and thus could travel freely only after electrons became bound into atoms. The photons released at the end of the era of nuclei make up the cosmic background radiation.

FIGURE 19.7 Spectrum of the cosmic background radiation from COBE. A theoretically calculated thermal radiation spectrum (smooth curve) for a temperature of 2.73 K perfectly fits the data (dots).

of this static is due to photons in the cosmic background radiation.

The cosmic background radiation came from the heat of the universe itself and therefore should have an essentially perfect thermal radiation spectrum [Section 6.3]. When this radiation first broke free 300,000 years after the Big Bang, the temperature of the universe was about 3,000 K, not too different from that of a red-giant star's surface. Thus, the spectrum of the cosmic background radiation originally peaked in visible light, just like the thermal radiation from a red star, with wavelengths of a few hundred nanometers. However, the universe has expanded by a factor of about 1,000 since that time, stretching the wavelengths of these photons by the same amount [Section 17.4]. Their wavelengths should therefore have shifted to about a millimeter, squarely in the microwave portion of the spectrum and corresponding to a temperature of a few degrees above absolute zero. In the early 1990s, a NASA satellite called the *Cosmic Background Explorer (COBE)* was launched to test these ideas about the cosmic background radiation. The results were a stunning success for the Big Bang theory. As shown in Figure 19.7, the cosmic background radiation does indeed have a perfect thermal radiation spectrum, with a peak corresponding to a temperature of 2.73 K. In a very real sense, the temperature of the night sky is a chilly 3 degrees above absolute zero.

TIME OUT TO THINK *Suppose the cosmic background radiation did not really come from the heat of the universe itself but instead came from many individual stars and galaxies. Explain why, in that case, we would not expect it to have a perfect thermal radiation spectrum. How does the spectrum of the cosmic background radiation lend support to the Big Bang theory?*

COBE achieved an even greater success mapping the temperature of the cosmic background radiation in all directions. It was already known that the cosmic background radiation is extraordinarily uniform throughout the universe. Conditions in the early universe must have been extremely uniform to produce such a smooth radiation field. For a time, this uniformity was considered a strike against the Big Bang theory because, as we discussed in Chapters 17 and 18, the universe must have contained some regions of enhanced density in order to explain the formation of galaxies. The COBE measurements restored confidence in the Big Bang theory because they showed that the cosmic background radiation is *not quite* perfectly uniform. Instead, its temperature varies very slightly from one place to another by a few parts in 100,000 (Figure 19.8). These variations in temperature indicate that the density of the early universe really did differ slightly from place to place—the seeds of structure formation were indeed present during the era of nuclei.

COBE's discovery of density enhancements bolstered the idea that some of the dark matter consists of WIMPs (weakly interacting massive particles [Section 18.4]) that we have not yet identified and that the gravity of this dark matter drove the formation of structure in the universe. Regions of enhanced density can grow into galaxies because the extra gravity in these regions draws matter together, even while the rest of the universe expands. The greater the density enhancements, the faster matter should have

FIGURE 19.8 This all-sky map shows temperature differences in the cosmic background radiation measured by COBE. The background temperature is about 2.73 K everywhere, but the brighter regions of this picture are slightly less than 0.0001 K hotter than the darker regions—indicating that the early universe was very slightly lumpy. We are essentially seeing what the universe was like at the surface marked "300,000 years" in Figure 19.2. (The central strip of this map, which corresponds to the disk of the Milky Way, has been masked out because the brightness differences there stem primarily from radio noise in the Milky Way.

collected into galaxies. Detailed calculations show that, to explain the fact that galaxies formed within a few billion years, the density enhancements at the end of the era of nuclei must have been significantly greater than the few parts in 100,000 suggested by the temperature variations in the cosmic background radiation. Because WIMPs are weakly interacting and do not interact with photons, we do not expect them to influence the temperature of the cosmic background radiation directly. However, the gravity of the WIMPs can collect ordinary baryonic matter into clumps that *do* interact with photons. Thus, the small density enhancements detected by COBE may actually echo the much larger density enhancements made up of WIMPs. Careful modeling of the COBE temperature variations shows that they are consistent with dark-matter density enhancements large enough to account for the structure we see in the universe today.

Synthesis of Helium

The discovery of the microwave background radiation in 1965 quickly solved another long-standing astronomical problem: the origin of cosmic helium. Everywhere in the universe, about one-quarter of the mass of ordinary matter is helium. The Milky Way's helium fraction is about 28%, and no galaxy has a helium fraction lower than 25%. A small proportion of this helium comes from hydrogen fusion in stars, but most does not. The majority of the helium in the universe must already have been present in the protogalactic clouds that preceded the formation of galaxies. In other words, the universe itself must once have been hot enough to fuse hydrogen into helium. The current microwave background temperature of 2.73 K tells us precisely how hot the universe was in the distant past and exactly how much helium it should have made. The result—25% helium—is another impressive success of the Big Bang theory.

A helium nucleus contains two protons and two neutrons, so we need to understand what protons and neutrons were doing during the era of nucleosynthesis in order to see why the primordial fraction of helium was 25%. Early in this era, when the universe's temperature was 10^{11} K, nuclear reactions could convert protons into neutrons, and vice versa. As long as the universe remained hotter than 10^{11} K, these reactions kept the numbers of protons and neutrons nearly equal.

As the universe cooled from 10^{11} K to 10^{10} K, neutron–proton conversion reactions began to favor protons. Neutrons are slightly heavier than protons, and therefore reactions that convert protons to neutrons require energy to proceed (in accordance with

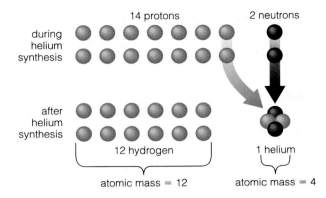

during
helium
synthesis

14 protons   2 neutrons

after
helium
synthesis

12 hydrogen   1 helium

atomic mass = 12   atomic mass = 4

FIGURE 19.9 During helium synthesis, protons outnumbered neutrons 7 to 1, which is the same as 14 to 2. The result was 12 hydrogen nuclei (individual protons) for each helium nucleus. Thus, the hydrogen-to-helium mass ratio is 12 to 4, which is the same as 75% to 25%.

$E = mc^2$). By the time the temperature of the universe fell to 10^{10} K, protons began to outnumber neutrons because the conversion reactions ran only in one direction: Neutrons changed into protons, but the protons didn't change back.

At 10^{10} K, the universe was still hot and dense enough for nuclear fusion to operate. Protons and neutrons constantly combined to form *deuterium*—the rare form of hydrogen nuclei that contain a neutron in addition to a proton—and deuterium nuclei fused to form helium. However, during the early parts of the era of nucleosynthesis, the helium nuclei were almost immediately blasted apart by one of the many gamma rays that filled the universe.

Fusion finally created long-lasting helium nuclei when the universe was about 1 minute old, at which time the destructive gamma rays were gone. Calculations show that the proton-to-neutron ratio at this time should have been about 7 to 1. Moreover, virtually all the available neutrons should have become incorporated into nuclei of helium-4. Figure 19.9 shows that, based on the 7-to-1 ratio of protons to neutrons, the universe should have had a composition of 75% hydrogen and 25% helium by mass at the end of the era of nucleosynthesis.

Thus, the Big Bang theory makes a very concrete prediction about the composition of the universe: It should be 75% hydrogen and 25% helium by mass (at least for ordinary matter). The fact that observations confirm this predicted ratio of hydrogen to helium is another striking success of the Big Bang theory.

TIME OUT TO THINK *Briefly explain why it should not be surprising that some galaxies contain a little more than 25% helium, but it would be very surprising if some galaxies contained less. (Hint: Think about how the relative amounts of hydrogen and helium in the universe can be affected by fusion in stars.)*

19.4 Inflation

The Big Bang model relies heavily on our knowledge of particle physics, which has been tested to temperatures of 10^{15} K, corresponding to the end of the electroweak era. Our knowledge of earlier times rests on weaker foundations because we are less certain of the physical laws at work. In fact, the best laboratory for studying the laws of physics at such high temperatures is the Big Bang itself. Different guesses as to how matter might behave at such high energies predict different outcomes for the universe we see today. If a particular model predicts that our universe should look different from the way it really does, then that model must be wrong. On the other hand, if a new model of particle physics explains some previously unexplained aspects of the universe, then it may be on the right track. The grand unified theories of particle physics discussed in Section 19.2 have not yet been proved, but many scientists believe these theories are on the right track for just this reason.

Prior to the early 1980s, scientists had identified several major aspects of our actual universe that were unexplained by the Big Bang model. Three of the most pressing unanswered questions were the following:

- *Where does structure come from?* According to the Big Bang theory, the process of galaxy formation required the density of the early universe to differ slightly from place to place. Otherwise, the pull of gravity would have been equally balanced everywhere, prohibiting the collapse of protogalactic clouds. The subtle temperature differences seen in the cosmic background radiation tell us that regions of enhanced density did indeed exist at the end of the era of nuclei, when the universe was 300,000 years old. However, the standard Big Bang theory cannot explain where these density enhancements came from.

- *Why is the large-scale universe so smooth?* Not only is the slight lumpiness of the early universe a problem, but its large-scale smoothness is also a problem. Observations of the cosmic background radiation show that, overall, the density of the universe at the end of the era of nuclei varied from place to place by no more than about 0.01%. The standard Big Bang theory does not explain why distant reaches of the universe look so similar.

- *Why is the density almost critical?* The density of matter in the universe is somewhere around 20–100% of the critical density [Section 18.6]—remarkably close to the critical density considering that the standard Big Bang theory doesn't say anything about what the density should be. Why isn't it 1,000 times the critical density, or 0.0000001 the critical density?

Physicist Alan Guth realized in 1981 that grand unified theories could potentially answer all three questions. These theories predict that the separation of the strong force from the GUT force should have released enormous energy, causing the universe to expand dramatically, perhaps by as much as 10^{30} times in less than 10^{-33} second. This dramatic expansion is what we call *inflation,* and while it sounds outrageous it may have shaped the way the universe looks today.

Giant Quantum Fluctuations

To understand how inflation solves the structure problem, we need to recognize a special feature of energy fields. Laboratory-tested principles of quantum mechanics require that the energy fields at any point in space be always fluctuating. Thus, the distribution of energy through space on very small scales is slightly irregular, even in a complete vacuum. These tiny quantum "ripples" can be characterized by a wavelength that corresponds roughly to their size. In principle, quantum ripples in the very early universe could have been the seeds for density enhancements that later grew into galaxies. However, the wavelengths of the original ripples were far too small to explain density enhancements like those we see imprinted on the cosmic background radiation.

Inflation would have dramatically altered these quantum fluctuations. The fantastic growth of the universe during the period of inflation would have stretched tiny ripples in space to enormous wavelengths (Figure 19.10). Ultimately, inflation could have caused these quantum ripples to grow large enough to become the density enhancements that later formed large structures in the universe. Amazingly, all the structure we see today in the universe may have started as tiny quantum fluctuations just before the period of inflation.

size of ripple before inflation = size of atomic nucleus

size of ripple after inflation = size of solar system

FIGURE 19.10 During inflation, ripples in spacetime would have stretched by a factor of perhaps 10^{30}.

Equalizing Temperatures and Densities

The idea that different parts of the universe should be very similar shortly after the Big Bang may seem quite natural at first, but on further inspection the smoothness of the universe becomes difficult to explain. Imagine observing the cosmic background radiation in a certain part of the sky. You are seeing that region as it was only 300,000 years after it formed. The microwaves coming from it have taken some 10 billion years or more—almost the entire age of the universe—to travel all the way to Earth. Now imagine turning around and looking at the background radiation coming from the opposite direction. You are also seeing this region at an age of 300,000 years, and it looks virtually identical. The two microwave-emitting regions are billions of light-years apart, but we are seeing them as they were when they were only 300,000 years old. They can't possibly have had communication with each other then. A signal traveling at the speed of light from one to the other would barely have started its journey. So how do they come to be exactly the same temperature?

The idea of inflation solves this problem because it says that the entire observable universe was less than 10^{-35} light-second across before the episode of enormous expansion. Radiation signals traveling at the speed of light would have had time to equalize all the temperatures and densities in this region. Inflation then pushed these equalized regions to much greater distances, far out of contact with one another. Like criminals getting their stories straight before being locked in separate jail cells, all parts of the observable universe came to the same temperature and density before inflation spread them far apart. Thus, inflation accounts for how vastly separated regions can look so similar.

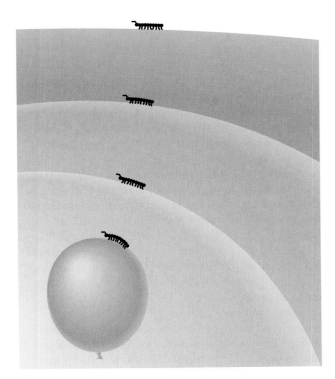

FIGURE 19.11 The flattening of the universe during inflation is similar to the flattening of a balloon's surface as you blow into it. To an insect on the balloon's surface, the balloon seems to become increasingly flat as it expands. The same is true of the universe during its period of inflation. Once inflation has ceased, the potential and kinetic energies of the universe are precisely balanced.

Balancing the Universe

Another way to say that the universe's density is close to critical is to say that the universe is remarkably flat. Recall that the density of a *flat universe* exactly equals the critical density. Its kinetic energy of expansion precisely balances its overall gravitational pull. According to Einstein's theory of general relativity, any imbalance in these energies causes curvature of spacetime, and deviations from precise balance grow more severe as the universe evolves. For example, if the universe had been 10% denser at the end of the era of nuclei, it would have collapsed long ago. On the other hand, if it had been 10% less dense at this time, galaxies would never have formed before expansion spread all the matter too thin. Thus, the universe had to start out remarkably balanced to be even remotely close to flat today.

Inflation balances the universe by ensuring that the mass-energy providing the universe's gravity precisely matches its kinetic energy of expansion. Once this balance is set, it should never waver. In terms of Einstein's theory, the effect of inflation on spacetime curvature is similar to the flattening of a balloon's surface when you blow into it (Figure 19.11). The flattening of space during the period of inflation

would have been so enormous that it would virtually have eliminated any curvature the universe might have had previously.

The fine balance created by inflation turns out to be both a success and a pitfall of the inflation theory. On one hand, it explains how the universe managed to stay balanced long enough to bring forth galaxies. However, our studies of dark matter tend to show that the current universe is slightly out of balance: The best current estimates put the actual density of the universe at only about 25% of the critical density [Section 18.6], suggesting that the universe has about four times more kinetic energy than gravity can overcome. If these dark-matter measurements hold up, inflation theory will have a tough time explaining this slight imbalance.

The Bottom Line

All things considered, inflation does a remarkable job of addressing the problems inherent in the standard Big Bang theory. Many astronomers and physicists believe that some process akin to inflation did affect the early universe, but the details of the interaction between high-energy particle physics and the evolving universe remain unclear. If these details can be worked out successfully, we face an amazing prospect—a breakthrough in our understanding of the very smallest particles achieved by studying the universe on the largest observable scales.

19.5 Did the Big Bang Really Happen?

You might occasionally read an article in a newspaper or a magazine questioning whether the Big Bang really happened. We may never be able to prove with absolute certainty that the Big Bang theory is correct. However, no one has come up with any other model of the universe that so successfully explains so much of what we see. As we have discussed, the Big Bang model makes at least two specific predictions that we have observationally verified: the characteristics of the cosmic background radiation and the composition of the universe. It also explains quite naturally many other features of the universe. And, at least so far, we know of nothing that is absolutely inconsistent with the Big Bang model.

You don't need to believe all you have read without question. The next time you are musing on the universe's origins, try an experiment for yourself. Go outside on a clear night and look at the sky. Notice how dark it is, and then ask yourself *why* it is dark. If the universe were infinite, unchanging, and everywhere the same, then the entire night sky would blaze as brightly as the Sun. Johannes Kepler [Section 5.3] was perhaps the first person to realize that the

The End of Time

According to the Big Bang theory, time and space have a beginning in the Big Bang. Do they also have an end? If the universe is closed, the answer seems to be a clear yes: Sometime in the distant future, the universal expansion will cease and reverse, and the universe will eventually come to an end in a fiery *Big Crunch*. But what if the universe continues to expand forever?

In an open universe, the end will come much more gradually. The star–gas–star cycle in galaxies cannot continue forever, because not all the material is recycled: With each generation of stars, more mass becomes locked up in planets, brown dwarfs, white dwarfs, neutron stars, and black holes. Eventually, about a trillion years from now, even the longest-lived stars will burn out, and the galaxies will fade into darkness. By this time, the expansion of the universe will have made the average distance between galaxy clusters about a hundred times greater than it is today.

At this point, the only new action in the universe will occur on the rare occasions when two objects—such as two brown dwarfs or two white dwarfs—collide within a galaxy. The vast distances separating star systems in galaxies make such collisions extremely rare. For example, the probability of our Sun (or the white dwarf that it will become) colliding with another star is so small that it would be expected to happen only once in a quadrillion (10^{15}) years. But forever is a long time, and even low-probability events will eventually happen many times. If a star system experiences a collision once in a quadrillion years, it will experience about 100 collisions in 100 quadrillion (10^{17}) years. By the time the universe reaches an age of 10^{20} years, star systems will have suffered an average of 100,000 collisions each, making a time-lapse history of any galaxy look like a cosmic game of pinball.

These multiple collisions will severely disrupt galaxies. As in any gravitational encounter, some objects lose energy and some gain energy. Objects that gain enough energy will be flung into intergalactic space, where they will drift ever-farther from all other objects with the expansion of the universe. Objects that lose energy will eventually fall to the galactic center, forming a gigantic black hole. The

night sky in such a universe should be bright, but this realization is now called **Olbers' paradox** after Heinrich Olbers, a German astronomer of the 1800s.

To understand how Olbers' paradox comes about, imagine that you are in a dense forest on a flat plain. If you look in any given direction, you'll likely see a tree. If the forest is small, you might be able to see through some gaps in the trees to the open plains, but larger forests have fewer gaps (Figure 19.12). An infinite forest would have no gaps at all—a tree trunk would block your view along any direction.

The universe is like a forest of stars in this respect. In an unchanging universe with an infinite number of stars, we would see a star in every direction, making every point in the sky as bright as the Sun's surface. Obscuring dust doesn't change this conclusion. The intense starlight would heat the dust over time until it too glowed like the Sun or evaporated away.

There are only two ways out of this dilemma: Either the universe has a finite number of stars, in which case we would not see a star in every direction, or it changes in time in some way that prevents us from seeing an infinite number of stars. For several centuries after Kepler first recognized the dilemma, astronomers leaned toward the first option. Kepler himself preferred to believe that the universe had a finite number of stars because it was finite in space, with some kind of dark wall surrounding everything. Astronomers in the early twentieth century preferred to believe that the universe was infinite in space but that we lived inside a finite collection of stars; they thought of the Milky Way as an island floating in a vast black void. However, subsequent observations showed that galaxies fill all of space more or less uniformly. We are therefore left with the second option: The universe changes in time.

The Big Bang theory solves Olbers' paradox in a particularly simple way: It tells us that we can see only a finite number of stars because the universe began at a particular moment. While the universe may contain an infinite number of stars, we can see only those whose light has had time to reach us. There are other ways in which the universe could change in time and prevent us from seeing an infinite number of stars, so Olbers' paradox does not *prove* that the universe began with a Big Bang. But we must have some explanation for why the sky is dark at night, and no explanation besides the Big Bang also explains so many other observed properties of the universe so well.

remains of the universe will consist of black holes with masses as great as a trillion solar masses widely separated from a few scattered planets, brown dwarfs, and stellar corpses. If the Earth somehow survives, it will be a frozen chunk of rock in the darkness of the expanding universe, billions of light-years from any other solid object.

If grand unified theories are correct, the Earth still cannot last forever. These theories predict that protons will eventually fall apart. The predicted lifetime of protons is extremely long: a half-life of at least 10^{32} years. However, if protons really do decay, then by the time the universe is 10^{40} years old the Earth and all other atomic matter will have disintegrated into radiation and subatomic particles, such as electrons and neutrinos.

The final phase in the future of an open universe is predicted to come with the disintegration of the giant, black-hole corpses of galaxies. Black holes are predicted to slowly evaporate through a process called *Hawking radiation*, finally disappearing in brilliant bursts of radiation. The largest black holes last longest, but even trillion-solar-mass black holes will evaporate sometime after the universe reaches an age of 10^{100} years. From then on, the universe will consist only of individual photons and subatomic particles, each separated by enormous distances from the others. Nothing new will ever happen, and no events will ever occur that would allow an omniscient observer to distinguish past from future. In a sense, the universe will finally have reached the end of time.

Lest any of this sound depressing, keep in mind that 10^{100} years is an indescribably long time. As an example, imagine that you wanted to write on a piece of paper a number that consisted of a 1 followed by 10^{100} zeros (i.e., the number $10^{10^{100}}$). It sounds easy, but a piece of paper large enough to fit all those zeros *would not fit in the observable universe* today. And, if that still does not alleviate your concerns, you may be glad to know that a few creative thinkers speculate about ways in which the universe might undergo rebirth, even after the end of time.

FIGURE 19.12 In a large forest (left), you'll see trees no matter where you look. In a small forest (below), you can see open spaces beyond the trees.

A poet once said "The whole universe is in a glass of wine." . . . It is true that when we look in a glass of wine closely enough we see the entire universe. There are the things of physics: the twisting liquid which evaporates depending on the wind and weather, the reflections in the glass, and our imagination adds the atoms. The glass is a distillation of the earth's rocks, and in its composition we see the secrets of the universe's age, and the evolution of the stars. . . . If our small minds, for some convenience divide this glass of wine, this universe into parts—physics, biology, geology, astronomy, psychology, and so on—remember that nature does not know it! So let us put it all back together, not forgetting ultimately what it is for.

RICHARD FEYNMAN, PHYSICIST (1918–88)

THE BIG PICTURE

Our "big picture" is now about as complete as it gets. We've discussed the universe from the Earth outward, and from the beginning to the end. When you think back on this chapter, keep in mind the following ideas:

■ Predicting conditions in the early universe is straightforward. The only real question is how matter and energy behave under such extreme conditions.

■ Our current understanding of physics allows us to reconstruct the conditions that prevailed in the universe all the way back to the first 10^{-10} second. Our understanding is less certain back to 10^{-35} second. Beyond 10^{-43} second, we run up against the present limits of human knowledge.

■ Although it may sound strange to talk about the universe during its first fraction of a second, our ideas about the Big Bang rest on a solid foundation of observational, experimental, and theoretical evidence. We cannot say with absolute certainty that the Big Bang really happened, but no other model ever proposed has so successfully explained how our universe came to be as it is.

Review Questions

1. Briefly explain why the early universe must have been much hotter and denser than the universe is today. How can we use this idea to determine the temperature and density conditions that prevailed at different times in the early universe?

2. What do we mean by the *Big Bang theory?*

3. What is *antimatter?* How were particle–antiparticle pairs created in the early universe? How were they destroyed?

4. Briefly summarize the characteristics marking each of the various eras shown in Figure 19.2.

5. What is the *Planck era?* Why can't our current theories describe the history of the universe during the Planck era?

6. What are *grand unified theories?* According to these theories, how many forces operated in the universe during the GUT era?

7. What do we mean by *inflation?* Briefly describe why inflation might have occurred at the end of the GUT era.

8. What characterized the universe during the *electroweak era?* What evidence from particle accelerators supports the idea that the weak and electromagnetic forces were once unified as a single electroweak force?

9. What characterized the universe during the *particle era?* What happened to the *quarks* that existed freely in the universe early in the particle era?

10. How long did the *era of nucleosynthesis* last? Why was this era so important in determining the chemical composition of the universe forever after?

11. Briefly explain why radiation was trapped for 300,000 years during the *era of nuclei* and why the cosmic background radiation broke free at the end of this era.

12. What do we mean by the *era of atoms* and the *era of galaxies?* What era do we live in? Explain.

13. Briefly describe the two key pieces of evidence that support the Big Bang theory.

14. Briefly describe how the *cosmic background radiation* was discovered.

15. Why does the Big Bang theory predict that the cosmic background radiation should have a perfect thermal radiation spectrum? How did the *Cosmic Background Explorer (COBE)* support this prediction of the Big Bang theory? What is the temperature of the cosmic background radiation?

16. Briefly explain why we expect the cosmic background radiation to be almost, but not quite, the same in all directions. What did COBE find regarding the smoothness of the cosmic background radiation?

17. Briefly describe how theoretical calculations allow us to predict the fraction of ordinary matter in the universe that should consist of helium as opposed to hydrogen. Do observations of the universe today agree with this theoretical prediction?

18. Describe three aspects of the universe that are not explained by the standard Big Bang model. How does inflation help answer these questions?

19. What is *Olbers' paradox?* Explain how this simple paradox provides at least some support for the idea that the universe began in a Big Bang.

Discussion Questions

1. *The Moment of Creation.* You've probably noticed that, in discussing the Big Bang theory, we never quite talk about the first instant. Even our most speculative theories at present take us back only to within 10^{-43} second of creation. Do you think it will *ever* be possible for science to consider the moment of creation itself? Will we ever be able to answer questions such as *why* the Big Bang happened? Defend your opinions.

2. *The Big Bang.* How convincing do you find the evidence for the Big Bang model of the universe's origin? What are the strengths of the theory? What does it fail to explain? Overall, do *you* think the Big Bang really happened? Defend your opinion.

3. *Forever.* If you could live forever, would you choose to do so? Suppose the universe will keep expanding forever, as the bulk of observations currently suggest. How would you satisfy your energy needs as time goes on? How would you grow food? What would you do with all that time?

Problems

1. *Life Story of a Proton.* Tell the life story of a proton from its formation shortly after the Big Bang to its

presence in the nucleus of an oxygen atom you have just inhaled. Your story should be creative and imaginative, but it also should demonstrate your scientific understanding of as many stages in the proton's life as possible. You can draw on material from the entire book, and your story should be three to five pages long.

2. *Web Project: New Tests of the Big Bang Theory.* The Cosmic Background Explorer (COBE) satellite provided striking confirmation of several predictions of the Big Bang theory. But new satellites are already being designed that will test the Big Bang theory further, primarily by observing the subtle variations in the microwave background radiation with a much higher sensitivity than COBE's. Use the Web to gather pictures and information about the COBE mission and its planned successors. Write a one- to two-page report about the strength of the evidence compiled by COBE and how much more we might learn from the upcoming missions.

3. *Research Project: Decay of the Proton.* One of the most startling predictions of grand unified theories is that protons will eventually decay, albeit with a half-life of more than 10^{32} years. If this is true, it may be possible to observe an occasional proton decay, despite the extraordinarily long half-life. Several experiments to look for proton decay are under way or being planned. Find out about one or more of these experiments, and write a one- to two-page summary in which you describe the experiment(s), any results to date, and what these results (or potential results) mean to the grand unified theories.

4. *Research Project: New Ideas in Inflation.* The idea of inflation solves many of the puzzles associated with the standard Big Bang theory, but we are still a long way from finding strong evidence that inflation really occurred. Find recent news or magazine articles that discuss some of the latest ideas about inflation and how we might test these ideas. Write a two- to three-page summary of your findings.

$\mathcal{I}$nterstellar $\overset{20}{\mathcal{T}}$ravel

$\mathcal{A}$CCORDING TO SCIENCE FICTION, OUR DESCENDANTS WILL SOON TRAVEL among the stars as routinely as we jet about the Earth in airplanes. They'll race around the galaxy in starships of all sizes and shapes, circumventing nature's prohibition on faster-than-light travel by entering hyperspace, wormholes, or warp drive. They'll witness incredible cosmic phenomena first-hand, including stars and planets in all stages of development, accretion disks around white dwarfs and neutron stars, and the distortion of spacetime near black holes. Along the way, they'll encounter numerous alien species, most of which will look a lot like us and share virtually the same level of technological advancement.

Real interstellar travel is likely to be limited by the speed of light. Moreover, given the billions of years over which the galaxy has existed, finding an alien species that shares our level of technological development would be an extraordinary coincidence; more likely, any other species we encounter will be either far ahead of us or far behind us.

Nevertheless, if our civilization survives long enough, it seems almost inevitable that our descendants will leave the confines of our solar system. In this final chapter, we will explore how that might happen. We will also consider whether other species might already have progressed beyond this point and, if so, how we might contact them.

20.1 Starships: Distant Dream or Near-Reality?

The vast distances between the stars mean that practical starships would have to travel near the speed of light in order to complete a journey within a human lifetime. Today's seemingly fast interplanetary spacecraft travel less than 1/10,000 the speed of light—meaning that starships will need to go 10,000 times faster. From this standpoint, we are as far from interstellar flight as cavemen were from the space age.

However, nearly all of the increase in speed we have achieved over our ancestors occurred within the past century or so, with the advent of flight and the dawn of the space age. If we can achieve a comparable speed advancement in the next century, our great-grandchildren may be building starships.

Regardless of how far in the future interstellar travel might be, we will have to overcome huge technological and social hurdles to achieve it. On the technological side, we will need entirely new types of engines to reach speeds close to the speed of light. We'll also need new types of shielding to protect crew members from instant death: When a starship travels through interstellar gas at near–light speed, ordinary atoms and ions will hit it in a flood of high-energy particles.

The social hurdles may be even more difficult to overcome. Building starships will almost certainly require construction facilities in space and probably cannot be achieved until we first learn to mine materials from the Moon or nearby asteroids. Thus, at least a few major space stations and outposts on the Moon are probably prerequisites to any era of starship exploration. Given global political realities, such colonization of space will probably not occur without far greater international trust and cooperation than we enjoy today.

Interstellar travel will pose further difficulties for starship crews. Einstein's theory of relativity offers high-speed travelers a "ticket to the stars": Time will run slow for the travelers, enabling them to make journeys across many light-years of space in relatively short amounts of time. For example, in a ship traveling at an average speed of 99.9% of the speed of light, the 50 light-year round-trip to Vega would take the travelers only about 2 years. But more than 50 years would pass on Earth while they were gone, which would make their lives very difficult when they returned. Family and friends would be older or gone, new technologies might have made their knowledge and skills obsolete, and it might take them many years just to understand the political and social changes that occurred in their absence.

Despite all these hurdles, we seem bound to become interstellar travelers unless we *choose* other-wise. Such a choice might be deliberate: Many people argue that money for space exploration would be better spent here on Earth, and others argue against space colonization on philosophical grounds. Or it might be the consequence of a catastrophe: Despite the lessening of international tensions since the end of the Cold War, more than 10,000 nuclear weapons remain in the global arsenal, and many nations possess the technology to build chemical and biological weapons of mass destruction. Our civilization might also succumb to disasters brought on by human activity, such as overpopulation, epidemic disease, or global warming.

If an alien civilization were watching us, they might well conclude that we are poised on the brink of the most significant turning point in human history. If we destroy ourselves, all our achievements in science, art, and philosophy will be lost forever. But if we survive and choose to continue the exploration of space, we may be embarking on a path that will take us to the stars.

20.2 Starship Design

So far we have spoken of starships that could carry humans across vast interstellar distances. But if we expand the definition to include robotic spacecraft, then we have already launched our first starships. Four interplanetary probes—*Pioneers 10* and *11,* and *Voyagers 1* and 2—are traveling fast enough to escape from the solar system. It will take them more than 10,000 years to cover each light-year of distance, and their trajectories will not take them on close passes of any nearby stars. Nevertheless, these spacecraft should suffer little damage during their journeys and are likely to look almost as good as new for millions of years. Each carries a greeting from Earth in case someone comes across one of them someday (Figure 20.1).

Current spacecraft, including both robotic probes like Pioneer and Voyager and crew-carrying spacecraft such as the Space Shuttle, are launched by *chemical rockets*. Their engines use energy from chemical reactions to drive hot gas out the back of the rocket, which causes the rocket to accelerate forward.

Chemical rockets will never be practical for interstellar travel because a limiting process prevents them from reaching very high speeds: Achieving higher speeds requires more fuel, but the added weight of additional fuel makes it more difficult to increase the rocket's speed. Fortunately, other technologies may allow for much faster travel.

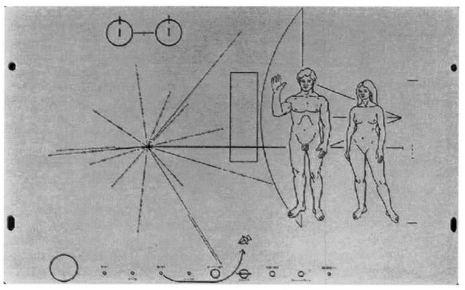

a *Pioneers 10* and *11* carry a copy of this etching as a greeting to anyone who might find them.

FIGURE 20.1 Messages aboard the Pioneer and Voyager spacecraft, which are bound for the stars.

b *Voyagers 1* and *2* carry a phonograph record—a 12-inch gold-plated copper disk containing music, greetings, and images from Earth.

Nuclear Rocket Propulsion

Rockets would be far more powerful if their engines used nuclear power rather than chemical power. Nuclear reactions generate far more power than chemical reactions because they convert some of the mass in atomic nuclei into energy in accord with $E = mc^2$. Fusing a kilogram of hydrogen, for example, generates more than a million times more energy than chemical reactions in a kilogram of hydrogen and oxygen.

One way to realize the benefits of fusion power would be to power starships with nuclear bombs. This idea, called *nuclear pulse propulsion,* involves generating energy with repeated detonations of relatively small H-bombs. One design, created under the name Project Orion, envisions the explosions taking place a few tens of meters behind the spaceship (Figure 20.2). The vaporized debris from each explosion would impact a "pusher plate" on the back of the spacecraft, propelling the ship forward. In principle, we could build an Orion spacecraft with existing technology, although it would be very expensive.

Another design for a fusion-powered rocket, called Project Daedalus, relies on generation of a continuous stream of energy from an on-board *controlled* nuclear fusion reactor (Figure 20.3). Because we have not yet achieved controlled nuclear fusion (as opposed to "uncontrolled" bombs), Daedalus remains beyond our current technological capabilities.

Fusion-powered starships such as Orion and Daedalus could probably achieve speeds of about 10% of the speed of light, completing journeys to nearby stars

FIGURE 20.2 Artist's conception of the Project Orion starship, showing one of the small H-bomb detonations that propel it. Debris from the detonation impacts the flat disk, called the pusher plate, at the back of the spaceship. The central sections (enclosed in lattice) hold the bombs, and the crew lives in the front sections.

FIGURE 20.3 Artist's conception of a robotic Project Daedalus starship. The front section (right) holds the scientific instruments. The large spheres hold the hydrogen for the central fusion reactor.

FIGURE 20.4 Artist's conception of a spaceship propelled by a solar sail, shown as it approaches a forming planet in a young solar system. The sail is many kilometers across; the scientific payload is at the central meeting point of the four scaffoldlike structures.

in a few decades. Such speeds are certainly sufficient for sending colonists on one-way voyages, particularly if the crew can be put into some sort of hibernation that prevents them from aging during the trip.

Matter–Antimatter Rocket Engines

In principle, there is an even more powerful energy source for rocket engines than fusion: matter–antimatter annihilation. Whereas fusion converts less than 1% of the mass of atomic nuclei into energy, matter–antimatter annihilation converts *all* the annihilated mass into energy. Starships with matter–antimatter engines could probably reach 90% or more of the speed of light.

However, while matter–antimatter engines may someday be the propulsion systems of choice for starships, this technology appears to be far in the future. The two major technological challenges to matter–antimatter engines are finding a way to produce sufficient quantities of antimatter to be used in the engines and finding a way to safely store the antimatter until it is needed. Neither challenge seems likely to be met soon.

Sails and Beamed Energy Propulsion

In principle, spacecraft with large, lightweight sails could be propelled by the solar wind and the radiation pressure of sunlight. Solar sailing may well prove to be a fairly inexpensive way of navigating within the solar system (Figure 20.4). Surprisingly, it may even be useful for interstellar travel. Energy flowing from the Sun would give an optimally designed sailing starship a small but continuous acceleration as it heads out of the solar system. By the time the ship reached the distance at which the solar wind could no longer provide acceleration, it might already be traveling at speeds of a few percent of the speed of light.

We might achieve even higher velocities by shining powerful lasers at the spacecraft sails. Such *beamed energy propulsion* theoretically could accelerate ships to substantial fractions of the speed of light—but only if we could build enormous lasers. For example, accelerating a ship to half the speed of light within a few years would require a laser that uses 1,000 times more power than all current human power consumption.

FIGURE 20.5 Artist's conception of a spaceship powered by an interstellar ramjet. The giant scoop in the front (left) collects interstellar hydrogen for use as fusion fuel.

Interstellar Ramjets

All the propulsion systems we've discussed so far have limited capabilities: Rockets can carry only a limited amount of fuel no matter what their fuel source, and beamed energy becomes too weak to accelerate starships when they get very far from home. One way around these difficulties would be to build a starship that collects fuel as it goes. Such ships, which in principle can accelerate continuously, are known as *interstellar ramjets* (Figure 20.5).

Imagine that an interstellar ramjet accelerates at 1*g* (the acceleration of gravity on Earth) for half the voyage to its destination and then turns around and decelerates at 1*g* until it arrives. The crew would find the trip quite comfortable, experiencing Earth-like gravity the entire way. During most of the journey, the ship would be traveling relative to Earth (and its destination) at a speed very close to the speed of light, so time on the ship would pass very slowly compared to time on Earth. For example, such a ship could make the trip to a star system in the Andromeda Galaxy in only about 29 years of ship time. Thus, the crew could travel to the Andromeda Galaxy, spend 2 years studying one of its star systems and taking pictures of the Milky Way Galaxy to bring home, and return in just 60 years of ship time. However, because the Andromeda Galaxy is about 2.5 million light-years away, the crew would find that about 5 million years had passed on Earth by the time they returned home.

Science Fiction

If you are a science fiction fan, this discussion of interstellar travel may be depressing. Interstellar tourism and commerce seem out of the question, even with interstellar ramjets that reach speeds very close to the speed of light. If we are ever to travel about the galaxy the way we now travel about the Earth, we will need spacecraft that can get us from here to there much faster than the speed of light.

Einstein's theory of relativity leaves little hope that we can ever find a way to travel *through* space faster than the speed of light, but science fiction writers have imagined all kinds of novel shortcuts that don't necessarily violate relativity or any other known laws of physics. Some of these go by names like hyperspace, wormholes, and warp drive. The bottom line is that our present knowledge does not allow us to say one way or the other whether any of these technologies are possible.

20.3 Radio Communication: Virtual Interstellar Travel

We cannot travel at the speed of light, but we can send information at the speed of light using radio waves, lasers, or other forms of light. Because radio waves penetrate interstellar gas and dust fairly easily, they probably will prove to be the most useful form of interstellar communication. Although we usually think of radio communication as a way of sending and receiving fairly simple messages, there is really no limit to how much information we could transmit or receive. In principle, we could become *virtual* interstellar travelers. Imagine, for example, that we sent a robotic probe to Alpha Centauri at 10% of the speed of light. It would take 40 years to get there, but it could send data back to Earth via radio in just 4 years. Nearby stars, at least, could be explored with robotic spacecraft in the same way we now explore the planets of our own solar system.

An even better way to gather data about distant worlds would be to receive signals from civilizations in other star systems. If such civilizations exist, they may already be broadcasting messages to us. The *search for extraterrestrial intelligence,* or *SETI,* is a project in which scientists use large radio telescopes to scan the skies for signals from other civilizations.

But how could we interpret these messages? After all, we could hardly expect alien civilizations to share any of our social customs, let alone to speak English. Fortunately, we know that all civilizations must share some things in common: the laws of mathematics and science that are the same everywhere. Imagine that we wanted to send a message to a distant civilization. As popularized in the movie *Contact,* we might begin by broadcasting radio pulses in the order of the prime numbers (1, 2, 3, 5, 7, 11, . . .)—clearly identifying the message as something sent by an intelligent species, since no known natural process can broadcast prime numbers. We could then represent the prime numbers with a binary code and use this same code to begin representing some of the laws of mathematics. Since alien civilizations would know these laws already, they could learn the intricacies of our code. Once they knew the full code, they could receive as much additional information about Earth and our civilization as we wished to send them.

Scientists working on SETI hope that someone out there is already broadcasting similar coded messages to Earth. Even if such messages are being continually broadcast to Earth, finding them may be rather like finding the proverbial needle in a haystack: There are billions of stars to search, and we don't know the radio frequencies at which other civilizations might choose to broadcast their messages. Nevertheless, we can make some guesses. For example, we might focus our attention on Sun-like stars and assume that the aliens would broadcast near some "natural" frequency, such as the frequency of 21-cm radio waves from hydrogen. In addition, new technologies allow us to search many frequencies simultaneously, increasing the odds of success. However, SETI efforts remain controversial because they require extensive time on powerful and expensive radio telescopes when we have no way of knowing whether anyone is even out there.

20.4 Civilizations

Whether we consider the virtual travel of radio communication or actual ships bound for the stars, a recurrent theme of interstellar travel is contact with other civilizations. Thus, we are led to one of the most exciting questions in science: Is anyone else out there?

How Many Civilizations Are Out There?

In principle, we could calculate the number of civilizations in the Milky Way Galaxy if we knew just a few basic facts. We would start with the number of planets that lie within habitability zones, which we'll call N_P (for "number of planets"). Then we would determine the fraction of those planets that actually have life; let's use f_{life} to stand for this fraction. Multiplying N_P by f_{life} would tell us the number of life-bearing planets in the Milky Way. Next, we'd multiply this number by the fraction of life-bearing planets on which intelligent beings evolve and develop a civilization, which we'll call f_{civ}. Finally, we'd multiply by the fraction of planets with civilizations that have those civilizations *now* (as opposed to the distant past or future), which we'll call f_{now}. Putting all these ideas together gives us the following simple equation (a variation on what is called the Drake equation) for the number of civilizations now inhabiting the Milky Way Galaxy:

$$\text{Number of civilizations} = N_P \times f_{life} \times f_{civ} \times f_{now}$$

The only difficulty in using this equation is that we don't know the value of any of its terms! In fact, the only term for which we can even make a reasonable guess is the number of planets, N_P. Our theories of star formation tell us that planetary systems ought to be common, and recent discoveries of protoplanetary disks around nearby stars—and of a few full-fledged planets—support this idea [Section 7.7]. Studies of our own solar system suggest that both Earth and Mars lie within the Sun's habitability zone [Section 11.6], and Venus is probably close to this zone. If other planetary systems are similar, we might expect the typical number of planets within the habitability zone to be between one and three. All in all, it seems likely that many, and perhaps even most, of the several hundred billion stars in the Milky Way have at least one planet within their habitability zone. The technology for detecting planets beyond our solar system is rapidly advancing, so we may be able to improve or confirm this estimate within the next one or two decades.

For the moment, we have no rational way to guess the fraction f_{life} of planets on which life actually arose. The fact that life arose quickly on Earth [Section 11.4] suggests that life develops fairly easily under the right conditions, but it is by no means conclusive. As far as we know today, f_{life} could be anywhere between near-zero (it happened only on Earth) and 1 (it happened on all planets suitable for life).

Similarly, we have little basis on which to guess the fraction f_{civ} of life-bearing planets that develop a civilization. However, the fact that life flourished on Earth for some 4 billion years before the rise of humans suggests that developing intelligent life is much more difficult than developing microbial life. The question is *how much* more difficult. Roughly half the stars in the Milky Way are older than our Sun, so if 4 billion years is a typical time for a civilization to develop then it should have happened already on about half the planets with life. On the other hand, if the Earth was "fast" and the typical time needed for intelligent life to develop is much longer, then most life-bearing planets may be covered with nothing more advanced than bacteria.

The huge uncertainties in f_{life} and f_{civ} make it impossible for us to determine the number of civilizations that have arisen in the Milky Way's history. If 1 in 1,000 suitable planets ends up with a civilization, then we are newcomers among a billion civilizations in the Milky Way. If 1 in 1 million develops a civilization, there would still have been some 100,000 civilizations out there in the Milky Way Galaxy alone. But if only 1 in 1 trillion planets ends up with a civilization, then we may be the first in the Milky Way—and the only civilization within millions of light-years.

The final step in our calculation—determining the number of civilizations that exist *now*—requires knowing how long a civilization typically lasts in a planet's history. Here we must define what we mean by *civilization;* let's use a definition that presumes a civilization advanced enough to travel in space. On the 4.6-billion-year-old planet Earth, the space age began only a little over 40 years ago. Thus, the fraction of Earth's history during which it has had a space-faring civilization is 40 in 4.6 billion, or about 1 in 100 million. If this fraction is typical, there would have to be some 100 million civilization-bearing planets in the Milky Way in order to have good odds of finding another civilization out there *now*. However, we'd expect this fraction to be typical only if we are on the brink of self-destruction, since the fraction will grow larger as long as our civilization continues to thrive. Thus, if civilizations are at all common, the key factor in whether any are out there now is their survivability. If most civilizations self-destruct shortly after achieving the technology for

space travel, then we probably are alone in the galaxy at present. But if most survive and become interstellar travelers, the Milky Way may be brimming with civilizations—most far more advanced than ours.

A Paradox: Where Are the Aliens?

Imagine that we survive and become interstellar travelers and that we begin colonizing habitable planets around nearby stars. As the colonies grow at each new location, some of the people may decide to set out for other star systems. Even if our starships traveled at relatively slow speeds—say, a few percent of the speed of light—we could have dozens of outposts around nearby stars within a few centuries. In 10,000 years, our descendants would be spread among stars within a few hundred light-years of Earth. In a few million years, we could have outposts scattered throughout the Milky Way and would likely have explored nearly every star system in the galaxy. We would have become a true galactic civilization.

Now, if we take the idea that *we* could develop a galactic civilization within a few million years and combine it with the reasonable (though unproved) idea that civilizations ought to be common, we are led to an astonishing conclusion: Someone else should already have created a galactic civilization. In fact, someone should have done it a long time ago.

Thus, we encounter a strange paradox: Plausible arguments suggest that a galactic civilization should already exist, yet we have so far found no evidence of such a civilization. There are many possible solutions to this paradox, but most fall into one of three basic categories:

1. There is no galactic civilization because civilizations are not common. Perhaps we are the first, or perhaps civilizations are so rare that many star systems, including ours, remain unexplored.

2. There is no galactic civilization because civilizations do not leave their home worlds—either because they are uninterested in interstellar travel or because they destroy themselves before achieving it.

3. There *is* a galactic civilization, but it has deliberately avoided revealing its existence to us.

We do not know which, if any, of these three explanations is the correct solution to the paradox, but each leads to intriguing possibilities. If the first is correct, then our civilization is an astonishing achievement—something that few if any other species have ever accomplished. From this point of view, humanity becomes all the more precious, and the collapse of our civilization would be all the more tragic. The second possible explanation is much less satisfying. Unless we "think differently" than other civilizations, it says that we will soon either lose our interest in interstellar travel or destroy ourselves—hardly a comforting thought. The third possibility is perhaps the most intriguing. It says that we are newcomers on the scene of a galactic civilization that has existed for millions or billions of years before us.

Perhaps members of this civilization are deliberately leaving us alone for the time being and will invite us to join them when we prove ourselves worthy. If so, our entire species may be on the verge of beginning a journey every bit as incredible as that of a baby emerging from the womb and coming into the world.

We, this people, on a small and lonely planet
Travelling through casual space
Past aloof stars, across the way of indifferent suns
To a destination where all signs tell us
It is possible and imperative that we learn
A brave and startling truth.

MAYA ANGELOU, excerpted from *A Brave and Startling Truth*

THE BIG PICTURE

Throughout our study of astronomy, we have taken the "big picture" view of trying to understand how we fit into the universe. Here, at last, we have returned to Earth and examined the role of our own generation in the big picture of human history. Tens of thousands of human generations past have walked this Earth, but ours is the first generation with the technology to study the far reaches of our universe and to travel beyond the Earth. It is up to us to decide whether we will use this technology to advance our species or to destroy it.

Take a long view, a view down the centuries and millennia from this moment. Imagine our descendants living among the stars, having created or joined a great galactic civilization. They will have the privilege of experiencing ideas, worlds, and discoveries far beyond our wildest imagination. Perhaps, in their history lessons, they will learn of our generation—the generation that history placed at the turning point and that chose life over death and destruction.

Review Questions

1. How do speeds of current spacecraft compare to the speed of light?

2. Describe several hurdles, both technological and social, that we must overcome in order to achieve interstellar flight.

3. Why do we say that *Pioneers 10* and *11* and *Voyagers 1* and *2* are our first interstellar spacecraft?

4. Why are chemical rockets inadequate for interstellar travel?

5. Briefly describe the prospects and limitations for each of the following alternative starship designs: nuclear propulsion, matter–antimatter engines, sails and beamed energy propulsion, and interstellar ramjets.

6. What is SETI?

7. Briefly describe and interpret the equation given in the text that allows us to estimate the number of civilizations in the Milky Way Galaxy. Why is the term f_{now} particularly important?

8. Describe several potential solutions to the paradox of "where are the aliens?"

Discussion Questions

1. *One-Way Trip.* From the sixteenth to the nineteenth centuries, many people left their homes in Europe on one-way journeys to the "new world" of America. In the future, similar one-way trips to colonies on Earth-like planets around other stars may be possible. If offered the opportunity, would you go? Do you know anyone who would? Why or why not?

2. *SETI Efforts.* SETI supporters argue that, despite the costs and the unknowns, the efforts are justified because contact with an extraterrestrial intelligence would be one of the most dramatic discoveries in human history. Do you agree? Defend your opinion.

3. *Aliens in the Movies.* Choose a science fiction movie (or television show) that involves alien species, and discuss what aspects of it are realistic and what aspects are not. Pay particular attention to the depiction of alien species and encounters between different species. Do you think such encounters are portrayed realistically?

Problems

1. *Research: Pioneers of Flight.* Write a one- to two-page biography of one of the following pioneers of flight: Otto Lilienthal, the Wright brothers, Charles Lindbergh, Amelia Earhart.

2. *Research: Development of Rocketry.* Write a one- to two-page paper about the development of rocketry. Be sure to include the roles of Konstantin Tsiolkovsky and Robert Goddard.

3. *Research: Starship Design.* Find more details about one of the proposals for starship propulsion and design discussed in this chapter. How would the starship actually be built? What new technologies would be needed, and what existing technologies could be used? Summarize your findings in a one- to two-page research report.

A FEW MATHEMATICAL SKILLS

This appendix reviews the following mathematical skills: powers of 10, scientific notation, working with units, and the metric system. You should refer to this appendix as needed while studying the textbook.

A.1 Powers of 10

Powers of 10 simply indicate how many times to multiply 10 by itself. For example:

$$10^2 = 10 \times 10 = 100$$

$$10^6 = 10 \times 10 \times 10 \times 10 \times 10 \times 10 = 1{,}000{,}000$$

Negative powers are the reciprocals of the corresponding positive powers. For example:

$$10^{-2} = \frac{1}{10^2} = \frac{1}{100} = 0.01$$

$$10^{-6} = \frac{1}{10^6} = \frac{1}{1{,}000{,}000} = 0.000001$$

Table A.1 lists powers of 10 from 10^{-12} to 10^{12}. Note that powers of 10 follow two basic rules:

1. A positive exponent tells how many zeros follow the 1. For example, 10^0 is a 1 followed by no zeros, and 10^8 is a 1 followed by eight zeros.

2. A negative exponent tells how many places are to the right of the decimal point, including the 1. For example, $10^{-1} = 0.1$ has one place to the right of the decimal point; $10^{-6} = 0.000001$ has six places to the right of the decimal point.

Table A.1 Powers of 10

Zero and Positive Powers			Negative Powers		
Power	Value	Name	Power	Value	Name
10^0	1	One			
10^1	10	Ten	10^{-1}	0.1	Tenth
10^2	100	Hundred	10^{-2}	0.01	Hundredth
10^3	1,000	Thousand	10^{-3}	0.001	Thousandth
10^4	10,000	Ten thousand	10^{-4}	0.0001	Ten thousandth
10^5	100,000	Hundred thousand	10^{-5}	0.00001	Hundred thousandth
10^6	1,000,000	Million	10^{-6}	0.000001	Millionth
10^7	10,000,000	Ten million	10^{-7}	0.0000001	Ten millionth
10^8	100,000,000	Hundred million	10^{-8}	0.00000001	Hundred millionth
10^9	1,000,000,000	Billion	10^{-9}	0.000000001	Billionth
10^{10}	10,000,000,000	Ten billion	10^{-10}	0.0000000001	Ten billionth
10^{11}	100,000,000,000	Hundred billion	10^{-11}	0.00000000001	Hundred billionth
10^{12}	1,000,000,000,000	Trillion	10^{-12}	0.000000000001	Trillionth

Multiplying and Dividing Powers of 10

Multiplying powers of 10 simply requires adding exponents, as the following examples show:

$$10^4 \times 10^7 = \underbrace{10,000}_{10^4} \times \underbrace{10,000,000}_{10^7} = \underbrace{100,000,000,000}_{10^{4+7} = 10^{11}} = 10^{11}$$

$$10^5 \times 10^{-3} = \underbrace{100,000}_{10^5} \times \underbrace{0.001}_{10^{-3}} = \underbrace{100}_{10^{5+(-3)} = 10^2} = 10^2$$

$$10^{-8} \times 10^{-5} = \underbrace{0.00000001}_{10^{-8}} \times \underbrace{0.00001}_{10^{-5}} = \underbrace{0.0000000000001}_{10^{-8+(-5)} = 10^{-13}} = 10^{-13}$$

Dividing powers of 10 requires subtracting exponents, as in the following examples:

$$\frac{10^5}{10^3} = \underbrace{100,000}_{10^5} \div \underbrace{1,000}_{10^3} = \underbrace{100}_{10^{5-3} = 10^2} = 10^2$$

$$\frac{10^3}{10^7} = \underbrace{1,000}_{10^3} \div \underbrace{10,000,000}_{10^7} = \underbrace{0.0001}_{10^{3-7} = 10^{-4}} = 10^{-4}$$

$$\frac{10^{-4}}{10^{-6}} = \underbrace{0.0001}_{10^{-4}} \div \underbrace{0.000001}_{10^{-6}} = \underbrace{100}_{10^{-4-(-6)} = 10^2} = 10^2$$

Powers of Powers of 10

We can use the multiplication and division rules to raise powers of 10 to other powers or to take roots. For example:

$$(10^4)^3 = 10^4 \times 10^4 \times 10^4 = 10^{4+4+4} = 10^{12}$$

Note that we can get the same end result by simply multiplying the two powers:

$$(10^4)^3 = 10^{4 \times 3} = 10^{12}$$

Because taking a root is the same as raising to a fractional power (e.g., the square root is the same as the 1/2 power, the cube root is the same as the 1/3 power, etc.), we can use the same procedure for roots, as in the following example:

$$\sqrt{0^4} = (10^4)^{1/2} = 10^{4 \times (1/2)} = 10^2$$

Adding and Subtracting Powers of 10

Unlike with multiplication and division, there is no shortcut for adding or subtracting powers of 10. The values must be written in longhand notation. For example:

$$10^6 + 10^2 = 1{,}000{,}000 + 100 = 1{,}000{,}100$$

$$10^8 + 10^{-3} = 100{,}000{,}000 + 0.001 = 100{,}000{,}000.001$$

$$10^7 - 10^3 = 10{,}000{,}000 - 1{,}000 = 9{,}999{,}000$$

Summary

We can summarize our findings using n and m to represent any numbers:

■ To *multiply* powers of 10, *add* exponents: $10^n \times 10^m = 10^{n+m}$

■ To *divide* powers of 10, *subtract* exponents: $\dfrac{10^n}{10^m} = 10^{n-m}$

■ To *raise* powers of 10 to other powers, multiply exponents: $(10^n)^m = 10^{n \times m}$

A.2 Scientific Notation

When we are dealing with large or small numbers, it's generally easier to write them with powers of 10. For example, it's much easier to write the number $6{,}000{,}000{,}000{,}000$ as 6×10^{12}. This format, in which a number *between* 1 and 10 is multiplied by a power of 10, is called **scientific notation.**

Converting a Number to Scientific Notation

We can convert numbers written in ordinary notation to scientific notation with a simple two-step process:

1. Move the decimal point to come after the *first* nonzero digit.

2. The number of places the decimal point moves tells you the power of 10; the power is *positive* if the decimal point moves to the left and *negative* if it moves to the right.

Examples:

$$3{,}042 \xrightarrow[\text{3 places to left}]{\text{decimal needs to move}} 3.042 \times 10^3$$

$$0.00012 \xrightarrow[\text{4 places to right}]{\text{decimal needs to move}} 1.2 \times 10^{-4}$$

$$226 \times 10^2 \xrightarrow[\text{2 places to left}]{\text{decimal needs to move}} (2.26 \times 10^2) \times 10^2 = 2.26 \times 10^4$$

Converting a Number from Scientific Notation

We can convert numbers written in scientific notation to ordinary notation by the reverse process:

1. The power of 10 indicates how many places to move the decimal point; move it to the *right* if the power of 10 is positive and to the *left* if it is negative.

2. If moving the decimal point creates any open places, fill them with zeros.

Examples:

$$4.01 \times 10^2 \xrightarrow[\text{2 places to right}]{\text{move decimal}} 401$$

$$3.6 \times 10^6 \xrightarrow[\text{6 places to right}]{\text{move decimal}} 3{,}600{,}000$$

$$5.7 \times 10^{-3} \xrightarrow[\text{3 places to left}]{\text{move decimal}} 0.0057$$

Multiplying or Dividing Numbers in Scientific Notation

Multiplying or dividing numbers in scientific notation simply requires operating on the powers of 10 and the other parts of the number separately.

Examples:

$$(6 \times 10^2) \times (4 \times 10^5) = (6 \times 4) \times (10^2 \times 10^5) = 24 \times 10^7 = (2.4 \times 10^1) \times 10^7 = 2.4 \times 10^8$$

$$\frac{4.2 \times 10^{-2}}{8.4 \times 10^{-5}} = \frac{4.2}{8.4} \times \frac{10^{-2}}{10^{-5}} = 0.5 \times 10^{-2-(-5)} = 0.5 \times 10^3 = (5 \times 10^{-1}) \times 10^3 = 5 \times 10^2$$

Note that, in both these examples, we first found an answer in which the number multiplied by a power of 10 was *not* between 1 and 10. We therefore followed the procedure for converting the final answer to scientific notation.

Addition and Subtraction with Scientific Notation

In general, we must write numbers in ordinary notation before adding or subtracting.

Examples:

$$(3 \times 10^6) + (5 \times 10^2) = 3{,}000{,}000 + 500 = 3{,}000{,}500 = 3.0005 \times 10^6$$

$$(4.6 \times 10^9) - (5 \times 10^8) = 4{,}600{,}000{,}000 - 500{,}000{,}000 = 4{,}100{,}000{,}000 = 4.1 \times 10^9$$

When both numbers have the *same* power of 10, we can factor out the power of 10 first.

Examples:

$$(7 \times 10^{10}) + (4 \times 10^{10}) = (7 + 4) \times 10^{10} = 11 \times 10^{10} = 1.1 \times 10^{11}$$

$$(2.3 \times 10^{-22}) - (1.6 \times 10^{-22}) = (2.3 - 1.6) \times 10^{-22} = 0.7 \times 10^{-22} = 7.0 \times 10^{-23}$$

A.3 Working with Units

Showing the units of a problem as you solve it usually makes the work much easier and also provides a useful way of checking your work. If an answer does not come out with the units you expect, you probably did something wrong. In general, working with units is very similar to working with numbers, as the following guidelines and examples show.

Five Guidelines for Working with Units

Before you begin any problem, think ahead and identify the units you expect for the final answer. Then operate on the units along with the numbers as you solve the problem. The following five guidelines may be helpful when you are working with units:

1. Mathematically, it doesn't matter whether a unit is singular (e.g., meter) or plural (e.g., meters); we can use the same abbreviation (e.g., m) for both.

2. You cannot add or subtract numbers unless they have the *same* units. For example, 5 apples + 3 apples = 8 apples, but the expression 5 apples + 3 oranges cannot be simplified further.

3. You *can* multiply units, divide units, or raise units to powers. Look for key words that tell you what to do.

- *Per* suggests division. For example, we write a speed of 100 kilometers per hour as

$$100\,\frac{km}{hr} \quad \text{or} \quad \frac{100\,km}{1\,hr}$$

- *Of* suggests multiplication. For example, if you launch a 50-kg space probe at a launch cost *of* $10,000 per kilogram, the total cost is:

$$50\,\cancel{kg} \times \frac{\$10,000}{\cancel{kg}} = \$500,000$$

- *Square* suggests raising to the second power. For example, we write an area of 75 square meters as 75 m^2.

- *Cube* suggests raising to the third power. For example, we write a volume of 12 cubic centimeters as 12 cm^3.

4. Often the number you are given is not in the units you wish to work with. For example, you may be given that the speed of light is 300,000 km/s but need it in units of m/s for a particular problem. To convert the units, simply multiply the given number by a *conversion factor:* a fraction in which the numerator (top of the fraction) and denominator (bottom of the fraction) are equal, so that the value of the fraction is 1; the number in the denominator must have the units that you wish to change. In the case of changing the speed of light from units of km/s to m/s, you need a conversion factor for kilometers to meters. Thus, the conversion factor is:

$$\frac{1,000\,m}{1\,km}$$

Note that this conversion factor is equal to 1, since 1,000 meters and 1 kilometer are equal, and that the units to be changed (km) appear in the denominator. We can now convert the speed of light from units of km/s to m/s simply by multiplying by this conversion factor:

$$\underbrace{300,000\,\frac{\cancel{km}}{s}}_{\substack{\text{speed of light} \\ \text{in km/s}}} \times \underbrace{\frac{1,000\,m}{1\,\cancel{km}}}_{\substack{\text{conversion from} \\ \text{km to m}}} = \underbrace{3 \times 10^8\,\frac{m}{s}}_{\substack{\text{speed of light} \\ \text{in m/s}}}$$

Note that the units of km cancel, leaving the answer in units of m/s.

5. It's easier to work with units if you replace division with multiplication by the reciprocal. For example, suppose you want to know how many minutes are represented by 300 seconds. We can find the answer by dividing 300 seconds by 60 seconds per minute:

$$300\,s \div 60\,\frac{s}{min}$$

However, it is easier to see the unit cancellations if we rewrite this expression by replacing the division with multiplication by the reciprocal (this process is easy to remember as "invert and multiply"):

$$300\,s \div 60\,\frac{s}{min} = 300\,\cancel{s} \times \underbrace{\frac{1\,min}{60\,\cancel{s}}}_{\substack{\text{invert} \\ \text{and multiply}}} = 5\,min$$

We now see that the units of seconds (s) cancel in the numerator of the first term and the denominator of the second term, leaving the answer in units of minutes.

More Examples of Working with Units

Example 1. How many seconds are there in 1 day?

Solution: We can answer the question by setting up a *chain* of unit conversions in which we start with 1 *day* and end up with *seconds*. We use the facts that there are 24 hours per day (24 hr/day), 60 minutes per hour (60 min/hr), and 60 seconds per minute (60 s/min):

$$\underbrace{1 \text{ day}}_{\substack{\text{starting}\\\text{value}}} \times \underbrace{\frac{24 \text{ hr}}{\text{day}}}_{\substack{\text{conversion}\\\text{from}\\\text{day to hr}}} \times \underbrace{\frac{60 \text{ min}}{\text{hr}}}_{\substack{\text{conversion}\\\text{from}\\\text{hr to min}}} \times \underbrace{\frac{60 \text{ s}}{\text{min}}}_{\substack{\text{conversion}\\\text{from}\\\text{min to s}}} = 86{,}400 \text{ s}$$

Note that all the units cancel except *seconds*, which is what we want for the answer. There are 86,400 seconds in 1 day.

Example 2. Convert a distance of 10^8 cm to km.

Solution: The easiest way to make this conversion is in two steps, since we know that there are 100 centimeters per meter (100 cm/m) and 1,000 meters per kilometer (1,000 m/km):

$$\underbrace{10^8 \text{ cm}}_{\substack{\text{starting}\\\text{value}}} \times \underbrace{\frac{1 \text{ m}}{100 \text{ cm}}}_{\substack{\text{conversion}\\\text{from}\\\text{cm to m}}} \times \underbrace{\frac{1 \text{ km}}{1{,}000 \text{ m}}}_{\substack{\text{conversion}\\\text{from}\\\text{m to km}}} = 10^8 \text{ cm} \times \frac{1 \text{ m}}{10^2 \text{ cm}} \times \frac{1 \text{ km}}{10^3 \text{ m}} = 10^3 \text{ km}$$

Alternatively, if we recognize that the number of kilometers should be smaller than the number of centimeters (because kilometers are larger), we might decide to do this conversion by dividing as follows:

$$10^8 \text{ cm} \div \frac{100 \text{ cm}}{\text{m}} \div \frac{1{,}000 \text{ m}}{\text{km}}$$

In this case, before carrying out the calculation, we replace each division with multiplication by the reciprocal:

$$10^8 \text{ cm} \div \frac{100 \text{ cm}}{\text{m}} \div \frac{1{,}000 \text{ m}}{\text{km}} = 10^8 \text{cm} \times \frac{1 \text{ m}}{100 \text{ cm}} \times \frac{1 \text{ km}}{1{,}000 \text{ m}}$$

$$= 10^8 \text{ cm} \times \frac{1 \text{ m}}{10^2 \text{ cm}} \times \frac{1 \text{ km}}{10^3 \text{ m}}$$

$$= 10^3 \text{ km}$$

Note that we again get the answer that 10^8 cm is the same as 10^3 km, or 1,000 km.

Example 3. Suppose you accelerate at 9.8 m/s^2 for 4 seconds, starting from rest. How fast will you be going?

Solution: The question asked "how fast?" so we expect to end up with a speed. Therefore, we multiply the acceleration by the amount of time you accelerated:

$$9.8 \frac{\text{m}}{\text{s}^2} \times 4 \text{ s} = (9.8 \times 4) \frac{\text{m} \times \text{s}}{\text{s}^2} = 39.2 \frac{\text{m}}{\text{s}}$$

Note that the units end up as a speed, showing that you will be traveling 39.2 m/s after 4 seconds of acceleration at 9.8 m/s^2.

Example 4. A reservoir is 2 km long and 3 km wide. Calculate its area, in both square kilometers and square meters.

Solution: We find its area by multiplying its length and width:

$$2 \text{ km} \times 3 \text{ km} = 6 \text{ km}^2$$

Next we need to convert this area of 6 km^2 to square meters, using the fact that there are 1,000 meters per kilometer (1,000 m/km). Note that we must square the term 1,000 m/km when converting from km^2 to m^2:

$$6 \text{ km}^2 \times \left(1{,}000 \, \frac{\text{m}}{\text{km}}\right)^2 = 6 \text{ km}^2 \times 1{,}000^2 \, \frac{\text{m}^2}{\text{km}^2} = 6 \, \cancel{\text{km}^2} \times 1{,}000{,}000 \, \frac{\text{m}^2}{\cancel{\text{km}^2}}$$

$$= 6{,}000{,}000 \text{ m}^2$$

The reservoir area is 6 km², which is the same as 6 million m².

A.4 The Metric System (SI)

The modern version of the metric system, known as *Système Internationale d'Unites* (French for "International System of Units") or **SI,** was formally established in 1960. Today, it is the primary measurement system in nearly every country in the world with the exception of the United States. Even in the United States, it is the system of choice for science and international commerce.

The basic units of length, mass, and time in the SI are:

- The **meter** for length, abbreviated m
- The **kilogram** for mass, abbreviated kg
- The **second** for time, abbreviated s

Multiples of metric units are formed by powers of 10, using a prefix to indicate the power. For example, *kilo* means 10^3 (1,000), so a kilometer is 1,000 meters; a microgram is 0.000001 gram, because *micro* means 10^{-6}, or one millionth. Some of the more common prefixes are listed in Table A.2.

Metric Conversions

Table A.3 lists conversions between metric units and units used commonly in the United States. Note that the conversions between kilograms and pounds are valid only on Earth, because they depend on the strength of gravity.

Example 1. International athletic competitions generally use metric distances. Compare the length of a 100-meter race to that of a 100-yard race.

Solution: Table A.3 shows that 1 m = 1.094 yd, so 100 m is 109.4 yd. Note that 100 meters is almost 110 yards; a good "rule of thumb" to remember is that distances in meters are about 10% longer than the corresponding number of yards.

Example 2. How many square kilometers are in 1 square mile?

Solution: We use the square of the miles-to-kilometers conversion factor:

$$(1 \text{ mi}^2) \times \left(\frac{1.6093 \text{ km}}{1 \text{ mi}}\right)^2 = (1 \, \cancel{\text{mi}^2}) \times 1.6093^2 \, \frac{\text{km}^2}{\cancel{\text{mi}^2}} = 2.5898 \text{ km}^2$$

Therefore, 1 square mile is 2.5898 square kilometers.

Table A.2 SI (Metric) Prefixes

Small Values			Large Values		
Prefix	Abbreviation	Value	Prefix	Abbreviation	Value
Deci	d	10^{-1}	Deca	da	10^1
Centi	c	10^{-2}	Hecto	h	10^2
Milli	m	10^{-3}	Kilo	k	10^3
Micro	μ	10^{-6}	Mega	M	10^6
Nano	n	10^{-9}	Giga	G	10^9
Pico	p	10^{-12}	Tera	T	10^{12}

Table A.3 Metric Conversions

To Metric	From Metric
1 inch = 2.540 cm	1 cm = 0.3937 inch
1 foot = 0.3048 m	1 m = 3.28 feet
1 yard = 0.9144 m	1 m = 1.094 yards
1 mile = 1.6093 km	1 km = 0.6214 mile
1 pound = 0.4536 kg	1 kg = 2.205 pounds

A.5 Finding a Ratio

Suppose you want to compare two quantities, such as the average density of the Earth and the average density of Jupiter. The way we do such a comparison is by dividing, which tells us the *ratio* of the two quantities. In this case, the Earth's average density is 5.52 grams/cm³ and Jupiter's average density is 1.33 grams/cm³ (see Table 7.1), so the ratio is:

$$\frac{\text{average density of Earth}}{\text{average density of Jupiter}} = \frac{5.52 \ \cancel{\text{g/cm}^3}}{1.33 \ \cancel{\text{g/cm}^3}} = 4.15$$

Notice how the units cancel on both the top and bottom of the fraction. We can state our result in two equivalent ways:

- The ratio of the Earth's average density to Jupiter's average density is 4.15.

- The Earth's average density is 4.15 times Jupiter's average density.

Sometimes, the quantities that you want to compare may each involve an equation. In such cases, you could, of course, find the ratio by first calculating each of the two quantities individually and then dividing. However, it is much easier if you first express the ratio as a fraction, putting the equation for one quantity on top and the other on the bottom. Some of the terms in the equation may then cancel out, making any calculations much easier.

Example 1. Compare the kinetic energy of a car traveling at 100 km/hr to that of a car traveling at 50 km/hr.

Solution: We do the comparison by finding the ratio of the two kinetic energies, recalling that the formula for kinetic energy is $\frac{1}{2}mv^2$. Since we are not told the mass of the car, you might at first think that we don't have enough information to find the ratio. However, notice what happens when we put the equations for each kinetic energy into the ratio, calling the two speeds v_1 and v_2:

$$\frac{\text{K.E. car at } v_1}{\text{K.E. car at } v_2} = \frac{\cancel{\frac{1}{2}}\cancel{m}_{\text{car}}v_1^{\ 2}}{\cancel{\frac{1}{2}}\cancel{m}_{\text{car}}v_2^{\ 2}} = \frac{v_1^{\ 2}}{v_2^{\ 2}} = \left(\frac{v_1}{v_2}\right)^2$$

All the terms cancel except those with the two speeds, leaving us with a very simple formula for the ratio. Now we put in 100 km/hr for v_1 and 50 km/hr for v_2:

$$\frac{\text{K.E. car at 100 km/hr}}{\text{K.E. car at 50 km/hr}} = \left(\frac{100 \ \cancel{\text{km/hr}}}{50 \ \cancel{\text{km/hr}}}\right)^2 = 2^2 = 4$$

The ratio of the car's kinetic energies at 100 km/hr and 50 km/hr is 4. That is, the car has four times as much kinetic energy at 100 km/hr as it has at 50 km/hr.

Example 2. Compare the strength of gravity between the Earth and the Sun to the strength of gravity between the Earth and the Moon.

Solution: We do the comparison by taking the ratio of the Earth–Sun gravity to the Earth–Moon gravity. In this case, each quantity is found from the equation of Newton's law of gravity. (See Section 5.4.) Thus, the ratio is

$$\frac{\text{Earth–Sun gravity}}{\text{Earth–Moon gravity}} = \frac{\cancel{G}\dfrac{\cancel{M}_{\text{Earth}}M_{\text{Sun}}}{(d_{\text{Earth–Sun}})^2}}{\cancel{G}\dfrac{\cancel{M}_{\text{Earth}}M_{\text{Moon}}}{(d_{\text{Earth–Moon}})^2}} = \frac{M_{\text{Sun}}}{(d_{\text{Earth–Sun}})^2} \times \frac{(d_{\text{Earth–Moon}})^2}{M_{\text{Moon}}}$$

Note how all but four of the terms cancel; the last step comes from replacing the division with multiplication by the reciprocal (the "invert and multiply" rule for division). We can simplify the work further by rearranging the terms so that we have the masses and distances together:

$$\frac{\text{Earth–Sun gravity}}{\text{Earth–Moon gravity}} = \frac{M_{\text{Sun}}}{M_{\text{Moon}}} \times \frac{(d_{\text{Earth–Moon}})^2}{(d_{\text{Earth–Sun}})^2}$$

Now it is just a matter of looking up the numbers (see Appendix C) and calculating:

$$\frac{\text{Earth–Sun gravity}}{\text{Earth–Moon gravity}} = \frac{1.99 \times 10^{30} \, \cancel{\text{kg}}}{7.35 \times 10^{22} \, \cancel{\text{kg}}} \times \frac{(384.4 \times 10^3 \, \cancel{\text{km}})^2}{(149.6 \times 10^6 \, \cancel{\text{km}})^2} = 179$$

In other words, the Earth–Sun gravity is 179 times stronger than the Earth–Moon gravity.

THE PERIODIC TABLE OF THE ELEMENTS

1																	2
H Hydrogen 1.00794																	**He** Helium 4.003

Atomic number — 12
Element's symbol — **Mg**
Element's name — Magnesium
Atomic mass* — 24.305

*Atomic masses are fractions because they represent a weighted average of atomic masses of different isotopes—in proportion to the abundance of each isotope on Earth.

3 **Li** Lithium 6.941	4 **Be** Beryllium 9.01218											5 **B** Boron 10.81	6 **C** Carbon 12.011	7 **N** Nitrogen 14.007	8 **O** Oxygen 15.999	9 **F** Fluorine 18.988	10 **Ne** Neon 20.179
11 **Na** Sodium 22.990	12 **Mg** Magnesium 24.305											13 **Al** Aluminum 26.98	14 **Si** Silicon 28.086	15 **P** Phosphorus 30.974	16 **S** Sulfur 32.06	17 **Cl** Chlorine 35.453	18 **Ar** Argon 39.948
19 **K** Potassium 39.098	20 **Ca** Calcium 40.08	21 **Sc** Scandium 44.956	22 **Ti** Titanium 47.88	23 **V** Vanadium 50.94	24 **Cr** Chromium 51.996	25 **Mn** Manganese 54.938	26 **Fe** Iron 55.847	27 **Co** Cobalt 58.9332	28 **Ni** Nickel 58.69	29 **Cu** Copper 63.546	30 **Zn** Zinc 65.39	31 **Ga** Gallium 69.72	32 **Ge** Germanium 72.59	33 **As** Arsenic 74.922	34 **Se** Selenium 78.96	35 **Br** Bromine 79.904	36 **Fr** Krypton 83.80
37 **Rb** Rubidium 85.468	38 **Sr** Strontium 87.62	39 **Y** Yttrium 88.9059	40 **Zr** Zirconium 91.224	41 **Nb** Niobium 92.91	42 **Mo** Molybdenum 95.94	43 **Tc** Technetium (98)	44 **Ru** Ruthenium 101.07	45 **Rh** Rhodium 102.906	46 **Pd** Palladium 106.42	47 **Ag** Silver 107.868	48 **Cd** Cadmium 112.41	49 **In** Indium 114.82	50 **Sn** Tin 118.71	51 **Sb** Antimony 121.75	52 **Te** Tellurium 127.60	53 **I** Iodine 126.905	54 **Xe** Xenon 131.29
55 **Cs** Cesium 132.91	56 **Ba** Barium 137.34		72 **Hf** Hafnium 178.49	73 **Ta** Tantalum 180.95	74 **W** Tungsten 183.85	75 **Re** Rhenium 186.207	76 **Os** Osmium 190.2	77 **Ir** Iridium 192.22	78 **Pt** Platinum 195.08	79 **Au** Gold 196.967	80 **Hg** Mercury 200.59	81 **Ti** Thallium 204.383	82 **Pb** Lead 207.2	83 **Bi** Bismuth 208.98	84 **Po** Polonium (209)	85 **At** Astatine (210)	86 **Rn** Radon (222)
87 **Fr** Francium (223)	88 **Ra** Radium 226.0254		104 **Rf** Rutherfordium (261)	105 **Db** Dubnium (262)	106 **Sg** Seaborgium (263)	107 **Bh** Bohrium (262)	108 **Hs** Hassium (265)	109 **Mt** Meitnerium (266)	110 **Uun** Unununilium (269)	111 **Uuu** Unununium (272)	112 **Uub** Unununbium (277)						

Lanthanide Series

57 **La** Lanthanum 138.906	58 **Ce** Cerium 140.12	59 **Pr** Praseodymium 140.908	60 **Nd** Neodymium 144.24	61 **Pm** Promethium (145)	62 **Sm** Samarium 150.36	63 **Eu** Europium 151.96	64 **Gd** Gadolinium 157.25	65 **Tb** Terbium 158.925	66 **Dy** Dysprosium 162.50	67 **Ho** Holmium 164.93	68 **Er** Erbium 167.26	69 **Tm** Thulium 168.934	70 **Yb** Ytterbium 173.04	71 **Lu** Lutetium 174.967

Actinide Series

89 **Ac** Actinium 227.028	90 **Th** Thorium 232.038	91 **Pa** Protactinium 231.036	92 **U** Uranium 238.029	93 **Np** Neptunium 237.048	94 **Pu** Plutonium (244)	95 **Am** Americium (243)	96 **Cm** Curium (247)	97 **Bk** Berkelium (247)	98 **Cf** Californium (251)	99 **Es** Einsteinium (252)	100 **Fm** Fermium (257)	101 **Md** Mendelevium (258)	102 **No** Nobelium (259)	103 **Lr** Lawrencium (260)

APPENDIX C

PLANETARY DATA

Table C.1 Physical Properties of the Sun and Planets

Name	Radius (Eq) (km)	Radius (Eq) (Earth units)	Mass (kg)	Mass (Earth units)	Average Density (g/cm³)	Surface Gravity (Earth = 1)
Sun	695,000	109	1.99×10^{30}	333,000	1.41	27.5
Mercury	2,440	0.382	3.30×10^{23}	0.055	5.43	0.38
Venus	6,051	0.949	4.87×10^{24}	0.815	5.25	0.91
Earth	6,378	1.00	5.97×10^{24}	1.00	5.52	1.00
Mars	3,397	0.533	6.42×10^{23}	0.107	3.93	0.38
Jupiter	71,492	11.19	1.90×10^{27}	317.9	1.33	2.53
Saturn	60,268	9.46	5.69×10^{26}	95.18	0.71	1.07
Uranus	25,559	3.98	8.66×10^{25}	14.54	1.24	0.91
Neptune	24,764	3.81	1.03×10^{26}	17.13	1.67	1.14
Pluto	1,160	0.181	1.31×10^{22}	0.0022	2.05	0.07

Table C.2 Orbital Properties of the Sun and Planets

Name	Distance from Sun* (AU)	Distance from Sun* (10⁶ km)	Orbital Period (years)	Orbital Inclination† (degrees)	Orbital Eccentricity	Sidereal Rotation Period (Earth days)‡	Axis Tilt (degrees)
Sun	—	—	—	—	—	25.4	7.25
Mercury	0.387	57.9	0.2409	7.00	0.206	58.6	0.0
Venus	0.723	108.2	0.6152	3.39	0.007	−243.0	177.4
Earth	1.00	149.6	1.0	0.00	0.017	0.9973	23.45
Mars	1.524	227.9	1.881	1.85	0.093	1.026	23.98
Jupiter	5.203	778.3	11.86	1.31	0.048	0.41	3.08
Saturn	9.539	1,427	29.46	2.49	0.056	0.44	26.73
Uranus	19.19	2,870	84.01	0.77	0.046	−0.72	97.92
Neptune	30.06	4,497	164.8	1.77	0.010	0.67	29.6
Pluto	39.54	5,916	248.0	17.15	0.248	−6.39	118

*Semimajor axis of the orbit.

†With respect to the ecliptic.

‡A negative sign indicates backward rotation.

Table C.3 Satellites of the Solar System*

Planet Satellite	Radius or Dimensions† (km)	Distance from Planet (10³ km)	Orbital Period‡ (Earth days)	Mass** (kg)	Density** (g/cm³)	Notes About the Satellites
Earth						**Earth**
Moon	1,738	384.4	27.322	7.349×10^{22}	3.34	Probably formed in giant impact.
Mars						**Mars**
Phobos	$13 \times 11 \times 9$	9.38	0.319	1.3×10^{16}	2.2	Probable captured asteroids.
Deimos	$8 \times 6 \times 5$	23.5	1.263	1.8×10^{15}	1.7	
Jupiter						**Jupiter**
Metis	20	127.96	0.295	9×10^{16}	—	Small moonlets within and near Jupiter's ring system.
Adrastea	$13 \times 10 \times 8$	128.98	0.298	1×10^{16}	—	
Amalthea	$135 \times 82 \times 75$	181.3	0.498	8×10^{18}	—	
Thebe	$55 \times ? \times 45$	221.9	0.6745	1.4×10^{17}	—	
Io	1,821	421.6	1.769	8.933×10^{22}	3.57	Most volcanically active object in the solar system.
Europa	1,565	670.9	3.551	4.797×10^{22}	2.97	Possible oceans under icy crust.
Ganymede	2,634	1,070.0	7.155	1.482×10^{23}	1.94	Largest satellite in solar system; unusual ice geology.
Callisto	2,403	1,883.0	16.689	1.076×10^{23}	1.86	Cratered iceball.
Leda	~8	11,094	238.7	4×10^{16}	—	
Himalia	92.5	11,480	250.6	8×10^{18}	—	Each probably a fragment of captured asteroid that broke apart.
Lysithea	~18	11,720	259.2	6×10^{16}	—	
Elara	~38	11,737	259.7	6×10^{17}	—	
Ananke	~15	20,200	−631	4×10^{16}	—	
Carme	~20	22,600	−692	9×10^{16}	—	Each probably a fragment of captured asteroid that broke apart.
Pasiphae	~25	23,500	−735	1.6×10^{17}	—	
Sinope	~18	23,700	−758	6×10^{16}	—	
Saturn						**Saturn**
Pan	10	133.570	0.574	4.2×10^{14}	—	
Atlas	$19 \times ? \times 14$	137.64	0.602	1.6×10^{16}	—	Small moonlets within and near Saturn's ring system.
Prometheus	$74 \times 50 \times 34$	139.35	0.613	5×10^{17}	—	
Pandora	$55 \times 44 \times 31$	141.7	0.629	3.4×10^{17}	—	
Epimetheus	$69 \times 55 \times 55$	151.42	0.694	5.6×10^{17}	—	
Janus	$97 \times 95 \times 77$	151.47	0.695	2.0×10^{18}	—	
Mimas	199	185.52	0.942	3.70×10^{19}	1.17	Small and medium-size iceballs, many with interesting geology.
Enceladus	249	238.02	1.370	1.2×10^{20}	1.24	
Tethys	530	294.66	1.888	6.17×10^{20}	1.26	
Calypso	$15 \times 8 \times 8$	294.66	1.888	4×10^{15}	—	Small moonlets sharing Tethys's orbit.
Telesto	$15 \times 13 \times 8$	294.67	1.888	6×10^{15}	—	
Dione	559	377.4	2.737	1.08×10^{21}	1.44	Medium-size iceball, with interesting geology.
Helene	$18 \times ? \times 15$	377.4	2.737	1.6×10^{16}	—	Small moonlet sharing Dione's orbit.

Name	Radius (km)	Distance	Orbital Period (days)	Mass (kg)	Density	Comments
Rhea	764	527.04	4.518	2.31×10^{21}	1.33	Medium-size iceball, with interesting geology.
Titan	2,575	1,221.85	15.945	1.3455×10^{23}	1.88	Dense atmosphere shrouds surface; ongoing geological activity possible.
Hyperion	$180 \times 140 \times 112$	1,481.1	21.277	2.8×10^{19}	—	Only satellite known not to rotate synchronously.
Iapetus	718	3,561.3	79.331	1.59×10^{21}	1.21	Bright and dark hemispheres show greatest contrast in the solar system.
Phoebe	110	12,952	−550.4	1×10^{19}	—	Very dark; material ejected from Phoebe may coat one side of Iapetus.
Uranus						
Cordelia	13	49.75	0.336	1.7×10^{16}	—	Small moonlets within and near Uranus's ring system.
Ophelia	16	53.77	0.377	2.6×10^{16}	—	
Bianca	22	59.16	0.435	7×10^{16}	—	
Cressida	33	61.77	0.465	2.6×10^{17}	—	
Desdemona	29	62.65	0.476	1.7×10^{17}	—	
Juliet	42	64.63	0.494	4.3×10^{17}	—	
Portia	55	66.1	0.515	1×10^{18}	—	
Rosalind	29	69.93	0.560	1.5×10^{17}	—	
Belinda	33	75.25	0.624	2.5×10^{17}	—	
1986U10	20	75	0.64	—	—	
Puck	77	86.00	0.764	5×10^{17}	—	
Miranda	236	129.8	1.413	6.6×10^{19}	1.26	Small and medium-size iceballs, with some interesting geology.
Ariel	579	191.2	2.520	1.35×10^{21}	1.65	
Umbriel	584.7	266.0	4.144	1.17×10^{21}	1.44	
Titania	788.9	435.8	8.706	3.52×10^{21}	1.59	
Oberon	761.4	582.6	13.463	3.01×10^{21}	1.50	
Caliban	30	7,170	−579	—	—	Too recently discovered for accurate determination of their properties.
1999U1	10	10,000	—	—	—	
Sycorax	60	12,214	−1,289	—	—	
1999U2	10	25,000	—	—	—	
1999U3	—	—	—	—	—	
Neptune						
Naiad	29	48.2	0.296	1.4×10^{17}	—	Small moonlets within and near Neptune's ring system.
Thalassa	40	50.0	0.312	4×10^{17}	—	
Despina	74	52.5	0.333	2.1×10^{17}	—	
Galatea	79	62.0	0.396	3.1×10^{18}	—	
Larissa	$104 \times ? \times 89$	73.6	0.554	6×10^{18}	—	
Proteus	$218 \times 208 \times 201$	117.6	1.121	6×10^{19}	—	
Triton	1,352.6	354.59	−5.875	2.14×10^{22}	2.0	Probable captured Kuiper belt object—largest captured object in solar system.
Nereid	170	5,588.6	360.125	3.1×10^{19}	—	Small, icy moon, very little known.
Pluto						
Charon	635	19.6	6.38718	1.56×10^{21}	1.6	Unusually large compared to its planet; may have formed in giant impact.

*Note: Authorities differ substantially on many of the values in this table.

†$a \times b \times c$ values for the dimensions are the approximate lengths of the axes for irregular moons.

‡Negative sign indicates backward orbit.

**Masses and densities are most accurate for those satellites visited by a spacecraft on a flyby.

STELLAR DATA

Table D.1 Stars Within 12 Light-Years

Star	Distance (ly)	Spectral Type		RA h	RA m	Dec °	Dec ′	Luminosity (L/L_{Sun})
Sun	0.000016	G2	V	—	—	—	—	1.0
Proxima Centauri	4.2	M5.5	V	14	30	−62	41	0.0006
α Centauri A	4.4	G2	V	14	40	−60	50	1.6
α Centauri B	4.4	K0	V	14	40	−60	50	0.53
Barnard's Star	6.0	M4	V	17	58	+04	42	0.005
Wolf 359	7.8	M6	V	10	56	+07	01	0.0008
Lalande 21185	8.3	M2	V	11	03	+35	58	0.03
Sirius A	8.6	A1	V	06	45	−16	42	26.0
Sirius B	8.6	DA2		06	45	−16	42	0.002
Luyten 726-8A	8.7	M5.5	V	01	39	−17	57	0.0009
Luyten 726-8B	8.7	M6	V	01	39	−17	57	0.0006
Ross 154	9.7	M3.5	V	18	50	−23	50	0.004
Ross 248	10.3	M5.5	V	23	42	+44	11	0.001
ε Eridani	10.5	K2	V	03	33	−09	28	0.37
Lacaille 9352	10.7	M1.5	V	23	06	−35	51	0.05
Ross 128	10.9	M4	V	11	48	+00	49	0.003
EZ Aquarii A	11.3	M5	V	22	39	−15	18	0.0006
EZ Aquarii B	11.3	M6	V	22	39	−15	18	0.0004
EZ Aquarii C	11.3	M6.5	V	22	39	−15	18	0.0003
61 Cygni A	11.4	K5	V	21	07	+38	42	0.15
61 Cygni B	11.4	K7	V	21	07	+38	42	0.09
Procyon A	11.4	F5	IV–V	07	39	+05	14	7.4
Procyon B	11.4	DA	—	07	39	+05	14	0.0005
Gliese 725 A	11.4	M3	V	18	43	+59	38	0.02
Gliese 725 B	11.4	M3.5	V	18	43	+59	38	0.01
Gliese 15 A	11.6	M1.5	V	00	18	+44	01	0.03
Gliese 15 B	11.6	M3.5	V	00	18	+44	01	0.003
DX Cancri	11.8	M6.5	V	08	30	+26	47	0.0003
ε Indi	11.8	K5	V	22	03	−56	45	0.26
τ Ceti	11.9	G8	V	01	44	−15	57	0.59
GJ 1061	11.9	M5.5	V	03	36	−44	31	0.0009

Note: These data were provided by the RECONS project, courtesy of Dr. Todd Henry. The luminosities are all total (bolometric) luminosities. The DA stellar types are white dwarfs. The coordinates are for the year 2000.

Table D.2 Twenty Brightest Stars

Star	Constellation	RA h	RA m	Dec °	Dec ′	Distance (ly)	Spectral Type		Apparent Magnitude	Luminosity (L/L_{Sun})
Sirius *ι*	Canis Major	6	45	−16	42	8.6	A1	V	−1.46	26
Canopus	Carina	6	24	−52	41	98	F0	Ib–II	−0.72	1,300
α Centauri	Centaurus	14	40	−60	50	4.4	G2	V	−0.01	1.6
							K0	V	1.3	0.53
Arcturus **2**	Boötes	14	16	+19	11	36	K2	III	−0.06	170
Vega **3**	Lyra	18	37	+38	47	26	A0	V	0.04	60
Capella **4**	Auriga	5	17	+46	00	46	G8	III	0.05	200
Rigel **5**	Orion	5	15	−08	12	820	B8	Ia	0.14	70,000
Procyon	Canis Minor	7	39	+05	14	11.4	F5	IV–V	0.37	7.4
Betelgeuse	Orion	5	55	+07	24	490	M2	Iab	0.41	52,000
Achernar	Eridanus	1	38	−57	15	65	B5	V	0.51	690
Hadar	Centaurus	14	04	−60	22	290	B1	III	0.63	30,000
Altair	Aquila	19	51	+08	52	17	A7	IV–V	0.77	10.5
Acrux	Crux	12	27	−63	06	390	B1	IV	1.39	33,000
							B3	V	1.9	11,000
Aldebaran	Taurus	4	36	+16	30	52	K5	III	0.86	220
Spica	Virgo	13	25	−11	09	260	B1	V	0.91	23,000
Antares	Scorpio	16	29	−26	26	390	M1	Ib	0.92	16,000
Pollux	Gemini	7	45	+28	01	39	K0	III	1.16	55
Fomalhaut	Piscis Austrinus	22	58	−29	37	23	A3	V	1.19	13
Deneb	Cygnus	20	41	+45	16	1,400	A2	Ia	1.26	52,000
β Crucis	Crux	12	48	−59	40	490	B0.5	IV	1.28	90,000

Note: Two of the stars on this list, α Centauri and Acrux, are binary systems with members of comparable brightness. They are counted as single stars because that is how they appear to the naked eye. All the luminosities given are total (bolometric) luminosities. The coordinates are for the year 2000.

STAR CHARTS

How to use the star charts:

Check the times and dates under each chart to find the best one for you. Take it outdoors within an hour or so of the time listed for your date. Bring a dim flashlight to help you read it.

On each chart, the round outside edge represents the horizon all around you. Compass directions around the horizon are marked in yellow. Turn the chart around so the edge marked with the direction you're facing (for example, north, southeast) is down. The stars above this horizon now match the stars you are facing. Ignore the rest until you turn to look in a different direction.

The center of the chart represents the sky overhead, so a star plotted on the chart halfway from the edge to the center can be found in the sky halfway from the horizon to straight up.

The charts are drawn for 40°N latitude (for example, Denver, New York, Madrid). If you live far south of there, stars in the southern part of your sky will appear higher than on the chart and stars in the north will be lower. If you live far north of there, the reverse is true.

Jan–March
© Sky Publishing Corp.

©1999 *Sky & Telescope*

Use this chart January, February, and March.

Early January — 1 A.M.	Early February — 11 P.M.	Early March — 9 P.M.
Late January — Midnight	Late February — 10 P.M.	Late March — Dusk

Apr–June
© Sky Publishing Corp.

©1999 Sky & Telescope

Use this chart April, May, and June.

Early April — 3 A.M.*   Early May — 1 A.M.*   Early June — 11 P.M.*
Late April — 2 A.M.*   Late May — Midnight*   Late June — Dusk

*Daylight Saving Time

July–Sept.
© Sky Publishing Corp.

©1999 *Sky & Telescope*

Use this chart July, August, and September.

Early July — 1 A.M.*   Early August — 11 P.M.*   Early September — 9 P.M.*
Late July — Midnight*   Late August — 10 P.M.*   Late September — Dusk

*Daylight Saving Time

Use this chart October, November, and December.

Early October — 1 A.M.* Early November — 10 P.M. Early December — 8 P.M.
Late October — Midnight* Late November — 9 P.M. Late December — 7 P.M.

*Daylight Saving Time

©1999 Sky & Telescope

GLOSSARY

21-cm line A spectral line from atomic hydrogen with wavelength 21 cm (in the radio portion of the spectrum).

absolute magnitude A measure of the luminosity of an object; defined to be the apparent magnitude of an object if it were located exactly 10 parsecs away.

absolute zero The coldest possible temperature, which is 0 K.

absorption (of light) The process by which matter absorbs radiative energy.

absorption-line spectrum A spectrum that contains absorption lines.

acceleration The rate at which an object's velocity changes. Its units are velocity divided by time; standard units are m/s².

acceleration of gravity The acceleration of a falling object. On Earth, the acceleration of gravity, designated by *g*, is 9.8 m/s².

accretion The process by which objects gather together to make larger objects.

accretion disk A rapidly rotating disk of material that orbits a star-like object as the material (e.g., white dwarf, neutron star, or black hole) gradually falls inward.

active galactic nuclei The unusually luminous centers of some galaxies, thought to be powered by accretion onto supermassive black holes. Some active galactic nuclei outshine all the stars in their galaxy combined, and some appear to be firing powerful jets of material millions of light-years into space. Quasars are the brightest type of active galactic nuclei; radio galaxies also contain active galactic nuclei.

adaptive optics A technique in which telescope mirrors flex rapidly to compensate for the bending of starlight caused by atmospheric turbulence.

albedo Describes the fraction of sunlight reflected by a surface; albedo = 0 means no reflection at all (a perfectly black surface); albedo = 1 means all light is reflected (a perfectly white surface).

altitude (above horizon) The angular distance between the horizon and an object in the sky.

amino acids The building blocks of proteins.

angular momentum Momentum attributable to rotation or revolution. The angular momentum of an object moving in a circle of radius *r* is the product $m \times v \times r$.

angular resolution (of a telescope) The smallest angular separation that two pointlike objects can have and still be seen as distinct points of light (rather than as a single point of light).

angular size (or **angular distance**) A measure of the angle formed by extending imaginary lines outward from our eyes to span an object (or between two objects).

Antarctic Circle The circle on the Earth with latitude 66.5°S.

antimatter Refers to any particle with the same mass as a particle of ordinary matter but whose other basic properties, such as electrical charge, are precisely opposite.

aphelion The point at which an object orbiting the Sun is farthest from the Sun.

apogee The point at which an object orbiting the Earth is farthest from the Earth.

apparent brightness The amount of light reaching us *per unit area* from a luminous object; often measured in units of watts/m².

apparent magnitude A measure of the apparent brightness of an object in the sky, based on the ancient system developed by Hipparchus.

apparent retrograde motion (of a planet) Refers to the period during which a planet appears to move westward relative to the stars as viewed from Earth.

apparent solar time Time measured by the actual position of the Sun in your local sky; defined so that noon is when the Sun is *on* the meridian.

arcminutes (or **minutes of arc**) One arcminute is 1/60 of 1°.

arcseconds (or **seconds of arc**) One arcsecond is 1/60 of an arcminute, or 1/3,600 of 1°.

Arctic Circle The circle on the Earth with latitude 66.5°N.

asteroid A relatively small and rocky object that orbits a star; asteroids are sometimes called *minor planets* because they are similar to planets but smaller.

asteroid belt The region of our solar system between the orbits of Mars and Jupiter in which asteroids are heavily concentrated.

astronomical unit (AU) The average (semimajor axis) distance of the Earth from the Sun, which is about 150 million km.

atmospheric pressure The surface pressure resulting from the overlying weight of an atmosphere.

atomic number The number of protons in an atom.

atomic weight (or **atomic mass**) The combined number of protons and neutrons in an atom.

atoms Consist of a nucleus made from protons and neutrons surrounded by a cloud of electrons.

aurora Dancing lights in the sky caused by charged particles entering our atmosphere; called the *aurora borealis* in the Northern Hemisphere and the *aurora australis* in the Southern Hemisphere.

azimuth (usually called *direction* in this book) Direction around the horizon from due north, measured clockwise in degrees. E.g., the azimuth of due north is 0°, due east is 90°, due south is 180°, and due west is 270°.

bar The standard unit of pressure, approximately equal to the Earth's atmospheric pressure at sea level.

baryonic matter Refers to ordinary matter made from atoms (because the nuclei of atoms contain protons and neutrons, which are both baryons).

baryons Particles, including protons and neutrons, that are made from three quarks.

basalt A type of volcanic rock that makes a low-viscosity lava when molten.

belts (on a jovian planet) Dark bands of sinking air that encircle a jovian planet at a particular set of latitudes.

Big Bang The event that gave birth to the universe.

binary star system A star system that contains two stars.

biosphere Refers to the "layer" of life on Earth.

black hole A bottomless pit in spacetime. Nothing can escape from within a black hole, and we can never again detect or observe an object that falls into a black hole.

black smokers Structures around seafloor volcanic vents that support a wide variety of life.

blueshift A Doppler shift in which spectral features are shifted to shorter wavelengths, caused when an object is coming toward the observer.

bosons Particles, such as photons, to which the exclusion principle does not apply.

bound orbits Orbits on which an object travels repeatedly around another object; bound orbits are elliptical in shape.

brown dwarf An object too small to become an ordinary star because electron degeneracy pressure halts its gravitational collapse before fusion becomes self-sustaining; brown dwarfs have mass less than $0.08 M_{Sun}$.

bubble (interstellar) The surface of a bubble is an expanding shell of hot, ionized gas driven by stellar winds or supernovae; inside the bubble, the gas is very hot and has very low density.

Cambrian explosion The dramatic diversification of life on Earth that occurred between about 540 and 500 million years ago.

carbon stars Stars whose atmospheres are especially carbon-rich, thought to be near the ends of their lives; carbon stars are the primary sources of carbon in the universe.

carbonate rock A carbon-rich rock, such as limestone, that forms underwater from chemical reactions between sediments and carbon dioxide. On Earth, most of the outgassed carbon dioxide currently resides in carbonate rocks.

carbonate–silicate cycle The process that cycles carbon dioxide between the Earth's atmosphere and surface rocks.

CCD (charge coupled device) A type of electronic light detector that has largely replaced photographic film in astronomical research.

celestial coordinates The coordinates of right ascension and declination that fix an object's position on the celestial sphere.

celestial equator (CE) The extension of the Earth's equator onto the celestial sphere.

celestial navigation Navigation on the surface of the Earth accomplished by observations of the Sun and stars.

celestial sphere The imaginary sphere on which objects in the sky appear to reside when observed from Earth.

central dominant galaxy A giant elliptical galaxy found at the center of a dense cluster of galaxies, apparently formed by the merger of several individual galaxies.

Cepheid variables A particularly luminous type of pulsating variable star that follows a period–luminosity relation and hence is very useful for measuring cosmic distances.

charged particle belts Zones in which ions and electrons accumulate and encircle a planet.

chemical enrichment The process by which the abundance of heavy elements (heavier than helium) in the interstellar medium gradually increases over time as these elements are produced by stars and released into space.

chromosphere The layer of the Sun's atmosphere below the corona; most of the Sun's ultraviolet light is emitted from this region, in which the temperature is about 10,000 K.

circulation cells (also called Hadley cells) Large-scale cells (similar to convection cells) in a planet's atmosphere that transport heat between the equator and the poles.

circumpolar star A star that always remains above the horizon for a particular latitude.

climate Describes the long-term average of weather.

close binary A binary star system in which the two stars are very close together.

closed universe Refers to the case in which the density of the universe is greater than the critical density, so that the expansion will someday halt and the universe will begin to contract.

cluster of galaxies A collection of a few dozen or more galaxies bound together by gravity; smaller collections of galaxies are simply called *groups*.

CNO cycle The cycle of reactions by which intermediate- and high-mass stars fuse hydrogen into helium.

coma (of a comet) The dusty atmosphere of a comet created by sublimation of ices in the nucleus when the comet is near the Sun.

comet A relatively small, icy object that orbits a star.

comparative planetology The study of the solar system by examining and understanding the similarities and differences among worlds.

compound (chemical) A substance made from molecules consisting of two or more atoms with different atomic number.

condensates Solid or liquid particles that condense from a cloud of gas.

condensation The formation of solid or liquid particles from a cloud of gas.

conduction (of energy) The process by which thermal energy is transferred by direct contact from warm material to cooler material.

conservation of angular momentum (law of) The principle that, in the absence of net torque (twisting force), the total angular momentum of a system remains constant.

conservation of energy (law of) The principle that energy (including mass-energy) can neither be created nor destroyed, but can only change from one form to another.

conservation of momentum (law of) The principle that, in the absence of net force, the total momentum of a system remains constant.

constellation A region of the sky; 88 official constellations cover the celestial sphere.

convection The energy transport process in which warm material expands and rises, while cooler material contracts and falls.

convection cell An individual small region of convecting material.

convection zone (of a star) A region in which energy is transported outward by convection.

core (of a planet) The dense central region of a planet that has undergone differentiation.

core (of a star) The central region of a star, in which nuclear fusion can occur.

Coriolis effect Causes air or objects moving on a rotating planet to deviate from straight-line trajectories.

corona (solar) The tenuous uppermost layer of the Sun's atmosphere; most of the Sun's X rays are emitted from this region, in which the temperature is about 1 million K.

coronal holes Regions of the corona that barely show up in X-ray images because they are nearly devoid of hot coronal gas.

cosmic background radiation The remnant radiation from the Big Bang, consisting of photons that have traveled freely through the universe since the end of the era of nuclei.

cosmic rays Refers to electrons, protons, and atomic nuclei that zip through interstellar space at close to the speed of light.

cosmological horizon The boundary of our observable universe, which is where the lookback time is equal to the age of the universe. Beyond this boundary in spacetime, we cannot see anything at all.

cosmological redshift Refers to the redshifts we see from distant galaxies, caused by the fact that expansion of the universe stretches all the photons within it to longer, redder wavelengths.

critical density (of the universe) The precise average density for the entire universe that marks the dividing line between eternal expansion and eventual contraction; if the actual density is less than the critical density, the universe will continue to expand forever; if the actual density is greater than the critical density, the universe will someday stop expanding and begin to contract.

crust (of a planet) The low-density surface layer of a planet that has undergone differentiation.

dark matter Matter that we infer to exist from its gravitational effects but from which we have not detected any light; dark matter dominates the total mass of the universe.

daylight saving time Standard time plus 1 hour, so that the Sun appears on the meridian around 1 P.M. rather than around noon.

declination (dec) Analogous to latitude, but on the celestial sphere; it is the angular north-south distance between the celestial equator and a location on the celestial sphere.

degeneracy pressure A type of pressure unrelated to an object's temperature, which arises when electrons (electron degeneracy pressure) or neutrons (neutron degeneracy pressure) are packed so tightly that the exclusion and uncertainty principles come into play.

degenerate object An object in which degeneracy pressure is the primary pressure pushing back against gravity, such as a brown dwarf, white dwarf, or neutron star.

differential rotation Describes the rotation of an object in which the equator rotates at a different rate than the poles.

differentiation The process in which gravity separates materials according to density, with high-density materials sinking and low-density materials rising.

diffraction grating A finely etched surface that can split light into a spectrum.

diffraction limit The angular resolution that a telescope could achieve if it were limited only by the interference of light waves; it is smaller (i.e., better angular resolution) for larger telescopes.

dimension (mathematical) Describes the number of independent directions in which movement is possible; e.g., the surface of the Earth is two-dimensional because only two independent directions of motion are possible (north-south and east-west).

disk component (of a galaxy) The portion of a spiral galaxy that looks like a disk and contains an interstellar medium with cool gas and dust; stars of many ages are found in the disk component.

Doppler effect The effect that shifts the wavelengths of spectral features in objects that are moving toward or away from the observer.

dwarf elliptical galaxies Particularly small elliptical galaxies with less than a billion stars.

eccentricity A measure of how much an ellipse deviates from a perfect circle, which has zero eccentricity.

eclipse Occurs when one astronomical object casts a shadow on another or crosses our line of sight to the other object.

eclipse seasons Periods during which lunar and solar eclipses can occur because the nodes of the Moon's orbit are nearly aligned with the Earth and Sun.

eclipsing binary A binary star system in which the two stars happen to be orbiting in the plane of our line of sight, so that each star will periodically eclipse the other.

ecliptic The Sun's apparent annual path among the constellations.

ecliptic plane The plane of the Earth's orbit around the Sun.

ejecta (from an impact) Debris ejected by the blast of an impact.

electromagnetic field An abstract concept used to describe how a charged particle would affect other charged particles at a distance.

electromagnetic spectrum The complete spectrum of light, including radio waves, infrared, visible light, ultraviolet light, X rays, and gamma rays.

electromagnetic wave A synonym for light, which consists of waves of electric and magnetic fields.

electrons Fundamental particles with negative electric charge; the distribution of electrons in an atom gives the atom its size.

electron-volt (eV) A unit of energy equivalent to 1.60×10^{-19} joule.

electroweak era The era of the universe during which only three forces operated (gravity, strong force, and electroweak force), lasting from 10^{-35} second to 10^{-10} second after the Big Bang.

element (chemical) A substance made from individual atoms of a particular atomic number.

ellipse A type of oval that happens to be the shape of bound orbits. An ellipse can be drawn by moving a pencil along a string whose ends are tied to two tacks; the locations of the tacks are the foci (singular, focus) of the ellipse.

elliptical galaxies Galaxies that appear rounded in shape, often longer in one direction, like a football. They have no disks and contain very little cool gas and dust compared to spiral galaxies, though they often contain very hot, ionized gas.

emission (of light) The process by which matter emits energy in the form of light.

emission-line spectrum A spectrum that contains emission lines.

equivalence principle The fundamental starting point for general relativity, which states that the effects of gravity are exactly equivalent to the effects of acceleration.

era of atoms The era of the universe lasting from about 300,000 years to about 1 billion years after the Big Bang, during which it was cool enough for neutral atoms to form.

era of galaxies The present era of the universe, which began with the formation of galaxies when the universe was about 1 billion years old.

era of nuclei The era of the universe lasting from about 3 minutes to about 300,000 years after the Big Bang, during which matter in the universe was fully ionized and opaque to light. The cosmic background radiation was released at the end of this era.

era of nucleosynthesis The era of the universe lasting from about 0.001 second to about 3 minutes after the Big Bang, by the end of which virtually all of the neutrons and about one-seventh of the protons in the universe had fused into helium.

erosion The wearing down or building up of geological features by wind, water, ice, and other phenomena of planetary weather.

escape velocity The speed necessary for an object to completely escape the gravity of a large body such as a moon, planet, or star.

evaporation The process by which atoms or molecules escape into the gas phase from a liquid.

event Any particular point along a worldline represents a particular event; all observers will agree on the reality of an event but may disagree about its time and location.

event horizon The boundary that marks the "point of no return" between a black hole and the outside universe; events that occur within the event horizon can have no influence on our observable universe.

excited state (of an atom) Any arrangement of electrons in an atom that has more energy than the ground state.

exclusion principle The law of quantum mechanics that states that two fermions cannot occupy the same quantum state at the same time.

exosphere The hot, outer layer of an atmosphere, where the atmosphere "fades away" to space.

exposure time The amount of time for which light is collected to make a single image.

fall equinox (autumnal equinox) Refers both to the point in Virgo on the celestial sphere where the ecliptic crosses the celestial equator and to the moment in time when the Sun appears at that point each year (around September 21).

false-color image An image displayed in colors that are *not* the true, visible-light colors of an object.

fault (geological) A place where rocks slip sideways relative to one another.

fermions Particles, such as electrons, neutrons, and protons, that obey the exclusion principle.

filter (for light) A material that transmits only particular wavelengths of light.

fireball A particularly bright meteor.

flare stars Small, spectral type M stars that display particularly strong flares on their surfaces.

flat (or Euclidean) geometry Refers to any case in which the rules of geometry for a flat plane hold, such as that the shortest distance between two points is a straight line.

flat universe Refers to the case in which the density of the universe is equal to the critical density; as with an open universe, expansion will continue forever in this case.

focal plane The place where an image created by a lens or mirror is in focus.

focus (of a lens or mirror) The point at which rays of light that were initially parallel (such as light from a distant star) converge.

force Anything that can cause a change in momentum.

frame of reference (in relativity) Two (or more) objects share the same frame of reference if they are *not* moving relative to each other.

free-fall Refers to conditions in which an object is falling without resistance; objects are weightless when in free-fall.

free-float frame A frame of reference in which all objects are weightless and hence float freely.

frequency Describes the rate at which peaks of a wave pass by a point; measured in units of 1/s, often called *cycles per second* or *hertz*.

frost line The boundary in the solar nebula beyond which ices could condense; only metals and rocks could condense within the frost line.

fundamental forces The four fundamental forces in nature are gravity, the electromagnetic force, the strong force, and the weak force.

fundamental particles Subatomic particles that cannot be divided into anything smaller.

galactic fountain Refers to a model for the cycling of gas in the Milky Way Galaxy in which fountains of hot, ionized gas rise from the disk into the halo and then cool and form clouds as they sink back into the disk.

galactic wind A wind of low-density but extremely hot gas flowing out from a starburst galaxy, created by the combined energy of many supernovae.

galaxy A huge collection of any-where from a few hundred million to more than a trillion stars, all bound together by gravity.

gamma rays Light with very short wavelengths (and hence high frequencies)—shorter than those of X rays.

gamma-ray burst A sudden burst of gamma rays from deep space; such bursts apparently come from distant galaxies, but their mechanism is unknown.

genetic code The "language" that living cells use to read the instructions chemically encoded in DNA.

geocentric universe (ancient belief in) The idea that the Earth is the center of the entire universe.

geology The study of surface features (on a moon, planet, or asteroid) and the processes that create them.

giants (luminosity class III) Stars that appear just below the supergiants on the H–R diagram because they are somewhat smaller in radius and lower in luminosity.

global positioning system (GPS) A system of navigation by satellites orbiting the Earth.

global wind patterns (or **global circulation**) Wind patterns that remain fixed on a global scale, determined by the combination of surface heating and the planet's rotation.

globular cluster A spherically shaped cluster of up to a million or more stars; globular clusters are found primarily in the halos of galaxies and contain only very old stars.

grand unified theories (GUTs) Theories that unify the strong, weak, and electromagnetic forces into a single force.

granulation (on the Sun) The bubbling pattern visible in the photosphere, produced by the underlying convection.

gravitational constant The experimentally measured constant G that appears in the law of universal gravitation;

$$G = 6.67 \times 10^{-11} \frac{m^3}{kg \times s^2}.$$

gravitational contraction The process in which gravity causes an object to contract, thereby converting gravitational potential energy into thermal energy.

gravitational encounters Occur when two (or more) objects pass near enough so that they can feel the effects of each other's gravity and can therefore exchange energy.

gravitational equilibrium Describes a state of balance in which the force of gravity pulling inward is precisely counteracted by pressure pushing outward.

gravitational lensing The magnification or distortion (into arcs, rings, or multiple images) of an image caused by light bending through a gravitational field, as predicted by Einstein's general theory of relativity.

gravitational redshift A redshift caused by the fact that time runs slow in gravitational fields.

gravitational time dilation The slowing of time that occurs in a gravitational field, as predicted by Einstein's general theory of relativity.

gravitational waves Predicted by Einstein's general theory of relativity, these waves travel at the speed of light and transmit distortions of space through the universe. Although not yet observed directly, we have strong indirect evidence that they exist.

gravitationally bound system Any system of objects, such as a star system or a galaxy, that is held together by gravity.

great circle A circle on the surface of a sphere whose center is at the center of the sphere.

Great Red Spot A large, high-pressure storm on Jupiter.

greenhouse effect The process by which greenhouse gases in an atmosphere make a planet's surface temperature warmer than it would be in the absence of an atmosphere.

greenhouse gases Gases, such as carbon dioxide, methane, and water vapor, that are particularly good absorbers of infrared light but are transparent to visible light.

Gregorian calendar Our modern calendar, introduced by Pope Gregory in 1582.

ground state (of an atom) The lowest possible energy state of the electrons in an atom.

GUT era The era of the universe during which only two forces operated (gravity and the grand-unified-theory, or GUT, force), lasting from 10^{-43} second to 10^{-35} second after the Big Bang.

half-life The time it takes for half of the nuclei in a given quantity of a radioactive substance to decay.

halo (of a galaxy) The spherical region surrounding the disk of a spiral galaxy.

Hawking radiation Radiation predicted to arise from the evaporation of black holes.

heavy elements In astronomy, *heavy elements* generally refers to all elements *except* hydrogen and helium.

helium-capture reactions Fusion reactions that fuse a helium nucleus into some other nucleus; such reactions can fuse carbon into oxygen, oxygen into neon, neon into magnesium, and so on.

helium flash The event that marks the sudden onset of helium fusion in the previously inert helium core of a low-mass star.

helium fusion The fusion of three helium nuclei into one carbon nucleus.

hertz (Hz) The standard unit of frequency for light waves; equivalent to units of 1/s.

Hertzsprung–Russell (H–R) diagram A graph plotting individual stars as points, with stellar luminosity on the vertical axis and spectral type (or surface temperature) on the horizontal axis.

high-mass stars Stars born with masses above about $8M_{Sun}$; these stars will end their lives by exploding as supernovae.

horizon A boundary that divides what we can see from what we cannot see.

horizontal branch The horizontal line of stars that represents helium-burning stars on an H–R diagram for a cluster of stars.

hot spot (geological) A place within a plate of the lithosphere where a localized plume of hot mantle material rises.

Hubble's constant A number that expresses the current rate of expansion of the universe; designated H_0, it is usually stated in units of km/s/Mpc. The reciprocal of Hubble's constant is the age the universe would have *if* the expansion rate had never changed.

Hubble's law Mathematically expresses the idea that more distant galaxies move away from us faster; its formula is $v = H_0 \times d$, where v is a galaxy's speed away from us, d is its distance, and H_0 is Hubble's constant.

hydrogen compounds Compounds that contain hydrogen and were common in the solar nebula, such as water, ammonia, and methane.

hydrogen-shell burning Hydrogen fusion that occurs in a shell surrounding a stellar core.

hydrosphere Refers to the "layer" of water on the Earth consisting of oceans, lakes, rivers, ice caps, and other liquid water and ice.

hyperbola The precise mathematical shape of one type of unbound orbit (the other is a parabola) allowed under the force of gravity; at great distances from the attracting object, a hyperbolic path looks like a straight line.

hyperspace Any space with more than three dimensions.

ices (in solar system theory) Materials that are solid only at low temperatures, such as the hydrogen compounds water, ammonia, and methane.

imaging (in astronomical research) The process of obtaining pictures of astronomical objects.

impact basin A very large impact crater often filled by a lava flow.

impact cratering The excavation of bowl-shaped depressions (*impact craters*) by asteroids or comets striking a planet's surface.

inflation (of the universe) A sudden and dramatic expansion of the universe thought to have occurred at the end of the GUT era.

infrared light Light with wavelengths that fall in the portion of the electromagnetic spectrum between radio waves and visible light.

intensity (of light) A measure of the amount of energy coming from light of specific wavelength in the spectrum of an object.

interferometry A telescopic technique in which two or more telescopes are used in tandem to produce much better angular resolution than the telescopes could achieve individually.

intermediate-mass stars Stars born with masses between about 2 and $8M_{Sun}$; these stars end their lives by ejecting a planetary nebula and becoming a white dwarf.

interstellar cloud A cloud of gas and dust between the stars.

interstellar dust grains Tiny solid flecks of carbon and silicon minerals found in cool interstellar clouds; they resemble particles of smoke and form in the winds of red-giant stars.

interstellar medium Refers to gas and dust that fills the space between stars in a galaxy.

intracluster medium Hot, X ray–emitting gas found between the galaxies within a cluster of galaxies.

inverse square law Any quantity that decreases with the square of the distance between two objects is said to follow an inverse square law.

inversion (atmospheric) A local weather condition in which air is colder near the surface than higher up in the troposphere—the opposite of the usual condition, in which the troposphere is warmer at the bottom.

Io torus A donut-shaped charged-particle belt around Jupiter that approximately traces Io's orbit.

ionization The process of stripping an electron from an atom.

ionization nebula A colorful, wispy cloud of gas that glows because neighboring hot stars irradiate it with ultraviolet photons that can ionize hydrogen atoms.

ionosphere A portion of the thermosphere in which ions are particularly common (due to ionization by X rays from the Sun).

ions Atoms with a positive or negative electrical charge.

irregular galaxies Galaxies that look neither spiral nor elliptical.

isotopes Each different isotope of an element has the *same* number of protons but a *different* number of neutrons.

jets High-speed streams of gas ejected from an object into space.

joule The international unit of energy, equivalent to about 1/4,000 of a Calorie.

jovian nebulae The clouds of gas that swirled around the jovian planets, from which the moons formed.

jovian planets Giant gaseous planets similar in overall composition to Jupiter.

Julian calendar The calendar introduced in 46 B.C. by Julius Caesar and used until it was replaced by the Gregorian calendar.

Kelvin (temperature scale) The most commonly used temperature scale in science, defined such that absolute zero is 0 K and water freezes at 273.15 K.

Kepler's laws of planetary motion Three laws discovered by Kepler that describe the motion of the planets around the Sun.

kinetic energy Energy of motion, given by the formula $\frac{1}{2}mv^2$.

Kuiper belt The comet-rich region of our solar system that spans distances of about 30–100 AU from the Sun; Kuiper belt comets have orbits that lie fairly close to the plane of planetary orbits and travel around the Sun in the same direction as the planets.

latitude The angular north-south distance between the Earth's equator and a location on the Earth's surface.

length contraction Refers to the effect in which you observe lengths to be shortened in reference frames moving relative to you.

lenticular galaxies Galaxies that look lens-shaped when seen edge-on, resembling spiral galaxies without arms. They tend to have less cool gas than normal spiral galaxies but more gas than elliptical galaxies.

leptons Fermions *not* made from quarks, such as electrons and neutrinos.

life track A track drawn on an H–R diagram to represent the changes in a star's surface temperature and luminosity during its life; also called an *evolutionary track*.

light curve A graph of an object's intensity against time.

light gases (in solar system theory) Refers to hydrogen and helium, which never condense under solar nebula conditions.

light pollution Human-made light that hinders astronomical observations.

light-collecting area (of a telescope) The area of the primary mirror or lens that collects light in a telescope.

light-year The distance that light can travel in 1 year, which is 9.46 trillion km.

lithosphere The relatively rigid outer layer of a planet; generally encompasses the crust and the uppermost portion of the mantle.

Local Group The group of about 30 galaxies to which the Milky Way Galaxy belongs.

Local Supercluster The supercluster of galaxies to which the Local Group belongs.

longitude The angular east-west distance between the Prime Meridian (which passes through Greenwich) and a location on the Earth's surface.

lookback time Refers to the amount of time since the light we see from a distant object was emitted. I.e., if an object has a lookback time of 400 million years, we are seeing it as it looked 400 million years ago.

low-mass stars Stars born with masses less than about $2M_{Sun}$; these stars end their lives by ejecting a planetary nebula and becoming a white dwarf.

luminosity The total power output of an object, usually measured in watts or in units of solar luminosities ($L_{Sun} = 3.8 \times 10^{26}$ watts).

luminosity class Describes the region of the H–R diagram in which a star falls. Luminosity class I represents supergiants, III represents giants, and V represents main-sequence stars; luminosity classes II and IV are intermediate to the others.

luminosity–distance formula The formula that relates apparent brightness, luminosity, and distance:

$$\text{apparent brightness} = \frac{\text{luminosity}}{4\pi \times (\text{distance})^2}$$

lunar eclipse Occurs when the Moon passes through the Earth's shadow, which can occur only at full moon; may be total, partial, or penumbral.

lunar maria The regions of the Moon that look smooth from Earth and actually are lava-filled impact basins.

lunar month The time required for a complete cycle of lunar phases, which is about $29\frac{1}{2}$ days (also called a *synodic month*).

lunar phase Describes the appearance of the Moon as seen from Earth.

MACHOs Stands for *massive compact halo objects* and represents one possible form of dark matter in which the dark objects are relatively large, like planets or brown dwarfs.

magma Underground molten rock.

magnetic braking The process by which a star's rotation slows as its magnetic field transfers its angular momentum to the surrounding nebula.

magnetic field Describes the region surrounding a magnet in which it can affect other magnets or charged particles in its vicinity.

magnetic-field lines Lines that represent how the needles on a series of compasses would point if they were laid out in a magnetic field.

magnetosphere The region surrounding a planet in which charged particles are trapped by the planet's magnetic field.

main sequence (luminosity class V) The prominent line of points running from the upper left to the lower right on an H–R diagram; main-sequence stars shine by fusing hydrogen in their cores.

main-sequence fitting A method for measuring the distance to a cluster of stars by comparing the apparent brightness of the cluster's main sequence with the standard main sequence.

main-sequence lifetime The length of time for which a star of a particular mass can shine by fusing hydrogen into helium in its core.

main-sequence turnoff A method for measuring the age of a cluster of stars from the point on its H–R diagram where its stars turn off from the main sequence; the age of the cluster is equal to the main-sequence lifetime of stars at the main-sequence turnoff point.

mantle (of a planet) The rocky layer that lies between a planet's core and crust.

Martian meteorites Meteorites found on the Earth's surface that apparently were chipped off the surface of Mars.

mass A measure of the amount of matter in an object.

mass extinction An event in which a large fraction of the living species on Earth go extinct in a very short period of time.

mass-energy The potential energy of mass, which has an amount $E = mc^2$.

massive-star supernova A supernova that occurs when a massive star dies, initiated by the catastrophic collapse of its iron core; often called a Type II supernova.

mass-to-light ratio The mass of an object divided by its luminosity, usually stated in units of solar masses per solar luminosity. Objects with high mass-to-light ratios must contain substantial quantities of dark matter.

matter–antimatter annihilation Occurs when a particle of matter and a particle of antimatter meet and convert all of their mass-energy to photons.

mean solar time Time measured by the average position of the Sun in your local sky over the course of the year.

meridian A half-circle extending from your horizon (altitude 0°) due south, through your zenith, to your horizon due north.

metals (in solar system theory) Elements, such as nickel, iron, and aluminum, that condense at fairly high temperatures.

meteor A flash of light caused when a particle from space burns up in our atmosphere.

meteor shower A period during which many more meteors than usual can be seen.

meteorite A rock from space that lands on Earth.

Metonic cycle The 19-year period, discovered by the Babylonian astronomer Meton, over which the lunar phases occur on the same dates.

Milky Way Used both as the name of our galaxy and to refer to the band of light we see in the sky when we look into the plane of the Milky Way Galaxy.

millisecond pulsars Pulsars with rotation periods of a few thousandths of a second.

molecular bands The tightly bunched lines in an object's spectrum that are produced by molecules.

molecular clouds Cool, dense interstellar clouds in which the low temperatures allow hydrogen atoms to pair up into hydrogen molecules (H_2).

molecular dissociation The process by which a molecule splits into its component atoms.

molecule Technically the smallest unit of a chemical element or compound; in this text, the term refers only to combinations of two or more atoms held together by chemical bonds.

momentum The product of an object's mass and velocity.

moon An object that orbits a planet.

mutations Errors in the copying process when a living cell replicates itself.

natural selection The process by which mutations that make an organism better able to survive get passed on to future generations.

nebula A cloud of gas in space, usually one that is glowing.

nebular capture The process by which icy planetesimals capture hydrogen and helium gas to form jovian planets.

nebular theory The detailed theory that describes how our solar system formed from a cloud of interstellar gas and dust.

net force The overall force to which an object responds; the net force is equal to the rate of change in the object's momentum, or equivalently to the object's mass × acceleration.

neutrino A type of fundamental particle that has extremely low mass and responds only to the weak force; neutrinos are leptons and come in three types—electron neutrinos, mu neutrinos, and tau neutrinos.

neutron star The compact corpse of a high-mass star left over after a supernova; typically contains a mass comparable to the mass of the Sun in a volume just a few kilometers in radius.

neutrons Particles with no electrical charge found in atomic nuclei, built from three quarks.

newton The standard unit of force in the metric system; 1 newton $= 1 \frac{kg \times m}{s^2}$.

Newton's laws of motion Three basic laws that describe how objects respond to forces.

nodes (of Moon's orbit) The two points in the Moon's orbit where it crosses the ecliptic plane.

nonbaryonic matter Refers to exotic matter that is not part of the normal composition of atoms, such as neutrinos or the hypothetical WIMPs.

north celestial pole (NCP) The point on the celestial sphere directly above the Earth's North Pole.

nova The dramatic brightening of a star that lasts for a few weeks and then subsides; occurs when a burst of hydrogen fusion ignites in a shell on the surface of an accreting white dwarf in a binary star system.

nuclear fission The process in which a larger nucleus splits into two (or more) smaller particles.

nuclear fusion The process in which two (or more) smaller nuclei slam together and make one larger nucleus.

nucleus (of an atom) The compact center of an atom made from protons and neutrons.

nucleus (of a comet) The solid portion of a comet, and the only portion that exists when the comet is far from the Sun.

observable universe The portion of the entire universe that, at least in principle, can be seen from Earth.

Olbers' paradox Asks the question of how the night sky can be dark if the universe is infinite and full of stars.

Oort cloud A huge, spherical region centered on the Sun, extending perhaps halfway to the nearest stars, in which trillions of comets orbit the Sun with random inclinations, orbital directions, and eccentricities.

opacity A measure of how much light a material absorbs compared to how much it transmits; materials with higher opacity absorb more light.

opaque (material) Describes a material that absorbs light.

open cluster A cluster of up to several thousand stars; open clusters are found only in the disks of galaxies and contain young stars.

open universe Refers to the case in which the density of the universe is less than the critical density, so that the expansion will continue forever.

optical quality Describes the ability of a lens, mirror, or telescope to obtain clear and properly focused images.

orbital resonance Describes any situation in which one object's orbital period is a simple ratio of another object's period, such as 1/2, 1/4, or 5/3. In such cases, the two objects periodically line up with each other, and the extra gravitational attractions at these times can affect the objects' orbits.

outgassing The process of releasing gases from a planetary interior, usually through volcanic eruptions.

oxidation Refers to chemical reactions, often with the surface of a planet, that remove oxygen from the atmosphere.

ozone The molecule O_3, which is a particularly good absorber of ultraviolet light.

ozone hole A place where the concentration of ozone in the stratosphere is dramatically lower than is the norm.

pair production The process in which a concentration of energy spontaneously turns into a particle and its antiparticle.

parabola The precise mathematical shape of a special type of unbound orbit allowed under the force of gravity; if an object in a parabolic orbit loses only a tiny amount of energy, it will become bound.

parallax angle Half of a star's annual back-and-forth shift due to stellar parallax; related to the star's distance according to the formula

$$\text{distance in parsecs} = \frac{1}{p}$$

where p is the parallax angle in arcseconds.

parsec (pc) Approximately equal to 3.26 light-years; it is the distance to an object with a parallax angle of 1 arcsecond.

particle era The era of the universe lasting from 10^{-10} second to 0.001 second after the Big Bang, during which subatomic particles were continually created and destroyed and ending when matter annihilated antimatter.

peculiar velocity (of a galaxy) The component of a galaxy's velocity relative to the Milky Way that deviates from the velocity expected by Hubble's law.

penumbra The lighter, outlying regions of a shadow.

perigee The point at which an object orbiting the Earth is nearest to the Earth.

perihelion The point at which an object orbiting the Sun is closest to the Sun.

period–luminosity relation The relation that describes how the luminosity of a Cepheid variable star is related to the period between peaks in its brightness; the longer the period, the more luminous the star.

photon An individual particle of light, characterized by a wavelength and a frequency.

photosphere The visible surface of the Sun, where the temperature averages just under 6,000 K.

pixel An individual "picture element" on a CCD.

Planck era The era of the universe prior to the Planck time.

Planck time The time when the universe was 10^{-43} second old, before which random energy fluctuations were so large that our current theories are powerless to describe what might have been happening.

planet An object that orbits a star and that, while much smaller than a star, is relatively large in size; there is no "official" minimum size for a planet, but the nine planets in our solar system are all at least 2,000 km in diameter.

planetary nebula The glowing cloud of gas ejected from a low-mass star at the end of its life.

planetesimals The building blocks of planets, formed by accretion in the solar nebula.

plasma A gas consisting of ions and electrons.

plate tectonics The geological process in which plates are moved around by stresses in a planet's mantle.

plates (on a planet) Pieces of a lithosphere that apparently float upon the denser mantle below.

potential energy Energy stored for later conversion into kinetic energy; includes gravitational potential energy, electrical potential energy, and chemical potential energy.

power The rate of energy usage, usually measured in watts (1 watt = 1 joule/s).

precession The gradual wobble of the axis of a rotating object around a vertical line.

primary mirror The large, light-collecting mirror of a reflecting telescope.

primitive meteorites Meteorites that formed at the same time as the solar system itself, about 4.6 billion years ago.

processed meteorites Meteorites that apparently once were part of a larger object that "processed" the original material of the solar nebula into another form.

protogalactic cloud A huge, collapsing cloud of intergalactic gas from which an individual galaxy formed.

proton–proton chain The chain of reactions by which low-mass stars (including the Sun) fuse hydrogen into helium.

protons Particles found in atomic nuclei with positive electrical charge, built from three quarks.

protoplanetary disk A disk of material surrounding a young star (or protostar) that may eventually form planets.

protostar A forming star that has not yet reached the point where sustained fusion can occur in its core.

protostellar disk A disk of material surrounding a protostar; essentially the same as a protoplanetary disk, but may not necessarily lead to planet formation.

protostellar wind The relatively strong wind from a protostar.

protosun The central object in the forming solar system that eventually became the Sun.

pulsar A neutron star from which we see rapid pulses of radiation as it rotates.

pulsating variable stars Stars that alternately grow brighter and dimmer as their outer layers expand and contract in size.

quantum mechanics The branch of physics that deals with the very small, including molecules, atoms, and fundamental particles.

quantum state Refers to the complete description of the state of a subatomic particle, including its location, momentum, orbital angular momentum, and spin to the extent allowed by the uncertainty principle.

quantum tunneling The process in which, thanks to the uncertainty principle, an electron or other subatomic particle appears on the other side of a barrier that it does not have the energy to overcome in a normal way.

quarks The building blocks of protons and neutrons, quarks are one of the two basic types of fermions (leptons are the other).

quasar The brightest type of active galactic nucleus.

radar ranging A method of measuring distances within the solar system by bouncing radio waves off planets.

radial motion The component of an object's motion directed toward or away from us.

radiation pressure Pressure exerted by photons of light.

radiation zone (of a star) A region of the interior in which energy is transported primarily by radiative diffusion.

radiative diffusion The process by which photons gradually migrate from a hot region (such as the solar core) to a cooler region (such as the solar surface).

radiative energy Energy carried by light; the energy of a photon is Planck's constant times its frequency, or $h \times f$.

radio galaxy A galaxy that emits unusually large quantities of radio waves; thought to contain an active galactic nucleus powered by a supermassive black hole.

radio waves Light with very long wavelengths (and hence low frequencies)—longer than those of infrared light.

radioactive dating The process of determining the age of a rock (i.e., the time since it solidified) by comparing the present amount of a radioactive substance to the amount of its decay product.

radioactive element (or **radioactive isotope**) A substance whose nucleus tends to fall apart spontaneously.

recession velocity (of a galaxy) The speed at which a distant galaxy is moving away from us due to the expansion of the universe.

red giant A giant star that is red in color.

red-giant winds The relatively dense but slow winds from red-giant stars.

redshift (Doppler) A Doppler shift in which spectral features are shifted to longer wavelengths, caused when an object is moving away from the observer.

reflecting telescope A telescope that uses mirrors to focus light.

reflection (of light) The process by which matter changes the direction of light.

refracting telescope A telescope that uses lenses to focus light.

rest wavelength The wavelength of a spectral feature in the absence of any Doppler shift or gravitational redshift.

retrograde motion Motion that is backward compared to the norm; e.g., Venus is said to rotate retrograde because it rotates in the direction opposite to that of most other planets.

revolution The orbital motion of one object around another.

right ascension (RA) Analogous to longitude, but on the celestial sphere; it is the angular east-west distance between the vernal equinox and a location on the celestial sphere.

rings (planetary) Consist of numerous small particles orbiting a planet within its Roche zone.

Roche zone The region within two to three planetary radii (of any planet) in which the tidal forces tugging an object apart become comparable to the gravitational forces holding it together; planetary rings are always found within the Roche zone.

rocks (in solar system theory) Material common on the surface of the Earth, such as silicon-based minerals, that are solid at temperatures and pressures found on Earth but typically melt or vaporize at temperatures of 500–1,300 K.

rotation The spinning of an object around its axis.

rotation curve A graph that plots rotational (or orbital) velocity against distance from the center for any object or set of objects.

runaway greenhouse effect A positive feedback cycle in which heating caused by the greenhouse effect causes more greenhouse gases to enter the atmosphere, which further enhances the greenhouse effect.

saddle-shaped (or **hyperbolic**) **geometry** Refers to any case in which the rules of geometry for a saddle-shaped surface hold, such as two lines that begin parallel eventually diverge.

saros cycle The period over which the basic pattern of eclipses repeats, which is about 18 years $11\frac{1}{3}$ days.

satellite Any object orbiting another object.

scattered light Light that is reflected into random directions.

Schwarzschild radius A measure of the size of the event horizon of a black hole.

secondary mirror A small mirror in a reflecting telescope, used to reflect light gathered by the primary mirror toward an eyepiece or instrument.

sedimentary rock A rock that formed from sediments created and deposited by erosional processes.

seismic waves Earthquake-induced vibrations that propagate through a planet.

semimajor axis Half the distance across the long axis of an ellipse; in this text, it is usually referred to as the *average* distance of an orbiting object, abbreviated *a* in the formula for Kepler's third law.

shield volcano A shallow-sloped volcano made from the flow of low-viscosity basaltic (relatively "runny") lava.

shock wave A wave of pressure generated by gas moving faster than the speed of sound.

sidereal day The time of 23 hours 56 minutes 4.09 seconds between successive appearances of any particular star at its highest point on the meridian; essentially the true rotation period of the Earth.

sidereal month About $27\frac{1}{4}$ days, the time required for the Moon to orbit the Earth once (as measured against the stars).

sidereal year The time required for the Earth to complete exactly one orbit as measured against the stars; about 20 minutes longer than the tropical year on which our calendar is based.

silicate rock A silicon-rich rock.

singularity The place at the center of a black hole where, in principle, gravity crushes all matter to an infinitely tiny and dense point.

solar activity Refers to short-lived phenomena on the Sun, including the emergence and disappearance of individual sunspots, prominences, and flares; sometimes called *solar weather*.

solar circle The Sun's orbital path around the galaxy, which has a radius of about 28,000 light-years.

solar day Twenty-four hours, which is the average time between appearances of the Sun on the meridian.

solar eclipse Occurs when the Moon's shadow falls on the Earth, which can occur only at new moon; may be total, partial, or annular.

solar flares Huge and sudden releases of energy on the solar surface, probably caused when energy stored in magnetic fields is suddenly released.

solar maximum The time during each sunspot cycle at which the number of sunspots is the greatest.

solar minimum The time during each sunspot cycle at which the number of sunspots is the smallest.

solar nebula The piece of interstellar cloud from which our own solar system formed.

solar neutrino problem Refers to the disagreement between the predicted and observed number of neutrinos coming from the Sun.

solar prominences Vaulted loops of hot gas that rise above the Sun's surface and follow magnetic-field lines.

solar system (or **star system**) Consists of a star (sometimes more than one star) and all the objects that orbit it.

solar wind A stream of charged particles ejected from the Sun.

sound wave A wave of alternately rising and falling pressure.

south celestial pole (SCP) The point on the celestial sphere directly above the Earth's South Pole.

spacetime The inseparable, four-dimensional combination of space and time.

spacetime diagram A graph that plots a spatial dimension on one axis and time on another axis.

spectral resolution Describes the degree of detail that can be seen in a spectrum; the higher the spectral resolution, the more detail we can see.

spectral type A way of classifying a star by the lines that appear in its spectrum; it is related to surface temperature. The basic spectral types are designated by a letter (OBAFGKM, with O for the hottest stars and M for the coolest) and are subdivided with numbers from 0 through 9.

spectroscopic binary A binary star system whose binary nature is revealed because we detect the spectral lines of one or both stars alternately becoming blueshifted and redshifted as the stars orbit each other.

spectroscopy (in astronomical research) The process of obtaining spectra from astronomical objects

speed The rate at which an object moves. Its units are distance divided by time, such as m/s or km/hr.

speed of light The speed at which light travels, which is about 300,000 km/s.

spherical geometry Refers to any case in which the rules of geometry for the surface of a sphere hold, such as that lines that begin parallel eventually meet.

spheroidal component (of a galaxy) The portion of any galaxy that is spherical (or football-like) in shape and contains very little cool gas; generally contains only very old stars. Elliptical galaxies have only a spheroidal component, while spiral galaxies also have a disk component.

spin angular momentum Often simply called *spin*, it refers to the inherent angular momentum of a fundamental particle.

spiral density waves Gravitationally driven waves of enhanced density that move through a spiral galaxy and are responsible for maintaining its spiral arms.

spiral galaxies Galaxies that look like flat, white disks with yellowish bulges at their centers. The disks are filled with cool gas and dust, interspersed with hotter ionized gas, and usually display beautiful spiral arms.

spreading centers (geological) Places where hot mantle material rises upward between plates and then spreads sideways.

spring equinox (vernal equinox) Refers both to the point in Pisces on the celestial sphere where the ecliptic crosses the celestial equator and to the moment in time when the Sun appears at that point each year (around March 21).

standard candle An object for which we have some means of knowing its true luminosity, so that we can use its apparent brightness to determine its distance with the luminosity–distance formula.

standard time Time measured according to the internationally recognized time zones.

star A large, glowing ball of gas that generates energy through nuclear fusion in its core. The term *star* is sometimes applied to objects that are in the process of becoming true stars (e.g., protostars) and to the remains of stars that have died (e.g., neutron stars).

starburst galaxy A galaxy in which stars are forming at an unusually high rate.

stellar parallax The apparent shift in the position of a nearby star against distant objects that occurs as we view the star from different positions in the Earth's orbit of the Sun each year.

stellar wind A stream of charged particles ejected from the surface of a star.

stratosphere An intermediate-altitude layer of the atmosphere which is warmed by the absorption of ultraviolet light from the Sun.

stratovolcano A steep-sided volcano made from viscous ("sticky") lavas that can't flow very far before solidifying.

stromatolites Large bacterial "colonies."

subduction zones Places where one plate slides under another.

subgiant A star that is between being a main-sequence star and a giant; subgiants have inert helium cores and hydrogen-burning shells.

sublimation The process by which atoms or molecules escape into the gas phase from a solid.

summer solstice Refers both to the point on the celestial sphere where the ecliptic is farthest north of the celestial equator and to the moment in time when the Sun appears at that point each year (around June 21).

sunspot cycle The period of about 11 years over which the number of sunspots on the Sun rises and falls.

sunspots Blotches on the surface of the Sun that appear darker than surrounding regions.

superbubble Essentially a giant interstellar bubble, formed when the shock waves of many individual bubbles merge to form a single, giant shock wave.

supercluster Superclusters consist of many clusters of galaxies, groups of galaxies, and individual galaxies and are the largest known structures in the universe.

supergiants (luminosity class I) The very large and very bright stars that appear at the top of an H–R diagram.

supermassive black hole Giant black hole, with a mass millions to billions of times that of our Sun, thought to reside in the centers of many galaxies and to power active galactic nuclei.

supernova The explosion of a star.

Supernova 1987A A supernova witnessed on Earth in 1987; it was the nearest supernova seen in nearly 400 years and helped astronomers refine theories of supernovae.

supernova remnant A glowing, expanding cloud of debris from a supernova explosion.

synchronous rotation Describes the rotation of an object that always shows the same face to an object that it is orbiting because its rotation period and orbital period are equal.

tangential motion The component of an object's motion directed across our line of sight.

tectonics The disruption of a planet's surface by internal stresses.

temperature A measure of the average kinetic energy of particles in a substance.

terrestrial planets Rocky planets similar in overall composition to Earth.

theories of relativity (*special* and *general*) Einstein's theories that describe the nature of space, time, and gravity.

thermal emitter An object that produces a thermal radiation spectrum; sometimes called a "blackbody."

thermal energy Represents the collective kinetic energy, as measured by temperature, of the many individual particles moving within a substance.

thermal escape The process in which atoms or molecules in a planet's exosphere move fast enough to escape into space.

thermal pressure The ordinary pressure in a gas arising from motions of particles that can be attributed to the object's temperature.

thermal pulses The predicted upward spikes in the rate of helium fusion, occurring every few thousand years, that occur near the end of a low-mass star's life.

thermal radiation The spectrum of radiation produced by an opaque object that depends only on the object's temperature; sometimes called "blackbody radiation."

thermosphere A high, hot X ray–absorbing layer of an atmosphere, just below the exosphere.

tidal force A force that is caused when the gravity pulling on one side of an object is larger than that on the other side, causing the object to stretch.

tidal friction Friction within an object that is caused by a tidal force.

tidal heating A source of internal heating created by tidal friction. It is particularly important for satellites with eccentric orbits such as Io and Europa.

time dilation Refers to the effect in which you observe time running slower in reference frames moving relative to you.

timing (in astronomical research) The process of tracking how the light intensity from an astronomical object varies with time.

torque A twisting force that can cause a change in an object's angular momentum.

transmission (of light) The process in which light passes through matter without being absorbed.

transparent (material) Describes a material that transmits light.

Trojan asteroids Asteroids found within two stable zones that share Jupiter's orbit but lie 60° ahead of and behind Jupiter.

tropic of Cancer The circle on the Earth with latitude 23.5°N.

tropic of Capricorn The circle on the Earth with latitude 23.5°S.

tropical year The time from one spring equinox to the next, on which our calendar is based.

troposphere The lowest atmospheric layer, in which weather occurs. It is warmed by the greenhouse effect.

Tully–Fisher relation A relationship among spiral galaxies showing that the faster a spiral galaxy's rotation speed, the more luminous it is; it is important because it allows us to determine the distance to a spiral galaxy once we measure its rotation rate and apply the luminosity–distance formula.

turbulence Rapid and random motion.

ultraviolet light Light with wavelengths that fall in the portion of the electromagnetic spectrum between visible light and X rays.

umbra The dark central region of a shadow.

unbound orbits Orbits on which an object comes in toward a large body only once, never to return; unbound orbits may be parabolic or hyperbolic in shape.

uncertainty principle The law of quantum mechanics that states that we can never know both a particle's position and its momentum, or both its energy and the time it has the energy, with absolute precision.

universal law of gravitation The law expressing the force of gravity (F_g) between two objects, given by the formula

$$F_g = G \frac{M_1 M_2}{d^2}$$

$$\left(G = 6.67 \times 10^{-11} \frac{m^3}{kg \times s^2} \right).$$

universal time (UT) Standard time in Greenwich (or anywhere on the Prime Meridian).

universe The sum total of all matter and energy.

velocity The combination of speed and direction of motion; it can be stated as a speed in a particular direction, such as 100 km/hr due north.

virtual particles Particles that "pop" in and out of existence so rapidly that, according to the uncertainty principle, they cannot be directly detected.

viscosity Describes the "thickness" of a liquid in terms of how rapidly it flows; low-viscosity liquids flow quickly (e.g., water), while high-viscosity liquids flow slowly (e.g., molasses).

visible light The light our eyes can see, ranging in wavelength from about 400 to 700 nm.

visual binary A binary star system in which we can resolve both stars through a telescope.

voids Huge volumes of space between superclusters that appear to contain very little matter.

volatiles Refers to substances, such as water, carbon dioxide, and methane, that are usually found as gases, liquids, or surface ices on the terrestrial worlds.

volcanism The eruption of molten rock, or lava, from a planet's interior onto its surface.

wavelength The distance between adjacent peaks (or troughs) of a wave.

weather Describes the ever-varying combination of winds, clouds, temperature, and pressure in a planet's troposphere.

weight The net force that an object applies to its surroundings; in the case of a stationary body on the surface of the Earth, weight = mass × acceleration of gravity.

weightless A weight of zero, as occurs during free-fall.

white-dwarf limit (also called the *Chandrasekhar limit*) The maximum possible mass for a white dwarf, which is about $1.4 M_{Sun}$.

white-dwarf supernova A supernova that occurs when an accreting white dwarf reaches the white-dwarf limit, ignites runaway carbon fusion, and explodes like a bomb; often called a Type Ia supernova.

white dwarfs The hot, compact corpses of low-mass stars, typically with a mass similar to the Sun's compressed to a volume the size of the Earth.

WIMPs Stands for *weakly interacting massive particles* and represents a possible form of dark matter consisting of subatomic particles that are dark because they do not respond to the electromagnetic force.

winter solstice Refers both to the point on the celestial sphere where the ecliptic is farthest south of the celestial equator and to the moment in time when the Sun appears at that point each year (around December 21).

worldline A line that represents an object on a spacetime diagram.

wormholes The name given to hypothetical tunnels through hyperspace that might connect two distant places in our universe.

X rays Light with wavelengths that fall in the portion of the electromagnetic spectrum between ultraviolet light and gamma rays.

X-ray binary A binary star system that emits substantial amounts of X rays, thought to be from an accretion disk around a neutron star or black hole.

X-ray burster An object that emits a burst of X rays every few hours to every few days; each burst lasts a few seconds and is thought to be caused by helium fusion on the surface of an accreting neutron star in a binary system.

zenith The point directly overhead, which has an altitude of 90°.

zodiac The constellations on the celestial sphere through which the ecliptic passes.

zones (on a jovian planet) Bright bands of rising air that encircle a jovian planet at a particular set of latitudes.

CREDITS

Australian Observatory, photography by David Malin; (b) © IAC photo from plates taken with the Isaac Newton Telescope, photography by David Malin **17.3, 17.4** © Anglo-Australian Observatory, photography by David Malin **17.5** Wm. C. Keel and R. E. White III **17.6** © Anglo-Australian Observatory/Royal Observatory, Edinburgh, photograph made from UK Schmidt plates by David Malin **17.7** (**a, b**) © Anglo-Australian Observatory, photography by David Malin **17.8** © Anglo-Australian Observatory/Royal Observatory, Edinburgh, photograph made from UK Schmidt plates by David Malin **17.11** The Observatories of the Carnegie Institution of Washington **17.12** (**a**) J. Trauger/JPL and NASA; (**b**) Dr. Wendy L. Freedman/The Observatories of the Carnegie Institution of Washington and NASA **17.13** Megan Donahue/STScI **17.18** Robert Williams and the Hubble Deep Field Team (STScI) and NASA **17.21**

Hyron Spinrad et al. (University of California at Berkeley) **17.22** (**a**) Brad Whitmore/STScI and NASA; (**b**) Wm. C. Keel/University of Alabama **17.23** Robert Williams and the Hubble Deep Field Team (STScI) and NASA **17.24** Chris Mihos/Case Western Reserve University **17.25** © Anglo-Australian Observatory, photography by David Malin **17.26** Michael J. West/University of Hawaii at Hilo **17.27** (**a**) MPE/Max Plank Institute; (**b**) Wm. C. Keel/University of Alabama **17.28** John Bahcall/Institute for Advanced Study and NASA **17.29** NRAO/AUI **17.30** John Biretta/STScI **17.33** H. Ford (STScI/Johns Hopkins University); L. Dressel, R. Harms, A. Kochhar (Applied Research Corporation); Z. Tsvetanov, A. Davidsen, and G. Kriss (Johns Hopkins University); R. Bohlin, G. Hartig (STScI); B. Margon (University of Washington—Seattle); and NASA

Chapter 18   18.4 Caltech **18.5** (**a**) Omar Lopez-Cruz and Ian Shelton, courtesy of AURA/NOAO/NSF; (**b**) Steve Snowden (NASA/GSFC/USRA) **18.6** W. N. Colley and E. Turner (Princeton University), J. A. Tyson (Bell Labs/Lucent Technologies), and NASA **18.8** W. Couch/University of New South Wales, R. Ellis/Cambridge University, and NASA **18.9** Charles Alcock/Lawrence Livermore National Laboratory **18.10** Michael Strauss/Princeton University **18.11** Harvard-Smithsonian Center for Astrophysics **18.12** Frank Summers/American Museum of Natural History

Chapter 19   19.5 © Roger Ressmeyer/CORBIS **19.8** E. Ted Bunn/St. Cloud State University **19.12** (*left*) © John Kieffer/Peter Arnold, Inc.; (*right*) © Joel Gordon 1999

Chapter 20   20.1 (**a, b**) NASA/JPL

INDEX